Practical Capillary Electrophoresis

Practical Capillary Electrophoresis

Robert Weinberger
CE Technologies
Chappaqua, New York

ACADEMIC PRESS, INC.
Harcourt Brace Jovanovich, Publishers
Boston San Diego New York
London Sydney Tokyo Toronto

This book is printed on acid-free paper. ♾

ACADEMIC PRESS, INC.
1250 Sixth Avenue, San Diego, CA 92101-4311

United Kingdom Edition published by
ACADEMIC PRESS LIMITED
24–28 Oval Road, London NW1 7DX

Library of Congress Cataloging-in-Publication Data

Weinberger, Robert.
Practical capillary electrophoresis / Robert Weinberger.
p. cm.
Includes index.
ISBN 0-12-742355-9 (acid-free paper)
1. Capillary electrophoresis. I. Title.
QP519.9.C36W45 1993
547.7'04572—dc20 92–43461
CIP

Printed in the United States of America

93 94 95 96 BC 9 8 7 6 5 4 3 2 1

To Lisa, Julie and Jeremy

Contents

Preface

Capillary electrophoresis (CE) or high-performance CE (HPCE) is making the transition from a laboratory curiosity to a maturing microseparations technique. Now used in over one thousand laboratories worldwide, CE is employed in an ever-widening scope of applications covering both large and small molecules.

The inspiration for this book arose from my popular American Chemical Society short course entitled, as is this text, "Practical Capillary Electrophoresis." During the first eighteen months since its inception, nearly five hundred students have enrolled in public and private sessions in the United States and Europe.

I have been amazed at the diversity of the scientific backgrounds of my students. Represented in these courses were molecular biologists, protein chemists, analytical chemists, organic chemists, and analytical biochemists from industrial, academic, and government laboratories. Interestingly enough, CE provides the mechanism for members of this multidisciplinary group to actually talk with each other, a rare event in most organizations.

But the diverse nature of the group provides teaching challenges as well. Most of the students are well versed in the art and science of liquid chromatography. However, CE is not chromatography (usually). It is electrophoresis, and it is governed by the art and science of electrophoresis. For those skilled in electrophoresis, CE offers additional separation opportunities that are not available in the slab-gel format. Furthermore, the intellectual process of methods development differs from both slab-gel electrophoresis and liquid chromatography.

The key to grasping the fundamentals of CE is to develop an understanding of how ions move about in fluid solution under the influence of an applied electric field. With this background, it becomes painless to wander through the electrophoretic domain and explain the subtleties and permutations frequently illustrated on the electropherograms. Accordingly, a logical approach to methods development evolves from this treatment. This is the goal of my course, and I am hopeful that I have translated that message into this text.

Since I work independently, without academic or industrial affiliations, the writing of this text would have been impossible without the help of my friends and colleagues. In particular, I am grateful to Professor Ira Krull and his graduate student Jeff Mazzeo from Northeastern University for reviewing the entire manuscript; Dr. Michael Albin from Applied Biosystems, Inc., for providing his company's bibliography on HPCE; and the Perkin-Elmer Corporation, including Ralph Conlon, Franco Spoldi, and librarian Debra Kaufman and her staff, for invaluable assistance. I am also thankful to my associates throughout the scientific instrumentation industry for providing information, intellectual challenges, hints, electropherograms, comments, etc., many of which are included in this text. Lastly, I thank my students for helping me continuously reshape this material to provide clear and concise explanations of electrophoretic phenomena.

Finally, many of the figures in this text were produced by scanning the illustration in a journal article with subsequent graphic editing. While all efforts were made to preserve the integrity of the original data, subtle differences may appear in the figures produced in this book.

Robert Weinberger
Chappaqua, New York

Master Symbol List

A_{corr}	Corrected peak area
A_{raw}	Raw peak area
a	Fraction ionized
a	Molar absorptivity (Eq. (10.4))
α	Separation factor
B	Viscosity dependence on temperature
b	Detector optical pathlength
C, c	Concentration
C_c	Capacitance of the fused-silica wall
C_{ei}	Capacitance at the buffer–capillary wall interface
C_m	Coefficient for resistance to mass transfer in the mobile phase
C_s	Coefficient for resistance to mass transfer in the stationary phase
%C	Percent of crosslinker in a gel
D, D_m	Diffusion coefficient
D	Dynamic reserve (Eq. (10.5))
D	Capillary diameter (Eq. (9.10))
δ	Debye radius
ξ	Zeta potential
e	Charge per unit area
E	Field strength
E	Acceptable increase in H (Eq. (9.8))
E	Detector efficiency (Eq. (10.4))
ε	Dielectric constant
f	Frictional force (Stokes' Law)
g	Gravitational constant
γ	Field enhancement factor
H	Height equivalent to a theoretical plate
Δh	Height differential between capillary ends
dH/dt	Rate of heat production

I	Current
I_f	Fluorescence intensity
I_0	Excitation source intensity
k, κ	Conductivity
k'	Capacity factor
$\tilde{k}'$	Capacity factor in MECC
k_d	Desorption rate constant
K, λ	Thermal conductivity
K	Complex formation constant (Eq. (7.5))
L_d	Length of capillary to detector
l_d	Length of the detector window
L_f	Length of capillary from detector to fraction collector
L_t	Total length of capillary
l_{inj}	Length of an injection plug
m	Mass
M	Actual mass (Eqs. (10.6), (10.8))
N	Number of theoretical plates
N	Number of segments in a polymer chain (Eq. (5.3))
n	Number of charges
η	Viscosity
ΔP	Pressure drop
Φ	Quantum yield
ρ	Density (Eq. (9.11))
ρ	Resistivity (Eq. (9.13))
Q	Quantity of injected material
q	Ionic net charge
Φ	Overlap threshold
Φ_f	Fluorescence quantum yield
R	Resistance
R	Peak ratio (Eq. (11.2))
R	Displacement ratio (Eq. (10.5))
R_s	Resolution
r	Ionic radius (Stokes' Law)
r, r_c	Capillary radius
S/N	Signal-to-noise ratio
σ	Peak variance
σ_L	Peak variance in units of length
t_a	Adsorption time to a stationary phase

t_d	Desorption time from a stationary phase
t_L	Lag time
t_m	Migration time
t_{mc}	Migration time for a micellar aggregate
t_o	Migration time for a neutral "unretained" solute
t_r	"Retention" time in MECC
T	Temperature
%T	Percentage of monomer and crosslinker in a gel
μ	Ionic mobility
μ_{ep}	Electrophoretic mobility
μ_f	Mobility of uncomplexed solute (Eq. (7.5))
V	Voltage
υ	Ionic velocity
υ	Mean linear velocity (Eq. (8.2))
υ_{ep}	Electrophoretic velocity
υ_{eo}	Electroosmotic velocity
W	Power
w_{inj}	Width of an injection plug
w_s	Spatial width of a sample zone
w_t	Temporal width of a sample zone
x_i	Initial length of an injection plug
x_s	Zone length after stacking
Z	Number of valence electrons
z	Charge

"There are only three problems with capillary electrophoresis: injection, separation and detection" [1]

Chapter 1

Introduction

1.1 Electrophoresis

Electrophoresis is a process for separating charged molecules based on their movement through a fluid under the influence of an applied electric field. The separation is performed in a medium such as a semisolid slab-gel. Gels provide physical support and mechanical stability for the fluidic buffer system. In some modes of electrophoresis, the gel participates in the mechanism of separation as a molecular sieve. Non-gel media such as paper or cellulose acetate are alternative supports. These media are less inert than gels, since they contain charged surface groups that may interact with the sample or the run buffer.

A carrier electrolyte is also required for electrophoresis. Otherwise known as the background electrolyte or simply the run buffer, this solution maintains the requisite pH and provides sufficient conductivity to allow the passage of current (ions), necessary for the separation. Frequently, additional materials are added to the buffer to adjust the selectivity of the separation. These reagents, known as additives, can interact with a solute and modify its rate of electrophoretic migration. The theory and practice of electrophoresis have been the subject of many textbooks and conference proceedings (1–10).

Apparatus for conducting electrophoresis, such as that illustrated in Fig. 1.1, is remarkably simple and low-cost. The gel medium, which is supported on glass plates, is inserted into a Plexiglas chamber. Two buffer reservoirs make contact at each end of the gel. Electrodes immersed in the buffers complete the electrical circuit between the gel and the power supply. Many samples can be separated

[1] Comment from a conferee at HPCE '89, excerpted from a short course produced by Dr. Joseph Olechno, Dionex Corporation.

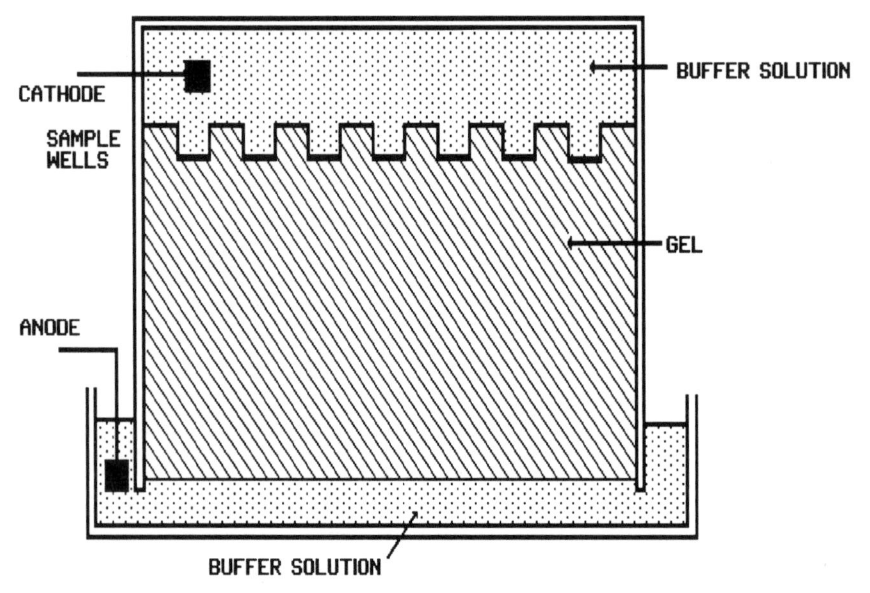

Figure 1.1. Drawing of an apparatus for slab-gel electrophoresis.

simultaneously, since it is possible to use a multilane gel. One or two lanes are frequently reserved for standard mixtures to calibrate the electropherogram. Calibration is usually based on molecular size or in isoelectric focusing, p*I*.[2]

Gels such as polyacrylamide or agarose serve several important functions: molecular sieving, reduction of diffusion and convection, and physical stabilization. The gel composition is adjusted to define specific pore sizes, each for a nominal range of molecular sizes. This forms the basis for separations of macromolecules based on size. With proper calibration, extrapolation to molecular weight is simple.

Reduction of convection and diffusion is another function of the gel matrix. The production of heat by the applied field induces convective movement of the electrolyte. This movement results in bandbroadening that reduces the efficiency of the separation. The viscous gel medium inhibits fluid movement in the electric field. Such a material is termed anticonvective. Since the gel is high-viscosity, molecular diffusion is reduced as well, further enhancing the efficiency of the separation.

Finally, the gel must be sufficiently viscous to provide physical support. Low-viscosity solutions or gels would flow if the plate were not held level. Immersion in detection reagents would be impossible, since handling or contact with fluid solutions would destroy the matrix and separation.

[2] Calibration of capillaries for isoelectric focusing and size separations is discussed in Chapters 4 and 5, respectively.

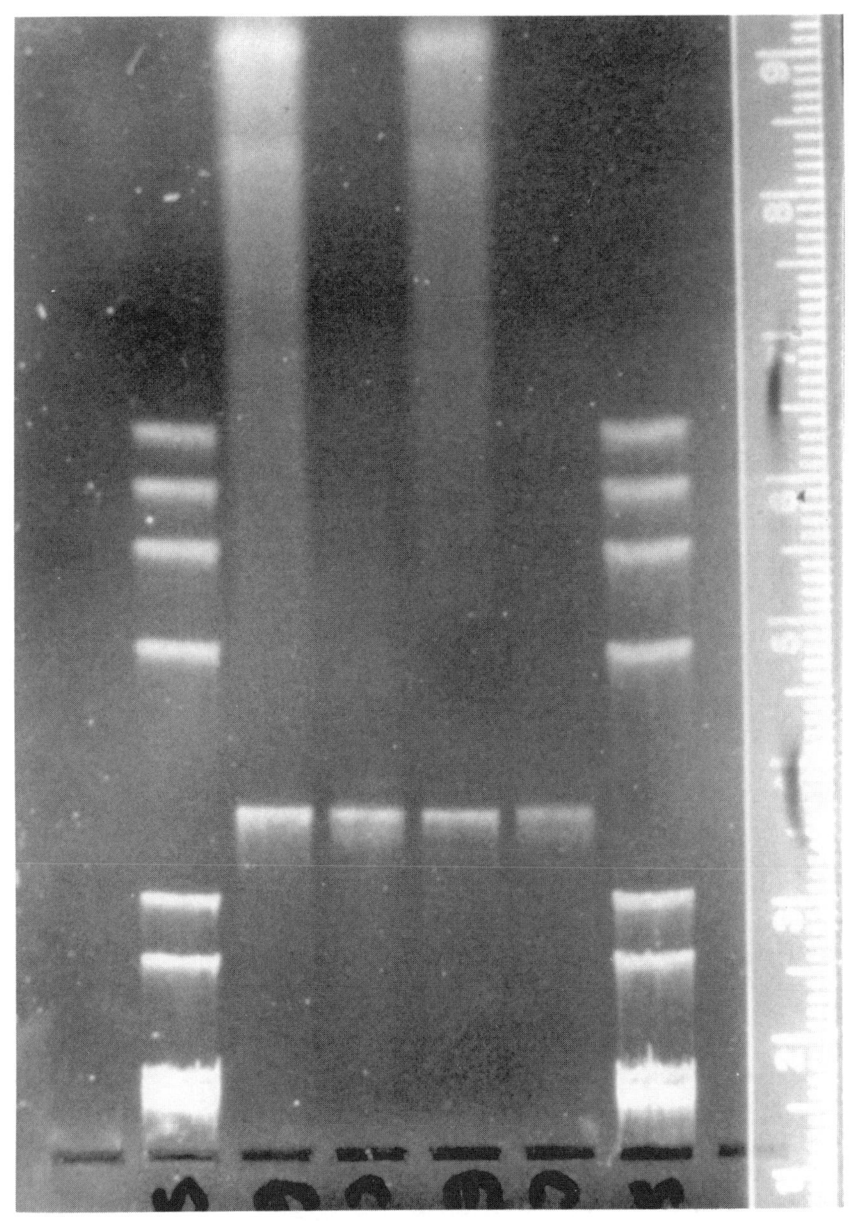

Figure 1.2. Slab-gel electrophoresis of a 500-mer double-stranded PCR reaction product in a 1.8% agarose ethidium bromide gel. Courtesy of Bio-Rad.

The basic procedure for performing gel electrophoresis is

1. prepare and pour the gel
2. apply the sample
3. run the separation
4. immerse the gel in a detection reagent[3]
5. photograph the gel for a permanent record[4]

The separation of some polymerase chain reaction (PCR) products is shown in Fig. 1.2. A restriction digest, used as a sizing standard, appears in the outer lanes. The middle three lanes of the gel show a triplicate run of a 500-mer double-stranded DNA PCR reaction. Quantitation for such a separation is difficult and often imprecise, but it can be performed with the aid of a gel scanner. Recoveries of material from the gel are performed using procedures such as the Southern blot (11). Sufficient material is recoverable for sequencing or other bioassays.

Separations of the sizing standard and 500-mer PCR product by HPCE using a size-selective polymer network (Chapter 5) are shown in Fig. 1.3. Quantitation is readily performed using peak area comparison to the standard. However,

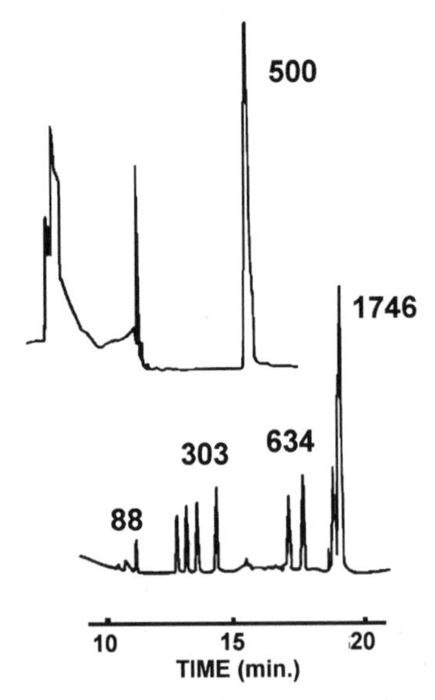

Figure 1.3. Capillary gel electrophoresis of a 500-mer (top) double-stranded PCR reaction product and a low molecular weight sizing standard (bottom). Capillary: 50 cm × 50 μm i.d. Bio-Rad coated capillary; buffer: 100 mM tris-borate, pH 2.3, 2 mM EDTA with linear polymers; injection: electrokinetic, 8 kV, 8 s; detection UV, 260 nm. Courtesy of Bio-Rad.

[3] On-line detection is performed on an instrument such as an automated DNA sequencer.

[4] Automated gel scanners can be used in the place of gel archiving or photography.

fraction collection is difficult, since only minuscule amounts of material are injected into the capillary.

1.2 Microchromatographic Separation Methods

The evolution of chromatographic methods over the last 30 years has produced a systematic and rational trend toward miniaturization. This is particularly true for gas chromatography, where the advantages of the open-tubular capillary displaced the use of packed columns for many applications.

Chromatographic separations all function based on differential partitioning of a solute between a stationary phase and a mobile phase. A packed column offers solutes "a multiplicity of flow paths, some short, the majority of average length, and some long" (12). Solute molecules select various paths through the chromatographic maze. The detected peak suggests this distribution and is broadened. In the open tubular capillary, the choices for solute transport are limited, so the solute elutes as a narrow band.

In order for the open tubular capillary to properly function, its diameter must be quite small. Larger-diameter capillaries present a problem, since solutes away from the walls do not sense the stationary phase in a timely fashion. However, a major problem with narrow i.d. capillaries is loading capacity. Injection sizes must be kept small to avoid overloading the system. In GC, this problem is overcome in part because sensitive detectors such as the flame ionization detector (FID), electron capture detector (ECD), and mass spectrometer are easily interfaced.

Improved efficiency is one of several advantages obtained through miniaturization. The most important of those is improved mass limits of detection (MLOD). Since dilution of the solute is minimized in the miniaturized system, improved MLODs compared with large-scale systems are obtained. This is particularly important when the available sample size is small, as often happens in biomolecule separations.

Miniaturization of GC has been exquisitely successful. These triumphs could not be directly transferred to liquid chromatography (LC) for several reasons. The most important is the lack of good detectors. Interface to the FID and ECD is not practical because of the incompatibility of the mobile phase with each detector. Pumping of the mobile phase at the low flow rates required by miniaturization is also more complex, particularly when gradient elution is required. Despite these problems, μ-LC systems are useful in sample limited situations. Several books have been devoted to this important field (13–15).

Most work with μ-LC employs 250 μm i.d. packed columns, so the advantages enjoyed by open-tubular GC are not realized in μ-LC. The instrumental problems posed by open-tubular LC have inhibited most people from using this technology.

1.3 Capillary Electrophoresis

The arrival of high-performance capillary electrophoresis (HPCE) solves many experimental problems of gels. Use of gels is unnecessary, since the capillary walls provide mechanical support for the carrier electrolyte.[5] The daunting task of automation for the slab-gel format is solved with HPCE. Sample introduction (injection) is performed in a repeatable manner. Detection is on-line, and the instrumental output resembles a chromatogram.

The use of narrow-diameter capillaries allows efficient heat dissipation. This permits the use of high voltage to drive the separation. Since the speed of electrophoresis is directly proportional to the field strength, separations by HPCE are faster than those in slab-gels. On the other hand, the relative speed of the slab-gel is enhanced, since multiple samples can be separated at once. HPCE is a serial technique; one sample is followed by another. This limitation may be overcome in the future if multicapillary instruments (16) are designed and introduced.

HPCE represents a merging of technologies derived from traditional electrophoresis and high-performance liquid chromatography (HPLC). Both HPCE and HPLC employ on-line detection. Developments in on-column micro-LC detection have directly transferred over to capillary electrophoresis. One of the modes of HPCE, micellar electrokinetic capillary chromatography (Chapter 7), is a true chromatographic technique. Electrically driven separations through packed columns (Chapter 8), while difficult in practice, have been reported from many laboratories. While there is much in common between these two techniques, the fundamentals of HPCE are based on electrophoresis, not chromatography.

Professor Richard Hartwick, from the State University of New York at Binghamton, starts off many of his lectures on capillary electrophoresis with a discussion of transport processes in separations (17). During a separation, there are two major transport processes that are occurring:

Separative transport arises from the free-energy differences experienced by molecules with their physicochemical environment. The separation mechanism may be based on phase equilibria such as adsorption, extraction, or ion exchange. Alternatively, kinetic processes such as electrophoresis or dialysis provide the mechanism for separation. Whatever the mechanism for separation, each individual solute must have unique transport properties for a separation to occur.

Dispersive transport, or bandbroadening, is the sum of processes of the dispersing zones about their center of gravities. Examples of dispersion processes are diffusion, convection, and restricted mass transfer. Even under conditions of excellent separative transport, dispersive transport, unless properly controlled, can merge peaks together.

[5] Gels are sometimes used in HPCE for running size separations.

According to Giddings as paraphrased by Hartwick, "Separation is the art and science of maximizing separative transport relative to dispersive transport." In this regard, capillary electrophoresis is perhaps the finest example of optimizing both transport mechanisms to yield highly efficient separations.

Fig. 1.4 and Fig. 1.5 illustrate this concept, using a series of barbiturate separations to compare HPCE and HPLC. The mode of electrophoresis used in Fig. 1.5 is micellar electrokinetic capillary chromatography (MECC), an electrophoretic technique that resembles reverse-phase LC. In the LC separation, amobarbital and pentabarbital coelute, but both are resolved by HPCE.

With some optimization work, amobarbital and pentabarbital can be separated by HPLC. But with HPCE, methods development often progresses rapidly because of the enormous peak capacity of the technique. Peak capacity simply describes the number of peaks that can be separated per unit time. With a couple of hundred thousand theoretical plates,[6] many separations occur without extensive optimization efforts. In addition, peak symmetry is excellent using HPCE unless wall effects

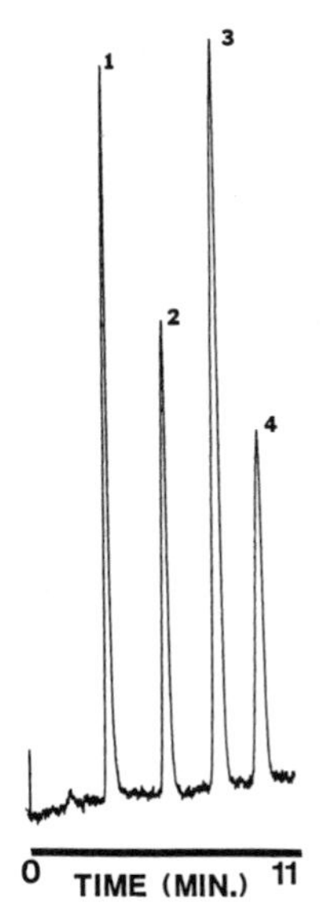

Figure 1.4. Reverse-phase liquid chromatography of barbiturates. Column: Econosphere C_{18}, 25 cm x 4.6 mm i.d.; mobile phase: acetonitrile:water, 55/45 (v/v); injection size: 20 µL; flow rate: 1.2 mL/min; postcolumn reagent: borate buffer, pH 10, 0.2 mL/min; detection: UV, 240 nm; solutes: 1) barbital, 2) butethal, 3) amobarbital and pentabarbital, 4) secobarbital; amount injected: 25 ng of each barbiturate from a 1.25 µg/mL solution.

[6] The theoretical plate (N) is a measure of the efficiency of a chromatographic or electrophoretic peak, $N = 16\ (t_m/\mathrm{PW})^2$, where t_m is the migration time and PW is the peak width at the baseline.

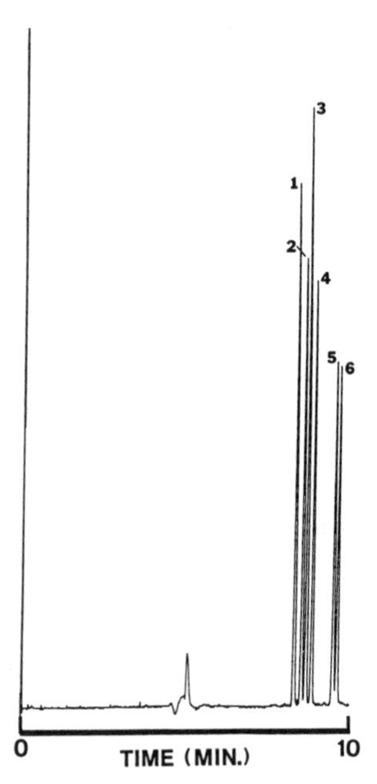

Figure 1.5. Micellar electrokinetic capillary chromatography of barbiturates. Capillary: 50 cm (length to detector) × 50 μm i.d.; buffer: 110 mM SDS, 50 mM borate, pH 9.5; injection: 1 s vacuum (5 nL); detection: UV, 240 nm; solutes: 1) phenobarbital, 2) butethal, 3) barbital, 4) amobarbital, 5) pentobarbital, 6) secobarbital; amount injected: 500 pg of each barbiturate from a 100 μg/mL solution.

(Section 3.4) occur. With the absence of a stationary phase, many factors that contribute to peak broadening and tailing are minimized.

It would be misleading to state that all separations are superior by HPCE, or that methods development will always be straightforward. It is realistic, however, based on the experiences of many separation scientists skilled in the art of both techniques, to predict that HPCE will provide the requisite speed and resolution in the shortest possible run time with the least amount of methods development, under most circumstances.

These same two figures illustrate the most important limitation of HPCE, the concentration limit of detection (CLOD). In Fig. 1.4, the LC separation requires a 1.25 μg/mL solution to give full scale peaks with 1–2 % noise (the postcolumn reagent merely alkalized the mobile phase, permitting sensitive detection at 240 nm). The CLOD is approximately 30-fold better by HPLC. The MECC separation shown in Fig. 1.5 required a solute concentration of 100 μg/mL for a similar response, although the noise was lower (0.5%). On the other hand, the MLOD by capillary electrophoresis exceeds HPLC by a factor of 100. The ideal detector for HPCE will be mass-sensitive and will not depend on the narrow optical pathlength defined by the capillary itself. Descriptions, advantages, and limitations of many HPCE detectors can be found in Chapter 10.

The preceding comparison is significant, since a μ-separation technique is compared to conventional HPLC using a 4.6 mm i.d. column. Would it be better to compare HPCE to μ-LC? Perhaps so from an academic standpoint, but this would not reflect the current usage and thinking in the real world. Chemists are contemplating using HPCE to replace or augment conventional HPLC as well as μ-LC. Table 1.1 provides a comparison of slab-gel electrophoresis, conventional LC, packed-capillary LC, and HPCE. It is easily concluded that the primary disadvantage of HPCE is sensitivity compared with conventional LC. The disadvantages of HPCE compared with slab-gel electrophoresis are difficulties of fraction collection and the inability to run multiple simultaneous lanes.

HPCE is a novel and alternative format for both liquid chromatography and electrophoresis. The unique properties of this technique include the use of:

1. capillary tubing in the range of 25–100 μm i.d.;
2. high electric field strength;
3. on-line detection in real time;
4. only nanoliters of sample;
5. limited quantities of mostly aqueous reagents.

Table 1.1. Comparison of Slab-Gel Electrophoresis, μ-LC, conventional LC, and HPCE

	Slab-Gel	μLC	HPLC	HPCE
Speed	slow	moderate	moderate	fast
Instrumentation Cost	low	high	moderate	moderate
Operating Cost	low	moderate	high	low
Sensitivity				
CLOD	poor	poor	excellent	poor
MLOD	poor	good	poor	excellent
Efficiency	high	derate	moderate	high
Automation	little	yes	yes	yes
Precision	poor	good	excellent	good
Quantitation	difficult	easy	easy	easy
Selectivity	moderate	moderate	moderate	high
Methods Development	slow	moderate	moderate	rapid
Reagent Consumption	low	low	high	minimal
Preparative Mode	good	good	excellent	fair
Ruggedness	good	good	excellent	fair
Separations				
DNA	excellent	fair	fair	excellent
Proteins	excellent	fair	fair	excellent
Small Molecules	poor	excellent	excellent	excellent

ular weight range of analytes separable by HPCE is enormous. A terature reveals applications covering small ions, small molecules, proteins, DNA, viruses, bacteria, blood cells, and colloidal particles.

1.4 Historical Perspective

A century of development in electrophoresis and instrumentation has provided the foundation for HPCE. Recent reviews describing the history of electrophoresis were published by Vesterberg (18) and by Compton and Brownlee (19). The highlights in the development of HPCE are given in Table 1.2.

A direct forerunner of modern CZE was developed by Hjertén in 1967 (20). To reduce the detrimental effects of convection caused by heat production, the 3 mm i.d. capillaries were rotated. While heat dissipation was unchanged, the rotating action caused mixing to occur within the capillary, smoothing out the convective gradients. In the 1970s, techniques using smaller i.d. capillaries were successfully developed (21, 22). Superior heat dissipation permitted the use of higher field strength without the need for capillary rotation. Jorgenson and Lukacs (23) solved, in 1981, the perplexing problems of injection and detection using 75 μm i.d.

Table 1.2. History of HPCE

Year	Technique	Detection	Other
1967	CZE in 3 mm i.d. rotating tubes (20)		
1971	CITP (35)		
1974	CZE in 200–500 μm i.d. glass capillaries (21)		commercial instrumentation for CITP
	Electroosmotic chromatography (36)		
1979	CZE in 200 μm i.d. teflon cappillaries (22)		stacking (22)
1981	CZE in 75 μm i.d. capillaries (23)	fluorescence (23)	
1983	CGE (24)		
1984	MECC (26)	ultraviolet (27)	
1985	CIEF (25)	laser fluorescence (28)	chiral recognition (28)
			coated capillaries (32)
1986			repulsion of proteins from capillary walls (33)
1987		mass spectrometry (29)	
		electrochemistry (30)	
1988		indirect fluorescence(31)	commercial instrumentation for HPCE
1990			field-effect electroosmosis (34)

capillaries. Their advances clearly defined the start of the era of HPCE. Fluorescence detection was required to record the electropherogram.

The 1980s proved ripe for invention. Adaptation of gel electrophoresis (24) and isoelectric focusing (25) to the capillary format was successful. In 1984, Terabe *et al.* (26) described a new form of electrophoresis called micellar electrokinetic capillary chromatography (MECC). Chromatographic separations of small molecules, whether charged or neutral, were obtained by employing the micelle as a "pseudo-stationary" phase.

Great advances in detection occurred during the 1980s. Walbroehl and Jorgenson (27) described improvements in UV detection to overcome, in part, the serious limitation of the short pathlength defined by narrow i.d. capillaries. Gassmann *et al.* (28) employed laser-induced fluorescence, improving detectability to the attomole range. Olivares *et al.* (29) interfaced CZE to the mass spectrometer via the electrospray. The importance of on-line mass spectrometry is significant because of the difficulty of carrying out fraction collection. Wallingford and Ewing (30) developed electrochemical detection sensitive enough to measure catecholamines in a single snail neuron. Kuhr and Yeung (31) employed indirect detection to measure solutes that neither absorbed nor fluoresced.

The problem of protein adherence to the capillary wall was addressed from several fronts. The use of treated capillaries was described by Hjertén (32) in 1985. Around the same time, Lauer and McManigill (33) employed alkaline buffers above the p*I* of the protein to effect solute repulsion from the anionic capillary wall. Based on these and related developments, wall effects have been substantially reduced.

In 1990, Lee *et al.* (34) described field-effect electroosmosis for externally controlling the electroosmotic flow within the capillary. This invention allows for decoupling of the buffer chemistry from the electroosmotic flow.

The first commercial instrument was introduced in 1988 by the late Bob Brownlee's company, Microphoretics. The following year, new instruments from Applied Biosystems, Beckman, and Bio-Rad were introduced. Later, SpectraPhysics, Isco, Europhor, Dionex, and Waters entered the fray. Modular systems from Lauer Labs, Groton Technologies, Jasco, and Europhor became available over the next few years.

The pace of development slowed during 1991–1992. Applications and methods development began to dominate the international symposia as users began to discover the merits and limitations of the techniques. Presently, HPCE is still regarded as a research technique, though more than 1,000 instruments have been sold through mid-1992.[7]

Most of the work described in this historical overview has been improved upon over the years. This includes the early work in capillary isotachophoresis (35) and

[7] Unpublished information from several industry sources.

electroosmotic chromatography (36). These advances are described in detail in the chapters of this text.

1.5 Instrumentation

The instrumental configuration for HPCE is relatively simple. Before 1988, all work was done on simple homemade systems of a design similar to Jorgenson and Lukacs's original work (23).

A schematic of a homemade system is shown in Fig. 1.6. The system consists of a high-voltage power supply, buffer reservoirs, an HPLC ultraviolet detector, capillary, and Plexiglas cabinet. A safety interlock can be employed to prevent activation of high voltage when the cabinet is open. The capillary can be filled with buffer by vacuum generated by a syringe. Samples are injected either by siphoning (elevating the capillary for a defined time at a specified height) or by electrokinetic injection (Section 9.3). While these simple systems provide good separations, precision may be poor because of the lack of temperature control and system automation.

Another common problem in homemade systems is excessive detector noise. The capillary is threaded through the detector and generally passes close to sensitive electronics where the high electric field frequently causes electrical disturbances because of inadequate grounding and shielding. This problem has been solved in commercial instrumentation. The advantage of homemade systems is primarily in the area of detection. It's easy to interface HPCE to fluorescence detection and, in particular, laser-induced fluorescence.

The arrival of commercial instruments has facilitated substantial growth in the field. These instruments provide the following basic features: a high-voltage power

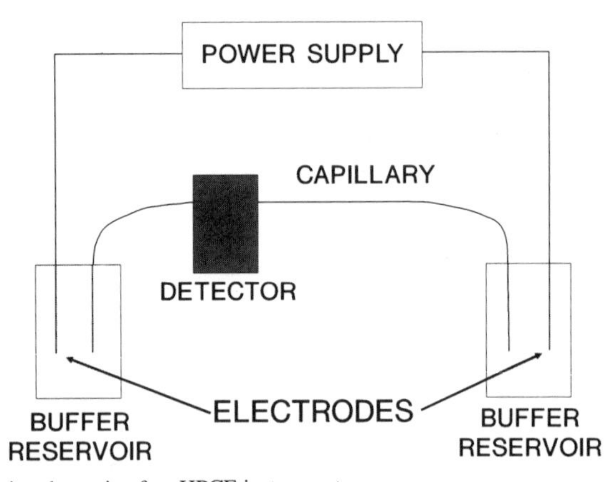

Figure 1.6. Basic schematic of an HPCE instrument.

chapter 1

"You can't just sing the song, Hudson. It's not enough to sing the song. You have to *own* it," said Holla Jones as she paced back and forth in front of her daughter, Hudson, on the stage of the Grand Ballroom in the Pierre Hotel. "Own the stage, own the song, and you'll own the crowd. And that, my dear," she said, pivoting to face Hudson, who stood half-hidden behind a curtain, "is how you become a star."

Hudson bit her full bottom lip. She would be playing her first show ever in just a few hours, and already her mom was using the *S* word. Then again, her mom used that word a lot. Actually, Holla Jones was much more than a pop star — she was a treasured piece of American pop culture. For the past twenty years, her songs had become instant hits all over the world. Her concerts sold out in minutes. Her albums went platinum. Her bubblegumpop-with-an-edge sound was copied by artists everywhere. And Hudson knew that she'd been waiting almost fourteen years to teach her only daughter everything she knew.

"So, you walk up to the mic like this," Holla said, taking short, quick steps on her stiletto-heeled booties toward an imaginary microphone at the edge of the stage. "The last thing you want to do is trip in front of an audience before you've even sung a note." She pretended to grab a microphone. "You slip it out of the stand, and then you hold it just a few inches from your lips, and then you back up just a little bit," she said, taking some steps backwards. "Then you say something to the crowd," she went on. "Be witty, but brief. And then, honey, you start to sing," she said, looking over her shoulder at Hudson and smiling.

At thirty-seven her mom was still beautiful, with flawless dark brown skin, lush lips, and straightened toffee-colored hair that fell past her shoulders. Her tight yoga jacket and pants showed off a body that was sculpted to the extreme: carved biceps, a rock-hard stomach, and slender, muscular legs. Her high, regal forehead didn't have one wrinkle, and she moved with a dancer's grace — shoulders thrown back, spine ramrod-straight. Hudson had inherited that grace, along with her mom's sweeping cheekbones and razor-sharp jawline. But her sea green eyes and wavy hair and French toast–colored complexion came from her dad — or at least she figured as much, based on the photos she'd seen of him. Michael Kelly had been Holla's backup dancer on her second concert tour. He was white and preppy-looking, with thick dark hair, a chiseled face, and soulful eyes, like Billy Crudup crossed with Mikhail Baryshnikov. In pictures he stood next to Holla, his head on her shoulder, smiling goofily into the camera. But they'd had a tumultuous relationship, and when the tour ended he broke up with her, just before she learned she was pregnant. He hadn't been heard from or seen since, and Holla, out of pride, had never tried to contact him. Sometimes Hudson wondered if he even knew he had a daughter. Holla didn't mention him too often, and most of the time it was almost as if he'd never existed at all.

"Mom, it's just the Silver Snowflake Ball," Hudson said. "It's not Radio City or anything."

"It doesn't matter," Holla said. "*Every* show is important. Your producer and your record-label executive are coming. They're going to want to see how you'll do this when it's time to go on tour. So come on out here. You can't hide behind those curtains all day."

Hudson stepped out of the wings, still wearing the ripped jeans and black sweater she'd worn to her last final exam. As of today, school was officially over for winter break, and all she really wanted to do right now was go home and take a nap. Besides, she and her mom had already spent hours talking about this, planning this, and rehearsing this. In a million years, she never would have guessed that she'd end up singing at Ava Elting's epic party. She hadn't even been sure that she would go. But then Carina Jurgensen, one of Hudson's best friends and the party planner for the event, had volunteered her as the night's entertainment, and she'd had no choice but to go along with it. Needless to say, Ava had been hoping for the Jonas Brothers or Justin Timberlake or some other big star she thought Carina could get, thanks to her billionaire dad and his A-list connections. But Ava had settled for Hudson. And now she needed to be prepared.

And her mom was right. In just six months her first album would drop, and then she would be playing shows all the time, at even scarier places. She needed to learn how to do this now. And even though she had a feeling that she hadn't quite inherited her mom's performance gene, at least she was getting a one-on-one tutorial that most other beginners would kill for.

"Okay, let's start the track," Holla said. "Jason?" she called out to the wings. "Can we have the music, please?"

Weeks ago, when Hudson was trying to decide on a song for the Ball, "Heartbeat" had seemed like the perfect choice. She'd written it about Kevin Hargreaves, who was four years older, a senior at Lawrenceville boarding school, and basically a complete stranger. But he was a Capricorn, which blended beautifully with Hudson's Pisces sign, and he had deep, bottomless gray eyes that had made her heart pound and her hands sweat every time she'd seen him. Which had been exactly twice — first on the beach in Montauk, and the second time by accident in the Magnolia Bakery near her house. Carina knew him and had practically pushed Hudson into Kevin's face both times. He'd barely made eye contact with her, and had pretty much said only "hey!" while Hudson stared at him, speechless. When she'd heard he was going out with Samantha Crain, a tenth grader at Lawrenceville, she was crushed. She'd gone straight to her piano, and two hours later she'd finished this song — a slow jazz- and soul-inflected number that she sang leaning over her piano, in her deep, smoky voice.

But the song had since gone through a transformation. A few months ago, Holla decided that Hudson's entire sound needed to change, that for the sake of her first album's sales she would need to go bigger, brighter, and more radio-friendly. It wasn't enough to have a small cult following — she needed to fill stadiums. So Hudson let her mom change studios. She let her take apart every track, layering it with digital beats and effects and backup voices. Until little by little, Hudson's music sounded exactly like hers.

Now, as the song came over the ballroom's speakers, Hudson fought the urge to cover her ears. It was bad enough that it sounded fake and manufactured. Now Hudson had to sing to it. She'd never tried to sing without sitting at her piano. She had no idea what she was supposed to do with her hands and arms and feet. Of course, Holla knew what to do.

"So, let's practice those dance moves, honey," Holla said, sidling up next to her. "First is the turn, like this," Holla said, executing a perfect, weightless spin on the toes of her boots. "You try it."

"Mom, I told you, I really don't want to dance," Hudson said.

"You've got to do *something*," Holla insisted. "Come on. Try it. You're such a good dancer."

Hudson threw herself to the left and barely did half a turn.

"You're not trying, Hudson," Holla said. "Come on. I know you can do better than this."

Hudson gazed out at the brightly lit ballroom, filled with tables and chairs yet to be moved out. At least nobody was watching them yet. *How much more fun would tonight be if I could just go to this party like everyone else?* she thought. *Just hang out with Carina and Lizzie and check out people's dresses and scope the room for cute guys?*

"Mom, I really can't do this," Hudson said after trying to imitate her mom's shimmy. "Do I need to dance? Why can't I just sing?"

"Oh, honey, don't be so negative," said Holla. "Don't you know what I always say about negativity?"

" 'Negative thoughts draw negative things,' " Hudson recited.

"That's right," Holla said, flipping her hair over her shoulder. "And you, my dear, are being *extremely* negative about this. Let's play that again!" she called over her shoulder to Jason in the wings.

Hudson waited for the music to begin. *This isn't right,* a voice said inside her. *Get out of this now. People will understand. Even Ava will have to understand.*

"Come on, Hudson, here we go," Holla said. "Let's do the turn, and then a shimmy to the right . . . that's it."

It was just one night, Hudson told herself. She'd get through this, somehow. After all, she was the child of two dancers. She had to have gotten some of their talent.

But inside, she wasn't so sure. Her mom was the star in the family. And something told her that it was always going to stay that way.

chapter 2

Several hours later Hudson was back behind the same curtain, trying not to hyperventilate. On the other side, the Silver Snowflake Ball was in full swing. Butterflies flew around her stomach as she clutched the scratchy silk fabric. At least she knew she looked good. Gino, her mom's hairstylist, had straightened her hair and then curled it into soft waves. Suzette, her mom's makeup artist, had dusted her face with shimmery powder and lined her eyes with a thick purple pencil. Her vintage black silk halter dress felt cool and soft against her skin. She looked like a star. Now all she had to do was act like one. And not pass out.

But first it would help to get a glimpse of her friends. She peeked out from the wings, ready to signal Carina or her other best friend, Lizzie Summers. And there, smack in front of her, was Carina, kissing some guy Hudson had never seen before. He was skinny, with spiky black hair and beat-up Stan Smiths, and he looked nothing like the guys Carina usually liked. It had to be

Alex, the cool downtown DJ Carina had been talking about non-stop for the past few weeks. Normally she would have given them space, but this was an emergency, so Hudson marched right up to them and tapped Carina on the shoulder.

"Sorry to interrupt," she said, "but I think I'm supposed to go on now."

"Oh my God, you look gorgeous!" Carina said, breaking away from the kiss. With her beachy-blond hair, cocoa brown eyes, and freckled nose, Carina usually looked like the picture-perfect surfer girl. But in her emerald green minidress and gold heels, she was stunning.

"Oh my God, I'm so happy I made you do this," Carina said, jumping up and down. Then she remembered that they weren't alone. "Oh, and by the way, this is Alex."

Hudson turned to the guy. He was definitely cute, with large, liquid brown eyes and sharp cheekbones. "Hey, it's nice to meet you," Hudson said. "I've heard a lot about you."

"Hi there," Alex said, shaking her hand. "Um, sorry to change the subject, but is that Holla Jones standing back there?" he asked, pointing into the wings.

Hudson barely turned around. She knew that her mom was hovering nearby.

"Holla's Hudson's mom," Carina told him.

"Wow," Alex said. "This is some school you go to."

As they chatted, Hudson could see that Alex was head over heels for Carina, despite his cool exterior. But she was starting to get more and more nervous. The Silver Snowflake Ball was the most exclusive holiday party in the city. Ava had made sure to invite only the highest-ranking students from all the New York City private schools, and even some boarding schools. Hudson couldn't quite see the crowd below the stage, but she could picture them, milling around, too cool to dance, too jaded to be excited about anyone who'd be performing. She knew that if she didn't do a good job tonight, she'd be the laughingstock of New York. But she also knew that she just needed to get this over with, so she reminded Carina and Alex that it was time for her to start.

"Okay, fine, break a leg," Carina said to her.

As Hudson turned to walk backstage she saw her mom coming toward her. Holla had changed into a tight black top and leather jeans.

"You ready?" Holla asked, reaching out to touch Hudson's curls. "Oh my God, what did Gino do to your hair? It's so . . . unruly."

"Mom —"

"Are you going to be able to dance in that dress?" Holla asked, looking her up and down with a disapproving frown. "It doesn't look like your hips can move in that. I thought you were going to wear the blue dress with the Lycra in it."

"Mom," Hudson said, feeling her heart rate start to rise. "Everything's fine."

"Now just remember, when you get out there, there's this thing called the fourth wall," Holla said, putting her hands on her slender hips. "It's like an invisible barrier between you and the audience. But you have to break it, over and over. You have to reach out into the audience and let them *know* that you're there —"

Hudson began to tune her mother out as the butterflies flitting around in her stomach turned into baby dragons.

"—and make sure, whatever you do, that you project your voice, even with the microphone, and remember"—she paused for dramatic effect—"Richard is here from Swerve Records. Chris is here. Everyone's watching you tonight. This has to be good."

Hudson nodded. From out of the corner of her eye she saw Ava Elting approaching. "Okay, fine, I have to go," she said, slipping away from her mother's stare just as Ava bossily inserted herself in front of her.

"So are you ready?" Ava asked. She wore her auburn curls piled up on top of her head and an electric purple dress with a side slit that was cut way, *way* too high up her leg.

"Sure," Hudson said, because she knew from the way Ava was looking at her that she didn't have a choice. "Let's go."

"Just remember, it's only *one* song," Ava emphasized. "We don't have time for any more."

"Don't worry," said Hudson. "I wasn't planning to do a full concert."

Ava was oblivious to Hudson's sarcasm. "Good luck!" she called out. Then she strode out onto the stage and right up to the mic, grabbing it like a pro. "Thanks everyone for coming!" she yelled. "And now I'd like to introduce to you the next huge pop music sensation, in her debut performance, the daughter of my really good friend Holla Jones, Hudson Jones!"

"My good friend Holla Jones"? Hudson thought as the applause roared from the ballroom. Ava hadn't even met her mom.

But then the applause started to die, and Hudson knew that it was time to walk out onstage. Her heart began to race. She took her first, tentative steps. *Here we go,* she thought. *You can do this. You can* totally *do this.*

With her eyes on the mic stand, she took the shortest steps she could on her three-inch heels. *The last thing you want to do is trip in front of an audience before you've even sung a note,* her mom had said.

Hudson looked out into the audience and blinked. She'd expected to see Lizzie's and Carina's smiling faces out in the crowd, but thanks to the blinding spotlight, there was only darkness. She couldn't see anything or anyone. She felt her throat tighten but she took the mic out of the stand and took a few steps back. She had no idea what to say.

"Hi, everyone. It's great to see you all here," she half-whispered, holding the microphone an inch from her lips. Her heart was beating so hard she thought it might fly up out of her throat and onto the stage. "This is a song off my first album. It's called 'Heartbeat.' "

She started to turn around, instinctively going to her piano. Then she remembered that it wasn't there. She was all alone. She bowed her head, gripping the mic with her sweaty hand. And when the song finally came blasting out of the speakers, Hudson raised her head to the audience—only to realize that she couldn't remember the song's first line.

The music—the awful, cheesy music—went on, blasting through the speakers. She stared into the darkness. If she turned her head and looked offstage, she knew that she'd see her mom jumping up and down, trying to get her to do some of the dance moves she'd spent so many hours trying to teach her. And she couldn't handle that right now.

Finally, the words came to her, just in time for her cue. She brought the mic to her lips and opened her mouth. The words were there, thank God, ready to be sung. She took a breath...

You can't just sing the song, Hudson. You have to own *it.*

...And nothing came out. She couldn't sing. It was as if she'd been running to the end of the diving board, preparing her body for a perfect swan dive, and then had just come to a dead stop.

She opened her mouth, ready to try again, ready to own that stage even though she was starting to shake and sweat and was pretty sure that she would probably never own a stage as long as she lived...

Nothing. She had no voice.

The music rolled on. She looked out into the darkness. This couldn't really be happening, could it? For a second she floated outside her body and saw herself standing there, mute and sweating. She couldn't see anyone, but they were definitely out there, watching this happen to her, watching her completely freeze up. And she already knew what they were thinking. She could practically feel it.

Her hands shook. Her whole body trembled.

Please, God, this isn't happening, she thought. *This really can't be happening.*

Finally the mic fell out of her slippery hands and hit the stage. *BOOM!* went the sound through the speakers. It shocked her awake. Somewhere inside of her, a loud voice spoke up. And this time it wasn't Holla's. It was entirely Hudson's.

Get out of here...NOW!

So she turned and ran.

chapter 3

Alone in the hotel bathroom, Hudson tried to catch her breath. She looked into the mirror above the sink. The perfect waves that Gino had spent an hour making with a curling iron now stuck, deflated, to her sweaty neck. Her kohl eyeliner had bled into Goth-like purple circles under her eyes. One of her gold drop earrings was mysteriously missing. And what was that smell? She sniffed under her arm. *Yuck.*

She turned on the water and splashed her face. *Did I really have to* run *offstage?* she asked herself, pumping some liquid soap into her hands. *Couldn't I have just walked gracefully into the wings? Or at least just hummed along?* She knew what people upstairs were thinking. And texting. And posting on Facebook right this very minute. The one thing she had never expected people to say about her:

OMG!! Holla Jones's daughter can't SING!

She blotted her face with a towel, trying not to shudder. Maybe it wasn't that bad. Maybe people didn't really care. Maybe it just looked like she'd forgotten something, or like she had to go to the bathroom. *Really* had to go to the bathroom.

She gazed at her clean, bewildered face in the mirror and shook her head. She'd totally screwed up. She'd completely and unforgettably blown it. But had it been all her fault? Holla had turned her into a wreck. For days, her mom had picked apart her voice, her body, her dancing, even her hair—*especially* her hair. And who told her kid when she went out onstage to sing for the very first time that "this has to be good"?

She heard the door to the ladies room creak open.

"Hudson?" said a hushed, familiar voice. "You in here?"

Hudson stepped away from the sink to see Carina and Lizzie stepping hesitantly into the powder room. Both girls looked almost as worried and out of breath as Hudson felt. Lizzie's hazel eyes seemed even larger than usual, or maybe it was that her red curls had been twisted up into a knot, away from her face. Her strapless, smoky blue gown showed off her pale shoulders. Both of her friends looked so pretty.

"Hey, guys," Hudson said meekly.

"Holy *shnit,*" Carina said, rushing over and throwing her arms around her friend. "What happened up there? Are you sick or something?"

Hudson hugged Carina and felt the knot in her stomach slowly loosen. "I wish," she said. She stood on her tiptoes to hug Lizzie. "Sorry I'm a little sweaty."

"It's okay," Lizzie said, letting her go but holding her by the arms. "But are *you* okay?"

Hudson's face burned. She could barely look at Lizzie. Of the three of them, Hudson was supposed to be the performer, the professional. She stepped back and shrugged. "I just blanked out, you guys. I froze."

Lizzie and Carina traded a look, their faces strained with concern.

Hudson looked down at the moss green carpet. "I got out there and I couldn't do anything. All the stuff my mom's been saying the past few weeks — that I'm singing the wrong way, I'm dancing the wrong way, I'm holding my arms too stiff, I'm not 'selling' the song enough... I couldn't get it out of my head." She glanced up at her friends. "Was it bad? What are people saying up there?"

"Nothing," Carina said, a little too quickly.

"Alex is already spinning some songs." Lizzie brushed some red tendrils out of her face. "People've already forgotten."

"Yeah, I'm sure they have," Hudson said bitterly.

"Look, Hudson, you can't listen to your mom," Carina declared. "She'd drive anyone crazy. She'd drive *me* crazy."

"You were just nervous, that's all," Lizzie said. "I would have had a heart attack up there."

"It's my fault," Carina exclaimed, kicking off her gold shoes and stretching her toes out on the carpet. "*I* made you do this. *I* put you on the spot with Ava. I knew you didn't want to do it. It's my fault. I should be arrested by the friendship police or something."

"No, it's not your fault," Hudson said soberly. "It's my fault."

"How is this *your* fault?" Lizzie asked.

"Because I shouldn't even be trying to be a singer," Hudson said. "Why would I even try?"

"Because you're incredibly talented," Lizzie answered firmly.

"But what good is that if I can't sing on a stage?" Hudson said. *And if my mom is always going to make me do things like her?* she thought, but didn't say.

"If it makes you feel any better," Carina said, "Alex thought you were really cool."

Hudson smiled. "He's cute. My guess is he's an Aquarius. Which would be perfect for you." Hudson loved checking up on whether her friends were compatible with the guys they liked.

"So what are you gonna tell your mom?" Lizzie asked, bringing them back on topic.

"I don't know," Hudson admitted. "If anyone has any ideas, now would be a good time to share."

"You should just come back up to the party," Carina offered. "Just have fun for the rest of the night. Who cares about what happened onstage?"

"I do," Hudson said, walking over to the mirror and giving her damp curls one last shake off her shoulders. She tried to picture walking back into the ballroom upstairs, past all the people who'd just watched her run away. Maybe Carina could do that, but Hudson couldn't.

Lizzie put her arm around Hudson's shoulders. "You're gonna be okay. I promise you."

"Thanks, Lizbutt," Hudson said. "Where's Todd tonight?"

Lizzie opened her purse. "I'm gonna text him now," she said. "He wanted to stay home with his dad tonight. I guess his dad's been really depressed." Todd's dad, Jack Piedmont, had been released on bail after being arrested for allegedly stealing money from the company he ran. Even though Todd was going through the worst time of his life, he and Lizzie still seemed very much together. They'd even dropped the L bomb a couple of weeks earlier.

Carina opened the door. "You're totally welcome to join us up there," she said, wobbling a little on her heels. "If you want to put off the Holla fallout a little longer."

"Are you kidding?" Hudson said as they stepped out into the small, deserted foyer. "She'd find me in five minutes."

Just then, Hudson heard the unmistakable sound of stiletto heels hurrying across a marble floor.

"*There* you are!" yelled a voice, and Hudson whipped around.

It was her mom, running toward her with her arms outstretched and her silky, highlighted hair bouncing softly against her shoulders. "Oh, honey, come *here*," she cried, throwing her arms around Hudson and pressing her firmly against the collection of necklaces resting against her chest.

Holla's amethyst-encrusted owl pendant dug into Hudson's cheek.

"Thank *God*," she said, squeezing Hudson so hard she couldn't breathe. "I've been looking for you all *over* this place."

Hudson pried herself away from her mom's embrace. "I just went to the bathroom for a minute. I'm fine."

Little Jimmy, Holla's linebacker-sized bodyguard, caught up to them, huffing and puffing slightly. Behind him was Sophie, Holla's new, perennially frazzled assistant, her Bluetooth still secured to her ear. Hudson gave them both an embarrassed smile. They smiled back, before politely looking down at the gray marble floor.

"Oh, honey, look at you," Holla said tenderly, touching

Hudson's hair and then her cheek. "You're a mess." Holla pushed Hudson's hair off her shoulder. "What happened to your other earring? Did you know that you're missing an earring?"

"Yes," Hudson said.

"Do you want to tell me what happened up there?" Holla asked, her voice softening. "Can you at least tell me that?"

Hudson stared at her mom. *Do you really not know?* she wanted to say. *You drove me crazy.* "I'm not really sure," she finally said. "I think it was just stage fright." She couldn't get into the truth. Not with so many people standing around them.

Holla folded her arms and her expression changed from concerned to controlled. "Go back upstairs and tell everyone it was food poisoning," she said curtly to Sophie.

"*Food* poisoning?" Hudson asked.

"She had some bad sushi," Holla added, ignoring Hudson's question. "And we're all really sorry for the inconvenience."

"Do I say what kind of sushi?" Sophie asked, scrambling inside her purse for her pen and notepad.

"It doesn't matter," Holla said in a clipped voice. "Just go."

Sophie turned on her heel and dashed down the hall, back toward the ballroom. Hudson glanced at Lizzie and Carina. They'd seen Holla flex her amazing powers of spin before, but they seemed stunned. "Mom, are you sure?" Hudson asked.

Holla put her arm around Hudson's shoulder and hugged her again. "Don't you worry about a thing," she said firmly. "Let me take care of this."

Before Hudson could respond, she heard Chris Brompton call out, "There you guys are! We got a little turned around."

She turned to see Chris approaching from the other end of the hallway, followed by Richard Wu, the executive from her record label. In all the chaos, she'd completely forgotten about them. She would have given anything for these men not to have seen her run offstage. Now they were going to comfort her. *Ick.*

"Hudson, you okay?" Chris asked, coming to stand next to her and peering into her face with his bright blue eyes. He wore his usual Levi's and a black button-down, instead of one of his vintage concert T-shirts.

He dressed up for me, she thought.

"Is there anything I can do?" he asked.

Just having him standing next to her was making her feel dizzy. "No, I'm fine," she managed to say. "Just a little" — she glanced at her mom — "food poisoning," she said, wincing at the lie.

"Really?" Chris said, touching her back. "Do you need anything?"

His touch sent a lightning bolt down her spine. She wanted to just look up into his eyes and ask him to hug her, but she restrained herself. "I don't think so."

Richard Wu flipped open his cell phone. Hudson had never seen him without it. "I've got a doctor I can call," he said, already scrolling through his phone. "I think he's an internist."

"She's fine, Richard," Holla declared. "It was just a little bad tuna."

Richard's eyebrows shot up. "Really?" he asked, glancing at Hudson.

Hudson shrugged and nodded.

"Okay." He put his phone away, but he didn't seem convinced.

"I think I should probably get Hudson home now," Holla said. "You girls should go back to the party. Especially since you look so adorable."

"Thanks, Holla," they murmured, visibly uncomfortable.

"You sure, Holla?" Richard asked, scrutinizing Hudson as if she were a jigsaw puzzle he couldn't solve. "We're happy to help."

"I think I just need to take care of my little girl," Holla said sweetly, closing her hand around Hudson's arm. "But I'll let you know how she's doing."

"You got it," Richard said. "Feel better, Hudson."

"Thanks," she said, unable to look him in the eye. "I will."

Chris waved. "I'll e-mail you. Have a great holiday."

Hudson waved back. *I absolutely won't,* she thought.

Holla steered her in the direction of the lobby. "Tell Fernald we're coming out now," Holla told Little Jimmy, who pulled out a cell phone.

"Bye, H," Carina said. "We'll text you later."

"Someone has to get back to her lov-ah," Lizzie teased.

Carina rolled her eyes.

"Have fun, guys," Hudson said as they backed away down the hall. She wondered what they thought of her, going along with such a blatant lie. But they knew the deal: Nobody said no to Holla Jones. She just hoped that they didn't feel sorry for her.

She caught up to her mom as they trotted down a short flight of stairs to the lobby. Behind them, Little Jimmy lumbered, still huffing and puffing. In all the years they'd had a bodyguard, Hudson had never seen Holla actually follow one.

"I want you to look on the bright side," Holla said, leaning in to speak into Hudson's ear. "At least this happened here. And not somewhere important."

"But I thought this *was* somewhere important," Hudson said. "Wasn't that why you said I couldn't make one single mistake? Isn't that why you said 'this has to be good'?"

Holla fixed her almond-shaped eyes on her daughter. "Honey, what are you talking about?" she said, clearly puzzled.

Little Jimmy jogged up next to them. "Looks like we got a crowd," he said, gesturing to the lobby doors.

Outside, through the glass, they could see that a mass of people had formed on the street. Apparently word had gotten out that Holla Jones was at the Pierre. Word always got out that Holla Jones was somewhere. All it took was a call to one of the paparazzi agencies, who usually paid handsomely for the tip.

Holla pivoted on her toes to give Little Jimmy a stony look, and then he ran ahead.

"Mom? Can we talk about this?" Hudson said.

"Later," Holla said firmly. She paused for a moment just inside the doors. Hudson watched her mom form The Face — the cool, tough-as-nails, mysterious exterior that she always showed to her fans. Holla dug a pair of black sunglasses out of her bag and slipped them on to complete the look. Hudson knew their fight was already history.

Hudson let her mom go through the revolving doors first. When Holla emerged on the street, the crowd exploded.

"HOL-LA!" people screamed. "HOL-LA!"

Hudson pushed through the doors and then she was right behind her mom. Several hotel security guards rushed up to the crowd to keep them at bay.

"HOL-LA!" someone screamed. "I LOVE YOU!"

Holla gave the crowd a slight wave and they screamed even harder. Hudson darted over to the far side of the sidewalk, close to the hotel. Crowds always scared her a little. And she was still fuming. How could her mom have said this wasn't important? Hadn't three hundred people just seen the most humiliating moment of her life? How was she supposed to forget about it, when she knew that they never would?

She spotted their SUV down Sixty-first Street and broke into a run toward it, eager to escape the screams. But Holla took her time, lingering near the crowd, deliberately egging them on with her cool detachment. Holla never wanted to sign autographs or shake hands, but she also didn't like to rush past her fans. It was a little game she played with them — not wanting to leave, but not wanting to really do anything with them, either.

Suddenly a girl's voice rose out of the crowd: "I want to *BE YOU, HOLLA*!"

No you don't, Hudson thought, as she reached the car. *I tried it tonight, and it really doesn't work.*

chapter 4

The SUV snaked through the narrow, cramped streets of the West Village, going farther and farther west toward the river. Hudson leaned against the tinted window, listening to Nina Simone sing "Here Comes the Sun" on her iPod. The argument with her mom still hung in the air, but they hadn't said a word since they'd gotten into the car. Instead they sat in silence, Hudson with her iPod, Holla with her knitting needles. Knitting was Holla's new hobby. She liked to make long scarves that neither she nor Hudson would ever wear. Holla claimed the hobby relaxed her, but judging from how fast her hands were working, Holla seemed anything but relaxed.

Fernald, Holla's driver, zoomed right past the front door of their four-story redbrick Georgian-style mansion and turned the corner onto Perry Street. The mansion was more than a hundred and seventy-five years old, and supposedly Edgar Allan Poe had lived in it, once upon a time. But Hudson was pretty sure he wouldn't recognize it now. Holla had gutted the inside, leaving just the staircase, the fireplaces, and the crown moldings. She'd added a yoga studio, a fitness room, an underground parking garage, a screening room, and, on the roof, a swimming pool. *Architectural Digest* had called it "The Queen of Pop's Dream Palace." The only part of the house they didn't show in the magazine spread was the black iron fencing that surrounded it. Holla liked things to be secure, which was good, because every photographer in the world seemed to know they lived there.

As they coasted up to the garage door and waited for it to rise, several photographers leaped out of the shadows and aimed their zoom lenses at the car. They were always there, camped out across the street, ever watchful for an arrival or a departure. Hudson waved at them; she figured it was the polite thing to do. Holla didn't look up from her knitting. She only sighed as they drove past. "What do they think they're getting?" she asked. "The windows are tinted."

Fernald steered their car down the curving ramp and into the garage, right next to Holla's silver Mercedes and black Lexus. Holla owned three cars and didn't drive any of them — not because she didn't know how, but because she couldn't park and walk away. The last time Holla had tried to walk down the street, she'd been mobbed in under five minutes.

"Thanks, Fernald," Holla said when they'd parked.

"You're welcome, Miss Jones," he said.

"And how's your wife doing?" Holla asked. "Does she like the elliptical?"

"She loves it," Fernald said happily, turning around. "I think we've already lost five pounds apiece."

"Great!" Holla said, patting Fernald on the shoulder. "Keep up the good work!" Holla loved to give generous gifts to her staff, even if they did always seem to be tools for self-improvement — exercise equipment, a haircut, a free session of tooth-whitening. Fernald wasn't even overweight — he'd just had a little potbelly — but that didn't matter to Holla.

Hudson followed her mom out of the car and walked behind her on the way to the elevator. She shivered in the unheated space, pulling her unbuttoned coat closer around her bare shoulders. "Honey, just put that on," Holla said, turning around. "You're gonna catch cold."

"I'm fine," she said stiffly.

At the elevator, Mickey, one of Holla's iron-jawed security guards, held the door open for them. "Evening, Miss Jones," he said.

"Evening, Mickey," she murmured in response. Hudson and Holla squeezed up against the wall to make room for Little Jimmy, who jogged toward the elevator, panting. The doors began to close before he got there, but Holla kicked out one leg and forced the doors back with a bang. He scooted inside.

"Thanks, Miss Jones," he said as the doors rumbled shut.

"You know, Jimmy," Holla said. "You're welcome to join me in any of my exercise classes. I'll have Raquel give you a schedule."

"Thank you, Miss Jones," said Little Jimmy, and Hudson could hear the embarrassment in his voice.

Since she was twelve, Hudson had been expected to attend at least two fitness classes a week, which could be anything from yoga to power hula-hooping. Now poor Little Jimmy was going to get roped in, along with the rest of the staff. She just hoped that he wouldn't be caught eating meat or cheese in front of Holla; that would be enough to get him fired. Holla had been obsessed with being "healthy" for years. Diabetes and heart disease ran in the Jones family, so Holla had cut meat, wheat, white flour, and sugar from her diet. This left fish, vegetables, fruit, whole grains, and all kinds of tofu. Naturally, everyone around Holla — including Hudson — was expected to eat this way, too. Even though Hudson could technically eat whatever she wanted when she wasn't at home, she found that she stuck mostly to food that was Holla-approved. Eating anything sweet — even Pinkberry — actually felt a little dangerous.

The elevator opened and they stepped into Holla's spacious all-white chef's kitchen. It was lined with gleaming chrome appliances and glass cabinets, and it had two of everything — two dishwashers, two refrigerators, and two six-burner stoves. Holla's kitchen could feed at least a hundred people, not that they'd ever tried it — Holla didn't usually have parties. The kitchen also doubled as the headquarters for Holla's live-in staff. When they walked in, Holla's blond, rail-thin chef, Lorraine, was rolling out dairy-free pastry on the butcher-block table; Mariana, the curvy Brazilian housekeeper, breezed through with armfuls of fluffy white towels; and Raquel, the sweet-faced and frighteningly competent house manager, polished a stack of silver. There was more staff, of course — a publicist, several yoga and fitness instructors, a business manager, a dog walker — but Lorraine, Mariana, and Raquel were the skeleton crew. There was also Sophie, who'd somehow beaten them downtown from the Pierre and was now sitting in front of a large computer monitor in the corner, reading e-mails. Seeing Sophie, Hudson wondered if Ava had believed the food-poisoning excuse. She walked over to the marble island in the center of the room and grabbed a handful of cut-up raw vegetables. Maybe chewing would relieve some of her stress.

"So?" Raquel asked, looking up from the silver ladle she was polishing. "How did it go?" Raquel had always worn her long, thick black hair in a braid, until the previous month, when Holla had decided she needed a change. She'd sent Raquel off to a boutique salon in SoHo, where they'd given her a layered bob. It still didn't look quite right on her.

"It didn't go so great," Hudson said.

"She had stage fright," Holla said bluntly, removing her leather coat and draping it over a chair.

"Oh," Raquel said, her face crumpling. "I'm sorry."

"It's fine," Hudson said, suddenly mortified. "It wasn't so bad."

"Would you like a little tea?" Lorraine asked gently, putting the kettle on the burner.

"That's okay," Hudson murmured.

"I'll have some," Holla said, brushing a perfect curtain of hair over her shoulder. "And Sophie? What did you say to people?"

"That it was food poisoning," Sophie chirped, avoiding eye contact with Hudson. "From tuna."

"Good," Holla replied.

Hudson felt the staff's eyes on her. It was Holla's lie, but she felt like a liar, too.

"And where are we with the tour?" Holla asked.

"Wembley in London, Madison Square Garden, Slane Castle in Dublin, and Staples Center in Los Angeles all sold out," Sophie read off the computer screen. "Tokyo goes on sale tomorrow. Sydney's still a question mark."

"Hmmm," Holla said. In May, Holla would release her tenth album. This summer would be her fifth world tour. "And where are we with *Saturday Night Live*?"

"You're booked. March seventh."

"Wonderful!" Holla clasped her hands and turned to face Hudson. "Wait — I just had a thought."

"What?" Hudson asked cautiously. She didn't like the way her mom was looking at her.

"What if you did *Saturday Night Live* with me?" Holla asked. "A mother-daughter duet." She looked at Sophie and Raquel. "Don't you guys think so? It would be fun!"

"Are you... are you serious?" Hudson stammered.

"Don't worry, you'll be a pro at this by then," Holla said, accepting a mug of ginger tea from Lorraine. "Oh, Sophie, call them back, would you? Call them back and tell them that —"

"Can I talk to you upstairs?" Hudson asked.

"'Course, honey," Holla said, taking a sip of tea. "You go on ahead. I'll meet you in your room."

Hudson grabbed her coat and walked to the staircase, stomping with barely concealed rage. Forty-five minutes earlier she'd frozen onstage and fled in terror, and now her mom wanted her to repeat the whole thing on live television? Hadn't she seen what had just happened? As usual, her mom was in complete denial of reality, just because she wanted something. She'd have to be honest with her about the real reason she'd run off the stage; she would have to tell her *exactly* how much pressure she'd put on her. And this probably meant having a terrible, earth-shattering fight.

When she reached the third floor she heard the sound of tinkling metal and scuttling paws as Matilda, her brindle-colored French bulldog, ran to greet her.

"Hi there!" Hudson said, scooping the dog into her arms. "How's my little girl?"

Matilda gave Hudson's chin a good licking, and Hudson rubbed Matilda's stubby head. "Mommy totally blew it," Hudson said.

Matilda gave an uncertain snort.

"Nope, it's true," Hudson said, and then put her down. Sometimes she wished she could be Matilda and not have to worry about anything but finding a cozy place to lie down and sleep.

Hudson walked into her room, which was technically a suite. The first room was where she did homework and practiced piano, and the second was her bedroom and closet. It was lucky that she had two rooms because every square inch of both was stuffed. Hudson loved to collect clothes, furniture, albums, Barbies—anything, really. Traveling with her mom on tour over the years, Hudson had been able to find items from all over the world. In the living room were a sheepskin rug from Denmark, a mirrored vanity table from an antiques store in Paris, and a battered leather armchair from a flea market in SoHo. In her bedroom stood a full-length mirror with claw feet; a shabby-chic, whitewashed dresser; and a vintage wrought-iron daybed from London, which was in turn covered with silk cushions from India. She could spend hours at a flea market, rifling through what other people thought was junk. And even though she loved all fashion, vintage designs were her favorite. She liked to think that she was stepping back in time whenever she slipped on clothes from another era.

"Why would you want to have other people's furniture?" Holla would ask, slightly aghast, whenever Hudson lugged home a cool footstool or area rug. Hudson would just shrug and smile. That was the entire point: Other people's marks on her clothes and furniture always made them seem more real.

As Hudson padded across the sheepskin rug, she tried not to look at her most important secondhand item, standing in the corner: a baby grand Steinway piano. It had been her Grandma Helene's. It was the first piano Hudson had ever played, back when she was five. She'd climbed right up onto the bench and started picking out chords by herself as her grandmother watched, amazed. Grandma Helene could play anything by ear, and she'd tried to teach her two daughters, Holla and Jenny, from the time they were little. Neither of them had cared much for it. Hudson, though, was different. She'd gotten all of Grandma Helene's talent, and then some. Grandma Helene became her first teacher, and Hudson was a fast learner. At seven, she learned Beethoven's Moonlight Sonata. At eight, she could play Chopin's Minute Waltz. At nine, she began to write her own songs. Grandma Helene gave Hudson the piano shortly before she became too sick to play it anymore. When she passed away, Hudson had the piano brought to her home and put in her bedroom, where she covered the walls and floor with special sound-absorbing pads. She played every night before she went to sleep, imagining that her grandma was still listening.

"Lizzie writes stories, I surf, and you play piano," Carina would say, and it was true. If she didn't play for a couple of days, Hudson would feel herself start to get anxious, and she would toss and turn in bed at night, thinking about scary things. But as soon as she sat down at her piano and played, all of that would go away. Writing music calmed her down and helped her cope.

But now, as she walked into her bedroom, she ignored the beautiful Steinway. The piano was what had gotten her into this mess in the first place. She wouldn't even look at it, let alone play it.

Hudson unbuckled the straps of her heels and pulled off her dress, then changed into a pair of flannel pajamas. Even with her mom's expensive renovations, the heating system in the old house could be a little funky.

"Honey?" Holla called out from the other room. "I brought you some tea." She walked into Hudson's bedroom and placed the mug on a side table.

"I didn't really have food poisoning," Hudson said.

"I know," Holla said, playing with the chains of her necklaces. "But you've had a rough night." Holla leaned down and straightened some picture frames on Hudson's bedside table.

"And asking me to be on *Saturday Night Live* is gonna make me feel better?" Hudson asked, flopping down on the bed and grabbing one of her silk-covered pillows.

Holla looked at her. "If you're going to make this your career, honey, you have to learn to let things go."

"Right. Like when your whole class watches you blank out,"

Hudson said. "Those were people I *know*. People I go to school with. It wasn't some random audience. And they're not gonna let it go."

"You can't care what people think," Holla said more forcefully. "That's what being an artist means. Do you think *I* ever cared about what people thought?"

"So the solution is for me to do live television," Hudson said.

"You just got afraid up there," Holla declared. "By the time you do the show, that'll be over with."

"What if it's not?" Hudson asked.

"Why do you always have to look on the dark side of everything?"

"And why do you always have to freak me *out* about everything?" Hudson asked.

"What? How did I freak you out?"

"By picking me apart. By telling me I'm doing everything wrong, all the time."

"I was giving you advice," Holla said flatly. "Can't I do that?"

"But ever since I started this album, it's like I can't do anything right. You want me to do things exactly like you."

Holla furrowed her brow, the way she did whenever Hudson said something she thought was ludicrous.

"You changed everything on my album," Hudson went on. "You changed every song."

"Because I wanted your album to sell," Holla replied.

"But those were my songs!"

"So you *don't* want to be a success?" Holla said, letting her voice get loud. "You *don't* want to sell out stadiums and be on the radio and have little girls scream your name when you walk down the street?"

Hudson squeezed the pillow. *Here we go,* she thought. *The unanswerable questions.* "I just don't want to be told over and over that I'm doing something wrong."

"Is that what you think?" Holla folded her arms. "Honey, you are gonna have to get a thicker skin. That's all this business is, you know. Being criticized. For everyone who loves you out there, there's someone who thinks you're awful. But that's not what I was doing. I'm just trying to help you. Who believes in you more than I do? Who's been your biggest champion your whole life? Who got you the best piano teacher in the city? Who put you in dance classes when you were five because you wanted them?"

Hudson picked at a loose thread on her bedspread, waiting for her mom to finish.

"I wish someone had believed in *me* enough to cheer me on," Holla went on. "I had to beg my mother to give me dance lessons. I had to beg her to take me to talent contests in Chicago. I had to do everything myself. My mom didn't care. That's why your aunt Jenny still doesn't know what she's doing with her life. Running around the world, pretending she's some kind of fashion designer —"

"*Jewelry* designer," Hudson corrected.

"Jewelry designer, whatever." Holla snorted. "She could have been dancing *Swan Lake* at Lincoln Center, and now she's flailing around, going nowhere. God knows I tried, but there was only so much I could do, after all." She shook her head, as if the memory of her younger sister was too much. "So listening to you complain about my interest in your career... It just sounds a little ungrateful. And tonight wasn't my fault. You can't pin that on me. I'm sorry you got scared, but that wasn't my fault."

Hudson knew her mom was right. She was trying to blame Holla, when the real problem was that Hudson just didn't have what it took. She'd *thought* she did. She'd thought she could be a musician. But she'd been wrong. How had she ever thought she'd be able to do this? When she could barely talk in class?

"I don't think I can do this," she said. "I want to get out of it."

"*You're* the one who wanted to do this," Holla said. "You're the one who told me you were ready."

"I know," Hudson said quietly. "I've changed my mind." Just saying it was such a relief.

"Honey," Holla said, moving closer to Hudson and taking her hand. "You're just upset. Don't say things you don't mean."

"I do mean it," Hudson said, looking her mom straight in the eye. "I've never meant anything more. I don't want this," Hudson said. "I'm not like you. We both know that."

Holla's face grew serious. She stood up from the bed. "You're making a big mistake. One that you will regret the rest of your life."

"I'm fourteen," Hudson said. "I'll have another chance."

"Not like this," Holla said. "I'm not always going to be able to help you this way. You're throwing away a huge opportunity here, an opportunity other girls would kill for."

But I'm not like other girls, Hudson wanted to say.

"And what are we going to tell your label?" Holla asked, her voice rising again. "That you just want to scrap it all? After we made all those changes?"

"Tell them I'm so sorry. Tell them I have stage fright. I don't know. Tell them anything."

Holla tapped her foot on the wood floor. "It's a good thing you're my daughter. Otherwise they would sue you for breach of contract."

"I know I'm your daughter." Hudson sighed. "Believe me."

Holla stared at her for a few more seconds. "I never thought that you'd be the type to give up."

For a second Hudson felt a twinge of sadness, mixed with fear—a feeling of regret before it had actually become regret. "Well, I guess that's who I am," she said, and lay down on her bed, facing the wall.

For a moment there was silence, and then she heard Holla walk out of the room. Hudson listened as the door shut, her eyes still on the wall. It was over.

Hudson lay there, unable and unwilling to move. Outside a car alarm wailed. She shut her eyes and wondered if she could just stay in this position for the rest of her life. *So that's it,* she thought. The album. Her music. Her dream. Everything. It was all finished. She knew that she'd done the right thing, but it didn't feel right. She felt even worse now than she had in the bathroom at the Pierre. But she hadn't had a choice, really, when she thought about it. Tonight had shown her that.

After a few minutes, Hudson got up and walked into the next room. Matilda stared at her from her dog bed, tilting her blocky head as if to say, *That didn't go so well, huh?*

Hudson grabbed her laptop. In times of crisis, she always needed to do two things: text Carina and/or Lizzie, and check the next day's horoscope. First she logged on to signsnscopes.com and checked Pisces for the next day, December 21.

Congratulations! With Uranus, the planet of surprise, moving into your tenth house of career, expect a major work development that will have you smiling!

Hudson closed her laptop. She hated it when she did something to contradict her horoscope. But maybe astrology was just a bunch of nonsense, anyway. According to her chart—the one Aunt Jenny had given her for her last birthday—she was supposed to be incredibly successful. Famous, even. "Almost as famous as your mom," Jenny had whispered, with a wink.

Hudson went back to her desk and got out the chart. It was covered in one of those sheets of plastic used for term papers. The chart was a large circle, sliced into wedges and covered in weird hieroglyphics and waves and cut through with straight lines radiating in all directions. She didn't know enough to actually read charts yet, but she remembered the spot on the circle where her career was. It was a mess of squiggly lines and shapes.

"Whatever," she thought, and put it back in the drawer, shoving it in far enough that she wouldn't be able to look at it again without doing some serious spring cleaning.

Just then her phone, which was still in her purse, on the floor, chimed with a text. She reached down and pulled it out.

It was from Lizzie.

Why'd u leave me w/this??:)

Just below the text was a photo of Carina and Alex with their arms around each other at the Ball, smiling and looking goofily into each other's eyes.

Hudson wrote back:

Because I have food poisoning, remember??

Hope you've been throwing up for hours ☺ Lizzie wrote.

U know it.

At least she still had her friends, she thought later, as she climbed into bed. And no matter what happened, that would never change.

chapter 5

Hudson jumped back into yet another *chataranga,* and her elbows buckled as she lowered herself, push-up style, an inch from the floor.

"Now curl your toes into upward dog, and take a deep healing breath," said Niva, Holla's yoga instructor, who paced calmly in front of them. "Reach your heart to the front of the room."

Hudson moved easily into upward dog, balancing her weight on her hands, but reaching her heart anywhere was a tall order. It was Christmas morning, and it was snowing, and she wanted to be lying in bed, smelling Lorraine's vegan chocolate bread pudding as it baked in the oven. Or walking Matilda down the snow-dusted West Village streets. Or exchanging gifts with Lizzie and Carina. (They'd tried once to do "Secret Santa" but the "secret" part had lasted only a few minutes.) Instead she was on a mat, sweating, just inches from her mom, who, several days after their fight in Hudson's bedroom, still seemed to be angry.

"Now curl your toes for downward dog," Niva droned in her ethereal yoga voice.

Hudson grunted a little as she moved back into the pose. Christmas was always a little tricky in Holla's house. It was the one day of the year when all the people who usually swirled around them — Holla's publicist, record executive, and manager, to name a few — went home to pay attention to their own lives and families, and the house was unusually quiet. Sometimes Holla could get her yoga teacher or makeup artist to come over for Christmas dinner, but today it would be only Jenny, Holla's sister. And Hudson knew that this was the real reason she felt a little antsy.

"Feel your limbs get heavy," Niva said as they lay in resting pose at the end of the class. "Feel your arms sink into the floor."

Hudson fluttered her eyes closed. She thought of Lizzie and Carina, who'd been so sweet about her disastrous night at the ball. The next day the two of them had taken her out to Pinkberry for a quick pep talk and reality check. "Nobody even talked about it the whole night, I swear," Carina had said around a mouthful of Cap'n Crunch–topped yogurt.

"Uh, that's because you were on Planet Alex all night," Lizzie had teased, nudging Carina with her elbow.

Now Lizzie was at her grandparents' house down in northern Florida, and Carina was with her dad in Aspen. The city felt empty without them, even though they'd been texting one another the entire week.

"Breathe in through your nose," Niva droned on, her voice getting more and more unworldly. "Let go of all earthly worries..."

Sounds good, Hudson thought, drifting close to sleep, just as the door to the yoga studio opened with a loud *whoosh*. Heavy footsteps shook the bamboo floor.

"Merry Christmas!" said a high-pitched voice.

Hudson opened her eyes and craned her head. Aunt Jenny was standing in the middle of the room, clutching some red shopping bags and looking amazing, as usual. She wore a belted coat of what looked like lavender-dyed rabbit fur, a thick cashmere scarf in deep purple, and earrings that were actually safety pins dripping gold chains. Her cropped hair showed off her beautiful oval face, which was, as always, free of makeup. Her boots were part grandma, part sexy, with pointy toes and stiletto heels, and her nail color matched the inky black of her vinyl bag. Jenny had been the one to take Hudson to her first flea market, and she'd given Hudson her first handbag: a crimson alligator Ferragamo clutch from the sixties. Hudson still had it, but she never left the house with it; she was too scared of losing it. "Oh," said Jenny, letting her shopping bags drop to the floor. "I thought you guys would be done by now. Sorry!"

She was famous for being either too late or too early, and today her arrival had fallen on the way-too-early end of the spectrum. They hadn't expected her until lunch.

"That's okay," Holla said. She slowly sat up. "We're just finishing."

Hudson sat up. "Hi, Jenny!"

"Hey, Hudcap! You look great! Awesome headband."

"Thanks!" Hudson beamed as she got to her feet. She loved her stretchy headband, which was covered with stones that looked like diamonds.

"Sweetie, the *floors*," Holla said.

Hudson looked down, but then she realized that Holla was talking to Jenny. Jenny's spiky heels had made scuff marks on the floor.

"Oh, whoops, sorry about that," Jenny said, examining the heels of her boots with an embarrassed smile.

"Just take them off," Holla said, smiling tightly. "Please. We don't allow shoes in here."

Jenny unzipped her boots and slipped them off. "Sorry, again," she said.

"No problem," Holla said warmly once the boots were off. She walked over to Jenny and the two sisters shared a tepid hug. "How are you?" Holla asked.

"Jet-lagged," Jenny said. "Ever since I got home I've been getting up at the crack of dawn."

"You've gotten taller," Jenny said to Hudson. "When was the last time I saw you?"

"I think just before you moved," Hudson said. She grabbed Jenny's hand. "How's Paris?"

"Incroyable," Jenny said with a smile.

"I want to come visit!" Hudson exclaimed. "Maybe for spring break?"

Jenny's smile faded, but only a little. "Sure."

"Are you hungry?" Holla asked.

"Actually, breakfast would be great," Jenny admitted.

"I'll take you," Hudson answered. She led her aunt upstairs to the kitchen while Holla left them to shower and change.

Hudson made herself a bowl of steel-cut oats while Jenny devoured one of Lorraine's gluten-free muffins.

"So my jewelry studio in Paris was awesome," Jenny said as she pulled off another hunk of muffin. "I shared it with three French girls. They were really mean at first, but then I figured out how to talk to them. French girls definitely make you work for it, but then they open up and they're really cool. I'll miss them."

"So...how's that guy?" Hudson asked, trying not to sound too nosy.

"What guy?" Jenny asked, wiping off her fingers with a napkin.

"The guy you moved there for," Hudson said. "The one you said was The One."

"Oh," Jenny said, grimacing slightly. "Jean-Paul." She shrugged. "It didn't really work out. He was a Capricorn. Totally wrong for me." She waved her hand. "Next time I really need a Sagittarius. Or a Libra. Plus, Paris was different than I thought it would be. I don't know. It was a little cold. I missed New York."

Hudson nodded uncertainly. "So you're not staying there?"

"No," Jenny said, shaking her head. "It's better if I leave."

Hudson ate another spoonful of oats and tried not to worry. One thing was for sure: Her mom wasn't going to like hearing any of this. Jenny had lived in more places for less time than anyone else Hudson had ever heard of. And each time she moved, Holla got more frustrated.

"So you've moved back here then?" Hudson asked.

"Well, sort of," Jenny said. "I still have to go back to Paris and get my stuff. But let's not talk about it, okay? I want to hear all about what's going on with *you*. How are Carina and Lizzie? Are they coming over?"

"No, but they're fine," she said. "They both have boyfriends now."

"Really?" Jenny's eyes lit up. "So now we have to get *you* one."

Hudson shrugged. "You know me. I'm not really into high school guys."

"Smart. Always go for the older ones," Jenny said. "They're the ones who know how to treat you. Then again, Jean-Paul was fifty, and that didn't mean anything." She put down what was left of her muffin. "So how are things with your mom?" she asked in a more serious voice.

Hudson hesitated. She'd learned not to tell Jenny too much about stuff with her mom. "Things are fine," she fibbed, scraping the side of her bowl with her spoon.

"Does she let you do anything?" Jenny asked. "I can tell she doesn't let you *eat* anything," she said, looking at the oatmeal. "But does she let you go anywhere? Or do you still have to be driven around in that car?"

"She's strict," Hudson said simply, hoping to get off the topic.

"And how's your album coming along?"

Hudson stood up quickly. "You know, I think I should probably get in the shower," she said. "I don't like to sit around in wet workout clothes. It's gross."

Jenny stared at Hudson. "I didn't mean to pry, Hudcap. I'm just interested in you."

"I know, but really, everything's fine," she said. "I'll be right back."

In the shower, Hudson lathered her hair, feeling guilty about her rude departure. But there was something about Jenny that made her nervous. Talking to her felt like walking a tightrope — if Hudson told her too much, or said the wrong thing, she would fall, and she didn't know where she might land.

After showering, she dressed in a long-sleeved silver tunic that always made her think of tinsel, pulled on her electric purple skinny jeans, and thrust her feet into some high UGGs. When she walked downstairs, she found that Holla and Jenny had moved into the serene, all-white living room and were curled up on the low, long sofa. A gigantic painting consisting solely of a splash of pink and purple hung on one wall. Candles flickered on the white piano in the corner, and a fire crackled in the fireplace. Holla tried hard to make this house cozy, but it was always a little too big, a little too cold, and a little too white for Hudson to really get comfortable.

"Hey! You're just in time for the trunk show!" Jenny joked. "Sit down and pick something out." Several different racks of earrings and bracelets were set up on the glass coffee table next to an open, velvet-lined display case. The shopping bags Jenny had carried into the yoga studio sat at her feet.

Hudson plopped down next to her aunt. "Those are cool," she said, pointing to a pair of silver earrings in the shape of serpents, with garnets for eyes.

"Try them on," Jenny said, taking them off the rack and handing them to Hudson.

Hudson slid them into her ears. Jenny hadn't been designing jewelry for very long, but she already knew what she was doing.

Hudson grabbed the hand mirror on the table and looked at herself. "Wow," she said, turning her head this way and that, watching the red stones catch the light.

"They're all yours," Jenny said.

"I hope the *InStyle* mention helped," Holla said, leaning back against a pile of throw pillows. She looked relaxed and comfortable sipping from a steaming mug of chai tea, an oversized white blanket over her legs, but Hudson could detect a slight edge in her voice.

"Yeah, it was great, thanks," Jenny said, running a hand through her hair. "Of course, I wasn't really ready to fill orders yet, but it always helps to get your name out there. So, Hudson," she said, turning back to her niece, "do you want those, or do you want to try a few others on?"

"I'll definitely take these," Hudson said, putting down the mirror. "Thanks."

"So... are you moving back here?" Holla asked, taking another sip of tea.

Jenny turned back to her sister. "Yes. Probably right after the holidays. But I spent the time in Paris well, and I made a ton more pieces, and there are a few stores in SoHo now that I think would be just perfect for my stuff. So I'm really excited."

Holla nodded slowly. "And the guy... just didn't work out?"

"No," Jenny replied. "Not so much."

"Well, in that case," Holla said delicately, "I want to propose something."

"What?" Jenny asked, not moving.

"You remember Kierce, my stylist?" Holla asked.

"Uh-huh," Jenny said carefully.

"Well, you know I've always loved your style, and obviously Hudson does, too." Holla chuckled. "And I've been thinking that before I start prepping for the tour next summer, I really ought to try someone new."

Jenny stared at Holla. "You want me to *work* for you?" she asked, baffled.

"Just until you get back on your feet," Holla said.

Jenny just stared at her.

"I think it would be fun," Holla went on. "And honestly, what other prospects do you have right now? What other incredible offers? You've been designing this jewelry for two years, and it's not *exactly* taking the world by storm —"

Jenny's shoulders sagged just the slightest bit at this.

"And I just see you going in circles, honey. I know how talented you are, and instead you're just running around from one city to another —"

"Just stop, okay?" Jenny interrupted suddenly. "Please? Stop."

"What?" Holla asked, sounding genuinely confused.

"That's possibly the worst idea ever," Jenny said. "God, Holla. Sometimes I don't know what planet you're living on."

"What planet *I'm* on?" Holla asked. "While you're running off to Paris to live with some guy you just met? Pretending you're a jewelry designer? Or is it a photographer? Or a handbag designer? I've lost track."

Jenny started pulling the earrings off the racks and placing them in the velvet-lined box. "Right," she muttered. "I forgot. You have all the answers."

"I don't," Holla said. "I just want you to be happy."

"I *am* happy," Jenny shot back. "I like my life. It's exciting, okay? I *like* the way it is."

"Mm-hmm," Holla said. "What's not to like? You get to traipse around Buenos Aires and come home and have me pay your credit card bill."

"That happened *once*," Jenny said, shooting to her feet. She started to gather her things.

"Where are you going?" Holla asked.

"Out of here," Jenny said, grabbing the shopping bags.

"Don't," Holla said. "It's Christmas. Let's just forget it and start over."

"Too late," Jenny said bitterly. She glanced around the room. "Where'd they put my coat?"

Holla pointed to the kitchen. "Back there," she said with an air of defeat.

Jenny leaned down and gave Hudson a hug. "Merry Christmas, Hudcap."

Hudson hugged her aunt back. "Thanks for the earrings," she whispered helplessly.

"Oh, and here," Jenny said, pointing to one of the shopping bags. "These are for you guys." She crouched down and pulled out several presents wrapped in shiny bronze wrapping paper and tied with red velvet bows. She stacked them carefully on the coffee table. "Just a few things from Paris."

Holla eyed the gifts but didn't move to take them. Hudson knew what they were: deliberately out-of-focus photos Jenny had taken and had carefully framed. For Christmas, she always gave them photographs of whatever city she'd just been living in.

"Have them take you down the back way; the front door's locked," Holla said.

"Of course it is," Jenny said, walking toward the kitchen.

They listened to Jenny ask Raquel for her coat and then leave through the service door. Holla stared gloomily into the crackling fire. Several long minutes seemed to pass. Holla's hands tapped out a silent rhythm on the top of the couch while Hudson sat next to her mom in silence. She didn't know what to do. She never did at times like this. "Mom?" she finally said.

Holla swung her feet to the floor and stood up, brushing at her leggings with her hands. "Honey, are you letting that dog jump up onto this couch? I've got hair all over me."

Holla's moment of reflection was clearly over.

"She doesn't even come into this room," Hudson said.

"Well, make sure she doesn't," Holla said. "I'm gonna check on lunch." She padded off in the direction of the kitchen, leaving Hudson alone.

Hudson hugged her knees to her chest and stared into the fire. This kind of fight had happened before, but for some reason she'd felt that Christmas would defuse the normal tension. She was sad for her mom. She knew how much Holla loved Jenny and wanted her life to work out. It wasn't fair for Jenny to always shoot her down. But she could also see how being Holla's stylist wasn't exactly a dream job.

And maybe it was better that Aunt Jenny had left. Having her around only made things more confusing. And today, it had made Hudson feel scared. Jenny had turned her back on her dancing talent and had become, according to Holla, a world-class screwup. And right now, as she sat by herself in front of the cold fire, Hudson felt like she was headed for exactly the same fate.

chapter 6

"Is it just me, or did this hallway get longer over break?" Hudson whispered, doing her best not to make eye contact with the cluster of girls standing by the bathroom door. She knew they were staring at her. And whispering. And it was making the walk to the stairwell a little torturous, even though she was sandwiched between her two best friends.

"Don't worry," Lizzie said. "We'll ignore them."

"No we won't," Carina whispered, and then looked over at the girls. "Um, is there a problem?" she demanded, sending the girls scurrying into the bathroom.

"You're making it worse, C." Hudson groaned.

"Well, we have to do something," Carina said. "They can't just think that's *okay*."

Hudson shifted her book bag up her shoulder and headed into the stairwell. The first day back at school had been worse than she'd thought. After Christmas Hudson spent the rest of break holed up in her room reading a stack of fashion magazines and books she'd picked out in the New Age/Self-Help section, with titles like *Falling Down, Getting Up* and *When Bad Things Happen*. She convinced herself that no one would remember the Silver Snowflake Ball by the time they went back to school. The night before school started, she'd even checked her horoscope just to be prepared. She'd read:

Tuesday, January 6—Pisces

With Pluto sitting firmly in your seventh house, get ready to be the center of attention! You'll be on everyone's lips—like it or not!

"Oh, God," she muttered and clicked off before she finished reading. That wasn't good. But still. What were the chances that almost three weeks later, people would still care?

She got her answer as soon as she stepped onto the floor of the Upper School. Ken Clayman and Eli Blackman were leaning against the wall under the bulletin board. Their faces lit up when they spotted her. "Hey, Hudson!" Ken called out. "Did you have any sushi for breakfast?"

Hudson darted down the hall toward her locker, her ears on fire. On her way she passed Sophie Duncan and Jill Rau, who grabbed each other, burst into giggles, and kept walking.

They remember, Hudson thought. *And nobody, nobody, believes it was food poisoning.*

Even seeing her friends for the first time in weeks was no relief.

"You guys, everyone remembers," she whispered to them in

homeroom, just as Ava Elting walked in and sent a searing look at her from across the room.

"No they don't," Carina said. Then she noticed Ava's glare. "Oh, yeah," she said. "They totally do."

"Carina," Lizzie complained.

Todd just gave Hudson a sympathetic smile.

As Madame Dupuis called roll, Hudson focused on a heart she was drawing, over and over, on a page of loose-leaf paper. When her name was called, more whispers and giggles rose from the back of the room.

"Here," Hudson said meekly.

At school she'd always been an object of vague curiosity, a kid other people noticed when she rose her hand in class, or when she wore her feathered headdress to the school dance. She was used to a certain amount of attention because of her mom, and because of her clothes. But that had been positive attention. This was different. Now she felt like a freak.

When it was time for their first free period of the day, Hudson couldn't get down to the library fast enough. "The worst part is, I can't even tell people the truth," Hudson said, still between Lizzie and Carina as they walked down the stairs to the library. "Now everyone thinks I can't sing, and that I made up some excuse to cover it."

"But food poisoning's, like, totally believable," Carina said.

"The point is I should have just told everyone I had stage fright."

"Somehow I don't think that would have gone over any better," Lizzie said.

"And I couldn't have gotten stronger hate vibes from Ava Elting," Hudson said. "She must totally want to kill me."

They walked into the library and Hudson stopped in her tracks. Just inside the doors was an empty table, covered with bags. And in the middle of the table, holding court like its owner, was a familiar black and red Hervé Chapelier bag. Ava's bag.

"Are you sure Ava didn't say anything to you?" Hudson asked Carina. "Like, that I ruined her party or something?"

"Not a thing," Carina said.

Knowing Ava's sometimes-shaky hold on the truth — she'd actually said Todd had cheated on her in order to save face when he dumped her — Hudson was pretty sure that Ava was talking about her behind her back. Especially after that furious look Ava shot her in homeroom. "You guys, let's sit back there," Hudson said. She pointed to an empty table in the corner farthest from Ava's table. It was the least she could do.

"Is Todd coming?" Hudson asked.

"He's coming a little bit later," Lizzie said. Hudson wondered for a moment if everything was okay between them. "So you're still scrapping the album?"

"Yup," Hudson said determinedly. "The record label peeps weren't too happy, but I guess my mom convinced them."

"That's too bad," Lizzie said.

"What do you mean?" Hudson asked.

"Just that you're not going to finish it," Lizzie said, opening her History book.

"Well, you guys saw what my mom did to it. How she totally took it over. It wasn't even mine anymore. Scrapping it was the right thing to do." Hudson took out her Geometry book. "I think I need to forget about music."

"Forget about music?" Lizzie said. "Are you serious?"

"It's the one thing you really love," Carina protested.

"I love other things," Hudson said. "Like fashion. Like astrology." She knew she didn't sound convincing. "Whatever, I'm taking a break. Believe me, it's the best thing for my sanity."

"Speaking of sanity," Carina whispered, tilting her head, "your biggest fan looks just as unhinged as ever."

Hudson looked over. Across the room, hunched over what looked like the *New York Times* crossword puzzle, marking her answers with a fountain pen, was Hillary Crumple. To Lizzie and Carina, Hillary Crumple was pretty much a stalker, blatantly obsessed with Hudson and her mom. They even thought that Hillary had given one of the tabloids Hudson's cell number. But Hudson doubted it. For one, Hillary just didn't look clued-in enough to do that kind of thing. She just seemed a little . . . different. She wore her brown hair tied back in a ponytail, but most of it hung loose and floated, staticky, around her head, despite a few plastic barrettes. She wore acid orange roll-neck sweater had blue waves sewn on the front, and a sequined dolphin jumped through them. Her Chadwick kilt hung to her shins, well past the appointed "cool" length of above the knees. As Hillary filled in another box, unaware of being watched, she made a tiny victory pump with her free hand.

"Wow," Hudson said, genuinely impressed. "She's not even doing that in pencil."

"I don't know," Carina said, eyeing her. "I still think she's the one who gave that tabloid your number."

They watched as Hillary put down her pen, stood up, and walked out of the library.

"Do you think she heard us?" Carina asked.

Lizzie shook her head. "What did they want?" Lizzie asked Hudson. "The tabloids? You never told us."

"Oh, some rumor about my mom dating John Mayer or something." Hudson shrugged. "The usual."

Just then Ava breezed into the library. Her devil-horned knit cap and silver coat were both dusted with snow, and as she sipped from her Starbucks cup, she locked eyes with Hudson and approached the table.

Oh, great, Hudson thought, looking down.

"What?" Lizzie whispered.

"Ava," Hudson said. "Incoming."

"Hi, Hudson," Ava said, coming to stand right next to Hudson's chair.

"Hey, Ava," Hudson said, barely able to look her in the eye.

"Hey, Ava," Carina said.

"Hey, Ava," Lizzie murmured.

"So how was your break, you guys?" Ava asked, popping the lid off her cappuccino. "Did you go away anywhere cool?" Hudson could already smell Ava's Daisy perfume.

"Vail," Carina said.

"Florida," said Lizzie.

"I was here," Hudson said.

"I was back down in Mustique," Ava said, sighing grandly. "It is just sooo beautiful there."

Good to know, Hudson thought. *Now please leave.*

"So, what happened at the Ball?" Ava said, turning to Hudson. "I thought for sure I'd hear from you after...I mean, given your dramatic exit."

Hudson played with the loose spine of her textbook as she felt a blush heat her face. "I wasn't feeling that well," she said, because she couldn't think of anything else.

"Oh, riiiigghht," Ava said, drawing out the word. "I forgot. What was it, again? Some bad tuna?"

Hudson glanced at Carina.

"Have *you* ever had bad tuna?" Carina asked Ava.

"No, but I'm sure it's awful," Ava said, narrowing her big brown eyes. "Almost as awful as not being able to sing."

"I have to go to the bathroom," Hudson said, getting up.

"More food poisoning?" Ava asked with mock sympathy.

Hudson walked past her, toward the door, and out into the hall. *I hate Ava Elting,* she thought, speeding to the bathroom. *I know it's not right to hate people, but she is truly the most evil person on earth.*

Inside the bathroom, she locked herself in a stall and looked at her watch. Ten twenty-five. This was going to be the longest day of her life. She just needed to visualize getting through this, like she'd read over break in one of her New Age books. If you visualized what you wanted, the book said, most of the time you could make it happen. She closed her eyes and pictured herself walking proudly through the school halls, her head held high, immune to the stares and whispers...and then breaking into song when people least expected it...

And then the girls' room door opened.

"I mean, it's one thing to not be able to sing," a girl said in a familiar lockjaw drawl. "But to tell everyone you have food poisoning? I'm just so embarrassed for her."

It was Ilona Peterson, Ava Elting's head henchwoman and easily the meanest girl in the freshman class.

"Oh my God, totally," said Cici Marcus, in her harsh, brittle voice. "Did she really think that people would buy that? *Please.*"

"I think it's kind of awesome," Kate Pinsky chimed in. "I mean, she's had this giant ego ever since fourth grade, and now everyone can see that she's just a big fake."

Hudson felt her stomach shrink into a cold iron fist. Ava had sent the Icks in to talk about her on purpose. And now she couldn't do anything but stand there and listen to it.

A toilet flushed a few stalls away, drowning them out. Someone else was in the bathroom listening to this, too, Hudson realized. This was actually getting worse.

"I mean, talk about negative attention," Ilona went on, not caring about the unknown person in the still-closed stall. "But I guess if you're the daughter of Holla Jones, that's all you know anyway."

"Her music is just soooo cheesy," Cici put in.

The stall door opened with a sharp *thwack*. "Could you guys be *any* more jealous or pathetic?" said a small, squeaky voice. "I mean, listen to yourselves. I almost fell off the seat."

Hudson craned her head to peek through the door crack, but the speaker was out of her line of sight.

"Um, nobody's talking to you," Ilona said icily. "And nobody asked you to eavesdrop, either."

"Yeah," Cici said.

"Well, *I'm* talking to *you,*" said the voice, "and if you're gonna gossip about someone you don't even know, *don't* do it about someone who's got more talent and style than the three of you will ever have in your entire *lives.*"

Hudson's mouth fell open. *Nobody* talked like this to the Icks. Who was this person?

"And just for your information," added the stranger, "she doesn't have an ego."

"How do you know?" Ilona said thickly. "Hudson Jones would never even *talk* to you."

"Yeah," Cici repeated numbly.

"As if we care what you think," Kate said. "And nice sweater."

Hudson heard them walk to the door.

"I hope *you* guys get food poisoning!" the stranger yelled as they walked out.

Hudson unlatched the stall door with trembling fingers. Whoever this girl was, she couldn't wait to thank her. And promise her eternal friendship, and possibly her firstborn child.

She threw open the door and there, standing on her tippytoes at the sinks, applying sparkly pink lip gloss, was Hillary Crumple.

"Hillary?"

Hillary turned around. "Oh, hi," she said, as if she'd known Hudson had been in the bathroom all along. "How was your break?"

"Uh…my break was fine," Hudson stammered, eyeing the exit.

"Mine, too," Hillary said casually. She turned back to the mirror and spread more lip gloss on her lips. "We just stayed here. It was kind of boring. What did you do? Did you guys go away? I really like your sweater. Where'd you get it?"

"Uh, I don't remember," Hudson said, trying to follow Hillary's line of questioning. "Yours is nice, too."

"Yeah?" Hillary turned back to Hudson and looked down at her sweater proudly. "Thanks. I got it for Christmas. My mom's life coach says that orange is supposed to make you more productive. And blue's supposed to be calming. What's your favorite color?"

Hudson glanced at the door again. "Um... silver?"

"Silver," Hillary mused, capping the lip gloss. "I'm going to have to check with the coach about that one."

"So, Hillary, thanks for what you said," Hudson said. "But you don't have to defend me or anything. It's okay. It's not your job."

"I know," Hillary said, slipping the lip gloss into a side pocket of the boxy pink and blue backpack at her feet. "But you're my friend. And friends stick up for each other."

We're friends? "Right," Hudson said uncertainly.

"And it's not like I'm lying to them or anything," Hillary said, taking a plastic barrette out of her backpack. "*US Weekly* said that you have an amazing voice. Didn't they interview your producer or something?"

"I'm not really sure," Hudson said, running her hands through her wavy black hair. "But thanks again, Hillary. And if you need anything, ever, just let me know."

"Then let's go shopping," Hillary said, turning to the mirror and securing some of her floating hair with the barrette. "Weren't we supposed to do that together? A couple months ago?" Hillary snapped the barrette shut and turned around again. “Do you remember we talked about it?”

“Yeah,” Hudson said, feeling caught. She did remember Hillary asking her to go shopping at the Chadwick dance back in the fall. “When’s good for you?”

“What about Saturday?” Hillary asked. “I could meet you downtown. Like, in NoLIta somewhere. What’s your favorite store?”

Hudson tried to imagine hanging out with Hillary in the chic neighborhood of NoLIta and her mind went blank. “There’s Resurrection,” Hudson said. “But it’s a little expensive—”

“Cool,” Hillary said. “Let’s meet at noon. That way we can get lunch, too.”

“Lunch,” Hudson said, trying not to sound surprised. “Great.”

“Great,” Hillary echoed, hoisting her book bag onto her shoulders. “See you then.” A moment later she was out the door, the folded-up *New York Times* sticking out of her book bag.

Hudson washed her hands at the sink, trying to process what had just happened. Lizzie and Carina were probably going to freak out—they were convinced Hillary was dangerous. But Hillary had just chewed out the Icks for her. Who else at Chadwick would have done that? Not even Lizzie or Carina would have been that gutsy. A two-hour shopping date was a small price to pay for that kind of loyalty. Even if Hillary’s loyalty felt a little unearned.

She wet the corner of a paper towel and pressed it to her closed eyelids. *Please, God,* she thought. *Don’t let me be known as the Girl With the Huge Ego and No Talent.* If only she hadn’t run off the stage. If only she hadn’t let her mom make up such a goofy excuse.

She wished she could blame her mom, or Carina, or even Ava Elting, but she couldn't. She had no one to blame but herself, and frankly, it sucked.

I hereby promise myself, Hudson Jones, that I will never, ever get up onstage again, she thought as she pulled the girls' room door open. *Ever ever ever.*

chapter 7

"Right up here's fine, Fernald," Hudson directed as the SUV swung over to the curb on Houston Street. Resurrection, the vintage boutique she and Hillary had chosen for their shopping date, was around the corner on Mott Street. But Hudson always preferred getting out of the car at least a block from wherever she was expected to be. It was embarrassing for people to see the SUV dropping her off.

"Just call me when you're ready," Fernald called over his shoulder.

"'Kay," she said brightly, slamming the door behind her. She could have walked over to NoLIta, or at least taken the subway down Seventh Avenue and then the bus along Houston, but Holla didn't like Hudson taking public transportation when she was by herself.

"Do you think I'm going to be kidnapped or something?" she'd asked Holla at the breakfast table, after yet another Saturday-morning power-yoga class. "I don't need Fernald to take me to Mott Street."

"You should see some of the mail I get," Holla had said, sipping her glass of kale juice. She was on one of her monthly juice fasts; she did them religiously to get rid of toxins. "So, I want you home at four o'clock. And two hours of homework this afternoon. Right?"

"Mom," Hudson protested, digging in to her oatmeal.

"Do you want to cram it all in tomorrow night?" Holla asked, squeezing a lemon slice into her drink. "You have to start learning time management, honey. It's absolutely the key to a successful life."

"Fine," Hudson said.

"And this semester we really have to bring up your math grade," Holla said. "Brown doesn't like C's."

"I didn't get a C last semester," Hudson said, carrying her bowl to the sink. "I got a B minus."

Holla threw the lemon slice into her glass and scowled. "Same thing," she remarked.

After waving good-bye to Fernald, Hudson turned south down Mott Street. She'd dressed in her Russian Spy/British Punk outfit — black wool leggings, black knee-high boots, black Russian hat, and a tartan plaid dress with strategically placed tears in the fabric. Her coat was a black trench with a bright red sash — a find from one of the street markets in Rome.

"Just don't let her steal a lock of your hair," Carina had said when Hudson had told her and Lizzie about her shopping date with Hillary.

"Or one of your buttons," Lizzie added. "Voodoo dolls always have buttons."

"Do you guys really think Hillary Crumple has a voodoo doll of me?" Hudson asked them. "That doesn't even make sense."

Far down the street, past the boutiques and espresso cafés housed in the ground levels of old tenements, Hudson could see a tiny figure in a bright pink knit hat and scarf and a gigantic puffy down coat that skimmed the ground. It had to be Hillary. Hudson raised one hand and sped up her walk. She hoped this wouldn't take too long.

"Hey!" Hillary yelled as Hudson approached. "That's such a cool hat. Where'd you get it? Moscow?"

"No. Somewhere around here."

"I love this neighborhood," Hillary said, glancing around. "Every single place down here is cool. You know? No Duane Reades, no Gristedes. Just really cool places. For cool people."

Hudson nodded. "Yeah, pretty much," she said, because she didn't know what else to say.

"I want to live down here one day. It's on my list of things I need to do before I'm thirty. Because that's when you have to get married, and after that, your life isn't fun anymore. Until you get divorced and you get to start over and be single again," Hillary said, pensive. "Or at least that's what my mom's life coach says. So, you ready to go in?"

Hudson looked through the window of Resurrection. Just past the shiny aluminum mannequins decked out in shift dresses from the sixties Hudson could see a saleswoman with ice blond hair. She was already eyeing Hillary with distaste. "Sure," Hudson said tentatively, and rang the doorbell.

The door buzzed and they entered the quiet, dimly lit boutique. Racks of vintage clothing lined the cherry red walls, while long tables laid with scarves and purses and sunglasses filled the interior space. Hudson stood very still and breathed in the scent of old leather and carefully preserved silk. Jenny had brought her here for the very first time, and she'd treated it more as a museum than a store. "Not that I can afford anything in here," Jenny had said good-naturedly as they walked in. "But I use this place for inspiration." Hudson knew exactly what she meant.

"So, I need something for my cousin's bar mitzvah," Hillary said loudly, unzipping her puffy coat to reveal a chunky purple turtleneck. "It's gonna be up in Westchester, at this really fancy hotel. So it sorta needs to be dressy." She darted over to the rack and pulled out a Pucci print dress. "What about this?"

The swirling print in aqua blue and yellow made Hudson squint. "That might be a little too much," she said gently. "You're really into color, huh?"

"I don't know." Hillary shrugged as she put the dress back. "I mean, I may as well stand out in the crowd."

"What about something understated like this?" Hudson said, pulling an ivory silk slip dress with a black ribbon belt off the rack.

Hillary wrinkled her nose. "Ehh. Boring. So what happened at the Silver Snowflake Ball?" Hillary asked, moving back to the racks. "Give me the scoop."

Startled, Hudson hung the slip dress back on its rack. "The scoop?"

"I think you can tell me the real story," Hillary said, rifling through some Bill Blass suits. "After everything I said to those girls in the bathroom."

Hudson picked up a yellow satin clutch purse. "It wasn't food poisoning. It was stage fright," she admitted.

"But how?" Hillary asked.

"What do you mean, *how*?" Hudson asked, putting down the bag.

"How could someone like *you* get stage fright? It doesn't make sense."

"I don't know," Hudson said, examining a rhinestone-encrusted lipstick case. "It just happens to people."

"But you're not just *people,*" Hillary said, inspecting a thin alligator belt she'd picked up off the table.

"What do you mean?"

Hillary put down the belt and pointed at the window. "Look."

Hudson looked over. Three photographers stood across the street, half-hiding behind a parked Datsun. They were taking pictures of the store with their zoom lenses.

"They're taking pictures of you, right?" Hillary asked.

"Oh, God," Hudson whispered.

"Do you want me to go out there?" asked the saleswoman.

"No, that's okay," Hudson said. "I got it." She would just give them a few shots and then maybe they'd go away. Her mom wouldn't be thrilled, but Holla knew this happened to Hudson every once in a while.

She walked over to the window and pretended to look at a silk scarf for almost ten seconds, giving them plenty of opportunity to get their shot. It was always a little embarrassing to see photos of herself in the tabloids. Even when they praised her fashion sense as "trailblazing" and "avant-garde." Once one of them had written,

"Move over, Kate Moss! Hudson Jones is the real style icon for teenage girls in the know." When she'd read that, Hudson had been flattered, but it had also made her anxious. Being a "style icon" was way too much pressure for anyone, especially her.

After she'd given the paparazzi at least several good shots, she opened the door to the store. "Okay, guys, thanks so much," she called out. "I don't want the store to freak out or anything."

"Hudson!" they yelled, still shooting. "Step outside!"

"You guys know how my mom feels about this," Hudson said. "Don't make me call her right now."

At that they lowered their cameras and edged away down the street.

When she walked back into the store Hillary was waiting with a smirk on her face and her hand on her hip. "I think we can stop pretending that you're just like everyone else now," she said.

"I was just saying that of course I get nervous being onstage," Hudson said. "Anyone would."

"So why did you say it was food poisoning?" Hillary asked.

Hudson fiddled with the belt of her coat. "That was my mom's idea."

"And you let her do that?"

"Well... she was just trying to be helpful."

"Did you think that was helpful?" Hillary inquired.

Hudson gave Hillary an annoyed look. "Should we try the place down the street? I think they have better stuff for what you need."

"Sure." Hillary zipped up her coat and they walked outside. Hudson tried to think of a way to permanently change the subject.

All this talk about stage fright and her mom was starting to get embarrassing.

Suddenly the SUV swung over to the curb. “Where to?” Fernald yelled through the lowered passenger-side window. He’d been waiting for them to emerge from the store.

“Who’s this?” Hillary asked.

“Uh, my driver,” Hudson said, blushing again. “That’s okay, Fernald!” she yelled. “We’re going to walk. We’re just going down there.” She gestured down the street. “I’ll call you, okay?”

Fernald gave her a thumbs-up and then drove down the street, leaving them mercifully alone.

“Wow,” Hillary said as they watched the hulking vehicle drive off. “Now I get it. No wonder you freaked out up there.”

“What do you mean?” Hudson asked.

“You’re so . . . watched.”

Hudson thought about this as they passed a woman pushing a baby stroller draped with plastic to keep out the cold. “Well, it’s just the way things are,” she explained. “There’s nothing I can do about it. My mom is really scared about kidnappers and stuff like that.”

“But it seems like she’s everywhere.”

“She’s not *everywhere*,” Hudson argued.

They turned into another store, which was a large, airy space, painted an industrial white and hung with full-length mirrors. Over the sound system came a familiar pounding song. One of Holla’s first hits.

“Uh, right,” Hillary said, pointing to the speakers.

“Look, I can’t help it if my mom’s music is everywhere.”

“It has nothing to do with that,” Hillary said, unwrapping her pink scarf from around her neck. "I just think your mom kind of rules your life."

Hillary's words stung. "It's a little more complicated than that," Hudson said.

"My mom's single," Hillary said. "So I get it. What's she going to open on Christmas, you know? Who's she going to hang out with on Saturday night?"

"My mom doesn't need me to hang out with her on Saturday night," Hudson said.

"I'm just saying, having only one parent around is hard," Hillary said. "But it's your life, too. And you need someone to step in and show you how to make it your best life."

"Who are you, Oprah?" Hudson asked. "This is my life. There's nothing I can do about it. And up until now, you kind of thought it was cool."

"I *still* think it's cool," Hillary said, following Hudson deeper into the store. "I think *you're* cool. But *you* don't think you're cool. I know you're an amazing singer. I can just tell. So why would you let one dumb night ruin everything?"

Hudson stared at Hillary. It was a good question.

"I read about you dropping your album on PopSugar. That's a really bad idea, by the way. Just so you know." Hillary pulled a black strapless bustier dress with a tulle skirt off the rack. "What do you think of this?"

"Uh, no. And let's please stop the self-help session, okay?" Hudson pulled out a simple black velvet dress with cap sleeves. "Try this."

Hillary examined the dress. "Really?"

"Look, if there's one thing about myself I *do* believe in, it's my taste in clothes," Hudson said, trying not to sound too sarcastic. "Trust me. Try it on."

"Fine," Hillary said, trudging into a fitting room.

Hudson stood there, relieved to be alone for a moment. She listened to Holla's song playing over the speakers, about to end. It had been a huge hit when Hudson had just been born. As with most of her mom's songs, she'd never paid much attention to the lyrics. But now the words jumped out at her.

Oh, baby, you know how much it hurts to let you go
But one day I swear, you're gonna know
That I will love you 'til the end of time...

She'd never thought that not having a father around was weird—lots of kids she knew had divorced parents and lived with their moms. But maybe she and Holla were a little too close. Maybe she *did* need to break away a little bit. Maybe she'd never noticed how much space her mom took up in her life, even when they weren't together.

Hillary walked out of the fitting room. "What do you think?" she asked, twirling around in the dress. It fit her perfectly. Hillary almost looked like Audrey Hepburn.

"I love it," Hudson said.

Hillary stopped twirling and beamed. "Good. I'm gonna take it. But first, I have something to say."

"Oh, God," Hudson said with a smile. "You do?"

"I think you need a life coach."

"What?" Hudson blurted out.

"My mom got one, and it's really helped," Hillary said. "She was a total mess after she and my dad got divorced. Sitting on the couch, eating Rice Krispies treats all the time, watching Animal Planet —"

"I don't need a life coach, Hillary," Hudson interrupted.

"No, but I think you need some help. I think you need to learn how to be Hudson. Not Holla Jones's kid. Just Hudson. And I'm happy to do it."

"Hillary?" Hudson said firmly. "No."

Hillary put up her tiny hand. "Fine, fine. Don't freak out. It was just a suggestion." She headed back into the fitting room, leaving Hudson feeling a little shaken. *Life coach?* She glanced around the store, wondering if anyone had heard them. They definitely weren't having lunch.

She waited as Hillary paid for her dress, and then they walked back outside. "I love it," Hillary said, almost giddy. "Thank you."

"You're welcome," Hudson said as they came to a stop at the corner of Prince Street. "Well, um, I think I have to go do some errands now."

"Okay." Hillary fixed her with a stare. "Are you offended by what I said?"

"Not at all," Hudson said, almost truthfully.

"Okay," she said. She turned this way and that. "Which way is uptown?"

"That way," Hudson said, pointing to Houston.

"Got it," Hillary said, and then she walked up the street. Hudson watched her pink hat disappear up the block. Hillary may have meant well, but she was definitely rude. Hudson still couldn't believe some of the things she'd said. Things that she didn't even want to tell Lizzie and Carina. *Your mom kind of rules your life. You need to learn how to be Hudson.* Who said things like that?

Just then she saw the SUV glide down Mott Street. It was Fernald, circling the block, waiting to pick her up. For a moment she thought about turning the other way and ducking down Prince Street. Suddenly she didn't feel like being under the watchful eye of Holla's staff. But then the feeling passed, and she stepped out into the street and flagged him down, just like she was supposed to.

chapter 8

By late afternoon it had started to rain. Hudson watched people hurry down the street under bobbing umbrellas as they drove along Washington Street, the tires of the SUV making a whooshing sound on the wet asphalt. It was just before four o'clock, so she didn't have to worry about being late. Fernald had taken her up to Carina's apartment, where she'd spent the rest of the day watching *Across the Universe.* She hadn't told Carina any details about her shopping trip with Hillary, except that nothing had been stolen for a voodoo doll.

They finally reached their corner. The rain had cleared out all but the most hard-core photographers, who stood across the street looking miserable in their hooded nylon jackets. She gave them a slight wave as they turned into the garage.

"Hellooo? Anyone here?" Hudson called out as she walked into the brightly lit kitchen. "Where is everyone?"

"They're in the prayer room," said Lorraine, stirring something sludgy and green in a mixing bowl. "Your mom and your record producer."

"*My* record producer?" Hudson asked.

Lorraine nodded. "The one with the gorgeous blue eyes," she said, winking.

Hudson dropped her bag on the kitchen table. Chris Brompton was here. She hadn't been in touch with him since that horrible night at the Pierre, except for one e-mail over Christmas break, when she'd given him a slightly pathetic excuse for stopping work on her album. Something about needing to take a break, to take time to be a kid. She wondered if he'd even believed her. He'd written back, saying that he was disappointed but that he completely understood. The nice, completely mysterious kind of response. But he was there, in her house, right that second. Maybe he wanted to her to reconsider. Or maybe he just wanted to see her again.

She climbed the stairs two at a time. Up on the third floor, tucked into various niches along the wall and protected by alarmed glass, were Holla's numerous awards: Billboard Artist of the Year, Grammy for Record of the Year, People's Choice Award, the NAACP Image Award for Outstanding Female Artist. On the opposite wall hung her framed gold and platinum records. When she was younger, Hudson loved to gaze at these statues and plaques and records, or even ask her mom if she could hold them every once in a while. Now she felt the need to just walk past them as quickly as possible.

Back when she'd bought the house, Holla had been a practicing Buddhist who needed a room for her chanting and meditation. But then Holla had abandoned Buddhism for something she called "nonspecific spirituality," and now the prayer room was just another office. Hudson pushed open the door.

"Are you *kidding* me?" her mom was saying. Holla sat perched on the edge of a white chaise in a snug fuchsia warm-up suit, and she was so absorbed in what she was saying, or who she was saying it to, that she didn't even notice Hudson's entrance. Sitting just a few inches away, at the desk, smiling at her mom in a way that made her heart stop, was Chris Brompton. He looked exactly the same as he had all those days they'd spent together in the studio: shaggy strawberry blond hair, kind but sexy blue eyes, weathered Levi's and short-sleeved T-shirt. Neither of them noticed her for a moment, until Chris glanced over at the door.

"Hey, Hudson," Chris said, getting up from the chair in his easy, laid-back way. "How's it going? Happy New Year!"

"Hi, Chris," Hudson said. Her heart beat rapidly as they hugged. Behind him, on the computer screen, Hudson glimpsed a list of tracks. But she couldn't tell whose songs they were. "What are you doing here?"

"Chris is going to do some work on my album," Holla announced. Her eyes were still glued to Chris's face, and she hadn't moved from her spot on the couch. "I was just so impressed with the work he did on your album, honey, that I asked him to put the finishing touches on mine."

Hudson stood perfectly still. She looked from Holla to Chris and back again. For a moment, she couldn't speak. It was as if a golf ball covered in spikes had suddenly lodged itself in her throat. "On yours?" she asked.

"Well, since you'd decided to put *yours* on the shelf, and he's so good," Holla said, reaching out to give Chris a playful swat on the arm as he sat back down, "I couldn't help myself."

"You couldn't?" Chris asked, laughing. "I'm flattered."

Hudson watched them grinning at each other, trying to absorb this. Chris Brompton had believed in *her.* He'd been *her* producer. Now, just like that, he was gone. The unfairness of it sliced through her. Her mom had thousands of fans — did she really need another one?

"So, Chris just had a brilliant idea," Holla said to Hudson. "You want to tell her?"

Chris swiveled around to the computer. "Okay, listen to this." He clicked on one of the tracks.

From the first sped-up, over-synthesized beat, Hudson knew exactly what it was: her song "Heartbeat." The song she'd tried to sing at the Silver Snowflake Ball. "What about it?" she asked, fighting off a sense of panic.

Chris paused it and turned back around. "It's a really good song, Hudson. I always thought it was your best. And, well" — he looked at Holla — "I think it would be perfect for your mom."

Hudson blinked.

"Only if you're okay with that, of course," Holla said, still smiling at Chris. She padded over to the desk. "I don't want you to feel weird about it."

"But... don't you have songs already?" Hudson asked.

"Chris doesn't think I have a single. *Do* you?" she asked him, with mock seriousness.

"Not like this one," Chris confessed, running a hand through his hair. "Of course," he said, turning to Hudson, "if you're not cool with it, it's no big deal."

Hudson couldn't move. *Of course* she wasn't cool with it. Did he even have to ask? Why didn't he know that already? It was her song. One of her favorites. Even though her mom had ruined it. And now he wanted to give it to someone who'd never liked it that much in the first place? "I thought you didn't like my music," Hudson managed to say.

Holla raised an eyebrow. "Honey, I *love* your music," she said. "Which is why I think this song — tweaked, of course — would be perfect for me."

"But it's *my* song," Hudson argued.

Holla stood and wobbled on her feet, as if she'd lost her balance, then caught the back of Chris's chair.

"You okay?" Chris asked, leaping to his feet and taking her arm.

"Yes, yes, I think so," Holla said, touching her forehead. "It's just this juice cleanse. It always makes me a little light-headed."

Chris led Holla over to the window seat and carefully patted her arm, as if she were a delicate thing that might break. "That better?" he asked. "Do you need some water?"

"That would be great," she said.

"You got it," Chris said.

As soon as he left, Holla looked up at Hudson. "So, you were saying?"

Hudson gathered herself. "I was just saying that it's my song," she said.

"So?" Holla said. "Are you saying you've changed your mind about the album?"

"No."

"So you just don't want me to have it," Holla prompted.

"I... I didn't say that," Hudson stammered.

"Then you just want it to go to waste," Holla said.

"No, but..."

"Honey, what's the matter? You're going to get all the royalties. I thought you'd be thrilled to have one of your songs actually out there instead of on a shelf gathering dust."

As usual, her mom had found the one point that Hudson couldn't debate. She looked at the framed cover of *Vanity Fair* on the wall, the one her mom had posed for years earlier. THE PRINCESS OF POP, it read. Her mom stood on a beach in a ballgown, wearing a tiara. And in her arms, naked and squirming, was Hudson, just a year old. Her mom looked so pretty, so happy. Overjoyed about her new little girl. The little girl who now was being petty and trivial and hopelessly stubborn. "Fine," she said. "You can have it."

Chris walked back into the room with a glass of water.

"Really?" Holla asked, suddenly contrite. "Are you sure?" She took the glass from Chris and sipped from it.

"Yeah," Hudson said. "Why not?"

"Thank you, honey," Holla said. She got up and gave Hudson an overpowering hug. "I'll do it justice. I promise. Now, you have some homework to do, right?"

"Uh- huh," Hudson said.

"Thanks, Hudson," Chris said. "You might have just given your mom her next hit."

"Awesome," she said as she backed out of the room. She grasped the banister as she walked down the stairs, trying to keep her feet moving, one in front of the other.

This had to be a joke. All her mom had done before now was

tell her how unmarketable her music was, and now she wanted to record one of Hudson's songs because she needed a single? She should have just said no. Carina and Lizzie would have just said no. It was just one stupid syllable. But, as usual, something inside of her had just shut down. Whenever she got angry, or hurt, she couldn't speak, couldn't fight back. It was like being caught in a sandstorm that blotted out words and sight and thought. And the only way out of it all was to just say *Fine*.

When she got to her room, she pulled out her iPhone. She needed to text Lizzie and Carina. They would be on her side. They'd understand. They always did. But she stopped, her finger poised over the screen. Her friends would ask why she hadn't fought back. Carina would go on and on about Hudson's inner bee-yatch. Lizzie would just say *Why didn't you say no?* And Hudson wouldn't have an answer. *Because I didn't feel like I could? Because it's my mom and she gets everything she wants?* Those weren't good enough answers.

She threw herself on her bed and buried her face in a cushion. Carina was right: She needed to find her inner bee-yatch, ASAP. But she had no idea how. This was who she was: Sensitive. Sweet. Nice. More at home by herself at her piano than in a crowd. The exact opposite of her fiery, vocal, afraid-of-nothing-and-nobody mom. She would never be any different, and she couldn't hope to be. Because how did people change who they were?

She sat straight up.

Maybe a person *could* change who she was.

If she got a life coach.

chapter 9

"Your mom wants your song?" Carina asked incredulously as they walked up the street toward school. "But she told you your stuff sucked. That day in the studio. I was there!"

"I know," Hudson said simply as they reached the school doors.

"I don't get it," Lizzie said. "After everything she said?"

"She's a Leo with an Aries rising," said Hudson.

"What does that mean?" Lizzie said.

"That it's not that surprising," Hudson said, pulling the doors open.

"But you didn't have to say yes," Lizzie said as they walked into the school lobby. She pulled off her knit hat and shook out her red curls.

"And what about your inner bee-yatch?" Carina cried, blowing on her cold fingers. Carina was always losing her gloves. "What happened to her?"

"Maybe Hudson doesn't have an inner bee-yatch," Lizzie said to Carina.

"If you don't start saying no to your mom, like, *now,* then things are just gonna get worse," Carina said.

"You guys, I *know,*" Hudson said as they passed the library. "That's why I've decided to get a life coach."

"A life coach?" Carina asked, crinkling her brown eyes. "Are you kidding?"

"Lots of people use them," Lizzie pointed out. "It's like a therapist who actually does stuff."

"Who are you getting?" Carina asked. "Please tell me it's not some kooky astrologer."

"Heeeyy, guys!" called a lilting voice behind them on the stairs. "Wait up!"

They slowly turned around to see Ava Elting climbing the stairs with the Icks. She'd traded her devil-horned hat for a black stocking cap with orange and red zigzagging lines around it, like the ones on an EKG machine. Her nails had been painted a deep sky blue, and she wore a skinny lavender scarf that barely covered her long neck. "Did you guys have a good weekend?" Ava asked. "I had the *best* time. I went up to Vermont to go snowboarding. I even got a private lesson from this guy who used to be on the Olympic team."

"That's nice," Lizzie said in a way that said she couldn't care less.

"So Hudson, I saw that you've pulled your album," Ava said, sidling up to her on the stairs. "That must have been such a hard decision for you."

Ilona and Cici snickered quietly. Kate supplied the deathstare.

"It wasn't that hard," Hudson said, trying to reach the top of the stairs as quickly as possible. "I'm just really busy with school right now."

"Was it because of what happened at the Ball?" Ava pressed. "I hope not. Considering it was just food poisoning and everything."

Hudson pursed her lips. "No. It had nothing to do with that."

"I just think it's a shame that things aren't working out," Ava replied. "I mean, that must be a lot for you to live up to. *I'd* crack under the pressure."

Just as Hudson began to get angry, she turned and saw Hillary walking up the stairs behind them. She sported her usual messy ponytail, and pieces of brown hair fell against her forehead. Her square pink and blue backpack was strapped to her back, and today her sweater was a blinding shade of tennis-ball yellow. "You guys go ahead," Hudson said. "I'll see you in homeroom."

Ava reached the door to the Upper School, grinning triumphantly. "See you later, Hudson," she said, giggling, and pulled the door open. The Icks followed, each of them giggling as well.

"Where are you going?" Carina asked, but Hudson was already descending the stairs toward Hillary, eager to put the Ava encounter behind her.

"Hey, Hillary!" she called out. "Can I talk to you for a second?"

Hillary pushed some hair out of her eyes. "Sure," she said in her tiny voice. "Those are really cute earrings."

"Thanks." Hudson fell into step beside her on the stairs. "So I was thinking about what you said the other day. About the life-coaching thing. I think I need to do it. I think I *have* to do it. So . . . can you still be my life coach?"

Hillary stopped climbing the stairs. Her thin legs were bare despite the freezing weather, and her kilt hung unevenly below her knees, as if she'd misbuttoned it. "If I do this for you," she asked in a low, portentous voice, "can I ask you for something?"

"Sure, what?" Hudson said.

Hillary crossed her arms. "Will you go with me to my cousin's bar mitzvah on Saturday?"

Hudson paused. This was a curveball. "But... but I don't even know your cousin."

"I know," Hillary said, unfazed. "I just want you to go with me. My friend Zoe was supposed to, but now she has to go to New Jersey for her grandma's birthday or something like that. So can you go? Please?"

"But... but why do I have to go with you?" Hudson asked.

Hillary hesitated. "Because there's going to be this guy there that I like."

"Oh," Hudson said, surprised. She hadn't thought Hillary even cared about guys yet.

"And I need someone to be with me when I talk to him," Hillary went on. "And tell me if they think he likes me back. He's my cousin's friend — my older cousin, Ben — and he's a sophomore —"

"A *sophomore*?" Hudson asked.

"And they're starting this band and he plays saxophone and he's on the chess team and, well, I think he likes me, but I really need a second opinion."

"Oh. Okay." Hudson tried to imagine a sophomore really liking Hillary, but she found it a little hard to swallow. "I'll go."

"Just don't tell anyone, okay?" Hillary asked, her voice rising with panic. "Because everyone'll think it's just some stupid crush. You promise?"

"I promise," Hudson said. "But…can you still be my life coach?"

"Oh, sure," Hillary said. She started taking the steps two at a time. "But first we have to figure out a life goal. You always have to have a life goal when you're being life-coached. So what do you want to change about your life?"

They reached the fourth floor and walked into the crowded Middle School hallway. It had been only a year since Hudson had gone to school up here, but it already felt like a million years ago. Several eighth-grade girls stared at her with undisguised worship.

"Well, it's like you said," Hudson told her, "I need to be my own person. And I really don't want to say 'fine' anymore."

"What do you mean?" Hillary asked.

"I mean, I'm always just saying 'fine.' When I really want to say anything *but* 'fine.' Do you know what I mean?"

Hillary nodded. If she thought Hudson didn't make any sense, she didn't say so. "Okay. First step, then: I want you to write down everything you're afraid of. All your fears." She reached into her boxy backpack and took out a pencil with a heart eraser stuck on the end of it. "Here," she said, taking out a piece of loose-leaf paper and scribbling something on it. "Take this." She handed it to Hudson. At the top of the page she'd written HUDSON'S FEARS.

"Wait," Hudson said. "Fears? Do you mean like earthquakes?"

"Just write down everything you think of."

"But why do I have to write them down?"

"My mom's coach says that once you write down your fears,

they lose their power over you," Hillary said, sticking her pencil back into her backpack.

Hudson couldn't help but feel a little disappointed. This was life-coaching? She'd assumed that Hillary would just rattle off more of her blunt insights and then offer some concrete, practical solutions. Writing down random fears wasn't going to change anything. But she didn't want to be late for homeroom, and most important, she didn't want to be rude. "Okay, fine," she said, putting the piece of paper in her backpack. "I'll let you know what I come up with."

When she slipped into her seat next to Carina and Lizzie in homeroom just before the second bell, Carina looked up from her Spanish homework, which she was frantically finishing. Spanish was Carina's least favorite subject.

"What was that about?" Carina asked.

"I just went to see my life coach," Hudson said.

"Hillary?" Carina shouted. "Your *stalker's* your life coach?"

"She's not my stalker," Hudson said.

Lizzie looked up from her copy of *This Side of Paradise*. "Well, she is an odd choice, you have to admit," she said.

Later, during her study period, Hudson took out her math notebook and uncapped her favorite Bic liquid gel pen. She read the words *Hudson's Fears* over and over. She'd never thought of herself as someone who had lots of fears. So now she wondered, *What am I afraid of?* She started to write.

getting a C in geometry
getting a B in everything else
Not getting into Brown or, well, anywhere

Something bad happening to Lizzie and Carina
Roaches, waterbugs, and snakes (ugh . . .)
Plane turbulence ☹
Not being liked
Small planes (except for Carina's dad's plane, which isn't that small, but it still counts)
Cute older guys who are really smart and like good music, and who I become friends with
All cute guys
Class presentations
Lunar eclipses (especially in Virgo)
Getting caught eating junk food in front of my mom
Being laughed at (too late for that ☹)

Later, as they waited for Bio to start, she took out the sheet and wrote more. And then she added some more fears during lunch. Before she knew it, she had four loose-leaf pages, all of them covered in fears. She was stunned. She'd had no idea that she was afraid of so many things. It was amazing that she'd even made it to ninth grade.

She folded up the pages and stowed them deep in the middle of her Geometry notebook. She didn't even want to show Carina and Lizzie. She was sure that neither of them was afraid of half as many things as she was.

At the end of the day, Hillary accosted her in the lobby as she was leaving.

"So, did you make your list?" she asked, eyeballing Hudson's bag.

"You don't have to see them, do you?"

"No," Hillary said. "Just tell me about one of them."

Hudson tried to remember one that wasn't completely embarrassing. "I think I said I was afraid of eating junk food in front of my mom."

"Great!" Hillary said, almost jumping out of her bluchers. "Okay. I want you to go home tonight and have pizza for dinner."

"What?" Hudson asked, stupefied. "I can't do that."

"Why not?"

"Because it's dairy. And white flour. And possibly nitrates."

Hillary shook her head. "But do you *like* pizza?"

Hudson nodded. "Yeah. I love it."

"Good," she said. "Then have it tonight. It's just for tonight."

"Wait," Hudson said, shifting her book bag to her other shoulder. "How is eating a pizza going to help me be my own person?"

Hillary folded her arms and cocked her head. "I know what I'm doing, Hudson," she said bossily. "Trust me."

Hudson walked out of the lobby and caught up with Lizzie and Carina, who were on the corner talking. "Can you guys come over for dinner tonight?" she asked them.

"What's up?" Lizzie asked.

"Are you guys having flaxseed tacos again?" Carina asked, wrinkling her button nose.

"No, I'm having pizza," she answered. "And I need you guys to help me."

chapter 10

Lizzie, Carina, Todd, and Hudson sat in a circle on the sheepskin rug, staring intently at the black cordless phone in Hudson's hand.

"If you're gonna do it, you gotta do it now," Lizzie urged, stretching her long legs out on the rug. "It's six forty-five. You said dinner was at seven, right?"

"On the dot," Hudson said. She pressed the Talk button on the phone so that they could hear the low hum of a dial tone, and then she pressed it again so that the phone hung up. It was the fourth time she'd almost dialed. In her dog bed in the corner, Matilda lifted her head and looked at Hudson like she was certifiably nuts.

"Okay," Carina said, taking charge. "What's the worst thing that can happen? It's *pizza*. Your mom's not going to throw you out of the house. Right?"

Hudson didn't say anything.

"Right?" Carina asked, less confidently.

"Carina's right," Todd said. "Pizza's loads healthier than a lot of other stuff you could be eating. There was a guy in my class in London who only ate Curly Wurlys and Aero bars for every meal."

"What's a Curly Wurly?" Carina asked.

"It's basically chocolate-covered caramel," Todd answered. "It's amazing."

"But why do they call it a Curly Wurly?" Carina asked, giggling.

"Let's get back to the pizza-ordering," Lizzie broke in, twisting her hair into a bun on top of her head. "Just do it. Nothing bad is going to happen. And for the record, having pizza more often wouldn't kill you."

"What do you mean?" Hudson asked.

"Just that you're always eating so healthy," Lizzie said gently. "It's like you're always trying to eat the *right* thing."

"Yeah, like if there's a choice between a burger and a salad, you'll always eat the salad," Carina added. "Because it's the right thing."

Hudson looked at Todd.

"I really don't know," Todd said, shrugging. "But I'll take everyone's word for it."

"Well..." Hudson argued. "Isn't that how we're supposed to be?"

"We're not supposed to be perfect, H," Carina said. "That's the worst thing you can be. Perfect."

Hudson knew Carina was right. "Okay. Here goes." She pressed the Talk button again. This time she dialed the number she'd written down on a piece of paper.

A man picked up. "Ray's!" he yelled into her ear.

"Hi, I'd like to order a large plain pizza, please," Hudson said carefully, as if she were speaking another language. "Or do I say cheese? Is cheese extra cheese or is it just plain?"

"What?" the man said.

"Nothing," Hudson said. "I'll just take a plain pizza. For delivery. Seven-fifty Washington Street."

"Large plain — you got a doorman or a buzzer?" he yelled.

"A buzzer," she said.

"Fine. Twenty minutes."

Click.

Hudson hung up the phone.

"Good for you," Lizzie said, reaching out and grabbing Hudson's arm. "I'm proud of you."

"I still don't know why Hillary wants me to do this," Hudson said.

"I do," Lizzie said. "How are you ever going to start living your own life if you don't even eat what you want?"

"Yeah," Carina said. "Maybe Creepy Crumple is onto something."

"Dinnertime!" Lorraine yelled over the intercom in Hudson's room.

"Don't worry, H," Todd said, helping her to her feet. "I think you're going to do just splendidly with this junk food."

"Thanks, Todd," Hudson said. He was so adorably English sometimes.

The three of them walked down the stairs to the kitchen, while Hudson thought about her friends' advice. She couldn't help it if she was used to eating healthfully. It was the right thing, after all. But the right thing for whom? It was hard to know sometimes. Her mom always prided herself on being the healthiest person she knew. Maybe even the healthiest person in America. But maybe being the healthiest person in America wasn't so healthy, Hudson realized. Maybe the point was to be healthy, but to not let it rule your life.

Hudson led the way into the kitchen and they took their places around the empty table.

"Remember," Carina said, sipping from the glass of triple-purified water at her place setting, "no matter what happens, it's just pizza."

Behind them, Hudson heard the elevator door rumble open, and soon Holla strode into the kitchen in a shiny silver trench coat, with Sophie in tow. "I'm meeting him tonight at the Rose Bar," Holla said, speaking into her cell phone. "I'll let you know what he says." She clicked off. "Sophie? Get us that sofa near the fireplace. And let Mr. Schnabel know we'll be there if he wants to join us."

Holla unbuttoned her trench. Instead of her usual studio uniform of warm-up suit and ponytail, she wore a black sweater with a deep V-neck and blue jeans tucked into knee-high leather boots. Her hair was down around her shoulders and curled at the ends. It was Hudson's favorite look for her mom — smart and elegant, but casual.

"Hi, everyone," she said, smiling at her daughter's friends. "I didn't know Hudson was having guests tonight." Even though Holla liked her privacy, she never seemed to mind when Hudson had her friends over.

"It was sort of a spur-of-the-moment kind of thing," Hudson said. "Mom, this is Todd."

"It's a pleasure, Miss Jones," Todd said, standing up and offering his hand.

"You can call me Holla," her mom said, shaking Todd's hand. Hudson could see that she thought Todd was cute. "So nice to meet you. Hello, girls," she said, leaning down and giving Carina and Lizzie each a hug and a kiss.

"Hi, Holla," Lizzie said.

"Hi," Carina murmured.

"Oh, it feels good to be home," Holla said, tossing her trench coat over the back of one of the bar stools by the island. "Another long day of mixing."

"So...how's it going?" Hudson asked. She never knew how exactly to talk to her mom around her friends. Lizzie and Carina were always a little uncomfortable around her, which made Hudson uncomfortable, too.

"We're getting a lot done, and thank God Chris is on board," Holla said as she scooted into the booth next to Hudson. "I realize now I should have had him this whole time." She looked at Hudson and frowned. "What if you got some bangs? I think they'd look cute on you. Don't you guys think?" she asked Lizzie, Carina, and Todd.

"So, Todd is a writer," Hudson said, changing the subject before they could answer. "Just like Lizzie."

"Oh?" Holla asked as Lorraine brought her a tall glass of ice water with lemon.

"His story is up for best short story at the entire school," Hudson said.

"Really?" Holla said, smiling as she sipped her water. "That's wonderful. What's it about?"

Todd blushed. "It's not really anything," he said, waving it off.

"It's about Lizzie," Carina said, smiling.

Hudson was relieved to see her friends relaxing in front of her mom.

"Really? So are you guys dating?" Holla asked.

"Uh, yes," Lizzie said shyly.

"Ohhh, that's so cute," Holla said. "Is there anyone in Hudson's life? She doesn't tell me a thing."

Hudson gulped and looked down at her plate. "Mom, please," she said under her breath. Bangs were one thing to discuss in front of her friends. Her love life was another.

"Everything's ready!" Lorraine called over from the island. "Come and get it!"

Hudson shot up from her seat and walked to the kitchen island, where Lorraine had set out a buffet.

"Baked tofu with sea-vegetable slaw," Lorraine said, gesturing to a platter heaped with what looked like stringy black seaweed. "Plus, sesame broccoli and wild rice with beet chutney."

"Where's the pizza?" Carina whispered desperately.

"Just take some," Hudson whispered back, spooning giant gobs of wild rice onto her plate, and even some of the sea-vegetable slaw. "It's hijiki."

"Hi-*what*-ee?" Carina whispered back.

"It's not bad." Hudson went back to her seat and politely began picking at her food.

"Did Hudson tell you guys that I'm singing one of her songs?" Holla asked between tiny mouthfuls of wild rice.

Her friends nodded as they pretended to eat. "H is really talented," Carina said.

"I know," Holla said. "And at least now the world is gonna have a chance to see that."

Hudson gripped the fork tighter as she swirled food around

her plate. She knew that her mom meant this as a compliment, but inside she felt like cringing.

"Honey?" Holla asked. "What's wrong? You love hijiki."

"I'm just not in the mood for it tonight," Hudson said.

"Well, is there something else you'd rather have?" asked Holla. "How about a crispy seitan sandwich?"

There was a buzz at the back door. Hudson, Lizzie, and Carina all sat up. The pizza had arrived.

"Are you expecting something?" Raquel asked Holla as she walked out of the kitchen to the back door.

Holla shook her head. "No, I don't think so."

Raquel went to the elevator as Lizzie, Carina, Hudson, and Todd exchanged urgent looks. *This is it,* Hudson thought. *The moment of truth.*

Raquel returned, carrying a pizza box. "Did somebody order a *pizza*?" she asked, scrunching up her face in horror.

The smell of melted cheese and oregano filled the room, answering her question.

"I did," Hudson said, springing out of her chair. She ran to the back door to pay the delivery man. When she returned, her mom was staring at the pizza box with undisguised fury. Carina and Lizzie and Todd looked down, inspecting their plates. Even Lorraine looked terrified. But the room smelled delicious.

"Hudson, you're being rude to Lorraine," Holla said evenly.

"I know," Hudson began. "But I just felt like having pizza tonight."

"Throw it out," Holla said to Raquel. "And sit down and finish your dinner," she said to Hudson. "I don't know what kind of point you're trying to make, but this is ridiculous."

Carina glanced up from her plate and gave Hudson a meaningful eyebrow raise. Hudson understood. She couldn't back down. Her heart pounding, she opened the pizza box and grabbed a thin, greasy slice.

"Hudson," Holla said. She sounded like she was struggling to stay even-tempered. "What are you doing?"

Hudson took a gooey, cheesy bite. "Yum," she couldn't help saying. She couldn't remember the last time anything had tasted this good to her.

"Do you know how bad that is for you?" Holla asked.

Hudson swallowed. "That's really good." She turned to her friends. "Want some?"

Carina and Lizzie and Todd all nodded eagerly.

"Go ahead, you guys," she said. "Have some." She walked the pizza box over to the table and held it out.

Carina quickly grabbed a slice. Then Lizzie did. Then Todd. Now all four of them were eating pizza and licking their fingers.

Holla watched them for a moment, her brown eyes glittering with anger. Finally, she picked up her fork and knife and sliced into her tofu. Lorraine and Raquel went back to quietly moving around the kitchen.

Hudson couldn't believe what she was doing. She was eating three of her mom's most hated food groups—dairy, wheat, and grease—right in front of her. And Holla wasn't stopping her. As soon as she finished the first slice, she reached for another with her greasy, cheesy fingers. So did Carina and Lizzie, and even Todd. Before long, they'd eaten the entire pizza. The empty grease-streaked box lay open on the kitchen counter. Hudson still stood next to it. She'd been too scared to sit down next to her mom, afraid that Holla might rip the slice right out of her hands.

"Can we be excused?" Hudson asked, finally daring to look at her mother.

Holla chewed her food and stared at her plate. "Yes," she said quietly. "And please get rid of that."

Without a word, Hudson took the pizza box into the hallway near the elevator and threw it into one of the recycling bins. As she clamped the top onto the bin, a thrill went through her. *I did it!* she thought. Maybe her "point" had been ridiculous, but she'd definitely made one.

She walked out to the hallway, where her friends were waiting for her.

"Oh my God, your mom was shooting dagger eyes at us," Carina whispered as they walked up the stairs.

"Are you okay?" Lizzie asked.

"I'm great," Hudson said, grinning. To Holla, food was a very big deal. Food was control. And Hudson had just taken control in Holla's house, for the very first time.

chapter 11

The next morning Hudson walked downstairs very, very quietly. She hadn't seen Holla since the pizza showdown the night before, and she wasn't sure what to expect. Hudson knew that she was in uncharted territory here.

She placed her book bag on a chair in the hall and walked into the kitchen to grab something to eat, steeling herself for a confrontation. But instead of Holla at the kitchen table, Chris Brompton was sitting there, reading the *New York Post* and sipping a soy cappuccino, like he enjoyed breakfast at Hudson's house all the time.

"Oh, hey, Hudson," he said, putting down his coffee. "'Morning." He got up out of his seat halfway and then just stood there awkwardly. Instead of jeans he wore long plaid swimming trunks and a burgundy T-shirt that read BONDI BEACH.

"What are you doing here?" Hudson asked. She glanced at the kitchen clock. It was quarter to eight.

"Oh, your mom just wanted me to come take one of her yoga classes," he said with a sheepish smile. "I'm just booting up with some caffeine." He held up his cup and took another sip, which left foamed soy milk all over his upper lip.

Lorraine walked into the kitchen. "Anything I can get you, Hudson?" she asked.

"I'll just have this," Hudson said, opening the refrigerator and grabbing an orange.

Lorraine retreated into the pantry, leaving them alone again.

"So, yoga, huh?" Hudson said, starting to peel her orange. "You better be careful. It's not for beginners."

"I can handle it," Chris said, sitting back down. "And I hope you're not weirded out by anything. You know, about me working with your mom."

"No, I think it's great," she lied, tearing off a hunk of orange peel. Even though she was still annoyed with him, being this close to him again, all alone, brought her back to that dizzy, exhilarated feeling she had when she worked with him in the studio. Or used to have, in the days before her mom had barged in. He'd made her feel so talented, so special, so seen.

"Just so you know, I was really surprised when you wrote me that e-mail," he said, looking straight at her. "About quitting the album. You're an amazing talent, Hudson. I hope you know that."

"Then why were you okay with changing it all?" The question had just slipped out.

"What do you mean?" Chris's laid-back smile dissolved into a frown.

"When my mom came in and changed every song, you seemed

totally fine with it." She stared right into his eyes. "Why? If you thought I was so talented?"

Chris looked dumbfounded. He stared into his cappuccino, then stirred it with his spoon. "Just because I respect your mom's talent and opinion doesn't mean I don't believe in you," he said. "And I think your mom happened to be right."

So he had no real answer, Hudson thought. Only an excuse for her mom. Chris had completely gone over to the dark side. "Well, I guess I better be going," she said, just as her mom walked into the kitchen.

"Well, good morning," Holla said, glancing first at Chris and then at Hudson. Her canary yellow yoga top looked beautiful against her dark brown skin, and was cut short to reveal her formidable six-pack. "And you actually made it," she said to Chris.

"I know," Chris said, sitting up straight. "My friends would never believe this."

"Hi, Mom," Hudson said.

"Hi, honey," Holla answered warmly. "You sleep okay?"

"Great," Hudson said. Her mom didn't seem to be the slightest bit angry. Or maybe she just didn't want to show it in front of Chris.

Holla fixed her gaze on Hudson's feet. "They let you wear those boots to school?" she asked, pointing at Hudson's Russian Spy/British Punk knee-high boots. "What about the ankle boots? The heels are so much lower —"

"I gotta go," Hudson said, backing out of the kitchen. "I'll see you later."

"Bye, Hudson," Chris said, waving.

"Bye, Chris," Hudson said tonelessly.

"Get ready to have your butt kicked," she heard Holla tease Chris as she walked out to the hall, grabbed her book bag, and pushed the elevator button with a sick feeling in her gut.

They're into each other, she thought. She'd seen that look on her mom's face before, and heard the way Holla's voice went up an octave whenever she was around someone she liked. And Chris was just her type: young, hunky, and boyishly handsome. Just like Hudson's dad.

But instead of feeling angry, she felt oddly Zen about it all. Her crush on Chris Brompton was officially over. She'd asked him straight-out why he'd caved so quickly to her mom's demands, and instead of being honest, instead of just saying that he liked her mom, that he'd wanted to date her, he'd talked about "respecting her opinion." *Whatever,* Hudson thought as she got into the elevator. She'd done the right thing, backing out of that album. Except she wanted that feeling back—that feeling of being talented, of being special in her own right. And now it seemed like she would never feel that again.

*

A few hours later she and Lizzie and Carina headed to the library for their free period.

"You don't have any proof," Lizzie said definitively, toying with the chain link wristband of her watch. "All you know is that he was there for yoga."

"Yeah, like six hours after they hung out at some bar the night before," Hudson said.

"He's probably terrified of her," Carina said. "I thought she was going to murder us all last night. Good job, by the way. You totally rocked that pizza thing."

"Thanks," Hudson said. "But it just makes me sad, you know? Like, I'll never have that again. Someone who made me feel the way he did. And now he's going to be my stepfather."

"He's *not* going to be your stepfather," Lizzie said.

"You think that every time your mom gets involved with someone," Carina said. "And, well, you know what usually happens."

Hudson *did* know. Her mom's relationships never lasted long. The men Holla picked always seemed so enamored of her. Until something changed. Sometimes she found out that they were still married, for example. And sometimes, the guys would start to retreat, as if a few weeks of constant togetherness were all they could offer. And every time, Holla would be destroyed. Her last boyfriend had been twenty-six and a sound engineer, and even though Holla had dumped him, and not the other way around, she'd come to Hudson's room every night, crying, for a week straight.

"You just need to focus on school right now," Lizzie said as they took a seat at one of the library tables. "We all need to."

Lizzie was right; Miss Evanevski had just warned them of a quiz the next day in Geometry. But just as Hudson pulled out her notebook she spied Hillary across the room, sitting at a table, alone, and wearing a bulky knit sweater with blue and purple sequined butterflies on it. She was doing the *New York Times* crossword again.

"I'll be right back," Hudson said, getting up.

"Tell your life coach you had *two* slices," Carina said, a proud grin on her face.

Hudson walked over to Hillary's table and slid into the seat across from her. "So guess what? I did it. I ordered a pizza."

"But did you actually eat it?" Hillary asked, her pen poised in the air.

"Yep! Two whole slices!"

"Great," Hillary said, capping her pen. "Did it taste good?" Hudson noticed the magenta plastic barrettes clamped firmly on either side of her head, and wondered if Hillary's mom still did her hair.

"It tasted amazing," Hudson whispered. "Of course, my mom was a little annoyed, but it was worth it."

"Good," Hillary said. "You passed the first test. I'm really proud of you."

"So, e-mail me the details about the bar mitzvah," Hudson said, getting up from the table. "I gotta go study for a Geometry quiz."

"Uh, we're not finished yet," Hillary said. "Sit down."

Hudson sat down. "There's more?"

"Of course there's more," Hillary said. "You didn't think that was it, right? A few slices of pizza and you're your own person?" Hillary stared at Hudson and pursed her lips as she thought. "You said you have a quiz?"

"Yeah," Hudson said, playing with her silver hoop earrings. "Geometry, aka hell."

"Good. Don't study for it."

"What?" Hudson said so loudly that the librarian sitting nearby put her finger to her lips.

"Don't study for it," Hillary repeated.

"But I can't not study for it," Hudson said. "I'll fail."

Hillary shook her head. "You may not get an A, but you won't fail."

"I'll barely get a C."

"And that would ruin your month, right?" Hillary challenged.

"Can I ask what the point of this is?" Hudson said, trying not to be annoyed.

"The point is to know that if you do get a bad grade, you can *survive* it," Hillary said, reclipping one of her plastic barrettes. "This is step two. You have to stop being afraid of bad grades. And I can tell that bad grades really freak you out."

"Is that a bad thing?" Hudson asked. "I mean, I'm not going to ruin my life just because I have a fear of bad grades."

"But you're living your life in *fear*," Hillary said. "You need to show yourself that getting a not-so-great grade isn't the end of the world. And you are *not* going to mess up your transcript with one quiz."

"I'm sorry, Hillary," Hudson said. "But I have to study."

"Then just do an hour," Hillary shot back. "Study for it for one hour. That's it. No more."

"Fine," Hudson said reluctantly as she got up to leave.

"You just said 'fine' again," Hillary pointed out.

"Whatever," Hudson said grouchily, walking back to her friends. Eating pizza in front of her mom was one thing. Now Hillary might be going a little too far.

*

That night, at her desk, Hudson checked her antique French boudoir clock. She'd been trying to solve sample Geometry problems for the past fifty-eight minutes, and by now her stomach was clenched into a knot. Geometry tended to do that to her. In class, finding the area of this triangle and that polygon made perfect sense, but as soon as she faced a problem on her own, everything she knew just melted away like snow. She looked at the answers Miss Evanevski had given them. She'd only gotten one on the first try. To make matters worse, she couldn't stop picturing Chris and Holla in some kind of romantic clinch. They were probably out together right this minute, drinking champagne out of glasses they clinked together like cheesy honeymooners. Ugh.

She eyed her piano across the room. Whenever she'd been stressed about something in the past, she'd sat down at it. Maybe she just needed to play right now.

She walked over and carefully lifted the heavy lid from the keys, then sat down on the polished bench. Her book of Chopin nocturnes still leaned against the sheet-music stand. She breathed in the smell of polished wood and touched the soft, velvety keys. Over the years she had spent hours here every night, first practicing the classics and then writing her own songs. Music had been the only thing she thought about. Until lately. She hadn't played in almost a month. As soon as she'd decided to turn her love for music into a career, everything had changed.

She decided to play an old game. She closed her eyes and tried to clear her mind. When a song appeared, she didn't question it or try to think of another one. She just started to play and sing the words.

Birds flying high you know how I feel
Sun in the sky you know how I feel

She finished the song, then sat quietly on the bench, just breathing. The knot in her stomach was gone, and her lungs were no longer tight. *Thank you, Nina Simone,* she thought. She felt exhilarated and relaxed, like she'd just dived into the Atlantic with Carina in the middle of summer and bodysurfed for an hour. How had she gone so long without playing her piano? No wonder she'd been feeling so anxious lately.

She closed the lid and went back to her desk. Her math notes were still spread all over the place. She'd done only four of the eight practice problems Miss Evanevski had given her, but she closed her textbook and got ready for bed. Hillary was right: She could survive one stupid quiz. She was sure of it.

*

"H, you okay?" Lizzie asked from beside her as she got out her protractor and graph paper. "You look like death."

"Yeah, I'm fine," Hudson said, taking out her pencils. She'd decided not to tell her friends about this latest life-coaching move from Hillary. "Just a little nervous."

She'd barely thought about the quiz as she sat through English and History. But as soon as she walked into Geometry, everything changed. All around her, people were studying — frantically. They sat bent over their books, working out problem sets, erasing and starting over. Hudson sat down, her heart pounding in her chest as if she were about to bungee jump without a cord. She'd been *crazy* not to study. And now she was going to pay the price.

"Hey, relax," Carina said, leaning past Lizzie. "You know this stuff. You're gonna be fine."

I would if I'd studied last night, she thought.

Miss Evanevski placed the tests on people's desks. "Okay, you can start," she announced to the room.

Hudson turned the test over. It was five problem sets. She scanned the pictures of circles and rays and shaded triangles and her mind went stubbornly blank. She was going to fail this quiz, and it was going to be long and drawn out and painful.

She worked through the first problem, the pencil slipping around between her sweaty fingers. Somehow she came up with an answer, but she had no idea if it was right. She moved on to the next one. Then the next one, and then the next, letting herself erase and start over only once for each question.

"Time!" Miss Evanevski called.

Breathless, Hudson looked up at the clock, sure that there was no way forty-five minutes had passed, but they had. People put down their pencils and protractors. Hudson looked down at her graph paper. It was covered with frantic scribbling. She was sure she'd failed.

"All right, hand them over," Miss Evanevski said, pacing through the room, collecting the quizzes. She smiled encouragingly at Hudson as she gestured for her quiz.

Hudson placed her paper in Miss Evanevski's carefully manicured hand. *There's no turning back*, she thought.

"That wasn't so bad," Carina said as they walked out of the classroom.

Hudson didn't say anything. She could imagine the big red F on her quiz already, throbbing like a cheap neon sign. But by the time she'd gotten to her locker, she felt oddly okay. The world hadn't ended. She was still basically herself. And whatever her grade ended up being, she knew that she would probably survive it.

A small note was sticking out of one of the air vents of her locker. She yanked it out and opened it.

You rule!

HC

“What’s that?” Carina asked.

“Just a note from my life coach,” Hudson said, stuffing the note in her bag.

“Is that stuff actually working?” asked Lizzie.

Hudson opened her locker. She felt anxious and exhausted from the Geometry test. But she also felt brave, and that was something she hadn’t felt in a while.

“I think it actually might be,” she said, and slammed her locker closed with a smile.

chapter 12

The next day in Geometry, Miss Evanevski passed out the corrected tests. "All in all, everyone did well," she said, placing Hudson's quiz facedown on her desk. "Though there were a few surprises."

Hudson turned over the quiz. In bright red marker, at the top, was her grade: 77. A C plus.

"Yes!" Hudson said out loud. She hadn't failed. It was a miracle.

Miss Evanevski gave her a quizzical look from across the room. Hudson immediately got rid of her smile.

"And let's go through it, shall we?" Miss Evanevski said, giving Hudson one last weird look.

"I got a C plus," Hudson confided to Hillary later that morning, in the library. "I didn't fail!"

"Told ya," Hillary said. "Congratulations on completing step two."

"So what's step three?" Hudson asked.

Hillary pulled two lipsticks out of her square backpack. "Before we get into that, which one do you think I should wear tomorrow?" she asked, uncapping them. "This one's Frisky Fuchsia and this one's Blushing Berry."

"Wait. Tomorrow?" Hudson asked cluelessly.

"The bar mitzvah!" Hillary said, slightly annoyed. "You're still coming, right? You have to tell me what Logan thinks."

"Oh, right," Hudson said, drumming her fingers on the library table. "Hillary, don't read too much into this, but my mom can be a little weird about me leaving the city," she said. "Especially with people she doesn't know."

"What about summer camp?" Hillary asked. "Haven't you ever been to camp?"

Hudson shook her head. "My mom says there are too many crazy people in the world. But she lets me go to my friend's house out in Montauk. That's about as close as I've gotten to camp."

"Well, no one's going to kidnap you at my cousin's bar mitzvah," Hillary snapped. "I promise. Just ask her."

"I'll try, Hillary," Hudson said, "but I can't promise anything."

"That's why you're doing all this stuff," Hillary said. "We're trying to help you become your own person, remember?"

That night, when she was finished with her homework, Hudson got up from her desk and sat down at her piano. Holla was still at the studio, recording with Chris. Just thinking about them together in the studio made Hudson feel weird, almost as if they were at a party she hadn't been invited to. She needed to distract herself. She closed her eyes and a song popped into her head — an old Fleetwood Mac song she'd always loved. Tentatively, she touched the keys, and then she started to sing.

For you, there'll be no more crying
For you, the sun will be shining

A knock on the door made her stop.

"Honey?" Holla called, opening the door a tiny bit. "Can I come in?" She stepped into the room lightly, with a dancer's grace. She looked impossibly sleek and thin in an off-the-shoulder electric green top, black denim leggings, and ankle boots, and her hair fell in soft waves to her shoulders. "Did you have dinner?"

"Yes," Hudson said.

"Finish your homework?"

"Uh-huh."

"Great." Holla headed to Hudson's walk-in closet. "You mind if I borrow something?"

"Go ahead," Hudson said. She needed to ask Holla about the bar mitzvah, so it could only help to let her mom peruse her wardrobe. "How was your day?"

"Long. And that new guy from the record label has to second-guess everything. As if I've never done this before." Holla walked out holding a hanger. "What about this?"

Hudson stared at the orange sherbet–colored cotton shift dress. It was from the line Tocca had put out in the mid-nineties. But Hudson had worn it only twice because it was so short. "Um, that's a summer dress," she said.

“I’m wearing it inside,” Holla said casually, folding it over her arm. “And how are you doing, sweetie? How’s school?”

“Okay.”

“How’s Geometry?”

Hudson took a deep breath. “It’s fine, Mom.” She didn’t want to tell her about the C plus.

Holla glanced at Hudson’s piano. “What are you doing?”

“Just playing a little.”

“Well, keep going,” she said, perching on the arm of Hudson’s battered leather armchair and crossing her legs. “You know how much I love your voice.”

“That’s okay,” Hudson said. “My voice is kind of weird tonight.”

“No, go ahead,” Holla urged. “I want to hear. I think it’s great you’re playing again.”

Hudson knew exactly what would happen if she started to sing, but there didn’t seem to be any way out of it. “Okay.”

She found the opening chords and started playing.

For you, there’ll be no more crying
For you —

“Honey, *fuller,*” Holla cut in. “Don’t rush it. Fill out the word. *Youuuuuu.*”

Hudson stopped playing and swung the lid closed over the keys.

“What?” Holla asked. “What happened?”

“Nothing,” she said.

Holla frowned. "I just want you to get the most out of practicing. And correct phrasing is very important. Why do you have to be so sensitive? Don't you think I know a little bit about this?"

"My friend wants me to go to a bar mitzvah tomorrow," Hudson said, changing the subject. "Can I go? It's in Westchester."

"Westchester?" Holla asked, wrinkling her nose as if she'd just smelled something bad. "Where in Westchester?"

"Larchmont, I think."

Holla shook her head as if this were utterly ridiculous. "What friend is this?"

"Her name's Hillary Crumple," Hudson said. "I kind of owe her a favor. Fernald can take me in the car, and I'll be back by five." *And I promise, nobody's going to kidnap me,* she wanted to add.

"I'll have Little Jimmy go with you," Holla said. "I don't need him at the studio."

"I don't need a bodyguard to go to a bar mitzvah."

Holla walked to the door and paused with her hand on the knob. "Fine, Hudson," she said. "You can go, but Fernald will take you there and bring you home. And there's something I'd like you to do for me."

"What?"

Holla shifted the dress on her arm. "I'd like you to check in on your aunt. Maybe spend the day with her on Sunday? I know she's finally back from Paris. For good this time," she added. "And I'm worried about her."

"So why don't you just go over there and see her?" Hudson asked. "I'm sure she'd love it."

"No, she wouldn't," Holla said, in a way that implied she'd given all of this a great deal of thought.

"I'm sure she's sorry about Christmas. You should just call her."

Holla shook her head. "Will you just do me that favor, please?" She walked over to Hudson and kissed the top of her head. "You're my partner in crime. You know that."

"Holla? Where'd you go?"

The voice calling from outside Hudson's bedroom door was unmistakably familiar. And unmistakably male.

"Are you up here? Where'd you go?"

Hudson's pulse raced. It was Chris.

Holla went to the door and opened it a crack. "I'll be down in a second," she said.

"Chris is here?" Hudson asked.

"Oh, we're just going to do some more work," Holla said matter-of-factly. "We'll be up in the office if you need us."

"Mom?"

"Yes?" From the way her face was glowing Hudson could tell that her mom had already fallen in love. "What, honey?" she asked. "What is it? You look so worried."

"Nothing," Hudson said. "I'm fine."

"Thanks for the dress," Holla said, and then she was gone.

chapter 13

"Okay, there he is," Hillary said, smoothing the front of her velvet dress and smacking her fuchsia-painted lips. "He's over there, checking out the dessert table."

Hudson looked across the spacious pink and blue–lit banquet hall, past the tables dressed with kelly green tablecloths and imitation roulette wheel centerpieces, and past the band onstage grinding out a cover of "Hey Ya!" For most of the ceremony, which had taken place at a synagogue a mile away, and then for most of the delicious meal they'd just eaten, Hudson had been waiting to spring into action as Hillary's wingwoman, on alert for Logan the Mysterious. But Hillary had refused to point him out. Whenever Hudson asked, Hillary would whisper, "You can't look now. He's right *there*."

But now Hudson could see two boys checking out the dessert buffet. One was tall and skinny, with a mop of curly brown hair and a dark blue suit that seemed a little short at the ankles. He kept trying to shake a sliver of chocolate frosting off the cake knife, but he wasn't having much luck. The boy next to him was shorter and stockier, with whitish blond hair. He was heaping his plate with everything on the table — brownies, cookies, and slices of three different cakes. "Which one is he?" Hudson asked.

"The cute one!" Hillary said.

"They've got their backs turned."

"The blond one!" Hillary exclaimed. "The other one's my cousin Ben. You met him at the synagogue."

"Oh, right," Hudson said, vaguely remembering Ben in the lineup of Hillary's extended family.

"Okay, I'm gonna go talk to him. How's my hair?"

Hudson looked at Hillary, who was still patting her hair and her dress. She looked good. Someone had mercifully whisked all of Hillary's flyaway hair into a neat, chic bun at the top of her head, and the black velvet dress looked perfect on her, even though she'd chosen to wear a strand of big white pearls with it. "You look great," Hudson said. "Go get 'im."

"Uh-uh, I'm not going alone," Hillary said, grabbing Hudson's hand. "Let's go."

As the two of them marched across the banquet room, Hudson tried not to be embarrassed. She never liked approaching guys. That was Carina's thing. Hudson's thing was to wistfully stare at them and wonder about them and look up their astrological signs. But never guys her age. To her, it was hard to get excited about a guy who was just doing homework every night.

The boys turned around and watched them approach. The tall, skinny boy — Hillary's cousin — smiled and waved at them with his fork. He looked shy and a little bit awkward, as if he were still getting used to being so tall. His blond, stocky friend didn't smile at them or wave at all. He was definitely cute, and he knew it. He had narrow, smoldering eyes, and he stared at Hudson in a familiar, unblinking way that she had seen before. *Uh-oh,* she thought, her heart sinking. *He likes me.*

"Hey, nerd," Hillary said to the tall, skinny boy, carefully ignoring his friend. "What are you guys doing?"

"Just having some cake," Ben said, just as a glob of it fell off his fork and onto the floor.

"This is my friend Hudson," Hillary said. "You guys already met."

"Hi," Hudson said.

"Oh, uh, hi there," Ben said, momentarily distracted by the cake on the carpet but smiling at her anyway.

"So . . . are you gonna go up there and jump on the bass?" Hillary said, pointing to the stage. "Ben's starting a jazz band," she explained to Hudson. "He's really good."

"Not really," Ben said shyly, waving off Hillary's compliment.

"This is my friend Hudson," Hillary said to the other boy. "This is Logan."

"Hey," Logan said, taking a bite of brownie with studied mellowness.

"Hi," Hudson replied, careful not to meet his gaze for too long.

"Are you guys looking for a singer?" Hillary blurted.

"We haven't started yet," Ben said, looking at Logan. "We're still trying to figure out what kind of jazz band we're going to be. It might just be drums, bass, and sax."

"Well, Hudson sings," Hillary said. "And plays piano. She almost did her own al—"

Before Hillary could say *album* Hudson clamped her hand around Hillary's arm. "I just sing a little," she said, shooting Hillary a quick warning glance.

"Yeah?" Ben asked. "What sort of stuff do you sing?"

Hudson smiled. "A lot of stuff," she said. "Mostly my own songs."

"She has an amazing voice," Hillary said. "You should *totally* audition her for your band. You don't want to do that awful coffee-shop jazz. That stuff is just noise. Ugh."

Just then the lead singer's voice rang out over the sound system. "All right, people! It's that time we've all been waiting for—when you get up here and show us what *you* got!"

They all looked to the center of the room. The rest of the band had cleared off the stage, and a projection screen was being lowered from the ceiling. As the lights dimmed, Hudson realized what was happening. It was karaoke time.

"Okay, who wants to go first?" the lead singer yelled again with an almost diabolical smile. "Come up here and pick out a tune! And then be the star of your very own music video!"

On the dance floor, kids milled around, wondering who would be the first to go.

"Perfect timing," Hillary said to Hudson. "Go up there and show everyone what I'm talking about!"

"What?" Hudson asked, clamping down on Hillary's arm even tighter. "You're joking, right?"

"I'm dead serious," Hillary said. "We were just talking about your great voice."

"Um, can I speak to you for a second?" Hudson asked, yanking Hillary aside.

"So what you do think? Is he into me?" Hillary asked as soon as they were alone. "I could feel him looking at me. You know, when I talked."

"I really don't know," Hudson fibbed. "I'd have to see you guys together more. But what are you doing? Why are you trying to make me sing?"

"Because this is the perfect step three," Hillary said. "Singing in public. This is your biggest fear. You can tackle it right now!"

"I'm not tackling any fears today," Hudson said, trying not to snap. "I came here to be your wingwoman."

"And I hate to criticize, but you could be just a little more talkative around Logan," Hillary said. "I can't do all the work."

Hudson took a deep breath. "Well, I really don't want you to mention the album to people."

"Fine, but those guys wouldn't know your mom if she fell on them. They're total nerds. The only TV they watch is the Discovery Channel."

"And I'm not singing here. No way."

"It's karaoke!" Hillary said, loud enough that Ben looked over at them. "Nobody expects anyone to be good."

Hudson met Ben's eye and he politely looked away. The group of seventh-graders had started edging closer to the stage. Hillary's cousin Josh, the boy being bar mitzvahed, was being pushed to the front of the dance floor. Hillary was right: How serious could this be? "I'll do it if you do it," she dared.

Hillary snorted. "Of *course* I'm doing it," she said. "Watch me."

With that she yanked her arm free and charged through the crowd. "I'll go!" she yelled. "I'll go!" Hillary almost knocked her own cousin down as she ran up to the stage.

"We have our first performer, everyone!" the lead singer shouted with glee into the mic. "What's your name?" He tipped the mic to Hillary.

Hillary grabbed the mic. "Hillary Victoria Crumple," she said, matter-of-factly. "Just start the first song. I'll sing anything."

Hudson watched with amazement as Hillary positioned herself in the center of the bright white spotlight. Hudson hadn't seen this kind of confidence and flair since the last time she'd seen Holla in concert.

"This is dedicated to my cousin Josh!" Hillary shouted into the mic. "And my friend Hudson!" A moment later a familiar beat started to play.

"Oh, God, no," Hudson said under her breath. She knew what the song was, and it was the worst song Hillary could have picked. And when Hillary began to chant into the mic in a wobbly, screechy voice, Hudson's blood ran cold.

I wanna hold 'em like they do in Texas plays
Fold 'em let 'em hit me raise it baby stay with me

She snuck a look at Ben and Logan, who didn't move. The kids on the dance floor seemed shocked speechless. But that wouldn't last. Pretty soon they were going to be snickering, if not all-out laughing. *I have to save her,* Hudson thought. *Before she makes a worse spectacle of herself than I did at the Silver Snowflake Ball.*

She began to push through the crowd. A few boys had already started to laugh. Finally she made it up to the stage, just in time to join Hillary for the chorus. Hudson threw her arm around Hillary and kept singing. All she was thinking about was trying to drown out Hillary's voice, or at least trying to make it sound like it was on-key. She didn't dare look down.

Until, toward the end of the song, she finally did, and saw that everyone was dancing. A middle-aged man with a paunch — somebody's dad — stood beside his chair, clapping his hands over his head. An elderly woman with wispy white hair did the twist on the dance floor. Even Ben's brother and his friends had stopped snickering and were jumping up and down in time with the beat.

Hudson looked over at Hillary, who was grinning, as if to say, *Isn't this great?*

When the song was finally over, the lead singer pounced on them with the mic. "That was fan-TAS-tic! What's your name?" he asked, tipping the mic toward Hudson.

"Hudson," she said, still out of breath.

"Let's hear it for Hudson and Hillary, everyone!" he cried.

Everyone clapped, hooted, and hollered. Hudson linked arms with Hillary and, grinning crazily, they both took a bow.

"That was amazing!" Hillary said as soon as they'd jumped down from the stage. "We totally killed!"

Hudson felt someone tap her on her shoulder. She turned around to see Ben staring at her with a radiant, awestruck smile. "Hillary was right," he said. "About your voice."

"Thanks."

"I hate it when she's right," he joked.

"Me, too," she joked back.

He ran a hand through his unruly hair. "So . . . this might be kind of sudden, but . . . would you be into being our lead singer?"

Hudson blinked. *Yes,* a voice inside her said. *Say yes.* "I'd love to," she said.

"Cool. Just come over tomorrow for rehearsal. Around two." He looked at Hillary. "I bet Hillary will want me to pay her a commission or something."

Hillary sidled up beside them. "What's going on?" Hillary asked.

"Hudson's gonna be our lead singer," said Ben. "Turns out you've got an eye for talent, Hil."

"Um, *obvi,*" Hillary said.

Ben glanced at the stage, where his brother was trying to sing an Eminem song. "I think I better get over there. Before the girl he likes never speaks to him again."

As he gave them an awkward wave and headed back toward the stage, Hillary started jumping up and down. "God, I love it when I'm right about stuff!" she shrieked. "You're going to be their lead singer!"

"But I can't do this," Hudson said, suddenly panicked. "I don't even live up here. How am I going to be the lead singer of a jazz band? Here? When my mom will barely even let me off the block?"

Hillary shook her head. "Don't you get it? This is just what you need! A second chance to do what you really want to do. And your mom has nothing to do with it!"

Hudson watched Ben onstage, backing up his brother on the Eminem song. The two of them looked so goofy that she had to smile. Singing up there had been fun, and now she had a feeling that being in a simple high school band might be even *more* fun. She wouldn't have to worry about concerts or clothes or *Saturday Night Live* appearances. Music would be fun again. And wasn't that what all this was about — fun?

Still, there was one last thing she had to do. She dashed back to her table, fished her phone out of her bag, and wrote a quick text.

Just got asked 2 b the lead singer of a bnd. Up in Larchmont. Yay or nay?

The answers came back right away, first from Lizzie, and then from Carina:

YAY!!!

With a smile spreading across her face, Hudson put the phone away and walked back to Hillary, who was at the dessert table, piling her plate with cookies. She was sure now. This was fate.

"I'm in," Hudson said to her. "As long as you don't tell Ben anything. You know, about my mom."

Hillary looked up. "No way," she said. "And you better thank me when you win your first Grammy."

chapter 14

The next morning Hudson sat bolt upright in bed and rubbed her eyes. She'd told Ben that she would be back up in Larchmont by two o'clock for rehearsal. But she'd also promised her mom that she'd see Aunt Jenny today. *Oops,* she thought. And it was already nine o'clock.

She grabbed her iPhone and texted her aunt.

Brunch at 11?

A moment later, Jenny wrote back:

Come over. 421 E. 76th Street.

Phew, Hudson thought. She would just be able to pull this off.

She showered and dressed in what she called her Seventies Urban Princess outfit: slim-fitting gray wool pants, an oversized cashmere cowl-neck sweater, and vintage platform boots. She drew back her curtains and looked out at the cloudless blue sky. The sidewalks were deserted — even the patch of sidewalk across from their house was empty of photographers. It had to be freezing outside. Hudson ran back to her closet and grabbed the thick black cape she'd picked up in Covent Garden during her mom's last tour and pinned it at her throat. Hopefully this wouldn't be too over-the-top for Larchmont.

When she opened her bedroom door she was surprised to find a big shopping bag waiting for her. There was a note pinned to the handles:

Honey plz bring this for Jenny. I thought she'd like it.
Mom

Hudson reached into the bag and pulled out the present: a thick white throw blanket, wrapped in red ribbon. It was exactly like the one they had in their living room. Hopefully Jenny wouldn't remember. Her mom meant well, but sometimes she could be a little dense. Hudson stuffed it back in the bag.

Soon she and Fernald were driving uptown. The few people out on the streets were bundled up with scarves wrapped over their mouths and noses and walked, heads bent, into the wind. Hillary had decided to spend the night in Larchmont after the bar mitzvah, so Hudson would be going up herself. Which meant that she'd be taking the train alone.

At Jenny's building, Hudson hopped out of the car with the shopping bag and the wind hit her face like a bunch of knives. "I'll text you in a bit!" she yelled to Fernald, her eyes watering from the cold.

He waved back to her, trusting as usual. Hudson felt a stab of guilt when she thought about her plan to elude him later — she'd need to get rid of him before she left for the train — but she pushed it aside.

Standing outside and shivering she pulled out her phone and called Carina.

"What are you doing today?"

"Staying the hell indoors," she replied. "It's like five below out."

"Can you come with me to Larchmont for my first band rehearsal?"

"*Larchmont?*" There was a long pause. Hudson could practically see Carina biting her lip, trying not to say no. "Sure," Carina finally said. "Just tell me where and when."

"Grand Central, one o'clock. Let's meet under the clock. We're taking the train."

Then she called Lizzie, who signed on right away. "As long as we're back by five," Lizzie said. "I'm trying to finish a story tonight."

"No problem." Hudson clicked off and went to Jenny's door. Thank God for her friends.

Jenny's building had no doorman, only a buzzer. Hudson rang the bell and a few seconds later the front door unlocked with a loud buzz. She pushed her way into a dingy, dim vestibule, then walked through another set of doors. There was no elevator, only a set of stairs. From above she heard a door open. "I'm up here!" Jenny's voice called. "Fourth floor!"

Hudson climbed the stairs, and when she turned the corner,

slightly out of breath — her platform boots didn't make it easy — Jenny was waiting for her at the door. "Hey, stranger!" she said, beckoning Hudson inside. "Come in and warm up!"

Jenny was more casual today in hole-ridden jeans, a faded red T-shirt, and a fuzzy Mr. Rogers–esque cardigan in a pale shade of yellowish brown. Hudson gave her a hug, and then walked inside. Her apartment was small, just one room, but, like Hudson's room, it was stuffed with eclectic pieces: a small chandelier suspended over a blond wood farm table and chairs, a Tiffany lamp on a skinny-legged end table, a beautiful red and purple dhurrie rug that stretched across the floor. But the two windows faced a brick wall, and it was so dark inside the apartment that it could have been nighttime.

"I haven't had a lot of time to finish decorating," Jenny said. "I've been so busy."

"That's okay," Hudson said, looking around. "It actually looks really cool in here."

"Do you want some tea?" Jenny asked. "I got the most amazing Earl Grey tea from this little tea shop in the Marais. And there's a French bakery around the corner from here that makes semi-decent croissants." Jenny put the kettle on to boil, then put the croissants on a plate and carried them to the table. "What's that?" she asked, pointing to the shopping bag.

"It's for you. From my mom."

Jenny gave Hudson a suspicious look. "What is it?"

"A housewarming gift," Hudson said, giving it to her.

Jenny reached into the bag and pulled out the throw blanket, letting it fall to its full length. "Isn't this the one you guys have in your living room?"

"Yes," Hudson said awkwardly. "I guess she thought you'd like it."

Jenny carefully spread the blanket over the back of her shabby chic–style couch. "Well, this time, I have to hand it to her. It's beautiful. Tell her I said thank you."

"I think she feels bad about Christmas," Hudson offered.

"So do I," Jenny said. The kettle began to whistle, so she slipped into the tiny kitchen and turned off the gas. "Every time I tell myself not to lose it, but she's just impossible to be around sometimes. You were there. You saw it." She poured the water into two mugs. "She just does that to me."

And me, too, Hudson thought. "I think she just worries about you."

"I know, but I like my life. I'm happy," she said, bringing the mugs of tea to the table.

"I think she just feels that you're a little . . ." Hudson let her voice trail off, aware that she was on dangerous ground.

"Lost?" Jenny said with a smile. "Look, I know I don't have Holla's discipline. Hardly anyone does. But I think I do okay." They sat down and Jenny stirred her tea, lost in thought. "She always wanted to be famous. That was always her thing. Did I ever tell you that?"

Hudson shook her head.

"She'd even talk about herself in the third person. She'd interview herself, pretend she was on Barbara Walters or something."

Jenny smiled as dunked her tea bag in and out of the mug. "I remember the day she won her first talent show. I was six, she was eleven. She sang and danced to some Madonna song. She was even better than Madonna."

"I'm sure," Hudson said, picking at her croissant.

"And I just wasn't like that," Jenny added. "I didn't need to be famous. I liked to dance. I was good at it. But when I blew my audition for the Martha Graham Company—"

"What? You blew an audition?" Hudson asked.

"Yep," Jenny said, taking a careful sip. "Just froze up. Forgot my routine. I practiced for weeks, and then when the time came I just stood there like an idiot, while a whole tableful of people stared at me."

"Why'd you freeze up?" Hudson pressed, clutching her mug.

"I'm not sure," Jenny said wistfully. "I was so self-conscious. At that point Holla was huge. She'd already won her first Grammy. She'd just had you. And there I was, trying to make it as a dancer. It was a lot of pressure."

Hudson stirred her tea, listening.

"When Holla heard what had happened, she called the company and talked them into giving me another chance."

"She did?" Hudson asked.

"I said no," Jenny said, looking down into her mug.

"Why?"

"Because I don't think I really wanted it. And I knew that no matter what, I would always be compared to her. How are you just a dancer when your older sister's insanely famous? You know what I mean?"

Hudson nodded. She knew exactly what Jenny meant, but she didn't want to say anything.

"The same thing happened with your father," Jenny said.

"My father?" Hudson asked, her curiosity piqued.

Jenny looked guilty. "I suppose you haven't heard a lot about what really happened with him," she said. "But I don't think he could handle it, either. Your mom's fame. Always being linked to someone like that. And your mom isn't exactly easy to deal with, either. But that's another story."

Hudson put down her mug. "Do you know where he is?" she asked, looking down at the table, trying to sound casual.

Jenny shook her head. "I heard he was in Europe for a little while, but now, I don't know. He was the type of guy who liked to stay under the radar."

Hudson let this sink in. She wondered if Jenny had ever seen him.

"But saying no to that audition was the best decision I ever made."

"But you loved dancing," Hudson argued.

Jenny gave Hudson a gentle smile. "I love a lot of different stuff. I'm an air sign. And you, my dear, are all water. Very creative, very sensitive."

"Yeah, so far it's been great," Hudson said wryly.

"Well, don't be surprised if you end up more famous than your mom. I told you it's in your chart, right?"

Hudson smiled weakly. "Too bad I pulled the album."

"What?" Jenny asked. "What happened?"

"The same thing that happened to you. Right before Christmas, I was supposed to sing at this big dance, and I froze up. Right there onstage. In front of three hundred people."

"Was your mom there?"

Hudson nodded.

"What'd she do?" Jenny asked, slightly horrified.

"Told everyone I had food poisoning—what do you think?" Hudson said, with an ironic smile. "But before it happened, she was driving me crazy. I was doing *this* wrong, I was doing *that* wrong. I needed to learn all these dance moves, and sing to this track. And before that she made my producer change my entire sound. To hers."

Jenny listened with a somber look on her face.

"So I just decided, forget it. It's not worth it. No matter what I do, it'll never be right for her. She wants me to *be* her."

Jenny put her hand on Hudson's wrist. "You can be whatever you want, Hudson." Her brown eyes were soft but also battle-scarred, as if she were a fellow survivor of something.

"But I've been thinking...what if I just joined a regular high school band? No record deal. No concert dates. No promotion. Just playing a couple shows here and there and jamming. Just for fun?"

Jenny raised one eyebrow. She seemed to sense that this wasn't exactly hypothetical. "I'd say it sounded great," she said. "Not everything has to be so serious, you know. You're allowed to have fun in your life."

Hudson glanced at her watch. It was almost twelve thirty. She needed to go meet her friends at Grand Central. "I have to go. I told my friends I'd meet them."

Jenny smiled knowingly. "Your friends, or your band?"

Hudson smiled back. "You should come over to the house sometime. My mom really does feel bad. She loves you. She really does."

Jenny rolled her eyes. "Sometimes I think we're just not meant to be friends."

"Maybe we can throw a little birthday party for you next month," Hudson said, undeterred. "Your birthday's February seventeenth . . . what about that Saturday night? The twenty-first?"

"Whoa," Jenny said, holding up her hand. "Are you sure that's a good idea?"

"Of course," Hudson said, getting to her feet. "I'm sure she'd want to do it."

"We'll see," Jenny said, hugging her niece good-bye. "I'm always here for you, you know. And it looks like I'm actually staying here for a while. So come hang out with your crazy aunt."

"I will," Hudson said as she left.

Downstairs, in the dim vestibule, Hudson remembered what she had to do. She called Fernald.

"Hey, Fernald, I think I'm gonna be staying here awhile," she said lightly. "Jenny'll drop me home in a cab." Her heart raced with the lie.

"Sure thing, Hudson," he said, before she clicked off.

She had no idea how she was going to do this every time she had rehearsal, but she told herself she'd worry about that later.

chapter 15

"If my dad could see me right now he'd freak," Carina said, settling into her seat across from Hudson and pulling a warm chocolate-chip cookie from a brown Zaro's bag. "But this definitely beats reading *Macbeth* all day."

"When was the last time we were on a train together?" Lizzie said, taking her battered copy of *Nine Stories* out of her purse.

"Sixth-grade field trip to D.C.," Hudson said. "Remember when Eli threw up in the club car?"

They burst out laughing as a bell sounded and the doors shut. They'd chosen a group of four seats that faced one another in the back of the train, and Carina propped her Chuck Taylors up on the cracked Naugahyde seat next to Hudson.

"Yeah, I remember," Carina said. "Some of it got on my foot."

"Ugh!" Lizzie yelled.

The train lurched forward, and soon they were rolling along

through a dark tunnel. Hudson still couldn't believe what they were doing, but being with her friends made her less anxious.

"Thanks so much for coming with me, you guys," Hudson said. "Who knew that doing karaoke at a bar mitzvah could be so eventful?"

"What'd you sing?" Lizzie asked, putting her book away.

" 'Poker Face,' " Hudson said. "But it was a duet. With Hillary."

Carina crumpled the brown bag. "*Hillary?*" she asked. "Your stalker Hillary?"

"She's not my stalker," Hudson said.

"Are you guys becoming friends now?" Lizzie asked warily, looking over the top of her book.

"She asked me to go and I went," Hudson said. "She's a little weird, but she's actually really cool."

"No one with that kind of backpack can be cool," Carina said.

"Be nice, C," Lizzie said, nudging her in the arm.

Suddenly the train shot out of the tunnel and into sunlight. Hudson looked out through the smudged window. They were on a track above Park Avenue, passing through Harlem. The cloudless sky glowed a deep cornflower blue, and sun glinted off the windshields of parked cars below. She'd never seen the city from this angle before.

"Wait," Lizzie said, pulling one unruly curl behind her ear. A moment later, it sprang back again. "Do the guys in the band know about your mom?"

"It never came up. And as long as they don't ask, I'm not telling them."

"But don't you think they'll find out?" Carina asked. "You're in the tabloids, like, once a month."

"These guys don't read *US Weekly,*" Hudson said. "They're into jazz and the Discovery Channel."

"But you don't have to be looking for info," Carina said. "All they have to do is see a photo of you somewhere —"

"And then I'll tell them," Hudson interrupted. "It's not the end of the world if they find out. But right now, it feels good to just be . . . nobody."

"And you're definitely not telling your mom," Lizzie prompted.

"No." Hudson played with her cowl-neck collar. "She'd never understand why I'd choose a high school band in the suburbs over Madison Square Garden."

"Is this guy at least cute?" Carina asked. "What does he look like?"

"Curly hair. Tall. Skinny. He's really nice."

"Sounds like a dork."

"He's not a dork," Hudson said.

"And there's nothing wrong with a dork," Lizzie said. "Todd's kind of a dork."

"Todd is definitely *not* a dork," Hudson replied. "And how's Alex?"

"Oh, I found out his birthday," Carina said. "It's September twenty-fourth."

"A Libra," Hudson said approvingly. "That's just what we want. His air balances out your fire."

"I wish you could figure out if he's compatible with my dad," said Carina. "Alex is supposed to come over for dinner next week.

And something tells me they're not gonna bond over music and subtitled movies."

"Don't even expect it to go well," Lizzie put in. "Todd's polite, he's a writer, he calls people 'sir,' and my dad still doesn't know his name. He calls him Brad. It's like he's mentally blocked him out or something."

"It's so funny; I think my mom would love for me to go out with someone," Hudson said, yawning. "That way she could tell me how to do that, too."

Lizzie and Carina laughed.

As they crossed the bridge into the Bronx, Hudson felt herself start to get drowsy from the gentle rocking of the train. A short time later, Hudson felt Lizzie's foot nudge her leg. She opened her eyes to see bare trees and power lines and a church steeple whizzing past the window. "I think we're the next stop," Lizzie said, as Carina rubbed her eyes.

The train began to slow down, and they passed a white sign that said LARCHMONT in big black letters. A few moments later they screeched to a stop. "Let's go," Hudson said, getting to her feet. The doors opened and they stepped out onto the platform as a gust of wind seeped in underneath her cape. "Hillary gave me directions. She said it's a quick walk."

"Uh, no," Carina said, heading for the line of black cabs waiting by the platform. "This may be the suburbs, but we're still taking a cab."

They got into a cab, gave the driver the address, and pulled out of the train station parking lot. Soon they turned onto a picturesque main street with an old-fashioned movie theater and a barbershop.

"This place is so cute," Hudson said. "Can you imagine living up here?"

"It's too quiet," Carina said bluntly.

"Our lives would be so different," Lizzie said. "We'd have to learn how to drive. And go to a school with a football team. And there would be cheerleaders."

"Cheerleaders," Hudson said, trying to picture it. "Do you ever think that we're gonna end up total weirdos, growing up in the city? Not driving, not going to football games and stuff like that?"

"I think if we end up total weirdos, it's gonna be for other reasons," Lizzie said.

They turned off the main street onto a rural road. "So I hope this isn't Silver Snowflake Ball, the Sequel," said Hudson.

"But do you know if this band is any good?" Lizzie asked. "You're so talented, H. It'd be good to know if they're up to your standards."

"You sound a little like my mom," Hudson said.

"No, seriously," Carina said. "How do we know that these guys even know how to play music?"

"Well, I guess we'll find out," Hudson said.

They passed several three-story homes, Victorians mostly, until they turned into a gravel drive. The half-melted remains of a snowman stood frozen on the front lawn. Beyond it was a Victorian shingle house. It looked old and friendly. A gray Ford Windstar was parked in the driveway. A few bikes lay on their sides in the snow.

They paid the driver and walked up the steps to the house. "Can you imagine having a front lawn?" Carina asked in a whisper as Hudson rang the doorbell.

"You have, like, three of them," Lizzie said.

"I mean, *all* the time," Carina said.

Hillary opened the front door. She wore jeans and a surprisingly muted navy blue sweater. It was the first time Hudson had ever seen Hillary in pants. And in a sweater that wasn't a blindingly bright color.

"Oh, hey," Hillary said, blinking her yellow-green eyes. "I didn't know you all were coming."

"I asked them to come for moral support," Hudson said. "You know Lizzie and Carina. You guys remember Hillary."

As they stepped into the house there were murmured greetings.

"Everyone's downstairs in the basement," Hillary said, leading them past a bench covered in coats and scarves and mittens. Hudson heard the whirring, tumbling sound of a dryer in the distance and, upstairs, the muffled blare of a TV. Hillary pulled Hudson aside. "Logan and I have been talking this *whole* time," she whispered excitedly into her ear.

"That's great!" Hudson said encouragingly.

"I really feel like he's *this* close to asking me out," Hillary said excitedly. "Should I have a date planned or should I leave it up to him?"

"I think we should just leave it up to him," Hudson said. "And by the way, you look really nice."

"Thanks. Do you guys want something to eat?" Hillary asked her friends in a louder voice.

"Sure," Carina said, going straight to the fridge and opening it. "Are these enchiladas?" she asked, taking a Tupperware container from the fridge. "Score."

"C, put that back," Lizzie said.

"No, she can have some," said a voice, and Hudson turned to see Ben walk into the kitchen. At first Hudson barely recognized him. He looked so different than he had the day before. Then Hudson realized that he was wearing clothes that fit: dark jeans, a black T-shirt that read STOP THE ROBOTS, and a pair of wire-rimmed glasses that were a little too square-shaped to be cool. "Hey, Hudson," he said shyly. "Nice... cape."

"Hey, Ben," Hudson said. "These are my friends Carina and Lizzie. They came with me for... well, because they'd never seen Larchmont." She quickly unpinned her cape and took it off.

"Oh, hey," Ben said, awkwardly shaking their hands. "Nice to meet you."

Hudson watched Carina and Lizzie size him up. Lizzie was always polite, but sometimes Carina could take her time. "Nice to meet you, too," Lizzie said.

"You sure I can eat these?" Carina asked, popping the lid off the Tupperware.

"Oh, sure. My mom would be flattered. Does anyone want something to drink?" he asked.

"Water would be great," Hudson said. She tried to imagine her mom making enchiladas—at least, with real cheese. It was impossible.

"So, lemme ask you something, Ben," Carina said, folding her arms and walking away from the Tupperware. "What *exactly* are your plans for this band?" She sounded just like she had the day they'd barged into Andrea Sidwell's photo studio to find out about Lizzie's modeling opportunities.

"My friends are a little protective," Hudson explained.

"No, I get it," Ben said, going straight to the sink to get Hudson a glass of water. "I saw you sing. You don't want to be around a bunch of deadweights." He grinned and pushed his glasses up the bridge of his nose. "Well, we're definitely not deadweights."

"No, it's not that," Hudson said, feeling the heat in her face.

"I think, Marina —"

"It's *Ca* -rina," Carina interrupted.

"Carina, I love jazz," he said. "Are you into jazz?"

"Uh, no," Carina said decisively. "But Hudson here is. Actually, she's got her own unique style. It's like a cross between Nina..." Carina turned to Hudson. "Nina, *who*?"

"Nina Simone," Hudson said. "Nina Simone and Abbey Lincoln. And a little bit of Julie London."

"And Lady Gaga," Ben said with a smile. "Let's go down to the basement. That's where we're all set up. And you can see for yourself if we're up to your standards." As he made his way out of the kitchen he stumbled on a broom handle but caught himself.

Hudson saw Carina almost giggle. "Be nice," she whispered.

"I am!" Carina whispered back.

Hillary tagged along. "Logan looks really cute today," she whispered to Hudson. "And we talked for, like, three whole minutes."

"Then I'm sure he's really into you, Hil," Hudson said.

Hillary wrinkled her nose. "You sure I look okay? Do you think I look too boring?" she asked, yanking on her sweater.

"I think you look perfect," Hudson said, making Hillary beam. "Very agnès b."

"Who's agnès b.?" Hillary asked.

"Oh, just...no one," Hudson said, knowing that Hillary probably wouldn't have heard of the super-influential French designer from the eighties. "So, have you guys picked a name?" Hudson asked Ben as she followed him down a back hallway.

"Right now we're the Stone Cold Freaks," Ben said. "But that's just temporary."

Thank God, Hudson thought. Hudson and Carina and Lizzie all flashed looks at one another.

"And I have to tell you, the other guys aren't *mad,*" he said, looking over his shoulder, "but they're a little weirded out. They knew we needed someone on piano, but they didn't know we were gonna have a lead singer. I spoke a little too soon."

Hudson cast a worried look at her friends.

"But don't worry. Let's just play one of your songs," he said, pushing up his glasses again. "And if it's as good as Hillary says, then great."

"No problem," she said cheerfully. So this was going to be more like an audition after all. In that case, she was extra glad she'd brought her friends.

Then Ben opened the door to the basement stairs, and she heard the music. Or at least what sounded like music. Someone was pounding mercilessly on the drums while a saxophone whined and warbled over the beat. Hudson knew that this was supposed to be the kind of hectic, free-form "coffee-shop jazz" that Hillary had referred to at the bar mitzvah. But this wasn't even that. This was just noise.

"Yikes," Lizzie said under her breath.

"Oh, God," Carina whispered. "It's *this* kind of jazz?"

"Just hold on, you guys," Hudson said as they walked down the stairs behind Ben. But she felt something inside her deflate and sink to the ground. The Stone Cold Freaks definitely weren't the studio band she'd used to record her album.

They reached the bottom of the stairs and walked into the basement, which had been converted into an old-fashioned rec room. There was a Ping-Pong table, a refrigerator, an upright piano, and a brown and pink plaid sofa that faced an old-fashioned wall-unit TV. Brown acoustic paneling covered the walls. And in the corner were the other Stone Cold Freaks—Logan sitting in a plastic folding chair, playing his sax, and behind him a freckled boy with bright red hair pounding his drum kit. Hudson almost had to plug her ears.

"Hey guys, Hudson's here," Ben said, waving at the two of them to stop. "Hudson, you know Logan. And this is Gordie," he said, waving to the redheaded guy on drums. "These are Hudson's friends Marina and—"

"Lizzie," Lizzie said.

"And it's *Ca*-rina," Carina said.

"Hi, guys," Hudson said, waving.

Ben turned to face her. "Wait. What'd you say your last name was?"

Hudson thought fast. Jones couldn't be a more common last name, and from what Hillary had said about Ben and his friends, they probably wouldn't make the Holla connection if she just told the truth. "Jones," she said.

"Hudson Jones," Ben said, oblivious. "So, guys, Hudson writes her own stuff. Right, Hudson?"

"I do," she said, feeling her heart start to race. What if they didn't like her music? She noticed Logan looking at her with a very different expression than yesterday; he seemed to be scowling.

Ben pointed to the upright piano. "Go ahead," he said. "Do one of your songs and we'll join in."

Hudson glanced at her friends. This was definitely an audition. Carina gave her a small thumbs-up as she sat down on the couch. Lizzie winked at her. "Okay," Hudson said, swallowing.

I can do this, she thought as she sat down on the creaky old piano bench. She touched one of the keys. The piano was horribly out of tune.

"Whenever you're ready," Ben said. "Just go for it."

Hudson looked down at the keys. Her heart was beating in her chest like she was about to run a marathon. This wasn't like jumping onstage with Hillary and singing a silly karaoke song. This was real. This was supposed to be good. And on top of that, she could practically feel Logan's scowl burning into her back. *I am going to make this guy like my music,* she thought. *No matter what.*

"Okay, this is a song called 'Heartbeat,'" she said. Her fingers found the familiar chords on the piano. *Go ahead,* she thought. *Just sing the first line. That's all you need to do.*

Her voice wavered at first. She hadn't tried to sing this song since the Silver Snowflake Ball. For a second she was back there, onstage, in front of all those people, knowing that her mom was just a few feet behind her in the wings, watching…

And then she remembered: *Nobody knows who I am.* Something inside of her swung open, like a gate being unlocked. She sang the first two lines.

I love the way you talk to me on the line
I love the way you tell me that you're mine

Before she knew it she was singing it the way she had the day she'd written it, slowly and passionately and smokily, letting her voice wrap around each syllable.

And then Ben started to play his bass—*thump a thump a thump a thump*—setting the perfect rhythm. He was good. She could tell right away.

Then Gordie started on drums, nothing too hard or distracting, just following Ben's lead.

Then, at the bridge, Logan blasted his sax, making Hudson jump. It was way too loud, and all over the place, like a manic foghorn.

When she was done, she sat and faced the keys for a moment, letting herself settle. Finishing a song was always a little like coming out of a trance—time would jump forward again and she'd suddenly become aware of her surroundings. The room was eerily quiet. That usually meant one of two things: People either loved the song or they hated it.

She turned around. Gordie sat with one hand still touching a cymbal, faintly smiling. Ben rested his bass on the floor and was blinking busily behind his glasses, his Adam's apple jumping up and down. Even Logan looked semi-impressed as he cradled his sax. Carina and Lizzie sat on the couch, clutching each other's hands. Even Hillary, standing against the wall, seemed to be moved.

"Did you really write that?" Ben finally asked.

"She totally did," Carina asserted from her spot on the couch.

"Wow," Ben said. "Can you give us a minute?" He glanced at his bandmates.

"Sure." Hudson practically leaped to her feet. "We'll just go upstairs."

Hudson took the stairs two at a time, with Hillary, Lizzie, and Carina behind her. "I knew it!" Hillary said when they got upstairs. "You blew them away!"

"Really?" Hudson asked.

"You *crushed* it!" Carina said, hugging Hudson.

"That was incredible," Lizzie said. "I got chills."

"They're totally going to want you," Hillary said.

"You think so?" Hudson asked.

Hillary nodded.

For the first time, Hudson realized that she didn't just *want* to be the lead singer of the Stone Cold Freaks. She *needed* to be.

Just then she heard the creak of Ben's feet on the stairs. "Hudson?" he called out. "Can you come down here?"

Hudson and Hillary exchanged a worried look. "Just go," Hillary said, swatting her on the arm.

Hudson walked back down the stairs. The Stone Cold Freaks had assembled themselves on the plaid couch. Gordie was smiling, but Logan's eyes were on the muted TV.

"So," Ben began, grinning, "can you come up for rehearsal on Wednesday? Around four thirty?"

"Really?" she cried, hopping up and down. "Yes, I definitely can!"

"Okay then," Ben said. He stood up and held out his hand. "Welcome to the Stone Cold Freaks."

She shook his hand, said good-bye, and then walked back up to the kitchen, not even feeling the stairs underneath her feet. She'd sung that song in front of total strangers, and they'd loved it. They'd loved it so much they'd asked her — for real, this time — to front their band.

She said good-bye to Hillary, and thanked her for everything, and then she and Lizzie and Carina walked out to the gravel driveway in the freezing cold, where they waited for the cab they'd called.

"You did it!" Carina squealed. "How do you feel?"

"I don't know why, but I'm more psyched about this than I was about my album," Hudson said. "Isn't that weird?"

"Maybe this is what you were supposed to do all along," Lizzie said, as a bright green cab turned into the driveway. "Maybe this is going to be more fun than your album."

"Maybe," Hudson said. "But first I really need to change their name."

chapter 16

That night Hudson sat at the kitchen table, picking at a plate of mushy boiled dandelion greens and feeling giddy. She was the lead singer of a *band*. *Outside* of the city. She had a whole new life. A life that her mom had no clue about. Hudson glanced over at Holla, who was across the room giving Sophie a list of orders. She almost felt a little guilty, but it was also undeniably thrilling.

"Tell them I want to be at the final callbacks," Holla said, pulling her sweaty hair up into a ponytail. She yawned slightly into her hand. "The last time I let Howard do it he picked the laziest dancers I've ever seen."

"You got it," Sophie said.

"Why is this so hard?" Holla grumbled. "Why do I have to do everything myself, all the time?"

Sophie stared at her, biting her full lips. "I'm not sure," she said helplessly.

"That's it 'til tomorrow," Holla said, turning back to the kitchen table. "You can go home now."

"Thanks, Holla," Sophie whispered as she began to gather her things from the computer desk. So far Sophie seemed to be dealing well with Holla's exacting demands, but Hudson was pretty sure she wouldn't last through the spring. Most of Holla's assistants quit — or were fired — after three months.

Holla walked over to the kitchen table and slid into a chair. Even though Hudson could see she was exhausted, her mom's posture was perfect. Lorraine placed a hot plate of mushed green vegetables in front of her. "It's always the same, every tour," Holla said to Hudson. "All these last-minute dramas. You're coming with me this summer, right?"

"Uh-huh," Hudson said. It seemed odd for her mom to ask, since that was what she did every time.

"So, honey, how was Jenny?" Holla asked. "And please, sit up straight."

"Jenny was great," Hudson answered, sitting up. "She loved the blanket."

"Really?" Holla asked, and Hudson could see the sudden interest in her eyes. "You guys spent an awfully long time together. What time did you get back here? Five thirty?"

Hudson remembered her lie to Fernald. "You know, I think she feels really bad about the fight you guys had," she said, ignoring Holla's question.

"She does?" Holla asked, putting down her fork. "What'd she say?"

"Just...well, that she feels bad," Hudson improvised. "I think we should do something nice for her. Just to let her know there are no hard feelings."

Holla zipped up her workout jacket. "I've done a lot of nice things for her, baby. It only makes things worse."

"I know, but maybe just this once we could do something to really show her that we're her family," Hudson said. "Like throw her a birthday party. Invite her friends. Sort of a welcome back to New York. Maybe then she'll stop running off all the time."

Her mom chewed her food slowly, considering this. "Where?"

"What about here?"

"*Here?*" Holla asked. "Why does it have to be here?"

"It doesn't," Hudson said. "It just might be more personal that way."

Holla sipped from her glass of coconut water. "Does she hate me?" she asked suddenly, the space between her eyebrows creased and her lips pressed together. "Just tell me the truth. I can handle it. Does she hate me?"

"She doesn't hate you, Mom."

"Because I promised Grandma that I would take care of her, that I would watch out for her," Holla said, her eyes growing soft and shiny. "We all knew that she was wild. And maybe I made it worse, being so successful, so quickly...Maybe it's my fault..."

Hudson touched her mom's arm. "It's okay. She's your sister. She loves you," Hudson reassured her.

Holla patted Hudson's hand, and just like that, the moment of insecurity passed. Her face hardened again, ever so slightly, and she picked up her fork. "Then let's plan it. But you make sure that she's okay with it. If I present it to her, she'll just say no."

"Fine." Hudson pushed the food around on her plate. "What if I go over to her place and hang out with her again? Like, Wednesday afternoon?" It was the perfect alibi for her Larchmont rehearsal date.

"Sure. And I'd like you to come down to the studio this week and give me some feedback. On your song."

Hudson stopped pushing a clump of dandelion greens around her plate. "Okay," Hudson said, too surprised to say anything else. "Let's do Thursday."

"Thursday it is," Holla said. "And honey, please sit up straight."

*

The next day Hudson was on her way out of Geometry class when Miss Evanevski said, "Hudson? Can you wait a moment?"

Hudson gave Carina and Lizzie a glance that said *Not a big deal, I'll meet up with you in a second* and whirled around. "Is something wrong?"

"I just wanted to discuss your quiz from last week," Miss Evanevski said. Hudson's Geometry teacher was tall and fragile-looking, with a pointy chin and a permanently disappointed expression, which intensified when she said, "I was a little alarmed to see your grade."

"I know," she said. "I'm sorry about that."

"Do you need some extra help?" Miss Evanevski sat down at her desk. "I'm happy to refer you to a tutor."

"It was just a onetime thing," Hudson explained.

Miss Evanevski frowned. "I'm supposed to inform your

parents if you get less than a C minus on a test," she said. "But this is so out of character for you, Hudson, that I'm tempted to call your mom anyway."

"Please don't," Hudson said. "Believe me, it won't happen again."

Miss Evanevski picked up a red pen and started shuffling piles of tests on her desk. "Why are you so sure?"

"Because I just decided not to study for it, sort of as an experiment," Hudson said, before she realized how that sounded.

Miss Evanevski frowned. "Experiment?"

"Well," Hudson said, "it's all part of my life-coaching."

"Your what?" Miss Evanevski asked, alarmed. "You're seeing a life coach? Are your parents aware of this?"

"Please," Hudson pleaded. "It's really nothing. Forget I said anything. I promise I'll study next time. You know I always do."

The sound of a footstep behind her made Hudson turn, and there on the threshold stood Ava Elting. From the smug smile on her face Hudson knew that she'd heard the entire conversation. "Oh, sorry," Ava said, tucking one of her auburn curls behind her ear.

"No, we're done, Ava. What is it?" Miss Evanevski said, beckoning her into the classroom.

Hudson headed for the hallway, her face burning. Of course Ava Elting had overheard. She replayed everything that she'd just said in front of Miss Evanevski. The C plus on her quiz. The life coach. It was hard to know which was more embarrassing. Both were enough ammunition to keep Ava busy for weeks. As Hudson passed her, she stared straight ahead and averted her eyes.

She sped down the hall to History and sat down next to Carina, Lizzie, and Todd.

"What was that all about?" Lizzie asked.

"I got a C plus on our last quiz," Hudson said, "and now I think Ava Elting knows that I have a life coach."

"Oh, jeez," Carina whispered.

It didn't take long for Ava to unleash this new tidbit. When Ava walked into History she took a seat right behind Hudson.

"So, is everything okay? I couldn't help overhearing that you're seeing a life coach," she drawled, making Hudson's stomach turn over.

"Uh-huh," Hudson said over her shoulder, holding her breath.

"I'm just soooo surprised," Ava went on. "I thought you were so together. At least, it seems that way. But I guess nothing is ever quite as it seems, is it?"

Hudson froze. Out of the corner of her eye, she could see Carina and Lizzie listening to this, too.

"Just so you know, I'm always, always here for you," Ava said. "I wouldn't want you to end up on *E! True Hollywood Story.*"

Hudson whirled around. "Don't you have more important, interesting things to think about than *my* life?" she asked.

Ava blinked her large brown eyes, as if she couldn't believe what she'd just heard. "Uh, *yeah,*" she finally said with her famous smirk. She tossed some of her auburn hair over her shoulder as she stood up. "I was just trying to be nice." Then she walked over to another empty chair as Mr. Weatherly sat on the edge of his desk.

"All right, people! Who wants to tell us about Alexander the Great?"

Hudson's heart pounded in her chest; she couldn't believe what she'd just said. Carina nudged Hudson's arm. "*You* are my new hero," she whispered.

Hudson knew that things with Ava probably weren't over. But she'd finally said exactly what had been on her mind. Maybe getting angry every now and again wasn't such a bad thing. Especially if it meant making Ava Elting get up and move seats in class. Maybe Hudson would never quite have an inner bee-yatch. But she could stick up for herself when the time came, and that was all that really mattered.

chapter 17

Hudson leaned back against the vinyl seat with her book bag on her lap and felt the wheels of the three thirty-five train roll underneath her. She'd told Fernald to drop her at Jenny's straight from school, and as soon as he'd driven away she'd taken the subway to Grand Central. She'd jumped on board just in time and scored a row of three seats all to herself. Looking at the cracked seats she wished Hillary or Lizzie or Carina could have come. Tonight it would just be her and the Stone Cold Freaks. *Note to self,* she thought. *Must change that name.*

At Larchmont she got off the train and got into a cab. As they pulled into the gravel drive at Ben's house, Hudson noticed that there were more cars in the driveway this time — a beat-up maroon hatchback and a forest green Saab. A couple of bikes lay on their side near the steps to the front door. Logan's and Gordie's, she assumed. She hoped Logan wouldn't still be acting weird. She'd tried to think of something to do or say to get him to like her but the things she'd thought of — bringing him cookies, or maybe getting him an iTunes gift certificate — just seemed a little desperate.

She paid the driver, then walked up the steps and rang the bell. The door opened and Hudson was greeted by a woman in her mid-forties with shoulder-length, softly layered brown hair and Ben's large brown eyes.

"Mrs. Geyer?" Hudson asked, remembering Ben's mom from the bar mitzvah.

"Call me Patty," she said, extending her hand. "Come on in, Hudson."

Hudson shook her hand. "It's nice to see you again," she said.

"You can leave your coat right there," said Mrs. Geyer, pointing to the bench in the entryway, which was again covered in coats and scarves. "You want some hot chocolate? Something to warm you up?"

"Hot chocolate would be great," Hudson said, following her into the kitchen.

"Hope instant's okay," Mrs. Geyer said, putting a kettle on the stove. "I know we're not supposed to be drinking stuff like that, but sometimes it's just so much easier."

"Yeah, I know," Hudson said. She didn't think her mom even had Hershey's chocolate in the house, let alone instant Swiss Miss.

"You know, I saw you sing with Hillary," she said. "You really are very talented."

"Thank you," Hudson said.

"And there's something so familiar about you," Mrs. Geyer said. "Maybe it's your voice. But you remind me of someone." She shook her head and shrugged. "So, you live in the city?"

"Yep," Hudson said, starting to get nervous. She scanned the kitchen for a cast-aside *US Weekly* or *Life & Style*. But before Mrs. Geyer could question her further, Ben walked into the kitchen. Or rather, tripped into the kitchen, over a hockey stick that lay diagonally on the ground. "Oh, hey, Hudson!" he said, grabbing a chair to steady himself.

"Honey," Mrs. Geyer said.

"Good news," Ben said, straightening up. "We have our first show. This Friday night."

"*This* Friday?" Hudson asked.

"Well, it's not really a show," Ben said, taking off his glasses and blowing on them to clean them. "It's my friend's birthday party. I told them we'd play at it. It's just at her house."

"Ellie is going to *love* you guys," said Mrs. Geyer, handing Hudson a mug of Swiss Miss.

"Are you sure we're ready for that?" Hudson asked as she followed Ben down the stairs to the basement. "We've barely had a rehearsal."

"Yeah, it'll be great," Ben said casually. "No worries."

"Okay," she said, unconvinced. He may not have been worried, but she was. And who exactly was Ellie?

When she reached the basement, she realized that the show was going to happen a lot sooner than Friday—it was about to happen now. There, draped on the pink and brown plaid couch and sitting cross-legged on the floor, were four girls and four guys. All of them seemed to be Ben's age. Two of the girls were identical twins. A couple of the guys watched the TV on mute. The rest seemed to be watching Gordie and Logan, who were in the middle of another experimental jazz set. Gordie wailed on his drum kit while Logan blasted away with his sax. Hudson didn't know how anyone could listen to them for long. She didn't want to say anything, but she thought that they definitely needed some quality rehearsal time before they played for a group.

"Okay, guys, Hudson's here," said Ben. "Everyone? This is Hudson Jones, our new lead singer."

"Hi, everyone," Hudson said, waving shyly.

"Hey," one of the girls said, getting to her feet. "I'm Ellie. It's my party on Friday. Oh, and don't worry—we have a piano. You guys are gonna be great." Ellie was Asian-American, and she had a friendly way of leaning in close when she talked. Her black hair splashed against her shoulders, and she wore an adorable purple ruffled blouse that Hudson immediately wanted to borrow.

"Thanks," Hudson said. "Are you guys here for our rehearsal?"

"Yeah, but we all promise to be super-quiet," Ellie said, dropping back down to the floor. "Okay, everyone," she announced to the group. "Time to zip it."

"Hudson, we definitely want to do the song you sang the other day," said Ben. "But we thought we'd just practice a few covers, too. Anything you want to start with?"

Hudson sat down at the piano. She could still feel the group of kids behind her, staring at her. She wished Ben had warned her about having an audience. This didn't seem fair. But then she reminded herself: *Nobody knows who you are.*

Gordie waved hello from behind the drum kit, and she waved back. Logan kept his head down, and busied himself by fidgeting with the keys of his sax. She couldn't tell if he was ignoring her.

"Let's do 'I'll Be Seeing You,'" she suggested. *At least it has lots of saxophone in it,* Hudson thought.

"Cool," said Ben.

She started playing, and when Logan joined in, his sax was so loud that it almost drowned out her voice. Hudson did her best to sing over him, but it wasn't easy.

"That was awesome!" Ellie yelled out when they were done. "You guys have to do that one!"

They followed with "Fly Me to the Moon," "At Last," and "Feeling Good," which Logan almost entirely blotted out with his sax. But Ben was phenomenal. He never missed one beat, and seemed to have practiced all of these songs for hours. Finally they ended with "My Baby Just Cares for Me," which had only a small part for Logan to ruin.

"That was soooo good!" Ellie said as the rest of the kids clapped. "Hudson, you have an amazing voice!"

"I told you, right?" Ben asked.

"Thanks," Hudson said, feeling herself turn red. Receiving praise always made her uncomfortable. When she happened to catch Logan glowering at her, she felt even more awkward. "But I probably should try to make the six-oh-seven train."

"Ellie can take you," Ben said, getting up. "She's got her license. She's a junior."

Hudson turned to Logan. "Good rehearsal," she said to him, smiling.

"Yeah," Logan replied, barely looking at her. Then he got up and went to sit down next to the twin girls, as if Hudson were no more than a stranger.

Yep, he hates me, Hudson thought as she followed Ben and Ellie out the back basement door. There was probably nothing she could do about it, but it gnawed at her, like one of her geometry problems.

Outside it was already dark, and a big frosty full moon hovered above the trees. Hudson craned her head back to look up at the night sky. "Wow," she said. "You can actually see stars."

"Uh, yeah," Ellie said, laughing, as she opened the door to her Saab.

"You can never see stars in the city," Hudson said. "There're too many lights."

"We could use a few more lights up here," Ben said, crawling into the back and flipping the seat back up so Hudson could sit in front.

Hudson got into the car and shut the door. She had never been driven by someone who was so close to her age. For a moment she was almost scared, but then Ellie turned the key in the ignition and expertly backed out of the driveway.

"What?" Ellie asked, catching Hudson watching her.

"I don't know any kids in the city who drive," Hudson observed.

"It kind of comes in handy up here," said Ellie, pulling out onto the street. "What part do you live in?"

"The West Village," she said.

"Cool," said Ellie, nodding with approval. "My dad works in the city. We go in once in a while, to shop or go to some restaurant that's way overpriced. Where do you go to school?"

"Chadwick," Hudson said. "Same place Ben's cousin Hillary goes."

"I've heard of that school." Ellie wrinkled her nose, as if she smelled something bad. "Isn't that where all those celebrities' kids go?"

Hudson started. "Um, I guess so," she said carefully.

"Wait," Ellie said. "Ben, didn't you ask your cousin once who went there? Who did she say again?"

"I can't remember. And I think you just missed the turn," Ben said, sitting forward. "It's back there."

"Oh, great," Ellie said, pulling over and then making a U-turn. "Um, how long have I lived here?" she joked. "Earth to Ellie."

Hudson smiled. She liked Ellie more and more. But she wondered if Hillary had ever told Ben who she was. It sounded like she had, a long time ago. What if he remembered?

"Where do *you* guys go to school?" Hudson asked quickly, before Ellie could return to her line of questioning.

"Mamaroneck High."

"It's in the next town over," said Ben.

"Do you guys have a football team?" Hudson asked.

"Of course," Ellie said. "Why?"

"That's something else we don't have," Hudson said.

"It's really not that great," Ben put in. "I'd rather go to Chadwick any day. At least there are cool music clubs downtown."

"Ben is *really* into jazz," Ellie said, as if Ben weren't in the car.

"I think that's great," Hudson said. She looked at Ben in the rearview mirror. He smiled.

"No. *Really* into it," Ellie repeated.

"Better watch out," he joked. "I'm pretty serious about this."

"Good," Hudson said, smiling back at him.

They pulled into the train station parking lot. A few people stood huddled on the platform.

"Thanks for the ride, Ellie." Hudson got out of the car and wrapped her scarf over her face. It was so cold her nose felt like it might fall off.

Ben got out of the backseat and moved to the front. "So be at my place by six o'clock Friday night," he said. "And Hudson?" Ben stuck his head out of the passenger side window. "You were awesome tonight."

"You really were," Ellie said.

"So were you, Ben," Hudson said. "I'll see you Friday."

She walked up the steps to the platform, touched by how nice they'd been to her. But she had to wonder if they would have been so nice if they'd known who she was. She remembered the way Ellie had wrinkled her nose as she said *celebrities' kids*. Eventually they were going to figure out who she was. But by then, she hoped, they'd be able to see her the same way as they had tonight: a shy girl with a big voice who loved jazz and Nina Simone. A girl, Hudson thought, watching Ellie's car drive out of the lot, who felt like she might just have a career in music after all.

chapter 18

"I just want you to be honest," Holla said as the service elevator creaked on the long ride up to the recording studio. "It's your song, so I want your true reaction."

"Sure, Mom," Hudson said uncertainly, shifting her book bag onto her other shoulder. No matter how often Holla said she wanted Hudson's "true reaction" to something—whether she wanted to know if her hands looked "veiny" or if the dance routine for a certain song was too *"Solid Gold"*—Hudson never told her mom what she didn't want to hear.

"I think you're going to be very proud," Holla said, taking off the black oversized sunglasses she'd put on to walk past the photographers and fans who'd been waiting outside the recording studio. "Chris thinks it's gonna be huge." She smiled nervously as she put the glasses in her ostrich-skin shoulder bag. "We'll see."

Hudson glanced at Little Jimmy and Sophie, and for a moment they made sympathetic eye contact with her. They, too, knew how

Holla felt about other people's opinions, even when she'd asked for them.

"And how's Jenny?" Holla asked.

"Jenny?" Hudson asked, momentarily thrown.

"You were with her for hours yesterday," she said. "Did you run the party idea past her? Does she want to do it?"

With a start, Hudson remembered lying about seeing Jenny so she could go to Larchmont for band rehearsal. "Oh, yeah, she thinks it's a great idea."

"And February twenty-first is fine with her?"

"Yup," Hudson asserted, as brightly as possible. Hadn't Jenny actually said that the other day? She couldn't remember. She needed to text her.

Hudson followed the group through a pair of glass doors into the studio's reception area, past the lounge and the kitchen, and down the hallway to a door marked CONTROL ROOM 2. Holla opened the door.

"Hey!" Chris said, standing up from behind the wraparound sound console. "Welcome back, Hudson!" He wore a blue knit hat that made his eyes seem extra vivid, and a T-shirt with Jimi Hendrix's face on it. He started to hug her, but then gave her a high five instead. "Thanks for coming down."

"Hey, Chris," Hudson said, feeling, as usual, a little awkward around him. "How's everything going?"

"Great. I can't wait for you to hear the song," he said. "Oh, this is Liam, the sound engineer," he said, gesturing to the older man sitting in a chair. He had big, sad eyes and a mustache, and seemed eager to get started. "Liam, this is Holla's daughter, Hudson. I used to be her producer."

"Hi," Liam said tersely, then went back to studying the knobs and dials of the immense console.

"Okay, who has a coconut water?" Holla asked, putting down her bag.

Chris opened a mini-fridge in the back of the room and pulled out a blue can. "Here you go, madam."

Holla popped open the can and gave him a meaningful smile. "*Merci*," she said with a giggle and took a couple of sips. "Okay!" she said sharply. "I'm ready. Let's do this." She turned to look at Hudson. "Remember, tell me what you think. Okay?"

Hudson nodded earnestly and Holla walked out. Hudson went to the couch in the back and dropped her book bag to the floor. She sat on the couch and spread her coat over her legs, staring at Chris as he manned the sound console. She still got a weird feeling about him and her mom. They were spending hours together in this tiny room every day. Then he was coming over at night. And he seemed extra amped-up right now, as if he were on his best behavior.

Chris leaned close to the intercom and said, "Okay, Holla, how are you doing in there?"

Through the glass they watched Holla walk into the vocal booth next door. She slipped on a pair of headphones and gave Chris a thumbs-up.

Chris turned around and looked at Hudson, his blue eyes sparkling. "Just wait, you're gonna love this," he boasted, and then turned back around. "And uh one," Chris said into the intercom. "Uh, two. Uh one, two, three, four..."

Liam hit a button and Hudson's song poured out through the speakers. It sounded even worse. They'd added even more tracks to it, which Hudson hadn't thought was possible. There were more crash cymbals and kick-drums, and now a chorus of digital voices screamed "Hey!" on every fourth beat. At one point, Hudson thought she could hear a sampled car alarm. *Oh, God,* she thought, hugging her knees to her chest. *How do I tell my mom that this is good?*

Through the glass she watched her mom in her stretchy purple sweater and yoga pants, swaying to the music.

And then Holla closed her eyes, tilted her chin, and started to sing.

Oooh, I love the way you talk to me on the line
Ooh, I love the way you tell me that you're mine
I love the way you won't let me go
And now I love that I'm telling you so . . .

Hudson had experienced many surreal moments in her life — being mobbed with her mom by screaming and crying Japanese girls in Harajuku, Tokyo; racing a Big Wheel through the winding tunnels below Madison Square Garden; being photographed by a cell phone camera in her dentist's waiting room — but this topped them all. Her thirty-seven-year-old mom was singing a song about a sixteen-year-old boy. A boy Hudson had once had a crush on. And she was practically writhing as she was doing it. She hugged her knees in tighter, cringing even more. The four-minute song seemed to last a lifetime.

"Perfect!" Chris yelled into the intercom when Holla was done. "That was just perfect!"

"Really?" Holla asked excitedly. She pulled off her headphones, looking flushed and giddy. Then she left the booth.

Chris swiveled around in his chair to face Hudson. "What'd you think?" he asked, his face bright with pride. "Sounds pretty good, huh?"

Hudson swallowed. "Yeah!" she chirped.

Holla bounded into the control room. "Phew!" she cried out, striking a dramatic pose, with the back of her hand pressed against her forehead. "So it was okay?" she asked Chris. "I didn't go too high at the end?"

"It was amazing," he said. "And someone else liked it, too." He jerked his thumb in Hudson's direction.

"You did? You liked it?" Holla asked, her brown eyes filled with delight.

Hudson bit her lip. "You guys really put a lot of work into it," she said carefully.

Holla's eyes narrowed, catching Hudson's hesitation. "What do you mean?" she asked, stepping closer to Chris.

"I just think it sounds a little . . . busy," Hudson said delicately.

Holla and Chris traded quizzical looks. "Busy?" Holla repeated.

"Forget it," Hudson said. "It's great. I'm sure it'll be a big hit."

"Wait, hold on," Holla said, holding up a hand. "You don't get away that easy. Explain what you mean."

"Um, just that it . . . well . . . it just doesn't sound like . . ." Hudson let her voice trail off as her mom continued to stare. "Can I talk to you in private?"

Holla looked at Chris and the sound engineer. Without a word they got up and went to the door. Sophie and Little Jimmy followed. The door shut behind them.

Alone with Holla, who seemed to be growing more annoyed every second, Hudson tried to figure out how to tell her mom the truth. "Your voice sounds great, Mom," she began, "but there's too much going on. He's done too much to the backup track. It's too layered. All the sampling . . . it just sounds a little cheesy."

At the word *cheesy*, Holla raised an eyebrow. "Sounds like you hated it."

"I didn't hate it," Hudson said. "I'm just saying that it could be better with less going on in it. That's all."

Holla nodded, seeming to mull this over. "Well, you haven't wanted me to do this song ever since we asked you about it," Holla said. "Admit it. You've had a problem with this from the beginning."

"You said you wanted my opinion, just now, in the elevator," Hudson said, getting angry. "And I'm giving it to you. I'm not trying to be mean or anything."

"This is supposed to be my next single," Holla said, her voice starting to get louder. "How could you tell me it's bad?"

"Because I'm just trying to be honest."

"Well, Chris seems to think it's great," Holla pointed out.

"Of course he does," Hudson muttered.

"What does that mean?" Holla asked.

"It means you're totally dating him," Hudson blurted out. "Right?"

Holla's angry expression lifted. "So that's what you're mad at me about? Chris?"

"I'm not mad," Hudson said.

"You're *always* mad," Holla spat. "You never like anyone I date. It's like I'm living with my mother."

"Maybe I'm just not looking forward to when you guys break up and you're in my room sobbing every night," Hudson said, letting her own voice get loud. "Since we all know *that's* coming."

Holla's eyes darkened, and she pointed to the door. "Go home," she said in a withering voice. "Right. Now."

Hudson picked up her book bag and went to the door. "I'm sorry," she mumbled as she reached for the knob.

"Just go," Holla repeated.

Hudson opened the door. Sophie, Chris, Liam, and Little Jimmy stood in the hallway, studying the carpet. She knew that they'd heard everything. Chris looked up at her. "Hudson?" he asked carefully, as if she were a mental patient who might have a seizure any moment. "Is everything okay?"

Hudson shook her head and trudged past him. As if she were going to pour her heart out to him, of all people. She pulled out her iPhone as she reached the elevators and typed two words:

Pinkberry. NOW.

*

"You did the right thing, H," Lizzie said a half hour later, digging her spoon into her cup of pomegranate Pinkberry topped with mochi. "You said how you felt. Your mom asked for your opinion and you gave it. That's all you can do."

"But what's the point if all she's gonna do is freak out?" Hudson took a bite of her yogurt with blueberries and let it melt on her tongue.

"It was a trap," Carina said, finishing her cup. "Whenever someone says they want your 'honest opinion,' it always, always means they don't."

Thankfully Carina had been hanging out at Lizzie's place, which made the Pinkberry on Columbus and Seventy-fifth Street superconvenient. Hudson looked out the window at Fernald in the black SUV, double-parked in front of the store. "But seriously, you guys," she said, cupping a blueberry with her spoon. "The song was awful. She needed to know. And then all that stuff with Chris..." She paused. "Obviously, I know we were never gonna be together or go out. And I feel dumb that I even had a crush on him. But why does *she* have to have him? Why does she have to have *everyone* love her, all the time? It's not fair."

"But that's the deal," Carina said, tossing her empty cup into the trash. "She's, like, the biggest star in the world."

"Do you think she *is* gonna end up in your room sobbing?" Lizzie asked, her hazel eyes locked on Hudson.

"I don't know," Hudson said, taking another small bite of her tart yogurt. "Maybe Chris is different. Maybe he's the one who'll really love her for who she is." She pushed the container away. "I'm just so sick of it. At least those guys get to leave one day. I'll never get to leave. I'm gonna have to deal with this the rest of my life."

Lizzie leaned closer and took Hudson's hand. "You're your own person, H. Really. Your mom has nothing to do with you." She chuckled. "I mean, believe me. I know how hard it is."

Hudson nodded. "But you had Andrea. I don't have someone

like that. I just have *me*. And I don't know if I can do it all on my own."

Lizzie and Carina put their hands on top of Hudson's. "Yes, you can," Carina said. "We know you can."

Hudson squeezed her friends's hands. She thought she might tear up for a moment so she pulled her hand away. "So, I forgot to tell you guys that we have a show. Tomorrow night. Up in Larchmont."

"You have a *show*?" Carina gasped. "Where?"

"At some girl's house. Her name's Ellie. She's really nice."

"Can we come?" Carina asked.

"We *have* to come!" Lizzie said. "Can I bring Todd?"

"Can I bring Alex?" Carina asked.

"You guys — I don't even know these people," Hudson hedged.

"And neither will we!" Carina cried. "Come on, you have to let us come."

Hudson tossed her melted Pinkberry in the trash. "Okay, fine. Let me check with Ben and I'll let you guys know."

Later, after she'd dropped Carina and Lizzie off at Lizzie's house, Hudson thought about what Lizzie had said in Pinkberry. She tried to imagine a world in which Holla didn't exist. Where she'd never seen billboards of her in Times Square or heard her songs as background music in taxicabs and restaurants and boutiques. Where she'd never seen her mom mobbed by fans screaming her name. It was impossible, like trying to imagine a world without sunlight or water. Her mom would always be there, taking up all the space in the room. Lizzie was wrong: Holla would always have *everything* to do with who Hudson was.

Later that night, as Hudson did her homework, she got an idea. Her mom was still at the studio, and probably would be for hours. So she picked up the phone and dialed.

"Hunan Gourmet!" said the voice on the other end of the line.

Hudson scanned the menu on her laptop. "Order for delivery, please," she said. "The chicken lo mein and the moo shu pork. With extra brown sauce."

Hunan Gourmet had a reputation for being the most organic and non-MSG Chinese takeout place in the West Village, but still, it was Chinese food. And Chinese food was very much not allowed in the Holla Jones Diet Plan. Ordering it felt like the ultimate way to rebel, and after Hudson's fight with Holla, it just felt like the right thing to do.

When it arrived, she pulled off the plastic tops, trying not to freak out about oozing toxins. Then she sat at the kitchen table, alone, slurping up her lo mein. Some of it was a little too salty, and some of it was too spicy, but most of it was delicious. To her credit, Lorraine didn't say a word. She just stood at the island, darting sly smiles in Hudson's direction as she chopped up some kale.

chapter 19

The next afternoon Hillary planted herself next to Hudson at the lockers. She wore a somber black sweater, and the absence of color made Hillary's skin look even paler than usual.

"How was rehearsal the other night?" Hillary asked.

"Great!" she said, deciding not to mention Logan and his angry glares. "And we have a show tonight. Wanna come?"

"I can't. I have to do something with my mom," Hillary said, her arms folded in front of her. "I'm so annoyed. But can you do me a favor? Can you ask Logan about me?"

From the little interaction she'd had with Logan so far, Hudson already knew that this wouldn't be easy. "I'll try," she said.

"Just mention my name, and maybe that I wish I could be at the party, and just see what he says. And watch his facial expression. Even if he's trying to play it cool, you'll be able to tell from his face."

Hudson pulled out her Geometry book. "I'll keep that in mind, Hil."

Hillary beamed. "Great!" she said. "And then tomorrow we can maybe go shopping again?"

Hudson needed to get to class. "Sure," she said. "Meet me at Kirna Zabete in SoHo. Noon."

As she watched Hillary walk away, Hudson wondered if it was a good or bad thing to be so superconfident. On one hand, people like her mom and like Hillary never let the word *no* stand in their way. But sometimes that meant that they also set themselves up for disappointment. Hudson wasn't sure, but she had the distinct feeling that Hillary was going to be headed for some real disappointment as far as Logan was concerned.

*

After giving Carina and Lizzie directions to the party in Larchmont and making plans to meet them at eight, Hudson left school and made the trip home to change. As Fernald navigated the traffic down Fifth Avenue, Hudson looked at the brittle winter sunlight slanting through the branches of Central Park and realized that she would need another alibi to get out of the house for the night. She still hadn't seen her mom since yesterday's fight at the recording studio, and with any luck, Holla and Chris would have plans tonight. Still, Hudson would have to say something to Raquel, so she thought quickly.

Jenny. She needed to ask her about the birthday party, anyway. She picked up her phone and called her aunt.

"Jenny?" she asked. "It's Hudson."

"Hey, Hudcap," came her aunt's cheery, but tired, voice over the line. "How's it going? I've missed you."

"So, I talked to my mom about that birthday party we'd like to throw you," she said. "My mom really wants to do it."

Jenny laughed. "Well, considering my sink just exploded and Barneys just told me they don't want to carry my line and I had the world's worst date last night," she said, "I'd say a party sounds kind of nice. But I'm gonna be out of town. Remember Juan Gregorio?"

"Who?" Hudson asked.

"The guy from Buenos Aires. He's invited me down for the week of my birthday. I think he misses me."

"Oh," Hudson said.

"So just tell her I can't do it the twenty-first. I'd call her myself, but I don't want to look pushy."

"Yeah, no problem," Hudson said. "And if my mom calls you tonight, will you say I'm with you?"

"O-kay," Jenny said cautiously. "What is it?"

"It's just this band I'm in. We have our first show tonight. Not at a club or anything. It's just a high school party. She still doesn't know about it."

"No problem," Jenny said. "I'll tell her. I hope it's going well."

"Okay, thanks!" Hudson said. "And let's hang out again soon!" *Thank God for Aunt Jenny,* she thought as she pressed the End Call button. She'd really lucked out when her aunt came home from France.

Back at home, Hudson showered, dried her hair, and put on her Leather Milkmaid dress, designed by Martin Meloy. Martin Meloy wasn't her favorite, since the whole debacle with Lizzie and his ad campaign, but she still loved the dress. She topped it off with a beat-up boy's motorcycle jacket she'd scored at a flea market in Florence and then slipped on her stretchy, rhinestone-covered headband for a little sparkle. This was going to be her first show, and she wanted to look good for it.

When she slipped downstairs, she almost walked right into Raquel, who was holding a tall, spindly orchid. "You look nice," she said. "Where are you going?"

"Aunt Jenny's taking me to a Broadway play," Hudson said, heading to the elevator. "We're having dinner before."

"Does your mom know this?" asked Raquel.

"I think so," Hudson said, trying hard to look Raquel in the eye.

"Just be home by eleven thirty," Raquel said, continuing past her down the hall.

Hudson gave Fernald the address of a restaurant on Forty-third and Sixth, which was about halfway between the theater district and Grand Central.

"Aunt Jenny will bring me home in a cab later," she told him as she got out.

Fernald nodded and she slammed the door. Then she ran down Forty-third Street, straight into the wind, amazed at how good she was getting at this. And she almost wanted to laugh. Probably every kid in Larchmont wished they could be in the city on a Friday night. And here she was, lying through her teeth so she could flee to the suburbs.

Ben had texted her that he and his mom would be waiting for her at the train station. As she stepped onto the platform at Larchmont, she saw his mom's car parked under a street lamp and recognized Ben's tall, lanky frame when he stepped out of the front seat. "Hudson!" he yelled. "Over here!"

She waved and ran to the car, relieved to see them. She still wasn't used to taking the train by herself.

Mrs. Geyer waved at her from behind the steering wheel. "Hi, Hudson!" she said.

"Thanks for picking me up," she said as she slipped into the backseat.

"How was your train ride?" Mrs. Geyer asked. "Does your mom know you're up here?"

"Oh, yeah," she said, trying to sound convincing. "She wanted to thank you for picking me up."

Ben got back into the passenger seat and closed the door.

"Because she should really have my phone number," Mrs. Geyer said. "Do you want to give it to her?"

"I'll give it to her later," Hudson said. "She's out right now."

This seemed to be enough for Ben's mom. They drove out of the parking lot and then turned in the opposite direction from town. "It's so quiet up here," Hudson marveled.

"*Too* quiet," Ben said. Hudson could see that Ben had tamed his hair with some kind of product, because it wasn't as springy as usual. He'd also switched his glasses for contact lenses, and she could smell some kind of spicy, musky scent that might have been aftershave.

"Ben thinks it's boring up here," said Mrs. Geyer, fiddling with the radio. "He'd love to live down in the city. What do your parents do in the city?" Mrs. Geyer asked.

"Mom," Ben objected. "Don't be rude."

"I'm not being rude, I'm just making conversation," Mrs. Geyer said.

"Well, it's just my mom and me, and she kind of... works from home." Hudson paused. "She's in the arts."

"Hmmm," Mrs. Geyer said as they turned into a driveway. "So she's a painter? Or a writer?"

"Kind of a cross between the two."

Mrs. Geyer drove up to what Hudson assumed was Ellie's house. It was a modest Tudor-style home with a stained-glass window in the front door. There were lights on in the windows downstairs, and already Hudson could see kids inside, milling around.

"All right, I'll come by to get you around eleven," Mrs. Geyer said. "Hudson, is that good for you? Or should we get you to the train a little earlier?"

"I'm getting a ride back to the city with my friends," she said. "But thank you."

"Have fun tonight," Mrs. Geyer said. "Ben, don't forget your bass."

"I know, Mom." Ben sighed.

Hudson and Ben got out of the car. He pulled the case containing his electric bass from the trunk, and as Mrs. Geyer pulled out of the driveway, he rolled his eyes at Hudson.

"Your mom's really cool," Hudson said. "I wish my mom was half as cool as that."

"She's okay," Ben said. "She's not that into this band thing. To her it's like this big distraction from what I should be doing."

"And what's that?" Hudson asked.

"Being on the chess team. Being on the physics team. Trying to get into MIT," said Ben. "Or Harvard. Or Johns Hopkins. Those are the only colleges that are officially approved."

"Really?" she asked.

"Both my parents are professors. And look at me. I'm, like, genetically engineered to be in a science lab somewhere, studying genomes or writing software."

Hudson laughed.

"But this is who I really am," he said, shaking the case of his bass. "I got big plans for this group. Just so you know."

"So I hear," she said, smiling. "What's your big plan?"

"That we play at Joe's Pub," he said simply, as if this were the most logical thing in the world. Joe's Pub was a famous club and cabaret space in New York. It booked all kinds of acts — jazz and pop and rock and even stand-up comedy — and everything from up-and-comers to the seriously famous.

"Joe's Pub?" Hudson asked.

"My dad took me there a couple of years ago," Ben went on. "He loves jazz. He's the one who got me into it. He gave me all his old John Coltrane and Miles Davis CDs. My mom wants to kill him." Ben smiled to himself and kicked at the gravel in the driveway. "So he took me to Joe's Pub to see Bill Frisell, who's probably the greatest jazz guitarist of all time, and I had kind of a flash-forward. I could just see myself doing the same thing, onstage there one day." Ben chuckled. "*Obviously* I know it's a long shot. Not to mention my parents would completely freak out if I actually got that far. But if I did play there, one day, then maybe my *not* doing what they want wouldn't be so hard for them to take."

Hudson listened, remembering the conversation she'd had with Richard Wu just a few months earlier. She'd wanted to book a show at Joe's Pub. Naturally, Holla had changed his mind about that. She'd wanted someplace bigger, more of a real concert hall, like Roseland. Hudson hadn't even fought her on it. "So you think that would do it? Playing at Joe's Pub?"

Ben shrugged. "It's just my goal right now. And maybe I could make it happen. The whole trick to this business is connections. That's how anyone gets anywhere. It's, like, ninety percent connections. I went to camp with this kid whose dad was some big-shot music exec guy. He could totally help us."

"I think it has a little more to do with talent," Hudson argued.

"Well, maybe," Ben said. "But what about all those kids who become movie stars because their parents are? You think they got the job just because of their talent?"

Hudson didn't want to answer that. "Look, if you *are* that serious about doing this, then what about changing the name?" she asked. "The Stone Cold Freaks isn't doing us justice. What do you think of... the Rising Signs?"

Ben didn't say anything.

"I'm kind of into astrology," Hudson explained.

"The Rising Signs," he murmured, looking off into the night. "That's kind of cool."

"Let's run it past Logan and Gordie and see what they think."

"The Rising Signs," Ben repeated as he leaned down to pick up his case. "Nice one."

They walked into Ellie's house, and as Ben stopped to chat with some friends, Hudson stepped slowly into the living room. She didn't usually go to parties alone, much less parties thrown by people she barely knew. Groups of girls walked past her, laughing and talking. All of them wore jeans and oxford shirts or sweaters.

Hudson looked down at her poufy leather and hot pink silk dress and felt a little self-conscious. A couple of girls eyed her dress from across the room. Hudson waved at them. They waved back, hesitantly, but kept talking. Hudson walked over to the piano and dropped her bag. At least she could hide out here for a while.

"Ooh, Hudson!" Ellie cried, coming toward her. She was more dressed up than her friends were, in jeans and a camisole lined with sequins along the neckline and straps. "Great dress! I love it!"

"Thanks," Hudson said, relieved. "And happy birthday."

"Hey, can I ask you something?" Ellie said in a low voice. "Ben just told me you're really into astrology."

"A little, yeah."

Ellie leaned closer, practically whispering. "Could you find out if me and this guy are compatible?" She nodded her head at Logan, who was walking toward them with his saxophone case.

"Him?" Hudson pointed. "Logan?"

"We sort of hooked up last weekend," Ellie admitted. "And I *think* he likes me. But I can't really tell. I can give you his birthday, though, and maybe you can tell me!"

"Uh...uh, sure," Hudson said.

"Awesome," Ellie said, getting distracted by a group of people walking into her house. "Have fun!"

She wandered off to greet the newcomers. Almost before Hudson could absorb this, Logan sat down next to her and opened his sax case. Hudson cast him a sidelong glance. If what Ellie had told her was true, and Hudson had the definite feeling it was, then she needed to figure out what to do about Hillary's unrequited crush.

Logan still seemed intent on ignoring her as he put his saxophone together.

"Do we have a set list yet?" she asked him, in a super-friendly voice.

Logan shrugged. "I guess we're just playing whatever you and Ben want," he muttered.

"No, we're not," Hudson said, confused. "I think we all should talk about it. Together."

"Yeah, just like we all talked about changing the name of the band," he said sarcastically, snapping the pieces of his sax into place. "Together."

"Wait. It's not changed. It was just an idea we had. Who told you that? Ben?"

"Whatever," Logan said under his breath. He slipped the sax onto its stand and walked away.

"Logan!" Hudson yelled, but he didn't turn around. She watched him pick his way through the crowd and disappear toward the kitchen.

She sat on the piano bench, glancing at the crowd to see if anyone had just overheard their fight. This band thing didn't feel right anymore. She felt like an intruder. She'd unwittingly caused a whole slew of problems just by showing up. *I should leave,* she thought. *I can't deal with this guy not liking me.*

But then she remembered: That had been one of her fears, something she'd written down on Hillary's list. *Not being liked.* And here it was, happening right now.

Just as she was about to go find the bathroom and take a timeout for a moment, she saw Lizzie and Carina and Todd and Alex wading through the crowd. She'd never been so glad to see them before. She ran over to them.

"Hey, guys!" she said.

"Hey!" Lizzie said brightly.

"We just walked right in!" Carina exclaimed. "There wasn't even a list or anything!"

Hudson had noticed that, too. In the city, the news of a party always spread dangerously fast. It wasn't rare to see the doormen of the buildings on Park Avenue holding guest lists and checking off names as they let people through to the elevator.

Alex checked out the crowd, tapping his feet to the music. "Nice playlist," he said. "But where's everyone supposed to dance? There's furniture everywhere."

"What's wrong, Hudson?" Todd asked, looking at her with his alarmingly blue eyes. "You look a little freaked out."

"Oh, nothing," she said, embarrassed. "I think this one guy in the band is a little bit mad at me."

"What for?" asked Carina.

"I kind of suggested we change our name."

"What is it now?" asked Todd.

"The Stone Cold Freaks."

"Oh, yeah," Alex asserted. "Definitely. Not even a question."

Ben and Ellie walked toward them, and Hudson waved for them to join. "Hey, you guys, meet my friends. This is Lizzie, Carina, Todd, and Alex. Everyone, this is Ben Geyer and Ellie Kim. Ellie's the one having the party."

After everyone said hi and spoke for a while, Ben pulled Hudson aside. "I think we better start. But I can't find Logan anywhere.

Hey, Gordie!" Ben pulled Gordie over as he walked by. "Where's Logan?"

"Haven't seen him." Gordie shrugged.

"I think he's mad at me," Hudson said. "You told him about the new name for the band, right?"

"Just for a second," Ben said. "I didn't tell him it was a done deal or anything."

"Let's try to find him," said Hudson, leading the way to the kitchen. Ben followed.

They looked everywhere. He wasn't in the kitchen or the dining room, or out on the back patio, where a group of kids hung out in the cold. Finally they spotted him in the laundry room, talking with a small group of people. "Can I talk to you for a second?" Hudson asked. She looked back at Ben and Gordie. "Alone, if you guys don't mind."

"No problem," Ben said, and he and Gordie retreated.

"What?" Logan asked, his eyes narrowed. The group of people he'd been hanging out with quietly walked back into the kitchen.

"I just want you to know that I'm sorry if you feel like I've come in and turned things around, because I haven't," she said. "Maybe Ben did speak a little too soon at the bar mitzvah, about having me be the lead singer, but then he gave you guys a chance to talk it out. And this whole thing with changing the band name — that was just something I mentioned to Ben for, like, two seconds. It's not a done deal."

Logan looked past her, fidgeting to get away.

"I know there's nothing I can do to make you like me," she went on, "and I really don't care if you do or not. But what I *do* care about is this band. And I don't want there to be drama before it's even started."

She had no idea where these words were coming from. This wasn't how she usually talked, and she'd never spoken to anyone like this before. Logan darted his eyes around the room, as if he were physically unable to make eye contact. And then he said, "Forget it. There's no drama. Are we going on now?"

"Uh, yeah," Hudson said. "Let's go."

They walked back into the kitchen, where Ben and Gordie waited for them beside the four-layer dip. "We're ready," she said simply, and the four of them walked back to the living room.

"What's going on?" Carina asked, brushing her blond hair back behind her ear.

"Just a little band drama," Hudson said, smiling. "Wish me luck."

"Break a leg, Hudson," Lizzie said, then leaned her head on Todd's shoulder. She looked like she was blissfully in love.

Hudson walked over to the piano. A few feet away, Gordie snapped his high-hat cymbal on above the snare.

"I say we start with the song you wrote, Hudson," Ben said. "Then 'My Baby Just Cares for Me,' 'Fly Me to the Moon,' and 'Feeling Good.' "

Hudson nodded and sat down at the piano.

Ellie called everyone into the living room. "Okay, you guys!" she yelled. "Here we go . . . the moment we've all been waiting for! May I present . . . well . . . Ben's jazz band! Take it away, guys!"

Hudson felt her heart leap into double-time as people quieted down. *You can't freak out now,* she thought. *Not now.* Her hands shook, but she still found the opening chords.

Nobody knows you here, she told herself. *Nobody expects you to be good.*

She started to play. After a few moments, she slipped into her music trance. Soon Ben joined her on his bass, then Gordie. And when she got to the bridge, and Logan started to play, he actually restrained himself. She didn't dare look up from the keys as she sang, but it didn't matter. Her voice was strong and elastic, hitting every note. And somehow, her talk with Logan in the laundry room had only filled her with more confidence.

When she reached the last chord, there was a moment of silence. And then the room broke into applause.

Hudson glanced back at the rest of the band and then, awkwardly, they all rose from their seats and took a small bow. *This is me,* Hudson thought, bowing. *This is what I'm good at. Like it or not.*

"Woo-hoo!" some guys called out.

The twins Hudson recognized from rehearsal jumped up and down.

"Hudson!" Carina called out. "We love you!"

At the end of their set, after they'd played four more songs and rocked every one of them, the Rising Signs, or whatever they were called, linked arms and took their final bows. Hudson locked eyes with Ben over Gordie's and Logan's heads and they traded an ecstatic smile. Their first official gig hadn't just been good. It had been great.

chapter 20

By the time they got off the "stage," it was almost time for Hudson and her friends to get in Carina's car and go back to the city. But Hudson wanted to linger just a little bit longer — especially because Ben and Ellie's friends kept coming up to her and telling her how much they'd loved the set.

"You have a really pretty voice," one girl said sweetly. "It's really strong."

"Where do you go to school?" asked another. "Do you live up here?"

"Hudson goes to Chadwick, in the city," said Ellie, who was now standing right next to Hudson in an almost proprietary way.

"Do you want to sing for a living?" asked one of the twins, who'd sidled up beside them.

"I don't know," Hudson said. "I think so."

"When's your next show?" Ellie asked excitedly. "Oh, and

Hudson, could you take a picture with me and my friends?" she asked, handing her iPhone to one of the twins.

Hudson looked at the camera. Getting her picture taken — and having it possibly be posted on someone's Facebook page, where a tabloid could find it — just wasn't worth the risk. "I can't. I look awful," she fibbed.

Luckily, Carina and Alex rushed over before Ellie could react. "You were amazing, H!" Carina said. "Just amazing!"

"When's your next show?" Alex asked.

"I don't know," Hudson admitted. "I don't think we have one booked yet."

"H, I think we gotta get back," Carina said. "My dad'll have a conniption if I get home after midnight."

"Okay. There's just one thing I still have to do," Hudson said, turning to look at Logan, who was talking with Ben and Gordie. She'd made a promise to Hillary, and she couldn't leave without at least trying to get some information for her.

"Hey, guys," Hudson said, approaching them. "Great show tonight."

"Everyone's asking when we're playing next," Gordie said, walking up to them. "What are we thinking?"

"I think I might be able to swing something at the Olive Garden," said Logan. "They sometimes do live entertainment."

"Um, Logan?" Hudson asked. "Can I talk to you for a moment?"

Logan ran a hand through his blond hair and glanced quickly at Ben.

"It's about something else," Hudson said. "Not the name of the band."

"Fine," he said.

She pulled him over toward the piano. "Um, so you know Ben's cousin, Hillary, right?"

"Yeah," he said, already looking both wary and confused.

"So, um... she's a really cool girl. And, well, I think you guys would, um..." *Oh, God, this is awful,* she thought. *Just say it.* "What do you think of her? Do you like her?"

Logan looked at her like she'd suddenly sprouted a third head.

"Because, um, I think, well..." she hedged.

Logan folded his arms. "*No,*" he said. "And please tell her to stop calling me."

He stalked off, leaving Hudson in shocked silence. *Oh, Hillary,* she thought. *This guy is such a jerk. He doesn't deserve you.*

Later, as Max, Carina's driver, drove them back to the city, Hudson tried to erase Logan's comment from her mind. She had no idea how she was going to tell Hillary about this. The old Hudson would have just decided not to tell her at all. But now Hudson wondered if it was such a good idea to spare Hillary's feelings.

"So when do you guys think you'll do another show?" Lizzie asked, holding Todd's hand in the backseat.

"I'm not sure, but someone mentioned the Olive Garden," Hudson said.

"You guys should play at Violet's," said Alex.

"Violet's?" Hudson asked. "How would we even get in there?" Violet's was a legendary music club in the East Village. It had been around for at least thirty years and everyone from Blondie to the Ramones had graced its small, dingy stage.

"This guy I work with at Kim's Video knows the booker," Alex said. "I'll have him talk you guys up. But if you put some songs on MySpace, then I think you have more than a shot at it."

"Really?" Hudson squealed. "You really think we have a chance?"

"If you're okay with being the opening-opening act," Alex said.

"C?" Hudson asked. "Can I hug your boyfriend?"

"Go ahead," Carina said.

Hudson leaned over her to Alex and hugged him. "Thank you!" she exclaimed.

"Well, don't thank me yet," Alex warned. "Let me talk to the guy first."

Hudson leaned her head on Carina's shoulder as they drove across the Triborough Bridge. In the distance the skyline of New York twinkled and glittered in the cold night. This was all so unreal. Everything was falling into place. Maybe the Rising Signs — because that's what they were to her now — actually had a shot at Joe's Pub, after all.

At eleven thirty, Carina's car pulled up to Hudson's house. "Here you go, Hudson," Max said in his gruff voice. "Last but not least. Have a good night."

Hudson scooted out of the backseat. "Thanks, Max."

He waited while she unlocked the front door and slipped inside. The lights were still bright in the kitchen, but from the relaxed, hushed feeling inside the house Hudson knew that her mom was still out. She waved to Raquel, who sat at the kitchen table by herself eating her dinner, and walked up the stairs. As she headed to her room, her iPhone dinged with a text. She pulled it out of her bag. It was from Carina.

Sorry in advance but thought you'd want to see this.

And then there was a link.

She hurried into her room, and clicked on the link.

It took only a second. Soon she was staring at the garish purple and yellow home page of a celebrity gossip site. And there was the headline, lurid and underlined and not at all surprising:

HOLLA'S GORGEOUS NEW GUY

Underneath it was a photo of her mom and Chris on a red carpet earlier that night. Holla stood next to Chris, her toned arm encircling his waist. Chris's blue eyes squinted at the camera in a sexy way as he pulled Holla in close to him. Hudson stared at the image for a moment. Up until now, their relationship had been something that existed only in her mind, a gross and annoying story to tell her friends. But now it was real. And it hurt, even though she knew that she and Chris would never have had a future.

She texted Carina back:

Looks like he might be my stepfather after all. Taking bets now.

Then she went to signsnscopes.com — now would be a good time to check her horoscope, she thought. But just before the page opened she closed her laptop. She thought of the car ride home with Alex, and Violet's, and how maybe she and the Rising Signs had a possible future after all. Whatever the future held for her, whatever was headed her way, maybe it was time to just let it surprise her.

chapter 21

The next morning her alarm went off at eight, but Hudson turned it off. The idea of running downstairs for yoga class seemed a little ridiculous. And ever since their fight in the studio, Hudson had been doing her best to stick to her room whenever Holla was home. So she closed her eyes and drifted pleasantly back to sleep. At ten, she opened her eyes with a start. She'd never, ever slept this late on a Saturday morning, except when she was sick. She jumped out of bed and got dressed.

When she got to the kitchen, Lorraine was washing her hands and Raquel was folding towels. "You're just getting up?" Lorraine asked. "Are you sick?"

"No, just tired," Hudson said, opening the fridge. "Do we have any bacon?" she asked.

Lorraine and Raquel both stopped what they were doing and looked at her. *"Bacon?"* Lorraine asked in disbelief.

"That's okay," Hudson replied. She opened the refrigerator and took out a quart of fresh-squeezed orange juice. "I was just in the mood for it."

Hudson poured herself a glass of juice and then made herself a bowl of oatmeal topped with berries. Then she walked over to the neat pile of daily newspapers and chose the *New York Post*. Holla mostly tuned the tabloids out, not bothering to read them or surf them or even acknowledge them. The *New York Post* was her one exception. It was in the kitchen every morning, stacked right next to the *Times* and the *Wall Street Journal* and *Women's Wear Daily*. Hudson wasn't sure why the *New York Post* was okay when all the other ones weren't, but then again, her mom was full of contradictions.

"Well, someone's finally up," Holla said as she glided into the kitchen in her workout clothes. Tendrils of sweat-soaked hair curled around her face. "Lorraine? Niva would love an açai smoothie before she leaves."

"Coming right up," Lorraine said, going straight to the blender.

Holla wandered over to the kitchen table. Hudson kept her eyes on her newspaper. "So what's wrong with you?" Holla asked. "Why'd you miss yoga?"

"I was just tired. I didn't think you'd really miss me."

Hudson looked up to see her mom peering at her. Hudson could see that she was trying to keep her patience. "What play did you see last night?"

"Play?" Hudson asked, unsure what her mom was talking about.

"Didn't you see a play last night, with Jenny?" Holla asked, puzzled.

"Oh, yeah," Hudson answered confidently. "*American Idiot*."

Holla seemed to expect more of an answer. "Did you like it?"

"Yeah. And Jenny's really psyched about the party, by the way."

"How's Jenny doing?" Holla asked, sitting down across from her. "Is she settled in finally?"

Hudson got up from the table and carried her bowl to the sink. She couldn't lie right to her mom's face. "I think she's doing better."

"Is she seeing someone?" Holla asked.

"Nope," she said, running water over her bowl. "By the way, I saw the pictures of you out with Chris last night."

Even though Hudson had her back turned, she could feel her mom stiffen at the mention of Chris's name. "Yeah, I took him to the Jay-Z documentary. It was fun."

"Why?" Hudson asked, turning around.

"Why?" Holla said, taken aback.

"I just thought that you had rules about this kind of stuff," said Hudson. "You know, waiting a month to take someone to a public event. That's what you always told me." Holla's rule about boyfriends was how she and Lizzie and Carina had come up with their own rule in the first place.

"Sometimes it's okay to break rules," Holla said. "And you're very opinionated these days."

"Opinionated?"

"I guess I'm not used to you being so . . . vocal," Holla said, as Lorraine placed a glass of coconut water in front of her.

"I just don't want to see you get hurt again," Hudson said.

Holla tilted her head, as if she didn't quite get what Hudson was saying. "Don't worry," she said with an edge in her voice. "Everything's going to be fine."

Hudson wiped her hands on a dish towel, then balled it up. She knew that she didn't quite believe Holla, but she didn't have any real reason for it. "I gotta go," Hudson said. "I'm meeting a friend at Kirna Zabete."

"If you see anything there that's cute, can you pick it up for me?" Holla said. "A dressy, third-date kind of top?"

"Sure, Mom." She started to walk out of the kitchen.

"You should really try to get something that's a little more fitted," Holla said, reaching out to touch Hudson's waist. "You have such a cute figure — why are you always trying to hide it?"

Hudson stepped out of her grasp.

"Hudson."

Hudson turned around.

"Please don't worry," Holla said, almost as if she were pleading with herself not to worry. "About Chris. Please, don't."

"I won't," Hudson said and walked out.

*

"Hey, superstar!"

Hillary was almost half a block away, but her high-pitched voice sounded loud and clear down the narrow SoHo street. Hudson waved her arms. "Hey!" she yelled back, trying hard not to laugh.

"I heard you rocked last night!" Hillary said, her face getting lost in her bulky pink knit scarf as she jumped up and down. "Ben said you guys totally killed."

"We did okay," Hudson admitted. "We didn't embarrass ourselves."

"Oh, come on!" Hillary said, as if that were the most ridiculous thing on earth. "I heard it was awesome. And Ellie worships you."

Hudson remembered what Ellie had told her about hooking up with Logan. "Well, she really likes you, too," she offered, opening the door to the boutique and walking inside.

Hillary unzipped her puffy down coat, and Hudson saw that she was wearing skinny dark-rinse jeans and a gray cashmere sweater. Her hair fell straight to her shoulders, and it looked magically thick and static-free. There were no plastic barrettes in sight.

"Wow, you look good," Hudson said.

"Thanks," Hillary said, blushing. "I think today I really need shoes." Hillary headed over to a shelf of shoes and picked up a suede bootie that looked like a cage for the foot. "So, what did Logan say? Did you ask him about me? I hope you weren't too obvious. Were you obvious?"

Hudson ran her hand over a strapless top with a fringed neckline. "I don't know if Logan is the best guy for you, Hil, to be perfectly honest."

"What do you mean?" Hillary asked.

"He got really mad at me for, like, no reason at the party, and . . ." Hudson hesitated. "I think he might be hooking up with someone."

Hillary dropped the bootie on the table with a thud. "Who?" she demanded, her eyes narrowed.

"I don't know," Hudson said. "Just someone."

"How do you know? Did he tell you that?"

"No," Hudson said. "Not really."

"Then how do you know?"

Hudson looked away and cringed. She should never have stepped into this. "I just think he's not even worth your time."

"*Who* did he hook up with?" Hillary demanded.

Hudson bit her lip and turned to face her. If she told Hillary the truth, then it might get back to Ellie. Or even to Ben. "I don't know," she said miserably. "I really don't. But the point is that he's just not that great a guy. You deserve so much more than him."

Hillary looked down at the floor, lost in thought. She shook her head. "I thought you were my friend," she murmured.

"I am. Of course I am." Hudson took a step toward her.

"You're not acting like it." Hillary zipped up her coat. "The least you could do is be honest with me. After everything we've been through."

"Fine. He said for you not to call him anymore," Hudson said bluntly. "That's what he said."

Hillary's already pale face turned even whiter.

Hudson felt an instant stab of remorse. "Hillary, I'm sorry," she said. "But that's what he said. You wanted to know."

Before Hudson could say anything more, Hillary turned and walked out of the store, letting the door swing shut behind her.

Hudson ran out of the store. "Hillary!" she yelled after her. "I'm sorry!"

But Hillary didn't turn around. She didn't stop. She trudged up the street in her gigantic puffy coat as if she couldn't get away from Hudson fast enough.

"Hillary! You wanted to know!" Hudson yelled.

Hillary picked up her pace, and when she reached the corner she hung a right and promptly disappeared.

Hudson went back inside and bought her mom a pretty floral-print Stella McCartney top that she would probably hate. She felt terrible. She wasn't the girl who didn't mince words; that was

Carina. And Carina probably wouldn't have thought that what she'd said to Hillary was that rude. So just to make sure, Hudson called her.

"Oh my God," Carina said. "She said she wanted to know."

"I know," Hudson said, crossing the cobblestone street. "And believe me, she would have said it to me."

"Don't even worry about it," Carina said. "You need to be a little more blunt. This is progress. And by the way, Alex says that his friend can definitely hook you up with a show at Violet's."

"What?" Hudson said, coming to a stop. "Are you serious?"

"Now you just need to create a MySpace page so they can hear you. Like, now."

Trembling with excitement, Hudson hung up the phone and called Ben.

"Violet's?" he asked. "Isn't that the place where the Ramones used to play, before they were the Ramones?"

"Yes," Hudson said. "And my friend can probably get us booked there. If we put up a MySpace page. And record some stuff for it."

"What are you doing tomorrow?" Ben asked.

"Nothing."

"Then come up here and let's get this done!"

*

The next day she told Raquel that she was going over to Carina's house, then she let Fernald drop her in front of Carina's building. As soon as he was out of sight she took a cab to Grand Central. On the train up to Larchmont she tapped her toes and sang her songs under her breath, getting ready. She knew that they would have to record mostly original songs for their MySpace page, and she knew just the ones she wanted: "For You," "Heartbeat," and a new song she'd just written.

She took a cab from the train station to Ben's house, and when she got there, Ben, Gordie, and Logan had Ben's mammoth laptop already set up. They recorded their songs on Ben's computer, and when they were finished, Ben fine-tuned the songs in GarageBand. It was impressive watching him work. Hudson could see that he definitely had a future in music.

"I think we're done," he finally said. "Wanna hear it back?"

He played back the three songs. The music sounded perfect, and her voice was as sultry as ever.

"Put it up," Hudson said. "I'll tell Alex's friend it's there. And I'll give them your number. Cool?"

"Absolutely," Ben said.

*

On Tuesday, just after Geometry, Hudson got the text from Ben: They were booked at Violet's to be the opening, opening act for a Monday-night show in a little less than a month. February twenty-third.

"We're booked!" Hudson shrieked in the hallway.

Mr. Barlow stepped his long, lanky frame out of his office and zeroed in on Hudson with his glacial blue eyes. "And I'm gonna be booking you in detention if you don't keep your voice down, Miss Jones."

"Sorry," Hudson said, running down the hall to tell her friends the good news.

chapter 22

During the days leading up to the Violet's gig Hudson could barely concentrate. She could already see herself and the band on Violet's notoriously small, crooked stage, playing their songs surrounded by the ghosts of rock-and-roll legends. Playing at Violet's meant that she was no longer in a high school band — she was in a *band*. She practiced her piano every spare moment. And at least once a week now, as she sat in Spanish or History or English, a song would come to her. After school she'd go home and head straight to her room, and for several hours she'd work on the song, letting it take shape as her hands slid over the keys.

She managed to escape up to Larchmont a few more times, using Aunt Jenny as her alibi. Whenever she was alone with Ben before or after rehearsal, she'd sometimes think about telling him who she was. It seemed a little weird to have "Hudson Jones" on their MySpace page. Someone, at some point, was going to figure out who she was. *Just tell him, already,* she'd think. *He deserves to know.* But Ben wasn't a boyfriend. A boyfriend deserved to know. Ben was just her bandmate, and if she told him about Holla Jones, he would probably want her to get her mom involved in the band. She remembered what he'd said that night about connections. And right now they seemed to be doing just fine the way they were.

Meanwhile, Holla spent long hours in the studio, putting the finishing touches on her album. She wouldn't get home until late. At night, Hudson would lie in bed, listening to the photographers rushing to snap the SUV as it drove into the garage. It was a relief to have a little break from her mom.

Hillary was someone else she barely saw. In fact, their friendship seemed to have dissolved altogether. Hillary no longer sat in the library in the morning, doing the crossword. Every once in a while Hudson caught glimpses of her trudging up the stairs to the Middle School. Hillary's fashion evolution still seemed to be in full swing. Her messy ponytail had been replaced by carefully blow-dried hair, and her pink knit scarf was gone, along with any hint of bright color. Her sweaters were blue, gray, or black, and even her pink and blue square backpack seemed to have been trashed in favor of a black messenger bag. It was as if her separation from Hudson had only made her more chic and fashionable.

And then there was her mom's ongoing love affair with Chris. When they weren't working in the studio, they took yoga classes together or and hung out in the prayer room, listening to music and talking about her tour. If she passed him on the stairs, he'd ask her a question: how she was, if she wanted to listen to any of her mom's tracks, if she'd had a good day at school. It seemed that after eavesdropping on her fight with Holla in the recording studio,

Chris was determined to be her friend. Sometimes she thought about telling Chris about her band. She wanted him to know that she hadn't completely given up on her music. But it seemed too risky, especially because he seemed surgically attached to Holla.

Sometimes it was a relief for Holla to have someone else to focus on. But it felt strange to be a third wheel in her own house, so soon. It had taken Holla weeks to allow her last boyfriend to spend any time at the house.

On the Saturday before the Violet's show, Hudson woke up late. The sun peeked through the crack in her velvet curtains, and it looked like one of the first days of spring. She got out of bed, showered, and dressed in a long-sleeved purple and gray striped dress layered over black tights and black ankle boots. She would need a new outfit for the Violet's gig, and today seemed like the perfect day to get something.

Down in the kitchen, Holla and Chris were digging into spelt pancakes and drinking kale–green apple smoothies. They were sweaty and flushed, and Hudson could see that they'd either been having a steamy makeout session or an intense power hula-hooping class. Or both.

"Hi, baby," Holla said casually. "You want some breakfast?"

"I'll just grab a muffin," Hudson said, going to the platter of them on the kitchen island.

"We missed you in yoga," Holla added. "Someone here has no idea how to do scorpion pose."

"Come on, do you really want me to be good at that?" Chris asked, as they played footsie under the table.

"So…how's the album?" Hudson asked, doing her best to ignore the lovefest.

"We finished last night," Holla said, clinking her smoothie glass with Chris's. "I think it turned out okay. The label's listening to it over the weekend."

"Just okay?" Chris asked, leaning in to kiss the top of Holla's nose. "I beg you to restate that."

"Okay, *great*," Holla said, kissing him back.

"Awesome," Hudson said, counting the seconds until she could leave.

"So, Hudson, I was thinking of stopping at Jeffrey today to get something for the party tonight," Holla said.

"Party?" Hudson asked.

"Jenny. I'm throwing a birthday party for her tonight. Remember?"

Hudson almost dropped her muffin on the floor. Today was February twenty-first. And then she remembered: Jenny was in Buenos Aires. In all her excitement about the Rising Signs and the Violet's booking, she'd completely forgotten to tell her mom that Jenny was going to be out of town.

"Raquel and Sophie have done an incredible job with the invitations," Holla was saying. "And everything's set. It's going to be about fifty people, most of them my friends, of course," Holla said to Chris. "My sister's social circle is…well, let's just say that it's not quite the kind of crowd you invite into your house."

Hudson couldn't move. The muffin still lay in her hand, dry and crumbly. She couldn't think straight.

"And tell me what you think of this," Holla said to Hudson. "You know how much Jenny loves macaroons. Well, I had about three hundred flown in from Ladurée." In answer to Hudson's blank stare she said, "They make the best macaroons in the world. They're a famous café in Paris. Don't you think she's going to love that?"

"Uh... uh, sure," Hudson said uncertainly.

"Just tell Jenny to be here at six. She doesn't need to help me set up." Holla noticed that Hudson was zoning out. "Hudson? Are you all right?"

"I'm fine," she said. "I'll be right back."

She ran up the stairs to her bedroom, where she'd left her phone. Maybe Jenny hadn't gone away after all, she thought desperately. Maybe her trip had been canceled. She called and waited. One ring. Two rings. Three rings.

Answer, she pleaded silently. *Please pick up.*

Finally she got Jenny's voice mail.

Hey! It's me! I'm out of the country but leave me a message or call me back later!

Hudson hung up and ran to her laptop. Trying hard to keep calm, she pounded out a note to Jenny over e-mail.

Forgot to tell mom about your party being moved. It's happening tonight. Can you come back in time for it?

She hesitated for a second and then added:

Can you call me ASAP?

She clicked Send, her stomach in knots. Her heart was beating so fast that she had to hold the edge of her desk and take deep breaths. She knew that there was the option to just go downstairs and calmly tell her mom the situation — that she'd forgotten, plain and simple, and that Aunt Jenny wasn't in town — but then Holla would want to know why it didn't come up when they went to see the Broadway play. And then Hudson would have to tell Holla that there'd been no play. And then Holla would start asking questions and find out about all the other times Hudson had lied about seeing Aunt Jenny. And then Holla would find out where she'd really been all this time... Hudson grabbed her purse and coat, left her room, and went down the stairs, still unsure of what to say. But only Chris was sitting at the kitchen table, wolfing down another plate of pancakes.

"Hey!" he said. "So what's the latest in the life of Hudson Jones?"

"Where's my mom?" Hudson asked.

"Oh, she rushed out of here. Had to go talk to a florist or something," he said.

"Oh." Hudson slung her coat over her shoulders. "I have to go out for a little bit."

"She's got her BlackBerry with her," Chris offered.

"That's okay," Hudson said, heading to the door. *Just call her right now and tell her the truth,* she thought. *Just tell her that you screwed up, that you're in this band, that you've been saying you're hanging out with Jenny as an excuse.*

But she pushed the thought away. She couldn't do it. Something about messing up that big in front of her mom made her want to run out of the house, head up to Grand Central, and leave the city for good.

She thought of calling her friends, but she knew that they

would make her fess up, and she just couldn't. Instead, she had Fernald drop her off at the Museum of Modern Art, where she walked through the halls of the permanent collection, barely registering the art on the walls. *Just call her,* she'd think, reaching for her phone. But if she did, she'd be saying good-bye to the Violet's gig. And she couldn't do that. Not to herself, and not to Ben.

That night Hudson watched, a smile frozen on her face, as the party guests arrived. Votive candles flickered in every corner of the living room, and black-jacketed waiters passed around platters of hors d'oeuvres. Holla weaved through the crowd, looking glamorous in a one-shoulder crimson dress that swept the floor. Hudson kept her eyes glued on her mother like a traffic accident, unable to look away. Holla was gracious and calm, the perfect hostess, but she glanced at the door repeatedly, waiting for the guest of honor to arrive. Finally, at six forty-five, she walked up to Hudson, who was still standing off to the side, contemplating a vegan dumpling but too filled with dread to eat it.

"Where's Jenny?" Holla demanded, one hand on her hip.

Hudson just shrugged.

"Call her right now and tell her to get down to this house," she snapped.

"No problem," Hudson said, and went up to her room. She sat down at her desk, trying to think. She'd known this moment was coming, of course. And she still didn't have any idea of how to resolve it. She opened her laptop. Jenny still hadn't written her back yet — not that that would have helped anything. Hudson bit one of her fingernails. It was awful to make Jenny look like even more of a flake than she already was, but her mom was used to this, after all.

Jenny had done worse things to her over the years. She'd even dated one of Holla's boyfriends right after Holla broke up with him. So blowing off her own birthday party almost wasn't that bad.

Hudson went back down to the party after a few minutes had passed. "She's not there," she said to Holla with as straight a face as she could.

Holla's eyes blazed. "What do you mean she's not there?"

"I mean, she's not answering her cell phone," Hudson said, cringing inside as she said it.

Holla shook her head as if she didn't quite understand, and then one of her party guests tapped her on the shoulder and drew her back into the crowd.

Hudson retreated into the corner. She reached for a mini veggie burger and popped it into her mouth even though she felt nauseous.

Holla asked Hudson to call Jenny again at seven, at eight, and then, one last time, at eight thirty. Each time Hudson went up to her room and just sat there, staring at the phone for a few minutes. *Just tell her the truth,* she'd think. But it was almost too late.

"She wasn't there," she'd say when she returned, as her mom's expression changed from disbelief to fury to quiet, enraged acceptance.

At the end of the night, after the last party guest had thanked Holla and wished her well, Hudson watched her mother shut the front door, then walk around the living room, blowing out each votive candle.

"Mom?" she asked, thankful for the gathering dark. "Are you all right?"

Holla didn't say anything.

"Mom? Are you okay?"

Her mom turned to her in the dark. "I don't want you to ever see her again," she said evenly. She left the room, leaving Hudson with an untouched tray of pastel macaroons.

chapter 23

"You go on at eight o'clock, you play six songs, and then you have two minutes to get off the stage," said Bruce, the manager and head booker at Violet's, wagging one thick, gnarled finger at them. He had watery blue eyes, a graying beard, and a very suspicious manner. "And no drinking. Do not go near the bar. You want some soda, you come ask me. You got that?"

"Don't worry. They won't be drinking," said Mrs. Geyer. Mrs. Geyer had agreed to be the band's official chaperone for the evening. Hudson was becoming more and more impressed with Mrs. Geyer every day. She'd helped the boys haul in equipment and then parked the car in a nearby garage. Now she sat in the corner with her handbag in her lap, quietly reading a magazine and doing her best to melt into the background. Hudson couldn't imagine her mom doing one of those things, let alone all of them. If Mrs. Geyer still had reservations about Ben's music career, she seemed to be getting over them.

"And anyone asks how old you are, just say eighteen," Bruce continued.

"That's no problem, sir," said Ben, with his usual politeness.

Bruce stared at him with narrowed eyes. "Don't be fresh with me," he said, pointing his finger. Then he walked out, leaving a trail of bad vibes behind him.

"For a guy who booked the Ramones, he seems a little uptight," Gordie said, adjusting his glasses.

"Dude has a killer beard," Logan observed.

"Anyone hungry?" Mrs. Geyer asked, reaching into her purse. "I've got beef jerky and Fruit Roll-Ups."

"Mom," Ben said, shaking his head. "I can't believe you brought Fruit Roll-Ups."

Hudson got up and paced around the small, cavelike dressing room. She'd heard and read so much about this place: famous brawls in the dressing room, police raids on the bathrooms. Of course, these days it was much tamer. The only reminder of Violet's wild past seemed to be here, on the dressing room walls, which were covered in mostly illegible graffiti.

"Hey, does anyone have a pen?" she asked.

Ben walked over and pulled one out of his back pocket. "You gonna add something of your own?"

"We have to leave our mark," Hudson said. She took his ballpoint pen and crouched low, careful not to let the hem of her flouncy burgundy chiffon dress get dirty. She pushed the pen into the peeling paint. *THE RISING SIGNS,* she wrote. And then the date.

"Hudson, is your mother coming?" Mrs. Geyer asked.

"Oh, she can't make it," Hudson said.

"Really?" Mrs. Geyer asked, surprised. "She can't?"

"She's out of town." She'd told Raquel to tell Holla that she was going over to Lizzie's house to study. It was a bit of a sloppy lie, but it was the best she'd been able to come up with, now that her mom and Jenny were so clearly on the outs. And Aunt Jenny was still out of the country. Every time she thought about Aunt Jenny she got a terrible knotted feeling in her stomach. But standing here now in this dressing room, she knew that she'd done the right thing by not telling her mom the truth.

"Your mother *does* know about all this, doesn't she?" Mrs. Geyer asked, trying not to sound too concerned.

"Oh, yeah, she's fine with it. Hey, do you want to come with me to the deli?" Hudson asked Ben. "I think I need some lozenges for my throat."

"Sure."

"Hurry back," said Mrs. Geyer. "You go on in twenty minutes."

As they walked out into the main room, Hudson realized that this was the first time she'd ever been in a real music club; her mom hadn't played places like this since before Hudson was born. Violet's was just one room, hardly any bigger than her bedroom at home. A cluster of tables stood on the floor, just a few feet from the stage. The bar on the side was barely longer than the island in Holla's kitchen. Above the bar hung a collection of old photographs. And there, against the far wall, was the small, cramped stage, bathed in soft reddish light.

"Wow," Ben said, looking around. "You have to thank that guy Alex for me. This is incredible."

"I know." Hudson laughed. "I can't believe we're here."

"I was thinking," Ben said, scratching his un-pomaded curly hair, "It might be kind of crazy, but we should try to get booked at Joe's Pub. What's the worst that can happen? They say no? Big deal."

"Right. I think we should."

They left the club and started walking to the corner. *Just tell him*, she thought. *It would be so easy.* Drops of rain were just starting to fall. A city bus wheezed its way up Bowery Street. Reggae music came from a taxi at the corner. Next door, a restaurant had just opened up. There was a velvet rope set up in front, manned by a large bouncer, and a throng of paparazzi jostling nearby, snapping the people walking in and out.

"I have this theory," said Ben, "that a velvet rope is basically all you need to make a place cool. That and a few weird guys with really big cameras just hanging out in front."

"I think you're right about that," Hudson said.

"Yeah. I mean, people will believe anything if they see a bouncer in front of a place —"

"Hudson!" someone yelled.

Hudson looked. It was a photographer. Before she could move he'd aimed his camera at her.

SNAP! The camera flash exploded in her face.

"Hudson!" another photographer yelled.

CLICK! went his camera as another flash blinded her.

"Where's your mother?" another one yelled. *SNAP!*

"What's going on?" Ben asked. "Why are they taking your picture?"

"Come here," she said, grabbing his arm and leading him into the deli.

"Do those people know you?" asked Ben. "How do they know your name? Why are they asking about your mom?"

She ran into the deli and down an aisle of potato chips and bags of mini-pretzels. She couldn't look at Ben. She looked at the potato chip bags, the shelves of candy, anywhere but his face.

"Hudson?" Ben asked. "What was that?"

Finally she met his worried gaze. "My mom is Holla Jones, Ben."

Ben blinked for a few moments. It was as if she'd just said she were a martian.

"Are you serious? Why didn't you tell me?"

"I should have," she said. "But I haven't told her about you guys yet, either."

"What?" Ben asked. "Why not?"

"Because I was supposed to put out my own album this spring."

"You *were*?" he said.

"And she'd never understand why I'm now in a jazz band up in Westchester."

"Thanks," Ben muttered.

"Sorry, that didn't come out right," Hudson said. "Look, I don't want my mom anywhere near this. She just takes over. She takes over everything. She took over my album and changed everything. She turned all my songs into this crazy sampled pop stuff, stuff I couldn't even recognize. It was awful. And then I tried to sing one of the songs in front of this big party and it was a disaster. And I figured that was a sign that it's just not meant to be after all. So I just decided to stop."

Ben nodded slowly, trying to understand.

"But then I found you guys," she said. "And you helped me

remember why I want to do this. Why I love music. And that I need to play the kind of stuff *I* want to play. No matter what." She shivered in the cold. "And being up in Larchmont, being in your basement, hanging out with Ellie, it's like I'm finally just me, you know? I'm not Holla Jones two-point-oh. For once."

Ben smiled. "Hudson," he said, stepping closer to her, "I know we don't know each other very well. But you're talented, okay? *Seriously* talented. And sometimes I feel like you don't really believe it. It's like you're embarrassed by it."

Hudson fidgeted with the bags of chips.

"And now I get it," he said. "I can't imagine having everyone in the world know my mom. And being compared to her. That would suck." He reached over and grabbed a pack of Ricola drops. "But you have to take up your own place in the world. And feel like you deserve to."

Hudson looked up at Ben. He gave her the Ricolas.

"And as far as not wanting your mom involved, that's fine, but think about the position you're in: You don't have to pound down doors. You don't have to stalk someone to listen to your tape. And there's nothing wrong with that."

"I know," she said, readjusting some pretzels on the rack. "I just didn't want you guys to expect that."

"I don't," Ben said. "But if this is what you really want to do, why are you making it so hard?"

She thought about that for a moment. "It's just how I want this to be right now."

Ben nodded. "Well, can I tell the rest of the guys?" he asked. "Would that be okay?"

"Not yet," she said. "Maybe later. But not now." She walked to the register.

"Okay." He touched her shoulder. "And Hudson?"

"Yeah?"

"Thanks for telling me the truth."

"You're welcome." She paid for the Ricolas and they walked out of the deli.

"So, are you gonna get your picture taken again?" he joked.

"It doesn't matter," Hudson said. "They don't usually end up anywhere." She watched the paparazzi milling around outside and was relieved to see that most of them had their backs to her. And then a couple walked out of the bar, past the velvet rope, and onto the sidewalk in front of them. They held their heads down and walked quickly by the paparazzi, who didn't seem to notice them. The guy was tall and skinny and his hair looked reddish-blond in the dim orange light of the street lamps. He held the hand of the woman with him. She was tiny and thin, with long brown hair that fell over the hood of her shearling coat.

Hudson watched as the man leaned down, kissed her, and then reached into the back pocket of his Levi's and pulled out a blue knit hat.

Hudson felt a shiver run through her. It was Chris Brompton. With another woman.

"Oh my God," she said.

"What?" Ben asked. "What is it?"

She watched Chris and his mystery woman cross the street.

"Hudson? What's going on?"

"I know that guy," she said, pointing, unable to say more.

"We gotta go," said Ben, glancing at his watch. "We have a show to play, remember?"

She stared at the man's back as he walked away. It could have been someone else. After all, she hadn't seen his face. But she knew with a sick certainty that it was him.

She was so distracted that she didn't even notice the bouncer in front of Violet's when they walked back inside.

"Wait. How old are you kids?" he asked.

"We're playing tonight," Ben explained, and showed the bouncer his bracelet.

They pushed through the crowd and headed toward the stage. The room had filled while they were gone. They reached the dressing room just as Gordie and Logan were about to walk out. "Where've you guys been?" Logan asked.

"Sorry!" Hudson said, digging in her purse for a comb and some lip gloss.

Bruce walked in, waving his arms. "What are you waiting for, kids? You're on!" he yelled. "Get out there!"

Hudson threw her things back in her purse and ran after her bandmates out onto the stage.

"Hey. We're the Rising Signs," Hudson said into the mic as she sat down at the piano.

A cheer rose up from the tables. A few people even whistled. Hudson pushed the image of Chris Brompton out of her head, leaned into the mic, and, thinking, *You're here, you're actually here,* she started to sing.

chapter 24

I have to tell her, Hudson thought as she lay in bed, staring at the ceiling. *I have to tell her what I saw last night.*

It was early — so early that her alarm clock hadn't even gone off yet. Outside her velvet curtains she could hear the whine of a garbage truck stopped outside their house. It was a school day, but Hudson felt too exhausted to even get up. She'd barely slept last night after the gig. The show at Violet's had been unbelievably and wonderfully great. The crowd had loved the songs, and even Bruce had seemed impressed. They'd even had some requests for CDs, even though they didn't have any yet.

But she couldn't get the picture of Chris kissing that woman out of her mind. It was clear that she had to say something. If she didn't, Holla would fall deeper in love with him. But if she *did* tell her mom, she would have to explain being in the middle of the East Village last night instead of at Lizzie's apartment.

"Don't say anything," Carina had warned her over the phone,

after the show. "Just pretend you never saw it. That's what I would do. No *way* do you want to get in the middle of that."

"You have to tell her!" Lizzie argued when Hudson called her after speaking to Carina. "Wouldn't *you* want to know?"

"But if I tell her, she's gonna want to know where I saw him," Hudson explained.

Lizzie seemed to be thinking. "Just say that you saw him near my house."

"But why would *we* be out on the street?" Hudson asked.

Lizzie didn't have any response to that.

"Ugh," she said into her pillow, remembering the conversation now, just as her vintage sixties alarm clock rang.

Hudson reached up and slammed the clanging bells with her hand. Then she grabbed her iPhone off the floor and, out of habit, checked her e-mail.

Jenny had finally written her back. Holding her breath, Hudson opened it.

> *Hey, Hudson, just got your message. I'm really shocked to hear about the party. I told you that I was going to be out of town. I was wondering why I never heard from you again. Guess you only needed me for an "alibi." I will be calling your mom as soon as I get home.*
>
> *Jenny*

With a piercing sensation in her chest, Hudson saw that Holla had been cc'd on the message.

Hudson threw the phone to the floor and walked to the bathroom on shaky legs. She was in serious trouble. Now her mom knew almost everything. Like a zombie, she showered, dried off, and put on her school uniform. When she opened the door, Matilda was there, excitedly circling her feet. Hudson picked her up. "I'm in trouble, Bubs," she whispered into the dog's ear. Matilda licked Hudson's nose almost sympathetically. Hudson walked down the stairs, passing the kitchen. If her mom were home, she knew where she would probably be, and there was no sense in dragging this out.

Hudson put Matilda down at the glass doors of the yoga studio. Inside, she could see Holla's pierced and dreadlocked hula-hooping instructor, Che, swiveling a hoop around her waist. Holla stood in the corner, conferring with Sophie, with her back to the door.

Hudson opened the door. *Here goes,* she thought. "Mom?" she asked.

Holla turned around before Hudson even finished speaking. Her eyes were dark and seemed even bigger than usual. Her chest heaved up and down.

"Mom, I saw the e-mail from Jenny—"

"You're grounded," Holla said. "Do you understand?"

"Mom, I'm sorry," Hudson said, feeling tears come to her eyes. "It was just a mistake."

"Oh, really? A mistake? Sophie?" Holla said, sounding like she was just barely keeping herself under control.

Hudson saw that Sophie held a rolled-up newspaper. It uncurled just enough to show the front page, and the *New York Post* masthead.

"Look at Page Six," Holla said.

Sophie handed Hudson the newspaper, without looking her in the eye. With shaking hands, Hudson found the page.

It was a photo of her with Ben. One of the photos that had been taken the previous night when they walked to the deli. And underneath was the caption:

Hudson Jones, daughter of icon Holla Jones, leaving Violet's, where she wowed the crowd last night with her own jazz- and soul-inspired songs.

"It's one thing if you're going to lie to me," said Holla, her voice eerily cool and controlled, "but your record label is gonna want to know why someone with crippling stage fright is singing at Violet's."

"I was gonna tell you," she began. "It's just this band I joined. Up in Westchester."

"In *Westchester*?" Holla exclaimed. "You've been going to *Westchester*?"

Hudson didn't say anything.

"Well, that's done," Holla said. "I'm calling your label today and telling them you're back on track. Do you understand me?"

Hudson didn't speak. She knew that her mom meant this.

"And about Jenny," Holla said in a withering voice. "Do you know how that made me look? In front of everyone? How could you let me be humiliated like that?"

Hudson swallowed again. "I'm sorry. I just didn't know what to say."

"And why would you want to play dumps like Violet's when you could be playing Madison Square Garden?" Holla asked, her voice ringing against the studio walls. Behind her, Che and Sophie seemed to cower. "When you have a finished album just sitting on the shelf? What's wrong with you?"

"I just wanted to do something myself," Hudson said feebly.

"Because I'm that horrible. Right? I'm that terrible." Holla shook her head. "I've done nothing but help you. I've given you music teachers and voice coaches and studios. And you throw it back in my face. Just like my sister."

"What if I don't *want* to play the Garden?" Hudson exploded. "What if I don't *want* to do things exactly the way you do?" Her voice was getting louder and louder. "What if I don't need to be a total egomaniac to be happy?"

Holla's face went slack. "You're done with that band. Today. And starting right now, you don't go anywhere but school and back. Do you understand me?"

Hudson turned and ran to the door, choking on her tears. She'd forgotten to tell her mom about Chris. But Holla didn't deserve to know. And when Holla did find out, Hudson wouldn't be there. She'd never be there for Holla again.

chapter 25

"Hudson, it's okay. Really. It's okay," Lizzie said, smoothing Hudson's hair with the palm of her hand.

"Come on, you're gonna make *me* cry," Carina pleaded, reaching out to pat Hudson on the back.

They weren't supposed to be inside the ladies room off the Chadwick lobby, because technically it was for visitors only. But Lizzie and Carina had pulled Hudson in there the moment they saw her, and Hudson had been only too grateful to follow them. She now stood with her head pressed up against Lizzie's shoulder, sobbing so hard she thought she might hyperventilate. As soon as she could breathe normally again, Carina wet a paper towel under the faucet and handed it to her.

"So what happened?" Lizzie asked. "Just tell us."

Hudson blotted her face with the paper towel. "My mom found out about the show. It's on Page Six. And I have to quit the band. *And* my aunt officially hates me."

"That sucks," Lizzie said.

"And now it's all over. Everything. After a great show last night, too." She sniffled, wiping her nose with the back of her hand.

"She was just angry," Carina reasoned. "Wait until she sees you at a show. She'll totally change her mind —"

"No, she won't, not now. That's all finished." Hudson grabbed a dry paper towel and wiped at her eyes. "It's like there's no talking to her. There's no reasoning with her. She said because I signed a contract I should pick up where I left off. But that album just isn't me. It's not even my music."

"Why don't you tell her that?" Carina asked. "Just say that. Tell her you want to make an album that reflects *you*."

"I tried," Hudson said. "That day when you guys were in the studio last fall. Remember how well that worked out?"

"Look, you've been brave, and you've put yourself out there, and you've gotten out from under your mom's thumb," Lizzie pointed out, her hazel eyes calm and reassuring. "And that's more than anyone else could do in your shoes. Maybe you can rejoin the band in a little while. After things calm down."

"And why does your aunt hate you?" Carina asked. "I mean, not to make things worse, but that part I still don't get."

"My mom threw her a party, and I knew my aunt couldn't be there, and I forgot to tell my mom. You know how controlling she is about stuff. I just couldn't tell her. I couldn't say that I'd made a mistake."

Her friends looked at her gently. "She's your mom, H," Carina said. "She doesn't expect you to be perfect."

"Yes, she does," Hudson said, feeling the tears start to come again. Hudson glanced in the mirror. Big red blotches spread out from her green eyes down to her cheeks, and her lips were swollen. "She can't be normal," Hudson said. "She can't eat like everyone else. She can't relax for a second. It's all about being the best, the biggest, the most amazing person in the world. She wants me to be like that, too. It's like I'll be some loser if I don't end up a superstar."

"Do you believe that?" Lizzie asked.

Hudson swallowed. "No."

"So then why does it bother you when she says that?" Lizzie asked.

Hudson played with the hair elastic around her wrist. "I guess a little part of me is afraid she's right," she said quietly.

Carina looked at her watch. "Oh, *shnit*. Madame Dupuis's gonna have a French cow if we don't get up there." She put her hands on Hudson's shoulders. "You okay?"

Hudson sniffled one last time and then splashed her face with more water. "God, I love school," she groaned, and then cracked a smile.

As they walked out of the bathroom, Hudson tried to believe what Lizzie had said. She *had* been brave. She'd tried to do her own thing. For just a few short weeks, she'd just been Hudson onstage. And it had been wonderful.

But that was all over now. And the sooner she accepted it, the better.

chapter 26

All day Hudson watched the clock. In every class, during every free period, she kept tabs on the time. Twelve o'clock. One o'clock. Two. Every minute that brought her closer to the end of the school day only increased her sense of dread. Spending the evening under the same roof as her mom was pretty much unthinkable. Right now she never wanted to look at or speak to Holla again. She almost asked Lizzie or Carina if she could sleep over. But whatever was waiting for her at home, she knew that she needed to face it. At least the worst seemed to be over.

When she walked out of school, the black SUV was waiting right at the curb, directly in front of the school doors. Holla had meant what she'd said about tightening her grip on Hudson's comings and goings.

"Bye, guys," she said to Lizzie and Carina as they hovered by the side of the school building, shielding themselves from the rain. "Say a prayer."

“You just have to get through tonight,” Carina said.

“Maybe your mom will reconsider the band thing,” Lizzie put in.

But as the SUV inched its way downtown, Hudson doubted it. For all she knew her mom had already booked her time in the studio to finish the album and get it released by summer. When she got home she took the service elevator up from the basement and walked into the kitchen, holding her breath.

“Your mother’s upstairs,” Raquel said sternly, arranging a spray of white flowers. “She wants to talk to you.”

Hudson felt an even bigger wave of dread. “Thanks,” she said listlessly, then took the elevator to the fourth floor, which belonged entirely to Holla’s suite of private rooms. Hudson walked down the hall, past the pale peach-colored bedroom and gym, where Holla ran on the treadmill and lifted weights in the afternoons. Then she turned into the closet. She slipped off her boots at the entrance — Holla didn’t like the idea of dirt so near her clothes — and entered.

Holla was standing on a pedestal in a magenta bandage dress, flanked by mirrored closet doors that magnified her reflection over and over, so that it looked like there were at least a hundred Hollas in the room. Kierce, her stylist, sat on a tufted ottoman to one side. He had a long black ponytail, ghostly pale skin, and a permanent, disapproving squint. Even though he wore only black, he was always trying to get Hudson to wear bright colors.

“We’re looking at stuff for *Saturday Night Live*,” Holla said curtly. “What do you think?” She twirled around on the pedestal with her hands on her waist, showing off the dress, which hugged every curve. "You like the color?"

"The color is not even a *question*," Kierce put in.

"I like it," Hudson said, relieved to talk about fashion.

"I don't know. I don't like what it does here," Holla said, gesturing to her flat stomach. "It gives me a pooch."

"No, it doesn't," Hudson said. "It looks really nice. I swear."

"And Kierce has a few things for you."

"For me?"

With a glum stare, Kierce took several clothes bags out of a cabinet and handed them to Hudson. "There's some Rodarte in there," he said. "Don't even ask me what I had to do to get that."

"But what is this for?" Hudson asked, giving Kierce a weak smile as she took the clothes.

"*Saturday Night Live*," Holla announced. "You're doing it with me. It's all set. March seventh. Two weeks." She turned to Kierce. "Would you unzip me, please?"

The bag of clothes slid out of Hudson's hands. "What?" she asked, her voice barely above a whisper.

"If you can try things on now that would be best," Holla said. "Kierce can take back whatever doesn't work."

"But, but..." she stammered. "Why am *I* doing *Saturday Night Live*?"

Kierce unzipped her dress. "Because I sent them one of your tracks," Holla said, "and they called back this afternoon saying they wanted you."

"So this was *your* idea?" Hudson asked, panicked.

"Honey, your label's thrilled. And I think in light of everything, this is the best thing for you." She stepped out of the dress and into the royal blue shirred silk dress that Kierce held open for her. "Most people have to wait 'til they have a huge hit to get on *Saturday Night Live*. But they want *you* now."

Her heart was pounding. "Mom, I can't," she said.

"Hudson..." Holla warned.

"I'm in a band now," she said bravely. "I need to be there for them. They're my priority."

"That's over," Holla snapped, turning her back so that Kierce could zip her up.

"It's not. I made a promise to them. And they like my music. They like me for me."

"Really?" Holla asked, turning around. "Are you sure about that?"

"What?" Hudson asked, picking up on her mom's sarcastic tone.

"Sophie heard from the publicist for Joe's Pub today," Holla said, suppressing a grin. "Apparently they'd been told — *promised*, actually — that I'd be playing a show there. In exchange for booking your band."

Hudson blinked. "That... that can't be right," she said.

"Sounds like a real loyal bunch of people," Holla said thickly. She assessed her reflection and turned back around. "Unzip me again," she said to Kierce.

Hudson stood there motionless. She didn't believe it. She *couldn't* believe it. But then she remembered what Ben had said that night in front of Ellie's house, about how getting ahead was ninety percent connections... and she started to feel a rumble of anger down deep inside of her.

"They're using you, honey," Holla said, stepping out of the dress. "I just thought you should know."

The sour taste of anger filled Hudson's mouth. "And your boyfriend's using *you*," she shot back. "I saw him leaving a bar with some woman. Holding her hand. And kissing her."

Holla froze, one leg out of the dress. "What?"

"I saw him last night," she said. "He was with someone. On a date. I saw them leaving some club." She knew she was being cruel, but she couldn't help herself. "Where'd he tell you he was last night?"

Holla still didn't move. "Visiting his family," she said hoarsely.

"Not from what I saw."

Kierce looked appalled. "Didn't you say he was in Poughkeepsie?" he asked Holla.

"So he's just like all the others, isn't he, Mom?" Hudson asked. "He *really* loves the spotlight. But you? Not so sure."

It was probably the meanest thing she'd ever said to her mother, but right now, the words were flying out of her mouth.

"Kierce, please hand me my phone," Holla said. She remained perfectly still. "I think that's enough now, Hudson. Why don't you just worry about your own life, okay?"

Hudson stormed out of the closet, grabbed her shoes, and went down a flight of stairs. Her heart was beating so fast it was all she could hear. When she reached her room, she pulled out her iPhone.

Ben picked up on the second ring. "Hudson? Hey, what's up?"

"Did you call Joe's Pub and tell them that my mom would play there?" she asked breathlessly.

"What? No," he said. "What are you talking about?"

"Well, someone did," she said. "Remember what you said about connections? How it's all about who you know? How I should use what I have?"

"What?" Ben sounded utterly thrown. "Hudson... what's wrong?"

"This isn't a game to me," she said. "The whole reason I joined your band was to prove to myself that I could do this, alone."

"You're *not* alone," Ben said. "There are three other people in this band besides you. You're not the only one who wants to get something out of this. And what are we all supposed to do? *Not* care that you're the daughter of Holla Jones?"

"So you told the guys," she said. "Great."

"I thought it was important," he said.

"And you told them at Joe's Pub," she said. "You told them who I was."

"It's not like we can hide it," Ben said.

"Thanks," she said. "That's just great. I trusted you."

"Hudson, hold on—"

"Good-bye, Ben," she said.

She slammed her finger on End Call before he could reply, then tossed the phone onto the bed as if it were on fire.

She sat down on the floor with her back to her bed and hugged her knees to her chest. *You did the right thing,* she said to herself. *This all had to end sometime. You knew things would change when you told them who you were. Better to get out now.*

She leaned her forehead against her knees, shutting her eyes against the tears. But she couldn't escape the feeling that she'd just made a gigantic mistake.

Then she got an idea. She stood up and went into the other room, where her laptop waited for her on the desk. She logged in to signsnscopes.com and clicked on Pisces.

Congratulations, little Fish! You achieved a massive breakthrough during the lunar eclipse! Everything you've ever wanted is finally within reach... Now all you have to do is go for it!

Hudson shut the laptop and went straight to bed.

chapter 27

"So, tomorrow's the big night," Lizzie said, picking some burnt crust off her BLT. "You feeling okay?"

"I can't believe you're doing *Saturday Night Live*!" Carina squealed, slamming her foot into the base of the diner table. "This is so cool!"

"I feel fine," Hudson said to Lizzie as she ate a forkful of coleslaw. "It's really not that big a deal." She picked up her Reuben and took a small bite.

"Are you sure I can't come?" Carina asked, grabbing Hudson's hand. "I'll sit in the audience, way in the back. You won't even see me, I *promise*."

"You guys, I wish you could," Hudson said gently. "But it's all part of my master plan: making sure nobody sees this. You haven't told anyone about it, right?"

Lizzie and Carina shook their heads.

"Good. And somehow my mom convinced *SNL* not to promote it," she said. "And I'm not doing any of the commercials with her, either."

"People *are* going to see this, you know," Carina reminded her. "Even if it *is* the first day of spring break."

"Thanks, C. You're really making me feel better."

"Then why did you say yes?" Lizzie asked.

Hudson rolled her eyes. "My mom's been a wreck ever since the breakup, which was basically my fault."

"It wasn't your fault, Hudson," Carina said.

"But I told her about it," Hudson said, taking another bite.

"And it was the right thing to do," Lizzie exclaimed. "Somebody had to!"

"I just wish it hadn't been me," she said.

"We know your mom broke up with some guy, but that's her life, Hudson, not yours," Lizzie said. "You shouldn't do this just because you feel like you have to."

"Well, what's my other option? Being in a band with a bunch of guys I don't trust?"

"So the guy used your name to get a booking," Carina said. "You know how people are. They don't think sometimes. And you really liked that band."

"You guys, please," she pleaded, putting down her fork. "I'm really nervous about this *SNL* thing. Don't make it worse for me, okay?"

Lizzie and Carina traded a look and then went back to eating their meals. Hudson resumed picking at her food. The breakup had hit her mother hard, just as Hudson had feared. Holla had confronted Chris about the mystery woman that same night, and after some denials, he confessed. She was his last girlfriend, with whom he'd never quite broken up. Luckily Holla's album was finished. She called Chris a few choice words and then hung up on him, and despite the frantic voice mails he'd left, she hadn't looked back.

Except Holla descended quickly into Breakup Mode—periods of manic activity followed by utter depression. The day after the breakup, Hudson came home to find Holla padding around the house barefoot, looking lost and remote. Hudson had seen her mom suffer through many a breakup, but she'd never actually been the instigator of one. Hudson told Lorraine to whip up a batch of her vegan chocolate-chip cookies and bring them up to the prayer room, where she and her mom ended up hanging out on the couch, watching *The Devil Wears Prada*. She told her mom that she was better off without Chris, and reminded her that she'd barely wasted any time on the guy. Still, Holla was heartsick.

"The whole time, he was with her," she kept saying, as if she didn't believe it. "The *whole time*."

When Holla eventually brought up the *Saturday Night Live* appearance again, it wasn't even a question that Hudson would do it.

"If you're out there, doing it with me, I'll feel strong," Holla said, squeezing Hudson's hand. "I won't be thinking about him."

"Sure, Mom," Hudson said, squeezing back. It was the least she could do.

Now the show was all Holla could talk about. They'd already had three hour-long rehearsals for a three-minute song. There had been sit-down meetings about hair and makeup with Gino and Suzette. Holla even wanted Hudson to do a dance solo, to which Hudson had reluctantly agreed.

Now Hudson sat with her friends in uncomfortable silence. She knew that she'd made a mistake by saying yes. And there was no way out of it.

"Have you heard from Ben?" Lizzie asked quietly.

Hudson put down her fork and shook her head. "I wonder if he has spring break for the next couple of weeks, too."

"Are they still doing the Joe's Pub show?" Carina asked.

Hudson shrugged. "No clue. I haven't heard from any of them. Not that I would."

"Are you sure you don't just want to call him?" Lizzie asked. "You *did* hang up on the guy."

"Call him and say what?"

"At least hear him out," Carina said.

"I think I know all I need to," she murmured.

"Ben just doesn't seem like the kind of guy who would do something like that," Lizzie said. "Did he seem like that to you?" she asked Carina.

Carina shook her head. "I didn't think so. But people do crazy, selfish stuff sometimes. Hey, speaking of crazy," she said, glancing over at the corner. "Is that Hillary coming over here?"

Hudson looked up from her plate. If the girl gliding toward them had ever been Hudson's eccentric, fashion-challenged friend, it was now impossible to tell. This was an astonishingly elegant stranger, wearing silver ballet flats, a leather backpack falling artfully from one shoulder, and dark lipstick. She'd traded her sequined sweaters for a silk camisole and a cropped black blazer — both against the school's uniform code — and her once-dowdy Chadwick kilt had been rolled up at least five inches. She even wore jewelry: a silver cuff bracelet and silver leaf earrings that caught the light as she walked.

"Did *you* do that to her?" Lizzie asked Hudson.

"I don't think so," Hudson said.

"Hey, you guys," Hillary said as she approached.

"Hi...uh...hi," they murmured, all of them in awe of Hillary's new look.

Hillary's gaze zeroed in on Hudson. "Can I talk to you for a sec? It's kind of important." She drummed her fingers on the strap of her book bag, and Hudson noticed that her nails were French-manicured. Almost like Ava's.

"Uh...sure," Hudson said, still a little stunned. "I'll be right back, you guys." She followed Hillary outside onto the street.

Hillary walked down the block and planted herself in front of Sweet Nothings, Carina's favorite candy boutique.

"So, what's up?" Hudson said.

"Why'd you quit the band?" Hillary asked, point-blank. "Ben told me you quit. He also told me you hung up on him, but we'll get to that later. Why'd you quit?"

"That's between me and Ben."

"You know how much this Joe's Pub show means to him," Hillary said. "How could you do that?"

"Is he still playing the show?" Hudson asked cautiously.

"Of *course* they're not playing. They can't do anything without you."

"Hillary, this might be hard to believe," Hudson said, "but I think Ben did something really gross."

Hillary put her tiny fists on her hips. "*What* are you talking about?" she asked snippily.

"I think he told the people at Joe's Pub that if they booked us, my mom would personally show up and do a few songs there."

Hillary just stared at her.

"My mom got the call. And this was after I'd finally told him everything—who my mom was, all the reasons I wanted to be in his band, and why I didn't want her involved," she went on. "He totally destroyed my trust in him, okay? That's why I quit. That's why I hung up on him."

"Ben wouldn't do that," she said.

Hudson shrugged. "He did it. He practically admitted it to me."

"No. I know my cousin. And I know that this is something he would never do." Hillary pulled her phone out of her book bag. "Call him up right now and apologize."

"Hillary! No," Hudson said, backing away from her. "I'm not apologizing!"

"But you have to," Hillary said. "I know he didn't do this. And maybe it's not too late to save the Joe's Pub show—"

"It *is* too late," Hudson said. "I have another show tomorrow night. With my mom. I'm doing *Saturday Night Live*."

"With your *mom*?" Hillary asked, aghast.

"Yeah. With my mom. Why are you looking at me like that?"

Hillary shook her head. "I'm just..." Hillary was quiet for a moment. "What about all the work we've done together?" she finally said. "What about doing stuff your own way?"

"I tried that," Hudson said, stepping away from her. "It didn't work."

Hillary went quiet again.

"I'm not the one who did anything," Hudson added. "This is all his fault. So if it's okay with you, I have to get back to my friends."

Hudson wheeled around and walked quickly up the block, not looking back. She was so angry that her chest hurt. None of this was Hillary's business. And she had no right to take her phone out like that and push it in Hudson's face.

Carina and Lizzie watched her carefully as she made her way back into the diner and to the table. "Everything okay?" Lizzie asked.

"Yeah, everything's great," Hudson fibbed, reaching for her bag.

"What happened?" Carina prodded.

"She's mad that I got into a fight with her cousin," Hudson answered. "She said that it wasn't like him to do something like that."

"Maybe she's right," Lizzie said.

"Well, she still had no right to get mad at me," Hudson said. "She's just upset with me about other stuff, too."

"What other stuff?" Lizzie asked.

"Nothing," Hudson murmured.

"I think you just got a text," Carina said, gesturing to Hudson's bag.

Hudson pulled out her phone. It was from her mom.

Rehearsal today at 4—don't be late!

"You guys, help," Hudson pleaded.

"What?"

"I really don't think I can do this. How do I get out of this?"

Carina and Lizzie looked back at her, stricken. Hudson knew what they were thinking: *Sorry, H. It's too late.*

chapter 28

"Now, just stand very still for a second," said Paula, the wardrobe lady, as she pulled Hudson's silvery purple dress tighter around her back. "You're so tiny, but this should do the trick."

Hudson stared in the full-length mirror as Paula secured the pin and the dress magically shrank a size. It was, Hudson had to admit, beautiful—metallic and sparkly and belted, with short sleeves and a V-neck. She'd actually bought it online, and Holla had taken one look at it and had decided to wear something in almost the exact same color.

"Okay, I think that should do it," Paula said, squinting at Hudson in the mirror. "You're going to look *adorable* next to your mom out there."

"Thanks," Hudson said, stepping off the pedestal. "Do you need me to send her in here?"

"She looked fine to me during dress rehearsal, but I'll check on her. And I think it's better if I go to her," she said, grabbing her sewing box.

Hudson followed Paula out of the wardrobe room and into the main hallway of the *Saturday Night Live* studios. It was jammed with people — writers, producers, NBC pages dressed in uniform and scurrying around in the five minutes before air. A few cast members ran by, already in makeup and costume for the opening sketch. A college-aged intern pushed a wardrobe cart up the hall, and a PA wearing headphones led some people into the greenroom. At the end of the hall was a set of double doors, and behind that was the studio, where the *SNL* house band was grinding out "Mustang Sally," warming up the audience. Through the doors Hudson could hear them clapping and hooting. Even though she'd been terrified of this night, now she looked around her in awe. It was hard not to get swept up in this.

Paula turned left at the double doors and then right, into Holla's dressing room. Hudson followed her but almost bumped right into Little Jimmy, who had been edged out of the room.

"Too many people in there," he said and pointed at the crowd spilling out of the narrow doorway and into the hall. Hudson had never seen most of these people before, but she could tell from their suits and dead-serious expressions that they were from Holla's label. Hudson pushed her way inside.

At the far end of the room, Holla sat in a director's chair in front of a mirror framed by tiny white bulbs, her head thrown back and her eyes closed, letting Suzette and Gino do their work. Suzette applied fake eyelashes while Gino ironed Holla's hair, which hung straight down on either side of her head, perfectly smooth. Her bangs — newly cut — just skimmed her thickly lined eyes. Her strapless, bandage-style dress in the same purple-gunmetal metallic sheen as Hudson's glinted under the lights.

Brendan, the music producer she'd met at the rehearsal, approached Hudson through the crowd. He was dressed in jeans and beat-up sneakers and held a folder under his arm. He was cute, with short, rumpled-looking black hair, but Hudson tried not to notice. "So we'll be coming to get you around ten after twelve," he said, "but before that someone'll come by to mic you and your mom."

"Great," she said, forcing a smile.

"I know you didn't want to rehearse, to keep everything under wraps," he said, "but we just have to know in advance . . . Are you all set? It *is* live television."

"I'm fine," she said, trying to sound like she went on *Saturday Night Live* once a month.

Brendan looked at Hudson just a bit longer than necessary, as if he were trying to figure this out himself. "Okay, I'll see you out there," he said, checking his watch. "Have a great show."

The TV looming above them in the corner of the room flickered, and a blue screen came on.

"What's that?" Hudson asked.

"It's the live feed from the studio," Paula explained. "As soon as the show starts, you'll be able to watch it in here." Paula approached Holla. "Just want to make sure you're okay with wardrobe?"

Holla held up her hand. "It's all good. But let me see you, Hudson," she ordered. "Come over here."

Hudson dutifully pushed her way up to her mother's chair.

Suzette and Gino moved aside and Holla beamed at Hudson in the mirror. "Adorable," she said, holding Hudson around the waist. "Just *adorable*. Look at us. This is *brilliant*." They practically looked like twins in their matching silvery purple dresses.

Kierce walked up to them and gave Hudson a sweeping head-to-toe squint. "It's perfect!" he decided.

"Hey, the show is starting!" somebody yelled. Hudson glanced up at the TV. A hush fell over the crowd as the screen went black, and the opening sketch began.

"Honey." Holla grabbed Hudson's wrist and pulled her in closer. Hudson almost collided with Suzette's mascara wand. "You *are* ready for this, right?" she asked.

The crowd in the room laughed uproariously at the TV.

"Do you remember the dance moves?" Holla asked, sounding impatient. "Shimmy, then double-turn, then head thrown back?"

"Yeah," Hudson said. "Of course." At least she thought she did. It would be hard to forget something so embarrassing.

"Good." Holla narrowed her eyes. "If you feel yourself getting anxious, just remember what I always say: *No negativity*."

"Right," Hudson said.

Holla released her grip, and Hudson stepped aside to let Suzette apply Holla's mascara. As she tried to make her way to the sofa, she repeated those words back to herself: *No negativity*. They did nothing to quell the feeling of doom that was starting to overtake her.

"Okay, people!" Sophie cried. "Holla would like everyone who doesn't need to be here to please move into the greenroom! Now!"

Reluctantly the men in suits started to head to the doors. Hudson sat on the leather sofa, staring at the yellow and red diamond-patterned carpet. Did she need to be here? This was a mistake. Maybe an even bigger one than the Silver Snowflake Ball.

"Good luck, Hudson," someone said to her, patting her on the shoulder. "We'll be watching."

She looked up. It was Richard Wu, her record label executive.

"We're so excited you're back. And we've got a great tour lined up for you already," he said with a smile. "If tonight goes well, we'll even get your mom to come out with you onstage a few times." He squeezed her shoulder. "Now just go out there and have fun."

He walked out the door before she could say anything. *My mom?* she wanted to yell. *On my tour? Come out onstage with me?*

And then she realized: Her mom didn't just want Hudson to be a mini-Holla, the perfect daughter following in her mom's footsteps. She wanted a way to keep reaching younger fans.

She had to get out of this. If she walked out onto that stage and did this show, it would kick off a chain of events that she would never be able to stop.

Minutes later, a knock on the door made her jump, and then a bearded PA walked in, holding a couple of mics and sound packs.

"I just need to get these on the two of you before we go out there," he said. He walked over to Hudson first. "This'll only take a second."

Hudson stood still as the PA looped the mic around the back of her dress. She hoped he couldn't hear her pounding heart.

"Okay, now you," he said to Holla.

Holla took her time getting out of her chair, never taking her eyes off her reflection as she fixed her hair. "Okay, fine, go to it," she said, and the PA clipped the mic to her dress.

Brendan walked into the room. "Okay, how're we doing? We all ready to go?"

"Just a second," Holla said, turning around in front of the mirror to check for any unsightly bulges. "I'm ready," she said.

"Mom?" Hudson suddenly asked. "Are you planning on doing more duets with me? You know, if I go on tour?"

Holla frowned. "What are you talking about, baby?" she asked, looking back in the mirror.

"Just what I said," Hudson said evenly. "Are we going to be doing this again? Is this something you talked about with my label?"

"Let's talk about this another time, shall we?" Holla said sternly.

"No," Hudson said, blocking her way. "I need you to tell me now."

Holla folded her arms. She looked more glamorous than Hudson had ever seen her, with her Cleopatra bangs and dramatically dark eyes. She took a deep breath, as if she were trying not to lose her temper. "We don't have time for one of your moods right now, honey. We have a show to do." She pushed past Hudson and began walking toward the door.

"Just follow me," Brendan said, trying to still sound excited.

Hudson followed them, quietly furious at being cut off. Of course it hadn't been great timing, but her mom could have answered her question. Brendan led them out of the dressing room and into another hallway, which seemed to be the entrance to the stage. Hudson spotted two or three *SNL* cast members, in costume, waiting to go on.

"Okay, we're going to have you wait out here," Brendan said, leading them up to another door.

Beyond it Hudson could hear the band playing. The show had to be at a commercial break. In just a few moments they would be back, and it would be time for her to walk out onstage.

She tried to picture what it might be like in there. The cameras moving around silently like ghosts on the studio floor. The people in the audience. The stage manager motioning for them to wrap it up, that time was running out...

She started to breathe fast. Her vision got darker. It was as if someone were pulling a blindfold over her eyes.

And then a small voice rose up inside of her: *This isn't right, Hudson. Don't fight it anymore.*

"Okay, here we go," Brendan said, grabbing the door handle and pulling the door open.

Hudson grabbed her mom's arm. "No. I can't."

Brendan and Holla turned around. "Hudson, come on," Holla said, trying to smile.

"I can't do this," Hudson said. "I can't be you. This isn't me. It never will be. As much as you want it to be. I just can't do it."

"Hudson," her mom warned, stealing a look at Brendan.

"I'm not quitting," she interrupted. "I'm saying no. I totally respect what you do, Mom. But this isn't what *I* do."

Brendan pulled the door open. Thunderous, earsplitting applause poured in. "I'm sorry, we've got to go," he yelled. "Now!"

Holla didn't move. Something rippled across her face—a

moment of understanding, of acceptance. Or maybe it was just that she couldn't argue anymore. She touched Hudson's cheek. "She's not coming," she said over her shoulder to Brendan. "Let them know out there."

Brendan pulled up the mic on his headset. "It's just Holla. The daughter isn't coming. *It's just Holla.* You got that?" Brendan pressed on his earpiece, then nodded, satisfied. "They're ready for you," he said to Holla. "Let's go." Then Holla turned and followed Brendan into the studio, and the doors swung shut behind them.

"Her tenth album comes out this Tuesday," she heard the guest host announce, "and she's here for the fourth time. Ladies and gentlemen... Miss Holla JONES!"

As the studio erupted in applause, Hudson looked up at the monitor hanging in the corner. Her mom stood on the stage, shimmering in the sparkling purple dress. It had only been a few seconds, and already she owned the room. She pulled the mic out of the stand and executed a perfect turn as the song started. On her first note, goose bumps rose along Hudson's arm. Hearing Holla sing this song wasn't weird anymore. Her mom was a star. She could sing anything. This was what she'd been born to do. She would always be a star first and a mother second. And maybe, Hudson realized, that was how things were meant to be.

chapter 29

"I need to ask you something," Hudson said carefully. "And I just want to say, in advance, that if I offend you or something, I'm really sorry."

Across from her, Jenny frowned slightly and rested her chin on her wrist. "Okay. Go ahead. Offend me."

Sitting across from Jenny at her wooden kitchen table, Hudson thought her aunt looked just as beautiful as ever. Her eyes were a little puffy from sleep, but her cropped hair had been highlighted with warm caramel streaks and her lips shone with clear gloss. When she'd called her that morning, Hudson hadn't expected an invitation for homemade crêpes suzette and tea. But Aunt Jenny had been incredibly gracious under the circumstances.

"First, I want to say I'm so sorry about the party," Hudson said. "I should have just told my mom in the first place that I forgot. I don't know why I didn't. I'm so sorry."

Jenny nodded, and then spooned some powdered sugar onto her

crêpe. "Obviously, it was worse for your mom," she said. "I can't believe you did that to her. And I can't believe I missed all those macaroons."

Hudson didn't say anything.

"But at least your mom and I started e-mailing again because of it," Jenny admitted. "And we're having lunch next week."

"Really?" Hudson asked, impressed. "You are?"

"Not at your house; at a restaurant," Jenny said, holding up one hand as she cut into her crêpe with the other. "Neutral territory. Of course we'll probably have to close down the restaurant so she doesn't get mobbed," she added.

Hudson smiled and then took a small bite of her crêpe. "Oh my God," she said. "This is incredible. You're a really good cook, you know that?"

"Thanks. But don't tell your mom I fed you white flour," Jenny said in a mock whisper. "So, what did you want to ask me?"

"When you decided not to audition for Martha Graham the second time," Hudson said carefully, "it was because you didn't want to compete with my mom, right?"

"*That's* what you want to know?" she asked.

Hudson nodded.

"Ye-es," Jenny said. "But I also don't think I wanted it bad enough. I didn't want that life."

"But don't you regret it?" Hudson asked. "Don't you wish you'd at least tried?"

Jenny reached across the table and took Hudson's hand. "Is this about what happened last night?"

"I told you how my mom changed my album because she said it wouldn't sell?"

Jenny frowned again and nodded.

"And at first I didn't care if it sold or not. I just wanted it to be my thing. *My* vision. But to my mom, it's like there's no point in even trying if you're not going to be huge."

Jenny nodded. "Right."

"And sometimes I think there's a part of me that believes that. I joined this band, up in Westchester, which you probably realize," Hudson said, embarrassed.

"Yeah, I got that," Jenny said knowingly.

"And I was finally doing my own thing again. But my mom found out and she's hurt. She thinks I'm crazy for wanting to be in some high school band and play these tiny clubs. She doesn't understand why I don't want what she has. We're all supposed to want that, right?"

"Oh, Hudson," Jenny said, shaking her head as she looked down at her plate. "I wish I could have been there for you a little more. I really do. It's my fault I wasn't." She leaned so close that Hudson could smell her fig-scented perfume. "Your mom is an amazing person. She's accomplished a lot. But you know how you're scared of being in the spotlight? She's scared of being *out* of it. She's been doing this since she was ten. It's all she knows. And sometimes having thousands of people love you from a distance is easier than living in the real world, where people can reject you and leave you and *see* you. And I don't think your mom knows how to be seen. As a real person. I think that scares her. More than anything else in the world."

Hudson bit her lip. It hurt to hear these things about her mom, but she knew that they were true.

"So my question is, do you really want to be like that?" Jenny asked. "Does anyone?"

Hudson shook her head.

"You don't have to be like your mom," Jenny said. "Not even if you want to do what she does. That night you got stage fright? That was your inner self, telling you that what you were doing didn't feel right. And so you ran off that stage. That was the bravest thing you could have done."

It had never occurred to Hudson that running off the stage at the Silver Snowflake Ball had been brave.

"We're living in a time where we're all told we're nothing if we're not famous," Jenny said. "Sometimes it's easy to forget how crazy that is."

"But you're the one who told me that it's in my chart, being famous," Hudson said. "You're always telling me that."

"I should have just said successful," Jenny said. "When you were little, you used to love to hold your mom's awards and sing her songs. I thought that's what you wanted. But there are many kinds of success. You can play music and put all your passion into it, but it doesn't have to be your whole life. There's a middle road out there, Hudson. Your mom had no idea what she was getting into, and now she's stuck. She doesn't have a choice. But you know what that life is like. *You* have a choice."

Hudson looked at the glass vase of early spring daffodils on Jenny's kitchen table. *A middle road.* She had never thought of it that way.

"Sometimes I wish I could talk to my dad about this stuff," Hudson said.

Jenny nodded. "I know. But you can always come see me if you need some reminding."

"And you'll have to let me know how lunch with my mom is."

"Well, one thing's for sure," Jenny said, taking another bite of her crêpe. "It's going to be very, very healthy."

Hudson smiled and picked up her fork. A middle road. She liked how that sounded.

chapter 30

"Well, you *almost* did *Saturday Night Live,*" Carina said later that afternoon as she dug her spoon into her pomegranate Pinkberry. "And how many people can say that? Seriously?"

"Thanks for looking on the bright side, C," Hudson said. She took a small bite of her yogurt with blueberries, lost in thought.

"*Go you,* that's what I think," Lizzie said, wagging a spoon of plain topped with mochi. "You totally listened to yourself, you spoke up to your mom, you realized that it wasn't right for you in the end —"

"And you may have finally cured her of her mini-me obsession," Carina added. "How is she now?"

"She's actually great," Hudson said. "This morning she seemed totally normal. I couldn't believe it."

She'd been expecting a stony glare, or at least a lecture on sleeping in so late when she walked into the kitchen that morning.

But her mom just gave her a warm smile and started talking about the awesome after-party at the Standard Hotel.

"Everyone *loved* the song. It's probably going to be a big hit. Shows how much I know."

Hudson looked out the window. Couples strolled up Bleecker Street in the sun, their coats open to the mild late-winter day. Everyone looked so happy that spring was almost here. "I know I did the right thing," she added. "I just wish the band wasn't over. First the album was over, now the band is over—"

"Wait," Carina interrupted, leaning forward. "What about the first album? The one you loved? What happened to that?"

Hudson shrugged. "I don't know. It still needs to be mastered, but the tracks are all done."

"Then why doesn't your label just release that one?" Carina asked, shaking her blond ponytail. "Just go in there and tell them that's the album you want to release. You could go out there and promote that one, right?"

"Exactly!" Lizzie exclaimed, hopping up and down in her chair. "Go in there and tell them that's your true sound and always has been!"

"And you'd be totally fulfilling your contract," Carina pointed out.

Hudson tapped her foot under the table to the beat of the music playing over the store's speakers. Why had she never thought of this idea before? For just a moment she pictured Ben onstage with her, but blocked it out. "I will," she said. "That's a great idea."

"I think that's the best idea I've ever had," Carina said proudly, polishing off her yogurt.

"But what about my mom?" Hudson asked.

"I think by now you probably have her blessing," Lizzie said gently, licking her full lips. "And there's no way you're ever going to do this totally on your own. Your mom will always be a part of it. You're just going to have to accept that."

Hudson nodded; she knew her friends were right. Maybe she'd been a little unrealistic this whole time: There was no way she could ever stop being Holla Jones's kid.

When they walked out of Pinkberry and onto Bleecker Street, Hudson heard her phone chime with a text. She pulled it out of her bag. It was from Hillary Crumple.

Hey can u meet me at Kirna Zabete? Need to talk to you.

"Who is that?" Lizzie asked.

"Hillary," said Hudson. "She wants to go shopping again."

"Personally, I don't think that girl needs any more clothes," Carina put in.

"I think she's still mad at me about the stuff with Ben," Hudson said, texting to Hillary that, yes, she'd meet her. "But maybe you guys are right. Maybe I shouldn't have gotten so angry at him."

"Guys," Carina said ruefully, shaking her head. "I will never, ever, understand them."

"Except Alex, of course," Lizzie said.

"No, even him. He wants to dye his hair blue. Can you believe it? If he does it, I'm gonna kill him."

"They set the court date for Todd's dad," Lizzie said quietly. "Todd's really upset about it. He didn't want his dad to have to have a trial. He wishes he would just fess up and go to jail. He thinks that's almost less humiliating."

"But you guys are still okay, right?" Carina asked.

"Yeah, he's not acting weird around you or anything, right?" Hudson asked.

"We're great," Lizzie assured them. "I just feel bad for him, that's all."

"Don't feel *too* bad for him," Carina said. "Guys don't like it when you feel sorry for them."

As Hudson listened to her friends talk about guys, she couldn't help but feel a little left out. It wasn't that she really wanted a boyfriend; right now her life felt full enough without one. But *not* having a boyfriend made her feel a little behind. She was technically the oldest of the three of them, but now she felt like the baby of the group. Both her friends were going out with guys and having experiences that she just couldn't relate to. Sometimes she found herself wondering if she ever would.

"Well, you guys, I gotta go meet my dad at his office," Carina said. "He wants me to look at this proposal for a new networking site or something."

"So you're giving the Metronome thing another try?" Lizzie asked.

Carina shrugged. "He begged," she said, grinning. "What was I supposed to do?"

Hudson smiled. She knew that things had changed between Carina and the Jurg — so much so that now when he asked her to do him a work-related favor, she actually did it.

“So I think I’m gonna go meet Hillary,” Hudson announced. “Thanks for letting me vent, you guys.”

“Congratulations, Hudson,” Lizzie said, tucking a red curl behind her ear. “I mean it. Even though I’m not your official life coach, you should know what a big deal last night was.”

“Thanks, Lizbutt. I know it was.”

“Be proud of yourself for that,” Lizzie said.

Carina waved good-bye to Hudson, and then she and Lizzie started walking up Sullivan Street toward Washington Square Park. Hudson tilted her face up to the sun and soaked in the feeble winter rays. Her old album was still out there. It hadn’t disappeared or gone away. All this time she’d thought of it as gone forever, when it was still intact, and waiting for her to return to it. It didn’t matter anymore what her mom thought of it. And maybe she could take the middle road to it, just like Jenny had said.

She wrapped her chunky knit scarf closer around her, and started walking to SoHo.

chapter 31

Hillary stood outside Kirna Zabete, tapping the toe of her tiger-striped ballet flats as Hudson walked up Greene Street. Hillary looked incredible—maybe too incredible. She'd traded in her puffy down coat for a belted swing coat with a fake-fur collar, and she'd pulled her hair back in a chic ballerina knot. Her bag looked like a knockoff of the Lizzie bag by Martin Meloy—bright white leather and gleaming silver buckles. And in her hands was a cluster of shopping bags.

"Hey," Hudson said, walking up to her. "Do you want to just get some coffee? I'm not really in the mood to shop."

Hillary shrugged and they started walking back toward Prince Street. Hudson didn't say anything; she was still a little scarred by Hillary's tongue-lashing the other day, and she didn't want to have another fight in her favorite store.

"So I noticed you weren't on the show last night," Hillary said as she maneuvered herself, shoulders first, past the tourists. She still walked as if she wore that gigantic backpack. "What happened? Did they cut you out at the last minute?"

"I decided not to do it," Hudson said, ignoring Hillary's slightly cruel remark. "It didn't feel right to me."

Hillary's shopping bags smacked against a lamppost. "Well, I think you seriously messed up with something else," she said. "I spoke to Ben, you know."

"Of course you did."

"And guess what? He *didn't* call Joe's Pub and make that deal." She shouldered her way past a dog walker. "But he found out who did."

"Who?" Hudson asked. But before Hillary spoke, she already knew who it was.

"Logan," Hillary said softly.

They stopped at the corner of Broadway. The signal read WALK but Hudson just stayed on the curb. "How did Ben find out?" she asked.

Hillary waited with her at the corner. "I guess Ben told Gordie and Logan who you were."

"Even though I told him not to."

Hillary sighed as if she wished Hudson wouldn't interrupt. "They promised to keep it to themselves. But Logan made some comment about Ben being lame for not trying to use your mom to get some gigs. And after you hung up on him, Ben found out that Logan had called Joe's Pub and promised them your mom. So Ben kicked him out of the band."

"He kicked him out?"

"Yep," Hillary said. "And they've been friends since, like, kindergarten."

Hudson winced. "What about the band now?" she asked cautiously. "Is it over?" Without a pianist and a saxophonist, how could they still have a jazz band?

"I don't know," Hillary said. "His parents are kind of happy it's over, I think. Come on. Let's cross the street."

Hudson followed Hillary across Broadway as the words *it's over* knocked around inside her head. It was bad enough that she'd kept Ben from playing at Joe's Pub. But now it seemed that she was responsible for the total demise of the band itself. Not to mention Ben's dream of being a jazz musician.

"So, the other thing I guess I need to say to you," Hillary said, turning around to face her, "is that you were right. As much as I hate to admit it."

"Right about what?" Hudson asked. So far it hadn't felt like she'd been right about much.

"About Logan being kind of a jerk." Hillary looked at Hudson and there was a flicker of sadness in her yellow-green eyes. "He hooked up with Ellie and then he hooked up with one of the McFadden twins, too." She wrinkled her nose with distaste. "But he really did give me his number back in January. Just so you know."

"Oh, Hillary," Hudson said, and without thinking, put her arms around her. "I'm so sorry." She squeezed Hillary's tiny frame. Eventually Hillary let her shopping bags drop to the ground and hugged Hudson back.

After a few moments Hillary pulled away and wiped her eyes with the back of her hand. "Whatever. It's not that important."

"Is he the reason you changed your look?" Hudson asked gently.

Hillary looked down at the sidewalk and nodded. "Why? Do you think that's lame?" she asked.

"Not at all," Hudson said. "I just think that I liked the old Hillary Crumple better."

Hillary looked up. "You *did*?" she asked.

"Yeah. Maybe she wasn't the trendiest girl on the face of the earth, but she was a true original. And this Hillary..." Hudson gestured toward Hillary's clothes. "Well, she looks nice and everything, but she's definitely not original."

A tentative smile spread across Hillary's face. "Yeah, I guess it's not really me," she said. "And God knows, it's expensive."

"Is Ben's family going away for spring break?" Hudson asked suddenly, changing the subject.

"Just for a couple of days," Hillary answered. "They go down to see his grandparents in Florida."

"When are they leaving?" Hudson asked.

"I think their flight is tonight."

Hudson checked her watch. It was almost two. "Hold on one second," she said, taking out her phone. She dialed Ben's number. It rang and rang and rang.

"Are you calling him?" Hillary asked.

"Yeah," Hudson said.

"Oh, he's not gonna answer," she said. "It's the Westchester chess championships today."

"Where are they?" Hudson pleaded.

"At the high school in White Plains," said Hillary.

"How long will they last?" Hudson asked.

“What are you gonna do? Just barge in there while he’s playing chess?” Hillary asked in reply.

“Yep,” Hudson said. She glanced back at the N and R subway station on the corner.

“So, no shopping?” Hillary asked with a wry smile. Then she laughed and said, “Just kidding.”

Hudson leaned down and hugged Hillary. “Have a great spring break. And thank you. For everything.”

Hillary hugged her back. “Good luck up there. Say hi to the nerds for me.”

Hillary pulled out of the hug, picked up her shopping bags, and almost knocked down a small child as she headed off down Broadway.

chapter 32

Hudson stared out the smudged train window at the parade of school buildings and churches and budding trees on the way to White Plains. *Stupid, stupid, stupid,* she thought. Of course Ben wasn't the one who betrayed her. Of course it had been Logan, who'd had it in for her from the beginning. And she'd hung up on Ben, on top of it. She flinched just thinking about it.

She owed Ben Geyer everything. He was the reason she was going to go back to her old album. He was the reason she'd finally discovered music on her own terms. And in return, she'd accused him of something terrible and hung up on him. Oh, and kept him from realizing his one dream in life: playing at Joe's Pub. *Great,* she thought. *Way to really screw up.*

By the time the train lurched to a stop at the White Plains station, Hudson was on her feet. She stepped onto the platform and headed toward the line of idling black cabs.

"The high school, please," she said as she opened the back door and threw herself into the seat.

White Plains was more of a city than Larchmont was. The car took her through downtown and then veered off into a more residential neighborhood. It finally pulled up in front of a large, squat brick building that looked shuttered and empty. "Can you wait here for a sec?" she asked, pressing a ten-dollar bill into the driver's hand. Then she took off at a run.

She threw open the main door of the school and ran down the empty hall. It smelled strongly of lemon soap, and her sneakers squeaked on the shiny linoleum. Being in school on a weekend always felt strange. She passed classroom after classroom until she came to the end of the hall and a pair of double doors. With all her might, she pulled one open.

She was in the school cafeteria. At least ten pairs of faces looked up from their chessboards. First Hudson saw Ellie, her hand poised over a pawn. And there, at the next lunch table, his brow knit in concentration, was Ben.

He looked up at her. "Hudson?" he said slowly. "What are you doing here?"

"Excuse me!" yelled the proctor. A tall, extremely thin older man with a bow tie and round, delicate glasses stood up from another table. "We're in the middle of a tournament here. Please wait outside—"

"I need to talk to Ben Geyer," Hudson managed to say, panting. "Please. Just for a second."

"You may absolutely *not,*" said Mr. Bow Tie.

Hudson ignored him and ran down the aisle of tables to get to Ben.

Now Ben and Ellie were both staring at her openmouthed.

"Hillary just told me the truth," she blurted out. "She told me that Logan was the one who called Joe's Pub and promised them my mom. I'm so sorry, Ben. I never meant to hang up on you. And I never meant to screw up the Joe's Pub thing. I'm so sorry."

Ben shook his head. "O — okay," he stammered.

"So I'm here to ask you to take me back," she said. "Or, actually, not that. I want you to be in my band now. I'm going to put out my first album. And I want you to play shows with me. Would you be my bassist? Please?"

"Excuse me, young lady, but you're going to have to leave!" the proctor yelled, walking up to her.

Ben looked at her and then at the proctor. He seemed utterly at a loss for something to say.

"But if you don't want to be in my band, I totally understand," she rambled. "If you don't want to be my friend anymore, I get that, too."

"Just so you know, Hudson, I don't do stuff like that," he finally said. "You're my friend. I don't betray my friends. Ever."

"Okay." She exhaled.

"And as far as being your bassist," he said, breaking into a smile, "I'd love to."

"Really?" she asked, her heart pounding. "And wait a second. I can't believe I never asked you this before, but what's your sign?"

His eyebrows shot up. "Virgo," he said. "Why?"

"I'm a Pisces, which means we're, like, a perfect match!" She looked over to see the proctor's face turning purple with rage. "Sorry. I guess I should go."

"Yeah, good idea," Ben said, glancing at the proctor. "We have a few more minutes, and I'm totally about to win this."

Awkwardly, she leaned down to hug him, and then stuck out her hand. He shook it. "I can't say we're still going to be called the Rising Signs," she said.

"The Hudson Jones Trio is fine with me," he said.

"Hudson Jones Trio," she said, mulling it over. "I like that."

She walked out of the cafeteria and back into the hallway, ignoring the glare from Mr. Bow Tie. So Ben didn't hate her, after all. And even if the Rising Signs were over, she had something even better in place now. As she pushed the school doors open, she realized that she even had a possible name for her album.

Hudson Jones: The Return.

chapter 33

"Do you want to do the talking or should I?" Holla asked as Fernald navigated the late-afternoon traffic on Fifty-seventh Street. April rain pattered against the windshield.

"I'll do it," Hudson said.

"You start, and I'll come in at the end," Holla suggested.

"Mom, let me be in charge of this, okay?" Hudson replied.

Holla put her hand on Hudson's back. "Fine, but sit up straight."

Fernald double-parked in front of the smoke-colored skyscraper that held the offices of Swerve Records and leaped out of the driver's seat with an umbrella.

Upstairs, the receptionist picked up the phone. "Richard, they're here," she said quietly and hung up. "You can go right in," she said. "Last office on the right."

Hudson and Holla walked down the hall as Little Jimmy walked behind them. At each open door they passed Hudson could see people at their desks, craning their necks to get a good look at them. Or rather at Holla, who once again had the top single on the *Billboard* charts.

At the end of the hall they made a sharp right, into the corner office. Several executives sat on the long gray couch. Richard Wu stood up from his chair and walked over to greet them. He took Hudson's hand first. "Hi, Hudson," he said.

"Hi, Richard," she answered.

"Hello, Richard," Holla said, bringing him in closer for a kiss on the cheek.

"Congratulations on the single. To both of you," he made sure to add.

"Thank you," Holla said. "She's not bad, huh?" she asked, circling an arm around Hudson's shoulders.

Richard introduced them to the other Swerve executives, including the man whose office they were in. Hudson shook everyone's hand, and then she and her mom made themselves comfortable on the couch.

"So, I have to admit, Hudson," Richard said, "I was surprised to hear from you. We all thought after *Saturday Night Live* that you'd decided not to do this anymore."

"I know," she said, looking each of the four men in the eye. "I'm sorry about that. But I've changed my mind. I want to put out my album. My *first* album."

"But I thought you and your mother agreed that you wanted to go in a different direction," Richard said smoothly.

Hudson looked at Holla, who gave her a quick nod of encouragement. "I've changed my mind," she said. "And so has my mom. We think the first album is the better one."

The record executives traded surprised glances. "I don't know," said one of them, a man with a shaved head and a fat gold ring. "We spent a lot of money redoing it to your exact specifications."

"To *my* specifications," Holla said in a throaty voice. "It was my idea to change it. But," she said, gazing at her manicured hands, "I've since realized I was wrong."

The phrase *I was wrong* echoed through the office. Two of the executives looked at each other, as if to say, *Did I just hear that?*

"Whatever it costs to finish the first album," Holla said, "I'll pay for it myself."

Hudson watched as there were even more surprised glances around the room.

"So you're sure you want to do this, Hudson?" Richard prompted. "You want to put out a smaller, more intimate album that's going to be marketed in a very different way? Not on the scale of anything we would do with your mother — not even close."

Hudson nodded. "That's what I want," she said. "And that's the one you guys signed me to do in the first place."

"I know," Richard said, "but we believe there's potential for you to be a much bigger star."

Holla held up her hand. "This is what my daughter wants. And may I remind everyone that she wrote the number one song in the country?"

The executives traded more glances.

"I want it to be real music, real musicians, all of us recorded together," Hudson said. "And no stadiums, no big venues. I want to do small, intimate shows and work my way up. No TV appearances — only live gigs. I want to sing my songs the way I want to sing them."

Richard cleared his throat. "How soon can you go back in the studio?"

"June," Hudson said. "So I can finish school first."

Richard nodded. "I think that sounds good to us. We would plan on a Christmas release if that's the case."

"Great," Hudson said. "I've already written a few more songs, too. But I just have to ask you guys for one thing. One thing I'd like you to promise me."

"Sure," Richard said, steepling his hands under his chin.

Hudson took a deep breath and looked at Holla. "Mom? Can I speak to them privately for a sec?"

Holla eased out of her chair. "Sure, honey," she said, smiling encouragingly. "I'll be just outside."

Hudson waited until her mom was gone. As soon as the door shut, there was a palpable relief in the room. One of the guys even loosened his tie. The executives looked at Hudson expectantly.

"What is it, Hudson?" asked Richard.

"From now on, I want you to think of me as Hudson Jones, not Holla. Not even her daughter. Just me. Just another one of your artists. And if I do well, great. If I don't, you can drop me."

The men looked at one another across the room yet again. Richard fiddled with his watch band and swallowed hard, as if he were suddenly embarrassed by something. "Sure, Hudson," Richard said. "I think we can do that."

"And there's just one more thing."

Richard sat back down in his chair. "We're listening," he said.

"Is there any way we can set up a gig at Joe's Pub? Something this summer, maybe?"

Richard looked at the other executives. "I'll see what we can do."

When Hudson walked out of the office, Holla was sitting in the lobby while Little Jimmy stood guard, reading a magazine while people walking in and out of the elevator openly gawked at her. It was the first time Hudson could remember her mom waiting for her in a public place. "How'd it go?" she asked, putting down the magazine. "Did you say everything you needed to?"

"Yup," Hudson said.

"Did they listen?"

"Yes."

Holla smiled. "I'm proud of you, honey." She brushed some hair off Hudson's forehead. "Honey, do you ever wear any eyeliner? A little purple liner would really bring out your green eyes —"

"Mom?" Hudson said, taking her hand.

Holla pressed her lips together and smiled. "Sorry. So how do you feel about ditching the car and just taking a walk in the rain?"

"Are you serious?" Hudson asked in disbelief. "We can't do that."

"I've got dark glasses, an umbrella, and *him,*" Holla said, pointing her thumb at Little Jimmy. "I think I can walk down the street."

Hudson pushed the elevator button. "That sounds great, Mom."

chapter 34

Hudson leaned close to the mirror and applied one last coat of mascara to her lashes. She stood back and fluttered her eyes open and closed. She wasn't wearing purple liner, but her sea green eyes shone, anyway. Her mom would have been proud.

She ran her hands through her wavy, barely styled hair. The black silk halter dress still fit perfectly. Wearing it again could have been a bad idea, but she was glad she'd chosen it. No matter what happened, tonight was going to be a better experience than the Silver Snowflake Ball. She knew that now.

She zipped up her makeup bag and headed to the bathroom door. A sign beside the door read JOE'S PUB—CALENDAR OF EVENTS. There, under today's date, June tenth, was THE HUDSON JONES TRIO, 8:00 P.M. She had to read it a few times to really absorb it: *The Hudson Jones Trio.* It had a nice ring to it. And she couldn't have asked for a better way to celebrate the last day of ninth grade.

She stepped out of the bathroom and back into the dressing room. Ben and Ricardo, the drummer, sat together over a miniature chessboard. "Checkmate," Ben said as he knocked Ricardo's queen off its square. "Sorry about that."

"Are you crushing my drummer?" Hudson asked him.

"Hey, I'm not being crushed," Ricardo said. "And those just came for you." He pointed to the extravagant arrangement of red roses in a glass vase on a dresser.

Hudson walked over to the flowers. The card was in a tiny envelope on the dresser. Hudson opened it and read:

To Hudson,
Good luck tonight. We miss you on tour. London isn't the same without you.
Love, Mom

Hudson folded up the card and slipped it into her purse. She never would have expected it, but she actually missed her mom. Her first gig at Joe's Pub wasn't going to be the same without Holla there.

There was a knock on the dressing room door, and then a voice asked, "Is it weird if I'm back here?"

Hudson turned around and saw Hillary standing on the threshold. She wore her dark-rinse jeans, but her pink shell and sweater set had embroidered hearts all over the front of it, and her hair was back to its usual messy ponytail, slightly tamed by plastic barrettes.

"I just wanted to wish you luck," she said in her small, rapid-fire voice. “Or tell you to break a leg and fall down some stairs. Whatever.”

“Thanks, Hil,” Hudson said. “And I have to say, you look great.”

“Really?” Hillary said, looking down at her outfit. “It definitely took a lot less time to get ready. You were right about those clothes, by the way. They weren’t me. And they weren’t gonna make some guy like me, either.”

“A guy not worth your time,” Hudson added. “I hope you remember that part.”

Hillary rolled her eyes. “Yeah, well, everyone’s entitled to a bad crush, right?”

“Hey, nerd,” Ben called out. He got up and checked his chunky black watch. “Nice of you to drop by, but I think we need to start.”

“Don’t embarrass the family,” Hillary said to him, punching him in the arm. “See you out there.” Then she left and walked back out to the restaurant.

Hudson turned to look in the mirror one last time. “So how do I look?”

“Really pretty,” Ben said shyly.

“Thanks,” Hudson said, turning to look at Ben. “I’m ready if you guys are.”

“Let’s go,” said Ricardo.

As she and Ben and Ricardo walked down the hall toward the stage, she could feel the old butterflies start to flit around her stomach. But she let them do their thing. She knew that they couldn’t hurt her.

“Can you believe we’re here?” Ben whispered in her ear.

Hudson shook her head. "No. Not one bit."

"So, ladies and gentlemen, without any further ado," said the announcer, "we introduce the Hudson Jones Trio!"

The three of them walked into the room and up onto the stage. There, at the tables nearest them, so close she could practically touch them, were all of her friends: Lizzie and Katia and Bernard; Carina, the Jurg, and Alex; Ellie Kim and her mom; Mrs. Geyer and Hillary. Everyone clapped and someone hollered, "Hudson!"

She made her way to the piano, praying she wouldn't trip. When she looked out at the room, past the stage, she could see people at the two-top tables. At one of them was Richard Wu, sitting with a colleague. He gave her a little wave, and Hudson smiled at him. This was nothing like the scary darkness of the Pierre Hotel's ballroom. She had friends here, people who were rooting for her. And as for the people she didn't know, they were probably here for the two acts that followed her band. In which case Hudson wasn't going to worry about what they thought.

"Hi, everyone," she said into the mic. "Thank you so much for coming. We're the Hudson Jones Trio, and we're going to start off with a song I wrote called 'For You.'"

She felt her heart flutter and the adrenaline shoot through her arms. But tonight she knew that she was going to be okay. She looked over at Ben. He smiled and nodded to say he was ready when she was.

She smiled back, took a deep breath, and started to sing.

acknowledgments

First off, I would like to thank my agent, Becka Oliver, for her unswerving support and encouragement. She is the best advocate, and friend, a writer could ask for.

I would also like to thank the following people: my wonderful editors at Poppy, Elizabeth Bewley and Kate Sullivan, who gave me valuable advice, suggestions, and feedback; Cindy Eagan, whose enthusiasm for this series makes me so happy; Matt Piedmont and Ethan Goldman, for schooling me in the basics of jazz; Janet Siroto, for her knowledge of Larchmont; JoAnna Kremer, for her impeccable copyediting; and Tracy Shaw, for her eye-catching cover design for this series. I am also indebted to John Lahr and his excellent article on stage fright, "Petrified," in *The New Yorker* (August 28, 2006.)

And, of course, endless thanks go to Edie, my family, and to Ido Ostrowsky, the coolest Gemini I know.

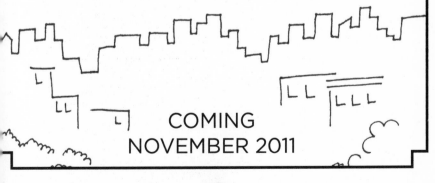

supply that can provide up to 30 kV, buffer reservoirs, electrodes, separation capillary, capillary temperature controller, UV detector, capillary-filling apparatus, and microprocessor control. Some instruments can also perform voltage programming, fraction collection, have alternative detectors such as fluorescence, and have cooled autosamplers. Data systems that are specifically designed for HPCE are beginning to appear in the field.

Automation is particularly important in HPCE. The best analytical precision is found when experiments are performed in a highly repeatable fashion. Table 1.3 lists the fully automated instruments along with their basic feature sets that are available in 1992. Manual and modular instruments are not included in this table.

Selection of the appropriate instrument need not be difficult.

Ask the following questions before placing an order:

1. Does the instrument feature set meet my present and future application needs?
2. Does the instrument company provide active support in my applied area of interest?
3. Can the instrument be upgraded as new technologies develop?

It is important during methods development to be able to vary conditions during a run. For example, can the voltage be increased, can pressure or vacuum be applied, can the detector wavelength be changed? An important application in the development phase is to determine if all components have passed through the capillary. This is simply accomplished by flushing the capillary with buffer and monitoring the detector output. Not all instruments have this capability.

Table 1.3. Instrumentation for HPCE

Instrument	Capillary	Detector UV/F	Sampler Positions	Cooling Capillary/ Sampler	Injection
Beckman P/ACE 2000	cartridge	filter/LIF	24	liquid[a]/none	P, E
Dionex CES 1	open	variable/variable	39	air[b]/none	P, E, G
ABI 270HT	open	variable/none	50	air/yes[a]	V, E
Isco 3140	open	variable/none	40	air[a]/none	V, E
Bio-Rad BioFocus 3000	cartridge	scanning/none	32	liquid[a]/yes[a]	P, E
Waters Quanta 4000	open	fixed/none	20	air[b]/none	G, E
Europhor IRIS 2000	cartridge	none/LIF	24	air[b]/none	P, G
SpectraPhysics 2000	cartridge	scanning/none	80	air[a]/none	V, E
Hewlett Packard HP[3D]CE	cartridge	PDA[c]/none	48	air[a]/none	P, E

[a]Sub-ambient temperature operation included.

[b]Ambient temperature operation only.

[c]Photodiode array.

P = pressure; V = vacuum; E = electrokinetic; G = gravity.

1.6 Sources of Information on HPCE

Keeping up with the literature in HPCE is no small task. Through July 1992, approximately 750 papers have appeared in the literature. The conference proceedings of the International Symposia on High Performance Capillary Electrophoresis that appeared in the *Journal of Chromatography,* Vols. 480, 516, 559, and 608, contain an impressive concentration of state-of-the-art results. Other journals containing numerous papers are *Analytical Chemistry, Journal of Microcolumn Separations, Chromatographia,* and the *Journal of High Resolution Chromatography.* The *Journal of Liquid Chromatography,* Vols. 12(13), 14(5), and 15(6,7), has good collections of papers on HPCE, and this series is expected to continue. For a comprehensive review of the literature, the biannual edition in *Analytical Chemistry* entitled "Fundamental Reviews" should be consulted (37, 38). For general information on the theory of electromigration techniques, see Ref. 39 for an excellent review.

1.7 Capillary Electrophoresis—A Family of Techniques

Capillary electrophoresis comprises a family of related techniques with differing mechanisms of separation.

These techniques, which are covered in the following chapters of this book, are:

capillary zone electrophoresis (CZE)
capillary isoelectric focusing (CIEF)
capillary gel electrophoresis (CGE)
capillary isotachophoresis (CITP)
micellar electrokinetic capillary chromatography (MECC)
capillary electroosmotic chromatography (CEC).

References

1. R. A. Mosher, D. A. Saville and W. Thormann, *The Dynamics of Electrophoresis* (VCH, 1992).
2. Z. Deyl, ed., *Electrophoresis: A Survey of Techniques and Applications: Part A: Techniques* (Elsevier, 1979).
3. D. Rickwood and B.D. Hames, *Gel Electrophoresis of Nucleic Acids: A Practical Approach,* 2nd Ed. (IRL Press, 1990).
4. D. Rickwood and B.D. Hames, *Gel Electrophoresis of Proteins: A Practical Approach,* 2nd Ed. (IRL Press, 1990).
5. A. T. Andrews, *Electrophoresis: Theory, Techniques and Biochemical and Clinical Applications* (Clarendon Press, 1981).

6. P. G. Righetti, *Isoelectric Focusing: Theory, Methodology and Applications* (Elsevier, 1983).
7. A. Chrambach, *The Practice of Quantitative Gel Electrophoresis* (VCH, 1985).
8. M. J. Dunn, ed., *Gel Electrophoresis of Proteins* (Wright, 1986).
9. O. Gaal, G. A. Medgyesi and K. Verczkey, *Electrophoresis in the Separation of Biomolecules* (Wiley, 1980).
10. J. W. Jorgenson and M. Phillips, Eds., *New Directions in Electrophoretic Methods* ACS Symposium Series 335 (American Chemical Society, 1987).
11. E. M. Southern, *J. Mol. Biol.* **98,** 503 (1975).
12. W. Jennings, *Analytical Gas Chromatography* (Academic Press, 1987).
13. M. Novotny and D. Ishii, eds., *Microcolumn Separations* (Elsevier, 1985).
14. D. Ishii, ed., *Introduction to Microscale High Performance Liquid Chromatography* (VCH, 1988).
15. F. J. Yang, ed., *Microbore Column Chromatography: A Unified Approach to Chromatography* (Marcel Dekker, 1989).
16. R. J. Zagursky and R. M. McCormick, *BioTechniques* **9,** 74 (1990).
17. R. A Hartwick, personal communication.
18. O. Vesterberg, *J. Chromatogr.* **480,** 3 (1989).
19. S. W. Compton and R. G. Brownlee, *Biotechniques* **6,** 432 (1988).
20. S. Hjertén, *Chromatogr. Rev.* **9,** 122 (1967).
21. R. Virtanen, *Acta Polytech. Scand.* **123,** 1 (1974).
22. F. E. P. Mikkers, F. M. Everaerts, and Th. P. E. M. Verheggen, *J. Chromatogr.* **169,** 11 (1979).
23. J. Jorgenson and K. D. Lukacs, *Anal. Chem.* **53,** 1298 (1981).
24. S. Hjertén, *J. Chromatogr.* **270,** 1 (1983).
25. S. Hjertén and M.-D. Zhu, *J. Chromatogr.* **346,** 265 (1985).
26. S. Terabe, K. Otsuka, K. Ichikawa, A. Tsuchiya and T. Ando, *Anal. Chem.* **56,** 111 (1984).
27. Y. Walbroehl and J. W. Jorgenson, *J. Chromatogr.* **315,** 135 (1984).
28. E. Gassman, J. E. Kuo, and R. N. Zare, *Science* **230,** 813 (1985).
29. J. A. Olivares, N. T. Nguyen, C. R. Yonker, and R. D. Smith, *Anal. Chem.* **59,** 1230 (1987).
30. R. A. Wallingford and A. G. Ewing, *Anal. Chem.* **59,** 1762 (1987).
31. W. G. Kuhr and E. S. Yeung, *Anal. Chem.* **60,** 1832 (1988).
32. S. Hjertén, *J. Chromatogr.* **347,** 191 (1985).
33. H. H. Lauer and D. McManigill, *Anal. Chem.* **58,** 166 (1986).
34. C. S. Lee, W. C. Blanchard, and C. T. Wu, *Anal. Chem.* **62,** 1550 (1990).
35. F. M. Everaerts and R. J. Routs, *J. Chromatogr.* **58,** 181 (1971).
36. V. Pretorius, B. J. Hopkins, and J. D. Schieke, *J. Chromatogr.* **99,** 23 (1974).
37. W. G. Kuhr, *Anal. Chem.* **62,** 403R (1990).
38. W. G. Kuhr, *Anal. Chem.* **64,** 389R (1992).
39. K. Kleparnik and P. Boček, *J. Chromatogr.* **569,** 3 (1991).

Chapter 2

Basic Concepts

2.1 Electrical Conduction in Fluid Solution

There are several simple concepts that are important to understand the physical processes that occur upon passage of an electrical current through an ionic solution.[1] These processes are far more complex compared to the passage of current through a metal. In metals, uniform and weightless electrons carry all of the current. In fluid solution, the current is carried by cations and anions. The molecular weight of these charge-bearing ions ranges from one for a simple proton to tens of thousands for large complex ions such as proteins and polynucleotides.

Conduction in fluid solution is still described by Ohm's Law,

$$E = IR, \tag{2.1}$$

where E is the voltage or applied field, I is the current that passes through the solution, and R is the resistance of the fluid medium. The reciprocal of resistance is conductivity. Kohlrausch found that the conductivity of a solution resulted from the *independent migration of ions*. When a current passes through an ionic solution, anions migrate toward the anode (positive electrode) while cations migrate toward the cathode (negative electrode) in equal quantities. Despite the passage of current, *electroneutrality* of the solution is always maintained as a result of ionization of water.

The conductivity of a solution is determined by two factors:

1. the concentration of the ionic species;

[1] See any basic text on physical chemistry for a thorough description of electrical conduction in fluid solution.

2. the speed of movement or mobility of the ionic species in an electric field.

In other words, highly mobile species are also highly conductive and vice-versa.

Mobility of ions in fluid solution is governed by their charge/size ratio. The size of the molecule is based on the molecular weight, the three-dimensional structure, and the degree of solvation (usually hydration). Data given in Table 2.1 (1) for alkali metals illustrate several of these important points: (i) The orders for the mobilities of the metal ions are the reverse of what is expected based on the metal or crystal radii data. These smaller ions are more hydrated than their larger counterparts. (ii) The current generated by 100 mM solutions of various acetate salts is proportional to the ionic mobility of the cation. This feature becomes important when selecting the appropriate counterion for preparing buffer solutions.

The forces governing this behavior are expressed by Stokes' Law,

$$f = 6\pi\eta r \nu \tag{2.2}$$

where η = viscosity; r = ionic radius; ν = ionic velocity. The competing forces of mobility (velocity) and viscosity are illustrated in Fig. 2.1 for an ion of radius r. Ionic size modifies mobility because of a solute's frictional drag through the supporting electrolyte. The frictional drag is directly proportional to viscosity, size, and electrophoretic velocity. An expression for mobility that contains these terms is

$$\mu = \frac{\nu}{E} = \frac{q}{6\pi\eta r}, \tag{2.3}$$

where q = the net charge and E = the electric field strength.

Table 2.1. Physical Properties of the Group 1A Metals

Metal	Metal Radius(Å)	Crystal Radius(Å)	Approx. Hydrated Radius(Å)	Hydration Number	Ionic Mobility[a]	Observed Current(μa)[b]
Li	1.52	0.86	3.40	25.3	33.5	111
Na	1.86	1.12	2.76	16.6	43.5	147
K	2.27	1.44	2.32	10.5	64.6	195
Rb	2.48	1.58	2.28	na	67.5	204
Cs	2.65	1.84	2.28	9.9	68.0	215

[a] At infinite dilution. Units are 10^5 cm^2/Vs.

[b] 100 mM buffer solution, 20 kV, 50 cm × 75 μm i.d. capillary at 20°C.

Data from *J. Liq. Chromatogr.* **13,** 2517 (1990).

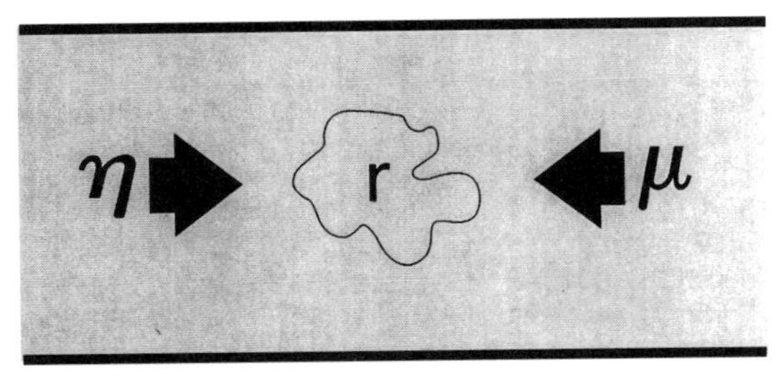

Figure 2.1. The forces acting on a solute of radius r during the course of electrophoresis.

2.2 The Language of Electrophoresis

There are several distinguishing differences between the terminology of chromatography and that of capillary electrophoresis. For example, a fundamental parameter in chromatography is the retention time. In electrophoresis nothing should ever be retained, so a more descriptive term is *migration time*. The migration time is the time it takes a solute to transit from the beginning of the capillary to the detector window. Expressions for some other fundamental terms are given in Eqs. (2.4)–(2.6). These include the *electrophoretic mobility* (μ_{ep}, cm^2/V·s), the *electrophoretic velocity* (v_{ep}, cm/s), and the *field strength* (E, V/cm):

$$v_{ep} = \frac{L_d}{t_m}, \tag{2.4}$$

$$\mu_{ep} = \frac{v_{ep}}{E}, \tag{2.5}$$

$$\mu_{ep} = \frac{L_d/t_m}{V/L_t}. \tag{2.6}$$

These equations define some fundamental features of HPCE:

1. Velocities are measured experimentally (Eq. (2.4)). They are determined by dividing the length of capillary from the injection side to the detector window (L_d) by the migration time t_m.
2. Mobilities are calculated by dividing the electrophoretic velocity, v_{ep}, by the field strength (Eq. (2.5)). The field strength is simply the voltage divided by the total capillary length (L_t).

   Mobility is the fundamental parameter of capillary electrophoresis. This term is independent of voltage and capillary length. Equations (2.5) and (2.6) define only the observed or relative mobility. To calculate the true mobility, a correction for a phenomenon known as electroendoosmotic flow (Section 2.3) must first be made.

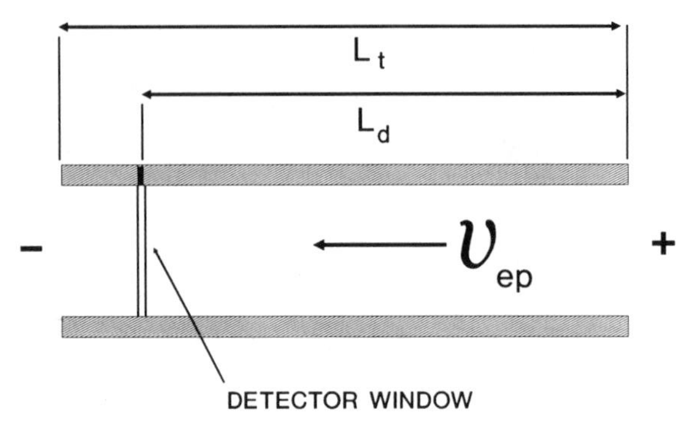

Figure 2.2. Illustration of a capillary defining the total length (L_t) and the length to the detector (L_d).

3. Two capillary lengths must be defined (Eq.(2.6)), the length to the detector (L_d) and the total length (L_t) as shown in Fig. 2.2. Ideally, $L_t - L_d$ should be as short as practical. Otherwise some system voltage (V) is wasted on maintaining field strength (E) over part of the capillary that does not participate in the separation because it lies beyond the detector window.

2.3 Electroendoosmosis

A. The Capillary Surface

Electroendoosmosis is one of the "pumping" mechanisms of HPCE. It occurs because of the surface charge, known as the *zeta potential,* on the wall of the capillary. Fused-silica is the most common material used to produce capillaries for HPCE. Technology developed for manufacturing capillary columns for GC is readily transferred over to HPCE. This material is a highly crosslinked polymer of silicon dioxide with tremendous tensile strength (2), although it is quite brittle. Other materials such as Teflon and quartz have been used (3), but performance and cost are less favorable.

Before use, capillaries are usually conditioned with 1 N sodium hydroxide. The base ionizes free silanol groups and may cleave some epoxide linkages as well. An anionic charge on the capillary surface results in the formation of an electrical double layer. The resulting ionic distribution is shown in Fig. 2.3 (4). Anions are repelled from the negatively charged wall region, whereas cations are attracted as counterions. Ions closest to the wall are tightly bound and immobile, even under the influence of an electric field. Further from the wall is a compact and mobile region with substantial cationic character. At a greater distance from the wall, the solution becomes electrically neutral as the zeta potential is not sensed.

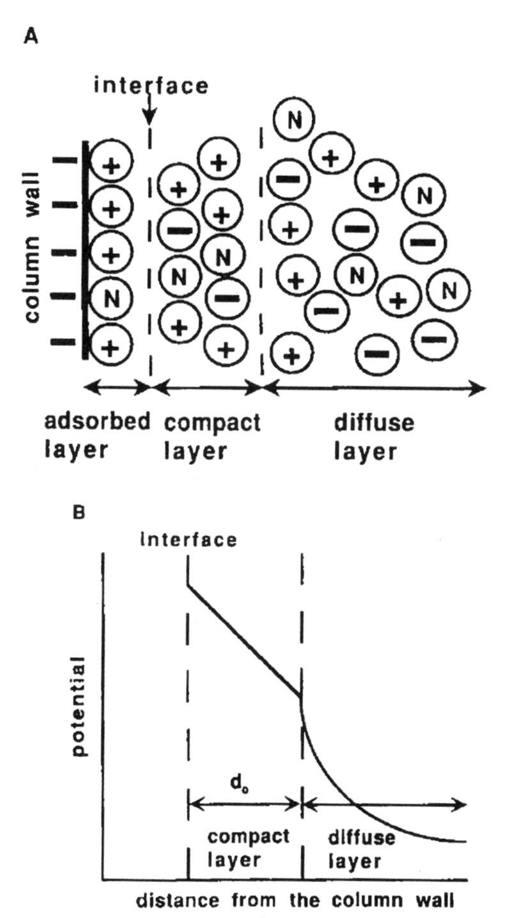

Figure 2.3. Representation of the electrical double layer versus distance from the capillary wall. Reprinted with permission from *J. Chromatogr.* **559,** 69, copyright ©1991 Elsevier Science Publishers.

Expressions describing this phenomenon were derived by Gouy and Chapman in 1910 and 1913, respectively. This diffuse outer region is known as the Gouy–Chapman layer. The rigid inner layer is called the Stern layer.

When a voltage is applied, the mobile positive charges migrate in the direction of the cathode or negative electrode. Since ions are solvated by water, the fluid in the buffer is mobilized as well and dragged along by the migrating charge. Although the double layer is perhaps 100 angstroms thick, the electroendoosmotic flow (EOF) is transmitted throughout the diameter of the capillary, presumably through hydrogen bonding of water molecules or van der Waals interactions between buffer constituents.

The electroosmotic flow as defined by Smoluchowski in 1903 is given by

$$\nu_{eo} = \frac{\varepsilon\zeta}{\eta} E, \tag{2.7}$$

where ε is the dielectric constant, η is the viscosity of the buffer, and ζ is the zeta potential of the liquid–solid interface. This equation is similar to Eq. (2.1) which describes mobility. Factors such as viscosity and temperature affect both phenomena in a similar fashion. The equation is only valid for capillaries that are sufficiently large that the double layers on opposite walls do not overlap each other (5).

B. Measuring the Electroosmotic Flow

To calculate the absolute mobility of a solute, the contribution of the EOF to the migration velocity must be accounted for. Routine measurement of the EOF is also necessary to ensure the integrity of the separation. If the EOF is not reproducible, it is likely that the capillary wall is being affected by some component in the sample or that a factor such as temperature is not properly controlled.

There are several practical methods to determine the EOF. One technique is to inject a neutral solute and measure the time it takes to transit the detector (6–8). Solutes such as methanol, acetone, benzene, dimethylsulfoxide, and mesityl oxide are frequently employed. In MECC, discussed in Chapter 7, a further requirement that the marker solute not partition into the micelle is also imposed. Conventional units for the EOF are mm/V/s, although migration time alone may be used when checking day-to-day reproducibility of a method under constant conditions.

Another means of measuring EOF is the current-monitoring method (9). The capillary and detector side buffer reservoir is filled with buffer while the other buffer reservoir (sample side) is filled with a 19/1 (buffer/water) solution. When the voltage is applied, the EOF drives the slightly diluted buffer into the capillary. Since the buffer is slightly diluted, the current will decrease as the buffer fills the capillary. The point of inflection (t_f) can be extrapolated as shown in Fig. 2.4. Then the EOF equals L_t/t_f. To convert to units of mobility, divide the measured velocity by the field strength. If there is no EOF or the direction of the EOF is different from that expected, there is no current change.

A similar method uses absorption detection to measure the change in absorption when the sample side buffer containing a solvent such as 0.5% acetone transits the detector. The use of either current monitoring or absorption monitoring is dependent on the minor changes in the sample side buffer causing insignificant changes in the EOF. Other less frequently used methods include weighing (10–12) and streaming potential measurements (10).

Once μ_{eo} has been calculated, μ_{ep} is found by subtracting (adding) the contribution of the EOF to the apparent electrophoretic mobility. The use of mobility as

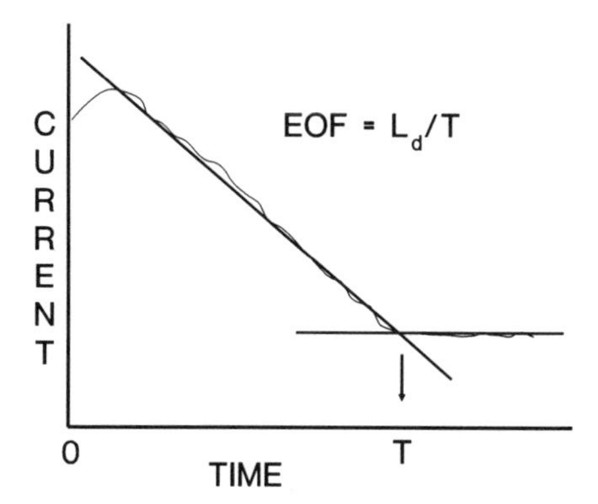

Figure 2.4. Use of the current monitoring method to determine electroosmotic flow. The electropherogram shows the current changes obtained when 20 mM phosphate buffer replaces 19 mM phosphate buffer in the capillary tube. Intersection of the regression lines designates the time used to calculate the EOF.

the "retention parameter" will frequently yield greater precision compared to electrophoretic velocity, since the impact of the EOF is factored out of the calculation.

C. Effect of Buffer pH

The impact of pH on the EOF and the mobility is illustrated in Figs. 2.5a and 2.5b. At high pH, the silanol groups are fully ionized generating a strong zeta potential and a dense electrical double layer. As a result, the EOF increases as the buffer pH is elevated (6, 13). A robust flow, typically around 2 mm/s at pH 9 in 20 mM borate buffer at 30 kV, 30°C, is realized. For a 50 µm capillary this translates to 235 nL/min. Since the total volume of a 50 cm × 50 µm i.d. capillary is only 980 nL, a neutral compound would reach the detector in 4.2 min. At pH 3, the EOF is much lower, about 30 nL/min.

The EOF must be controlled or even suppressed to run certain modes of HPCE. On the other hand, the EOF makes possible the simultaneous separation of cations, anions and neutral species in a single run. For example, a zwitterion such as a peptide will be negatively charged at a pH above its p*I*. The solute will electromigrate toward the positive electrode. However, the EOF is sufficiently strong that the solute migrates toward the negative electrode (Fig. 2.5a). At low pH, the zwitterion has a positive charge and will migrate as well toward the negative electrode (Fig. 2.5b). In untreated fused-silica capillaries, most solutes migrate toward the negative electrode unless buffer additives or capillary treatments are used to reduce or reverse the EOF (Sections 3.3 and 3.4).

The EOF is exquisitely sensitive to pH (6, 11, 14). Hysteresis effects have been reported (14) where the direction of approach to a particular pH value produces a

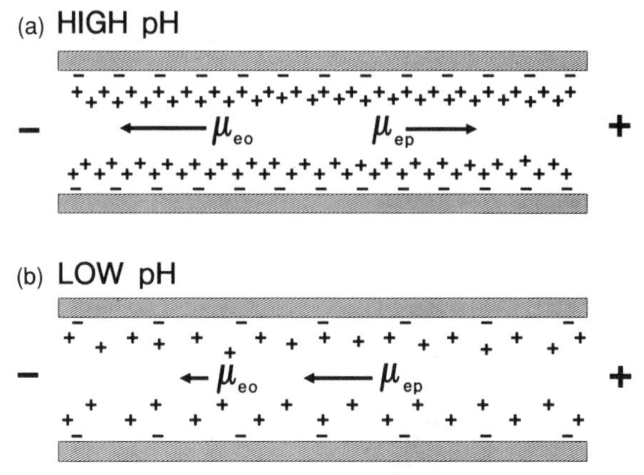

Figure 2.5. Behavior of electroendoosmotic flow and electrophoresis of a zwitterion (pI = 7) at (a) high and (b) low pH.

different EOF. When the pH is approached from the acid side, the measured EOF is always lower. Longer equilibration times would reduce hysteresis at the expense of total run time. Since the EOF will affect migration time precision, it is important to design experiments with these features in mind. The problems with EOF reproducibility are most severe in the pH range of 4–6 (14).

D. Effect of Buffer Concentration

The expression for the zeta potential is (15)

$$\zeta = \frac{4\pi\delta e}{\varepsilon}, \tag{2.8}$$

where ε = the buffer's dielectric constant, e = total excess charge in solution per unit area, and δ is the double layer thickness or Debye ionic radius. The Debye radius is $\delta = [3 \times 10^7 |Z| C^{1/2}]^{-1}$ where Z = number of valence electrons and C = the buffer concentration.

As the ionic strength increases, the zeta potential and similarly, the EOF decreases in proportion to the square root of the buffer concentration. This is confirmed experimentally (16) in Fig. 2.6 for a series of buffers where the EOF was found to be linear with the natural logarithm[2] of the buffer concentration. VanOrman *et al.* (16) also reported equivalent EOF for different buffer types as long as the ionic strength is constant.

[2] The linear relationship of EOF with the buffer concentration is probably a square root relationship, rather than the natural logarithum as indicated in the figure.

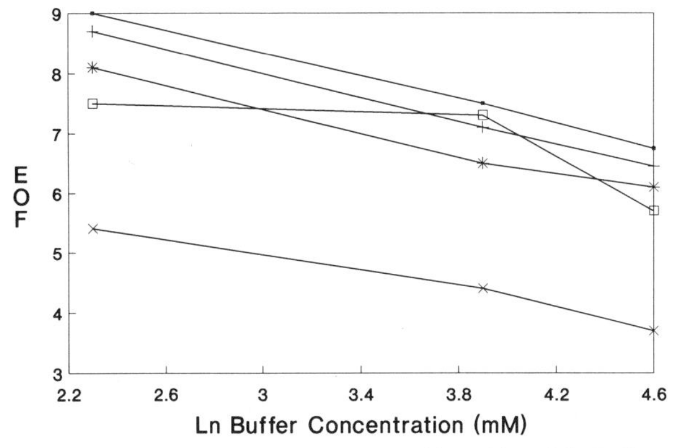

Figure 2.6. Electroosmotic flow at pH 7.5 vs. the natural logarithm of the buffer concentration. Key: ■ = ACES; + = HEPES; ✳ = HEPPSO; □ = phosphate; × = borate. Data from *J. Microcol. Sep.* **2,** 176 (1990).

E. Effect of Organic Solvents

Organic solvents can modify the EOF because of their impact on buffer viscosity (16) and zeta potential (17). Alcohols such as methanol, ethanol, and isopropanol and aprotic solvents such as acetonitrile, acetone, tetrahydrofuran, and dimethyl sulfoxide are useful to aid solubility and modify selectivity in CZE (Section 3.2) and MECC (Section 7.6). At high pH, increasing the organic solvent concentration usually decreases the EOF (17). In another report, acetonitrile was reported to increase the EOF (18). As a general guideline linear alcohols decrease the EOF while acetonitrile has a small effect, at least at low concentrations, *ca.* 10–20%.

F. Controlling the Electroosmostic Flow

Capillary isoelectric focusing (CIEF) and capillary isotachophoresic (CITP) separations are usually performed under conditions of very low or carefully controlled EOF using the combination of Teflon or coated capillaries plus the inclusion of an EOF modifier to the buffer (Chapters 4 and 6). Additives such as 0.5% hydroxypropylmethyl cellulose (19) or *s*-benzylthiouronium chloride (11) are effective in suppressing the EOF, particularly in conjunction with a coated capillary (19). Cationic surfactants such as cetyltrimethylammonium bromide can actually reverse the direction of electroosmotic flow (11). A new technique called *field effect electroosmosis* provides the opportunity to electrically tune the EOF (Section 12.2). While complete suppression of the EOF is unnecessary for most applications, control is critical to obtain reproducible migration times and resolution.

2.4 Efficiency

The high efficiency of HPCE is a consequence of several unrelated factors.

1. A stationary phase is not required for HPCE. The primary cause of band-broadening in LC is resistance to mass transfer between the stationary and mobile phases. For most modes of HPCE, this dispersion mechanism does not operate. Similarly, other dispersion mechanisms such as eddy diffusion and stagnant mobile phase are unimportant.
2. In pressure-driven systems such as LC, the frictional forces of the mobile phase interacting at the walls of the tubing result in radial velocity gradients throughout the tube. As a result, the fluid velocity is greatest at the middle of the tube and approaches zero near the walls (Fig. 2.7). This is known as *laminar* or *parabolic* flow. These frictional forces together with the chromatographic packing result in a substantial pressure drop across the column.

In electrically driven systems, the EOF is generated uniformly down the entire length of the capillary. There is no pressure drop in HPCE, and the radial flow profile is uniform across the capillary except very close to the walls where the flow rate approaches zero (Fig. 2.7).

Jorgenson and Lukacs derived the efficiency of the electrophoretic system from basic principles (20–22) using the assumption that diffusion is the only source of bandbroadening. Other sources of dispersion, including Joule heating (Section 2.6), capillary wall binding (Section 3.5), injection (Section 9.1), and detection (Section 10.5), lead to fewer theoretical plates than the theory predicts.

The migration velocity for a solute is as follows:

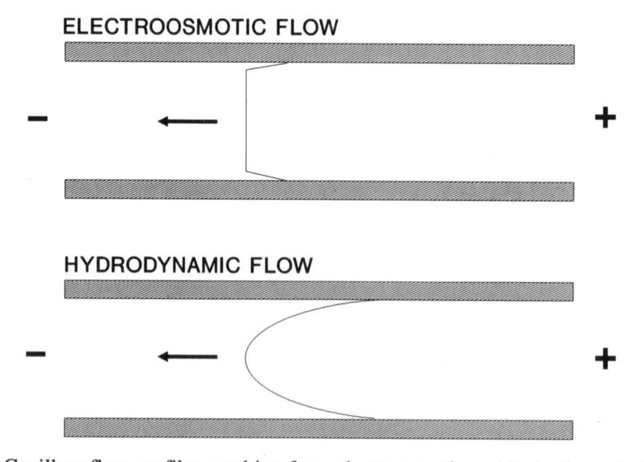

Figure 2.7. Capillary flow profiles resulting from electroosmotic and hydrodynamic flow.

$$v = \mu E = \mu V/L, \quad (2.9)$$

where μ = the apparent mobility, E = field strength, V = voltage, and L = capillary length. The time, t, for a solute to migrate the length, L, of the capillary is

$$t = L/V = L^2/\mu V. \quad (2.10)$$

Diffusion in liquids that lead to broadening of an initially sharp band is described by the Einstein equation:

$$\sigma_L^2 = 2Dt = \frac{2DL^2}{\mu V}, \quad (2.11)$$

where D = the diffusion coefficient of the individual solute. The number of theoretical plates, N, is given by:

$$N = L^2/\sigma_L^2. \quad (2.12)$$

Substituting Eq. (2.11) into Eq. (2.12) gives an expression for the number of theoretical plates,

$$N = \frac{\mu V}{2D}. \quad (2.13)$$

Some important generalizations can be made from this expression.

1. The use of high voltage gives the greatest number of theoretical plates since the separation proceeds rapidly, minimizing the effect of diffusion. This holds true until the point where heat dissipation is inadequate (Section 2.6).
2. Highly mobile solutes produce high plate counts because their rapid velocity through the capillary minimizes the time for diffusion.
3. Solutes with low diffusion coefficients give high efficiency because of slow diffusional bandbroadening.

Points 2 and 3 are contradictory. This is clarified by Fig. 2.8 and supplemented with some calculations in Table 2.2. Because of the indirect but inverse relationship

Table 2.2. Calculated Theoretical Plates for a Small and Large Molecule

Solute	MW	Mobility (10^{-4} cm^2/Vs)	Diffusion Coefficient (10^{-6} cm^2/s)	N
Horse heart myoglobin	13,900	0.65	1	975,000
Quinine sulfate	747	4	7	857,000

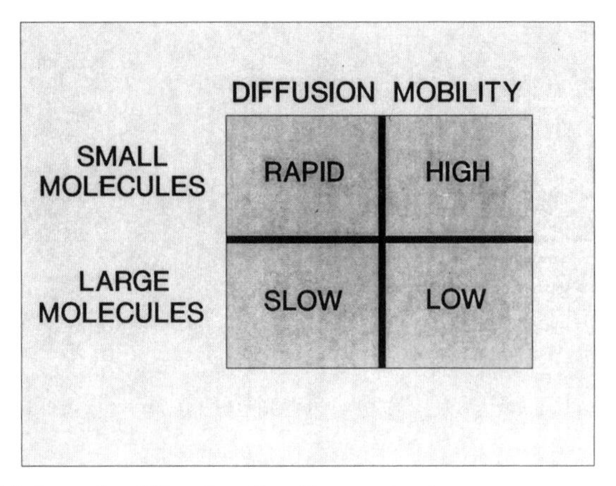

Figure 2.8. Diffusion and mobility of small and large molecules.

between mobility and diffusion, high-efficiency separations occur across a wide range of molecular weights.

HPCE is applicable for high-efficiency separations for both large and small molecules. The greatest number of theoretical plates is found in capillary gel electrophoresis (CGE). The use of an anticonvective gel matrix furthers the advantages of HPCE. The combination of HPCE in the gel format (Chapter 5) can yield millions of theoretical plates.

2.5 Resolution

The resolution (R) between two solutes is defined as

$$R_s = \frac{1}{4} \frac{\Delta\mu_{ep}\sqrt{N}}{\mu_{ep} + \mu_{eo}}, \tag{2.14}$$

where $\Delta\mu$ is the difference in mobility between the two species, μ_{ep} is the average mobility of the two species, and N is the number of theoretical plates. Substituting the plate count equation (2.13) where $\mu = \mu_{ep} + \mu_{eo}$ yields (20)

$$R_s = 0.177\, \Delta\mu_{ep} \sqrt{\frac{V}{(\mu_{ep} + \mu_{eo})\, D_m}}. \tag{2.15}$$

This expression suggests that increasing the voltage is not very effective for improving resolution. To double the resolution, the voltage must be quadrupled. Since the voltage is usually in the 10–30 kV range, Joule heating (Section 2.6) limitations are quickly approached.

To improve electrophoretic resolution, adjustments to $\Delta\mu_{ep}$, the selectivity of the separation, are best addressed. The control of selectivity is first accomplished through selection of the appropriate mode of HPCE. Next, proper selection of the separation buffer, buffer pH, and buffer additives are important. Both factors will be discussed in individual chapters. Each mode of HPCE has specific "tricks" for modifying selectivity and resolution.

Another means of improving resolution as predicted by Eq. (2.15) is to reduce the EOF or adjust the EOF to a direction opposite of electrophoretic flow. Under these conditions, the "effective" length of the capillary is increased, improving resolution at the expense of run time. Alternatively, the speed of separation can be enhanced by adjusting both electrophoresis and electroosmosis toward the same electrode. Then, resolution is sacrificed.

2.6 Joule Heating

The conduction of electric current through an electrolytic solution results in the generation of heat because of frictional collisions between mobile ions and buffer molecules. Since high field strengths are employed in HPCE, this *Ohmic* or *Joule* heating can be substantial. There are two problems that can result from Joule heating: (i) temperature changes due to ineffective heat dissipation; (ii) development of thermal gradients across the capillary.

If heat is not dissipated at a rate equal to its production, the temperature inside the capillary will rise, and eventually the buffer solution will degas or boil. Even a small bubble inside of the capillary will disrupt electrical continuity. At moderate field strengths, outgassing is not usually a problem, even for capillaries that are passively cooled.

The rate of heat production inside the capillary can be estimated by

$$dH/dt = IV/LA, \tag{2.16}$$

where L = capillary length and A = the cross-sectional area. By rearranging this equation using $I = V/R$, where R, the resistance = L/kA, and k = the conductivity, then

$$dH/dt = kV^2/L^2. \tag{2.17}$$

The amount of heat that must be removed is proportional to the conductivity of the buffer, and also to the square of the field strength.

Lacking catastrophic failure (bubble formation), the problem of thermal gradients across the capillary can result in substantial bandbroadening (23–25). This problem is illustrated in Fig. 2.9. The second law of thermodynamics states that heat flows from warmer to cooler bodies. In HPCE, the center of the capillary is

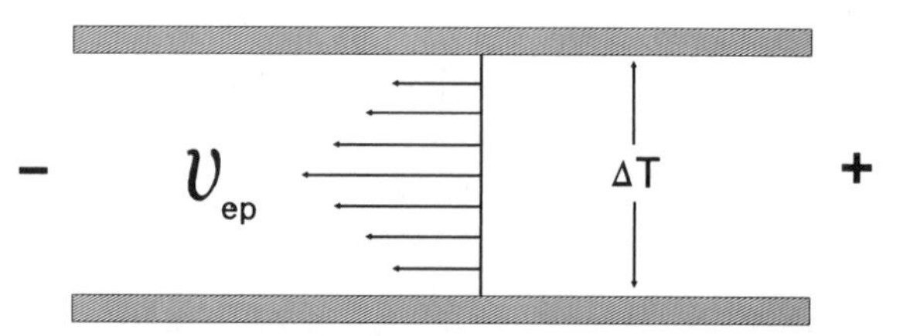

Figure 2.9. Impact of the radial temperature gradient on electrophoretic and electroosmotic flow.

hotter than the periphery. Since the viscosity of most fluids decreases with increasing temperature, Eqs. (2.3) and (2.7) predict that both mobility and EOF increase as the temperature rises.

This situation becomes similar to laminar flow, where the electrophoretic or electroosmotic velocity at the center of the capillary is greater than the velocity near the walls of the capillary. The temperature differential of the buffer between the middle and the wall of the capillary can be estimated from the following equation:

$$\Delta T = 0.24 \frac{Wr^2}{4K}, \tag{2.18}$$

where W = power, r = capillary radius, and K = thermal conductivity of the buffer, capillary wall, and polyimide cladding. A 2 mm i.d. capillary filled with 20 mM CAPS buffer draws 18 mA of current at 30 kV, giving a ΔT of 75°C. A 50 μm i.d. capillary filled with the same buffer draws only 12 μA of current, yielding a ΔT of 50 m°C. Since the thermal gradient is proportional to the square of the capillary radius, the use of narrow capillaries facilitates high resolution. On the other hand, the use of dilute buffers permits the use of wider-bore capillaries but the loading capacity of the separation is reduced (Section 11.4).

Unfortunately, narrow capillaries cause severe problems with sample loading capacity and optical detection. If a solution is injected equivalent to 1% of the capillary volume of a 50 cm × 50 μm i.d. capillary, the injection size is 9.8 nL. This small volume injection coupled to a 50 μm optical pathlength provides for concentration limits of detection (CLOD) that are about 50 times poorer than those of LC.

The compromise between sensitivity and resolution is illustrated in Figs. 2.10 and 2.11. Note in particular in Fig. 2.10, the cluster of peaks centered at a migration time of 31 min (26 min in Figure 2.11). With the 50 μm i.d. capillary, none of these peaks are baseline-resolved, but there is virtually no noise in the electropherogram. Separation of the same sample in a 25 μm i.d. capillary (Fig. 2.11) presents a different picture. The peaks are nearly baseline-resolved but there is substantial noise in the output. This presents one of several compromises that must be made in HPCE. In this case, sensitivity and resolution are competing analytical goals.

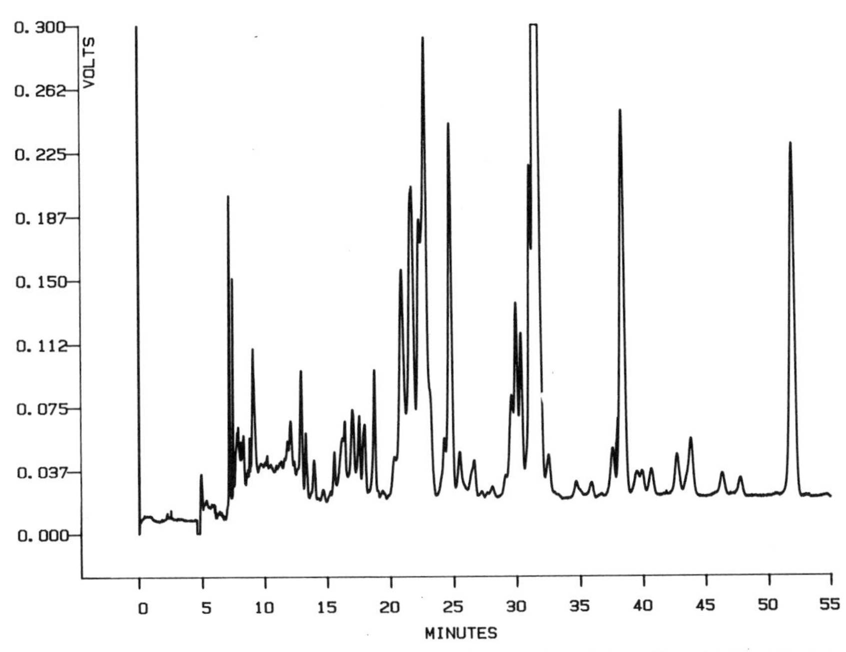

Figure 2.10. Separation of heroin impurities by MECC on a 50 μm i.d. capillary. Buffer: 85 mM SDS, 8.5 mM borate, 8.5 mM phosphate, 15% acetonitrile, pH 8.5; capillary: 50 cm (length to detector) × 50 μm i.d.; voltage: 30 kV; temperature: 50°C; detection: UV, 210 nm. Reprinted with permission from *Anal. Chem.* **63,** 823, copyright ©1991 Am. Chem. Soc.

The problem of Joule heating depends on the capillary diameter, the field strength and the buffer concentration. This combination of features is described in Figs. 2.12 and 2.13 (26). The data in Fig. 2.12 for a 50 μm i.d. capillary show a linear relationship between the EOF and the field strength for three buffer concentrations ranging from 10 to 50 mM. At the higher buffer concentrations, the EOF is suppressed as predicted by Eqs. (2.7) and (2.8). Fig. 2.13 contains data from the same experiments, except that a 100 μm i.d. capillary is used. A marked departure from linearity is found at the higher buffer concentrations.

Higher-concentration buffers are more conductive, draw higher currents, and produce more heat than more dilute solutions. In the 100 μm i.d. capillary, this heat is not properly dissipated. As a result, the internal temperature rises, reducing the viscosity of the buffer. Since Eq. (2.7), the basic expression for electroosmotic velocity, contains a viscosity parameter in the denominator, v_{eo} increases with decreasing buffer viscosity. Because the buffer viscosity is dependent on temperature, the capillary heat removal system plays an important role in deciding the maximum field strength, buffer concentration, and capillary diameter that can be

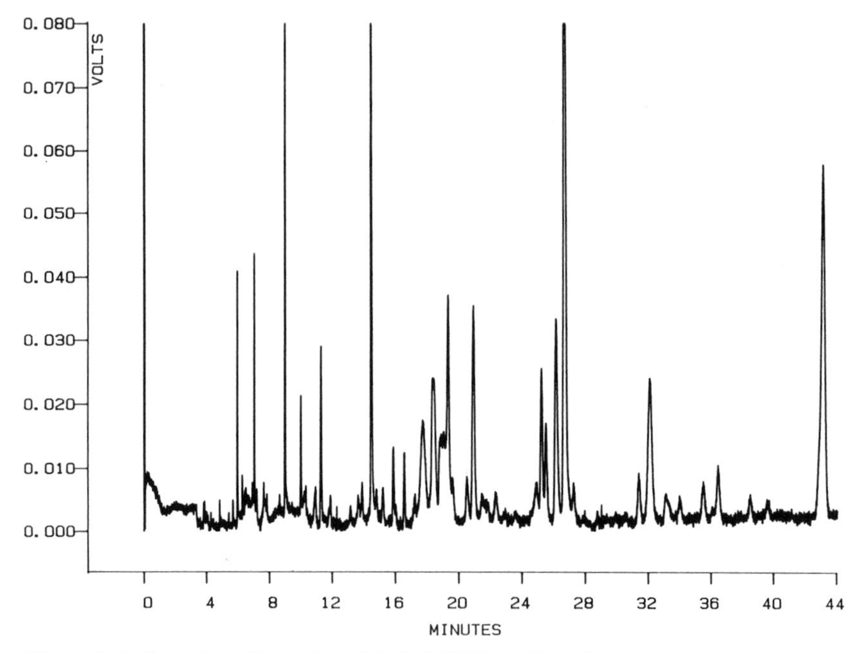

Figure 2.11. Separation of heroin impurities by MECC on a 25 μm i.d. capillary. Conditions as in Fig. 2.8 except for capillary diameter. Reprinted with permission from *Anal. Chem.* **63,** 823, copyright ©1991 Am. Chem. Soc.

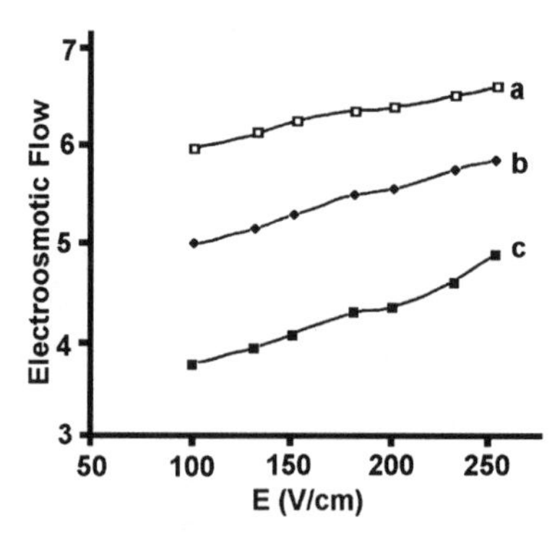

Figure 2.12. Effect of buffer concentration and field strength (E, V/cm) on the electroosmotic flow in a 50 μm i.d. capillary. Buffer: phosphate at a concentration of (a) 10 mM;(b) 20 mM; (c) 50 mM. Redrawn with permission from *J. Chromatogr.* **516,** 223 copyright ©1990 Elsevier Science Publishers.

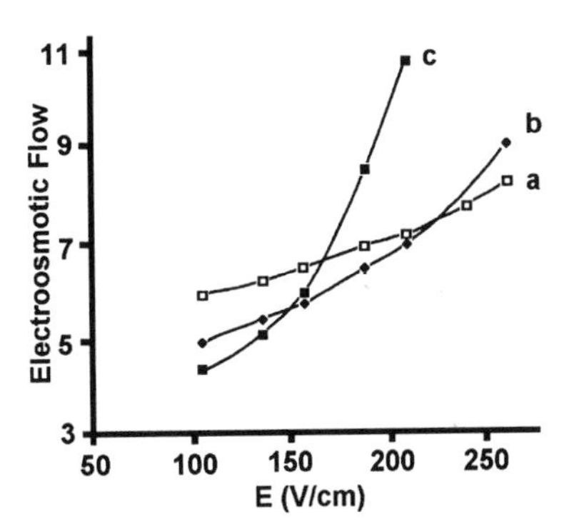

Figure 2.13. Effect of buffer concentration and field strength (E, V/cm) on the electroosmotic flow in a 100 μm i.d. capillary. Buffer: phosphate at a concentration of (a) 10 mM; (b) 20 mM; (c) 50 mM. Redrawn with permission from *J. Chromatogr.* **516,** 223, copyright ©1990 Elsevier Science Publishers.

successfully employed. Insufficient heat removal begins a vicious cycle leading to viscosity reduction, greater current draw, and higher temperature that further reduces the viscosity.

2.7 Optimizing the Voltage and Temperature

A. Ohm's Law Plots

A means of optimizing the voltage and/or the temperature despite the buffer concentration and capillary cooling system is very desirable. An *Ohm's Law plot* provides this tool with very little experimental work (27, 28). Simply fill the capillary with buffer, vary the voltage, record the current, and plot the results.

Some Ohm's Law plots are shown in Figs. 2.14 and 2.15 for an air-cooled and a water-cooled temperature control system, respectively. Whenever the graph shows a positive deviation from linearity, the heat removal capacity of the system is being exceeded. Operating on the linear portion of the curve will generally yield the highest number of theoretical plates. Sometimes separations can be run in a nonlinear section to optimize speed at the expense of plates.

Lowering the temperature below ambient can be used to extend the linear range of the Ohm's Law plot. This is useful when high ionic strength buffers are necessary. These strong buffers are particularly useful in micropreparative CE (Section 10.11), increasing the linear dynamic range (Section 11.4), and suppressing wall effects (Section 3.4). Increasing the temperature can also be employed to speed the separation since both v_{eo} and v_{ep} increase about 2%/K because of the decreased viscosity of the buffer medium.

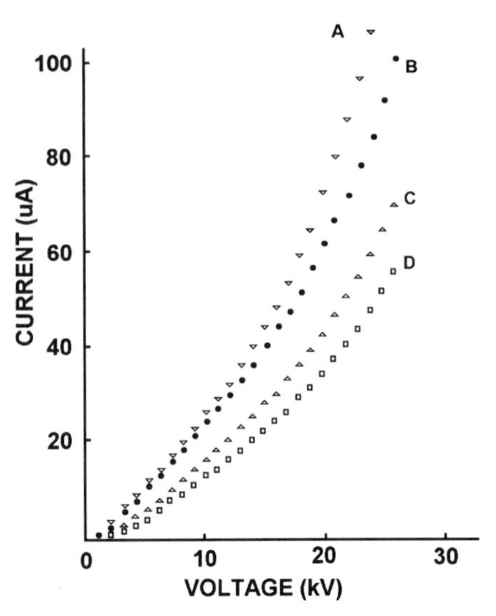

Figure 2.14. Ohm's Law plots for capillary temperature control by air circulation. (A) No control; (B) 25°C; (C) 10°C; (D) 4°C. Redrawn with permission from *J. High Res. Chromatogr.* **14**, 200, copyright ©1991 Dr. Alfred Huethig Publishers.

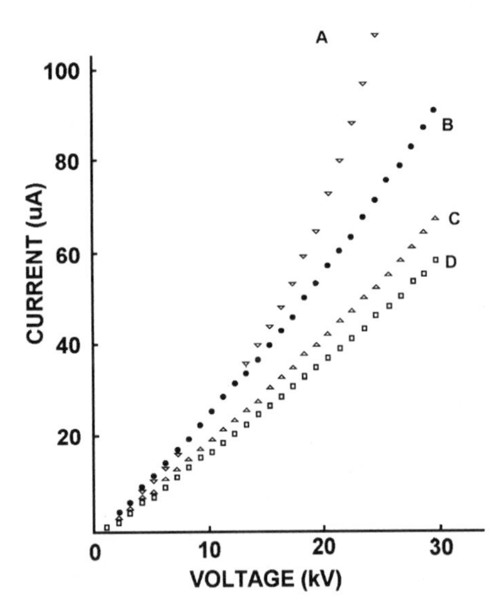

Figure 2.15. Ohm's Law plots for capillary temperature control by water circulation. (A) No control; (B) 25°C; (C) 10°C; (D) 4°C. Redrawn with permission from *J. High Res. Chromatogr.* **14**, 200, copyright ©1991 Dr. Alfred Huethig Publishers.

The Ohm's Law plot is an effective means of evaluating various instruments concerning the efficiency of their capillary cooling systems. Fluid-cooled systems are generally more effective than air-cooled systems, since the heat capacity of most fluids exceeds that of air.

B. Constant Voltage or Constant Current?

Power can be applied to the system in one of two ways. The voltage can be fixed, allowing the current to float based on the resistance of the buffer. Alternatively, the current can be fixed. Most published work in HPCE is in the constant-voltage mode. There is growing evidence (none published to date) that the constant-current mode may provide better reproducibility compared to constant voltage. Until this is better defined, both modes should be studied during methods development for CZE, MECC, and CGE separations. CITP is typically performed in the constant-current mode; otherwise, separation time becomes long. CIEF may benefit from the constant-current mode as well, although there is no evidence published to that effect.

2.8 Capillary Diameter and Buffer Ionic Strength

Some very subtle effects due to Joule heating can be seen when comparing separations run on capillaries with different inner diameters, or even the same capillaries run on various instruments with different capillary cooling systems. Some of these issues are illustrated in Figs. 2.14 and 2.15.

The ionic strength of the buffer influences not only the EOF and μ_{ep}, but indirectly the viscosity of the medium. More concentrated buffers have greater conductivity and generate more heat when the voltage is applied. The viscosity is dependent on the temperature, so there is also a dependence on the capillary diameter. This is shown for a series of runs in 50 and 75 μm i.d. capillaries (Figs. 2.16 and 2.17) (29).

With the 50 μm capillary, the migration times lengthen as the buffer concentration is increased. Ions in solution are always surrounded by a double layer of ions of the opposite charge. Since the migration of these counterions is in a direction opposite to the solute, increasing the concentration of the buffer reduces the mobility of the solute because of increased drag caused by counter-migration of the counterions.

With the 75 μm capillary, the solute migration times first increase as expected, but then they decrease. The decrease in migration time is a consequence of Joule heating becoming significant at higher buffer concentrations. Note, as well, the impact of buffer concentration on peak width. Sharper peak widths at the higher buffer concentrations are due to a phenomenon known as *stacking* (Section 9.6).

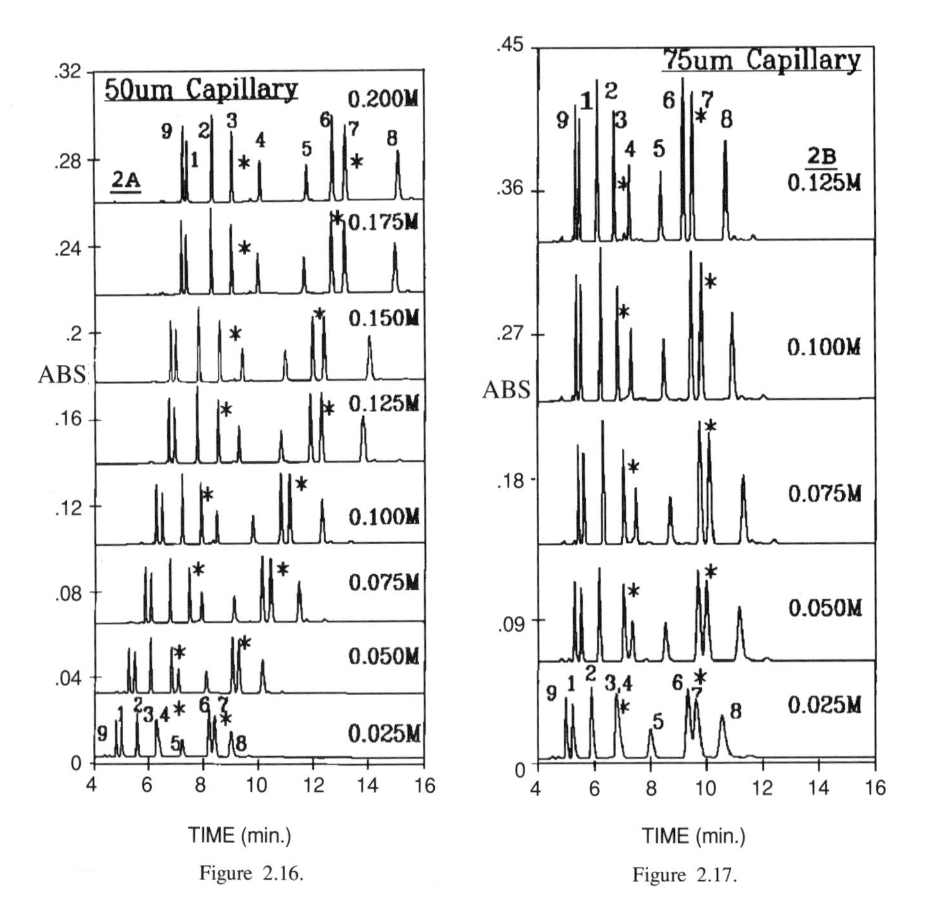

Figure 2.16.

Figure 2.17.

Figure 2.16. Effect of buffer ionic strength on peptide separations in a 50 μm i.d. capillary. Buffers: 0.025–0.200 M phosphate, pH 2.44; voltage: 30 kV; capillary: 50 cm to detector × 50 μm i.d.; key: (1) bradykinin; (2) angiotensin II; (3) TRH; (4) LRHR; (5) bombesin; (6) leucine enkephalin; (7) methionine enkephalin; (8) oxytocin; (9) dynorphin. Reprinted with permission from *Techniques in Protein Chemistry II*, 1991, Academic Press, pp. 3–19.

Figure 2.17. Effect of buffer ionic strength on peptide separations in a 75 μm i.d. capillary. Conditions as in Fig. 2.16 except for capillary diameter. Reprinted with permission from *Techniques in Protein Chemistry II*, 1991, Academic Press, pp. 3–19.

2.9 Optimizing the Capillary Length

The efficiency of the separation (theoretical plates) is directly proportional to the capillary length (30) provided the field strength is kept constant. The limitation here is available voltage. Most instruments produce a maximum of 30 kV. Once the capillary length reaches a certain point, the field strength must be reduced, and no further gains in efficiency are realized (3). Based on Eq. (2.14), the electrophoretic resolution is dependent on the square root of the number of theoretical plates, and thus on the square root of the capillary length (30). Increasing the capillary length beyond the limits imposed by the voltage maximum lengthens the separation time without any substantial benefits. Efficiency and resolution can deteriorate in extreme cases because of molecular diffusion.

2.10 Buffers

A. The Role of the Buffer

A wide variety of buffers (Table 2.3) are employed in CZE. The buffer is frequently called the *background or carrier electrolyte.* These terms are used interchangeably throughout this book. Other terms frequently employed are *co-ion* and *counterion.* A co-ion is a buffer ion of the same charge as the solute, and vice-versa for the counterion.

The purpose of the buffer is to provide precise pH control of the carrier electrolyte. This is important because both mobility and electroendoosmosis are sensitive to pH changes. The buffer may also provide the ionic strength necessary for electrical continuity. Usual buffer concentrations range from 10 to 100 mM though there are many exceptions. Dilute buffers provide the fastest separations, but the sample loading capacity is reduced.

Buffer solutions should resist pH change upon dilution and addition of small amounts of acids and bases. Concentrated buffer solutions do this well, but are too conductive for use in HPCE. The buffering capacity of a weak acid or weak base is limited to ± 1 pH unit of its pK_a. Operation outside of that range requires frequent buffer replacement to avoid pH changes (31). The buffer should also have a low temperature coefficient. Aromatic buffer constituents such as phthalates should be avoided if possible because of their strong UV absorption characteristics. This is generally less of a problem in HPCE because of the short optical pathlength. Strongly absorbing components such as carrier ampholytes (Section 4.1) prevent the use of the low UV region. Table 2.4 contains some of these data for a few common buffers.

Zwitterionic buffers such as bicine, tricine, CAPS, MES, and tris are also used, particularly for protein and peptide separations. They are all amines, and some

Table 2.3. Buffers for HPCE[a]

Buffer	pK_a	Mobility[b,c]
Goods Buffers		
MES	6.13	–26.8
ACES	6.75	–31.3
MOPSO	6.79	–23.8
BES	7.16	–24.0
MOPS	7.2	–24.4
TES	7.45	–22.4
DIPSO	7.5	
HEPES	7.51	–21.8
TAPSO	7.58	
HEPPSO	7.9	–22.0
EPPS	7.9	
POPSO	7.9	
DEB	7.91	–26.2
Tricine	8.05	
GLYGLY	8.2	
Bicine	8.25	
TAPS	8.4	–25.0
CHES	9.55	
CAPS	10.4	
Conventional Buffers[d]		
Citrate	3.12, 4.76, 6.40	
Formate	3.75	–56.6[e]
Acetate	4.76	–42.4[e]
Phosphate	2.14, 7.10, 13.3	
Borate	9.14	

[a]Data from *J. Chromatogr.* **545,** 391 (1991) except as noted.

[b]Effective mobility for fully ionized buffers at 25°C (10^{-5} cm^2/Vs). Buffer concentrations ranged from 5 to 6.75 mM.

[c]*Chem. Rev.* **89,** 419 (1989) except as indicated.

[d]Data from the Merck Index, 9th edition.

[e]Data from *J. Chromatogr.* **390,** 69 (1987).

buffers such as MES, tricine, and glycylglycine bind calcium, manganese, copper, and magnesium ions. PIPES, HEPES, and tris do not bind any of these metals, while BES binds only copper (32). Metal binding may be useful, or it may interfere with subsequent separations.

The advantage of the zwitterionic buffer is low conductivity when the buffer is adjusted to its p*I*. This provides for low current draw and reduced Joule

Table 2.4. Characteristics of Some Buffer Solutions[a]

Buffer	pH at 25°C	Dilution Value[b]	Buffer Capacity[c]	Temperature Coefficient[d]
50 mM $KH_2C_6H_5O_7$ (citrate)	3.776	+0.024	0.034	−0.0022
25 mM KH_2PO_4, Na_2HPO_4	6.865	+0.080	0.029	−0.0028
8.7 mM KH_2PO_4 + 3.0 mM Na_2HPO_4	7.413	+0.07	0.16	−0.0028
10 mM $Na_2B_4O_7$ (borate)	9.180	+0.01	0.020	−0.0082
50 mM Tris HCl + 16.7 mM Tris	7.382	na	na	−0.026

[a]Data from *pH Measurements*, Academic Press, 1978.

[b]pH change upon 50% dilution.

[c]pH change upon mixing 1 L of buffer with 1 gram equivalent of strong acid or base.

[d]Change in pH per degree Celsius.

heating, allowing higher buffer concentrations to be used. High buffer concentrations are sometimes useful to minimize interactions between the solute and the capillary wall (Section 3.4). Alternatively, high field strengths can be used to speed the separation.

Table 2.3 provides data describing the mobility of the fully ionized buffer component. Mobility matching between the buffer and solute has been proposed to improve the peak symmetry when the solute concentration is high (33). This problem will be discussed later (Sections 3.6A, 11.4).

B. Buffer Selection

For CZE and MECC separations, select a buffer from Table 2.3 depending on the desired pH. Use an initial buffer concentration of 20–50 mM. Buffers with low conductivity (mobility) are ideal to minimize Joule heating. Operating far from a buffer's pK_a or pI causes excessive conductivity because the buffer component may be fully ionized.

Selecting specific buffer anions can be useful for certain applications. Phosphate buffers are often used for low-pH protein separations (31, 34, 35). McCormick (35) suggested that phosphate ions bind to the capillary wall, reducing the impact of protein binding to anionic silanol groups. Pre-washing the capillary with pH 2.5 phosphate buffer was reported to reduce protein binding as well (36). Borate buffers are useful for separating carbohydrates (37–40) and catecholamines (41) because of specific complexation chemistry. Unless such specific interactions are identified, buffer selection based on the desired pH is usually satisfactory.

The selection of the buffer cation also plays a role in buffer conductivity (1). The correlation of the atomic radii and mobility was discussed in Section 2.1. In this regard, lithium and sodium salts are best used, since they contribute least to buffer conductivity.

Buffer Preparation

Titration of a buffer to the appropriate pH has some operational subtleties. In the trivial but ideal case, equimolar solutions of two different salts of the identical anion are blended to the appropriate pH. For example, to prepare a 50 mM pH 7 phosphate buffer, titrate a 50 mM monosodium salt with a 50 mM solution of the disodium salt. Under all possibilities, the final concentration must be 50 mM and the ionic strength must be consistent.

In other cases, the buffer is often titrated with acid or base to adjust the pH. Under these conditions, both pH and ionic strength are being adjusted (42). Unusual effects such as the reduction of EOF with increasing pH have been observed that are attributable to this problem. Selecting a buffer that requires only a minor titration will minimize these ionic strength effects. All buffers should be filtered before use through 0.45 μm filters.

2.11 Capillaries

A. Fused-Silica

Fused-silica, polyimide-coated capillaries similar to those used in capillary gas chromatography, is the material of choice for HPCE (Fig. 2.18). Internal diameters ranging from 25 to 100 μm are usually employed. Capillaries covering the range of 2–700 μm i.d. are commercially available.[3] Between 10 and 75 μm i.d., the i.d. tolerance is 3 μm. Fused-silica is a good (though not ideal) material because of its UV transparency, durability (when polyimide-coated) and zeta potential. Solute–capillary wall interaction is the most important limitation of the material. Functionalized capillaries and buffer additives (Section 3.4) are used to overcome this limitation. Gel-filled capillaries (Chapter 5) for CGE are also commercially available.

Today's instruments integrate the capillary in one of two fashions: with a capillary cartridge holder, or by direct insertion (open architecture) of the capillary into the unit. There are pros and cons for each approach. The cartridge concept provides a simple means of inserting a capillary and aligning the detection window with the instrument optics without excessive breakage. Some cartridges are integrated advantageously with the capillary cooling system. The disadvantages are cost, problems with incorporation of third-party capillaries, and slow changeover between capillaries within individual cartridges. Newer designs have overcome the latter two problems. The open-architecture systems permit rapid changeover and easy incorporation of third-party specialty capillaries. Refer to Table 1.3 to find which design is incorporated in various commercial instruments.

[3] Polymicro Technologies, Phoenix, AZ 85017.

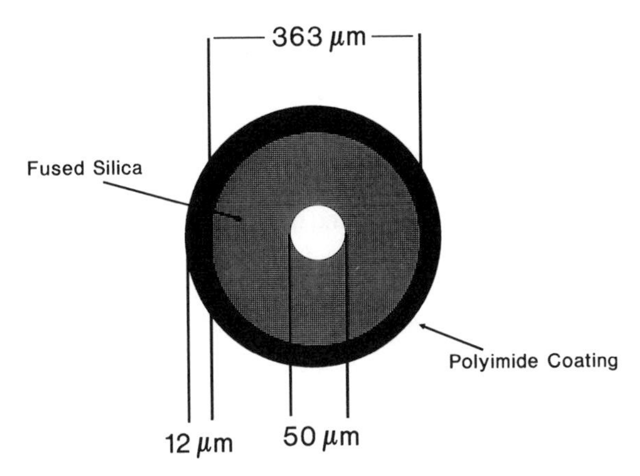

Figure 2.18. Cross-sectional view of a fused-silica capillary.

Since high-quality fused-silica capillaries can be purchased in bulk, it is advantageous to learn how to cut and condition a capillary. A new fused-silica capillary can be prepared and conditioned in half an hour with materials costs of under $5.00. At such a low cost, these home-cut capillaries are disposable items. Rather that attempt regeneration of a suspect capillary, it's wise to simply replace it.

B. Preparing A Fused-Silica Capillary

The procedure for preparing a bare silica capillary is given below. Only a ruler, a cutter (diamond cutter or silicon wafer), a butane lighter, methanol, and a tissue are required.

The method for introducing the detection window should *not* be used with gel-filled or surface-treated capillaries.

1. Nick the polyimide coating near the edge of the capillary as squarely as possible. Pull the capillary directly apart making sure not to pull at an angle. Measure from the cut end to the desired length, considering the length from the detector to the buffer reservoir, and cut again.
2. Measure the separation length of the capillary, and flame a length of about 2–3 mm with the butane lighter.[4] Clean the burnt polyimide with a tissue moistened with methanol. The capillary can now be inserted into the instrument. Take care not to bend the now-fragile detection window.[5]

[4] Heat burns off the coating without making the glass brittle. Alternatively, concentrated sulfuric acid at 130°C will remove the coating in a few seconds. This method is required when gel-filled or functionalized capillaries are used.

[5] For those not wishing to prepare a detection window, a fluorocarbon coated capillary (CElect-UVT) is available from Supelco, Bellefonte, PA. This capillary cannot be used with liquid cooling systems containing fluorocarbons because the coating will become brittle.

3. Wash the capillary for 15 min each with 1 N sodium hydroxide, 0.1 N sodium hydroxide, and run buffer. Change the detector side reservoir to run buffer. The system is ready to run.

The base conditioning procedure is important to ensure that the surface of the capillary is fully charged. For some methods it is necessary to regenerate this surface with 0.1 N sodium hydroxide, and in extreme cases, 1 N sodium hydroxide. This regeneration procedure is often necessary if migration times change on a regular basis. Regeneration may help if the zeta potential at the capillary wall is altered. Binding of solutes or sample matrix components may be the cause of this problem.

C. Storing a Fused-Silica Capillary

Not all attempts to store a fused-silica capillary will be successful. Capillaries are prone to clogging from buffer constituents, particularly if the capillary is left to dry out. While this is not too serious with bare silica, capillary damage can be costly with chemically modified capillaries. Clogging is more frequent in 25 and 50 μm i.d. capillaries. It is often better to discard an untreated capillary than store it.

The following procedure will maximize your chances of successfully storing a capillary.

1. Rinse the capillary with 0.1 N sodium hydroxide for several minutes.
2. Change both buffer reservoirs to distilled water.
3. Rinse for five minutes with distilled water.
4. Empty the appropriate buffer reservoir and draw air through the capillary for five minutes.
5. Remove the capillary from the instrument.
6. For capillaries not part of a cartridge assembly, slide some 0.5–1 mm i.d. Teflon tubing over the optical window and gently secure with tape or septa.

References

1. I. Z. Atamna, C. J. Metral, G. M. Muschik, and H. J. Issaq, *J. Liq. Chromatogr.* **13,** 2517 (1990).
2. W. Jennings, *Analytical Gas Chromatography*, pp. 29–31 (Academic Press, 1987).
3. K. D. Lukacs and J. W. Jorgenson, *HRC & CC* **8,** 407 (1985).
4. K. Saloman, D. S. Burgi, and J. C. Helmer, *J. Chromatogr.* **559,** 69 (1991).
5. T. A. A. M. van de Goor, P. S. L. Janssen, J. W. van Nispen, M. J. M. van Zeeland, and F. M. Everaerts, *J. Chromatogr.* **545,** 379 (1991).
6. T. Tsuda, K. Nomura, and G. Nakagawa, *J. Chromatogr.* **264,** 385 (1983).
7. H. H. Lauer and D. McManigill, *Anal. Chem.* **58,** 166 (1986).

8. Y. Walbroehl and J. W. Jorgenson, *Anal. Chem.* **58,** 479 (1986).
9. X. Huang, M. J. Gordon, and R. N. Zare, *Anal. Chem.* **60,** 1837 (1988).
10. F. M. Everaerts, A. A. A. M. Van de Goor, T. P. E. M. Verheggen, and J. L. Beckers, *HRC & CC* **12,** 28 (1989).
11. K. D. Altria and C. F. Simpson, *Chromatographia* **24,** 527 (1987).
12. K. D. Altria and C. F. Simpson, *Anal. Proc.* **23,** 453 (1986).
13. S. Fujiwara and S. Honda, *Anal. Chem.* **58,** 1811 (1986).
14. W. J. Lambert and D. L. Middleton, *Anal. Chem.* **62,** 1585 (1990).
15. T. Tsuda, K. Nomura, and G. Nakagawa, *J. Chromatogr.* **248,** 241 (1982).
16. B. B. VanOrman, G. G. Liversidge, G. L. McIntire, T. M. Olefirowicz, and A. G. Ewing, *J. Microcol. Sep.* 2, 176 (1990).
17. C. Schwer and E. Kenndler, *Anal. Chem.* **63,** 1801 (1991).
18. S. Fujiwara and S. Honda, *Anal. Chem* ***59,*** 487 (1987).
19. S. Hjertén, *J. Chromatogr.* **347,** 191 (1985).
20. J. W. Jorgenson and K. D. Lukacs, *Anal. Chem* **53,** 1298 (1981).
21. J. W. Jorgenson and K. D. Lukacs, *Clin. Chem.* **27,** 1551 (1981).
22. J. W. Jorgenson and K. D. Lukacs, *HRC & CC* **4,** 230 (1981).
23. A. E. Jones and E. Grushka, *J. Chromatogr.* **466,** 219 (1989).
24. E. Grushka, R. M. McCormick, and J. J. Kirkland, *Anal. Chem.* **61,** 241 (1989).
25. J. H. Knox, *Chromatographia* **26,** 329 (1988).
26. H. T. Rasmussen and H. M. McNair, *J. Chromatogr.* **516,** 223 (1990).
27. R. J. Nelson, A. Paulus, A. S. Cohen, A. Guttman, and B. L. Karger, *J. Chromatogr.* **480,** 111 (1989).
28. Y. Kurosu, K. Hibi, T. Sasaki, and M. Saito, *HRC & CC* **14,** 200 (1991).
29. G. McLaughlin, R. Palmieri, and K. Anderson, *Benefits of Automation in the Separation of Biomolecules by High Performance Capillary Electrophoresis* (J. J. Villafranca, ed.), Techniques in Protein Chemistry II (Academic Press, 1991).
30. G. M. McLaughlin, J. A. Nolan, J. L. Lindahl, R. H. Palmieri, K. W. Anderson, S. C. Morris, J. A. Morrison, and T. J. Bronzert, *J. Liq. Chromatogr.* **15,** 961 (1992).
31. A. D. Tran, S. Park, P. J. Lisi, O. T. Huynh, R. R. Ryall, and P. A. Lane, *J. Chromatogr.* **542,** 459 (1991).
32. C. Clark Westcott, *pH Measurements* (Academic Press, 1978).
33. V. Sustacek, F. Foret, and P. Boček, *J. Chromatogr.* **545,** 239 (1991).
34. M. Strickland and N. Strickland, *American Lab,* November, 60 (1990).
35. R. M. McCormick, *Anal. Chem.* **60,** 2322 (1988).
36. M. Zhu, R. Rodriguez, D. Hansen, and T. Wehr, *J. Chromatogr.* **516,** 123 (1990).
37. S. Honda, S. Iwase, A. Makino, and S. Fujiwara, Anal. Biochem. **176,** 72 (1989).
38. S. Tanaka, T. Kaneta, and H. Yoshida, *Anal. Sciences* **6,** 467 (1990).
39. S. Hoffstetter-Kuhn, A. Paulus, E. Gassmann, and H. M. Widmer, *Anal. Chem.* **63,** 1541 (1991).
40. S. Honda, S. Suzuki, A. Nose, K. Yamamoto, and K. Kakehi, *Carbohydrate Research* **215,** 193 (1991).
41. T. Kaneta, S. Tanaka, and H. Yoshida, *J. Chromatogr.* **538,** 385 (1991).
42. J. Vindevogel and P. Sandra, *J. Chromatogr.* **541,** 483 (1991).

Chapter 3

Capillary Zone Electrophoresis

Fig 7

3.1 Introduction

Separations by CZE are performed in a homogeneous carrier electrolyte. Otherwise known as free-solution capillary electrophoresis, CZE is further distinguished from other forms of HPCE by the absence of a gel (Chapter 5) or a pseudophase (Chapter 7). CZE is the simplest mode of HPCE.

The separation mechanism is illustrated in Fig. 3.1. Ionic components are separated into discrete bands when each solute's individual mobility is sufficiently different from all others. Separations of small ions, small molecules, peptides,

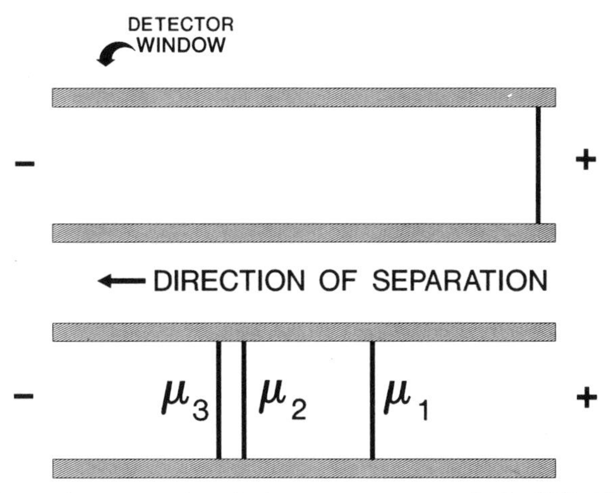

Figure 3.1. Schematic representation of a three-component separation by CZE at the moment of injection (top) and after separation (bottom).

proteins, viruses, bacteria, and colloidal particles have been reported. Zwitterions such as peptides are easily separated because the ionic charge can be fine-tuned by careful adjustment of the buffer pH.

Four fundamental features are required for good separations by CZE: (i) the individual mobilities of each solute in the sample differ from one another; (ii) the background electrolyte is homogeneous and the field strength distribution is uniform throughout the length of the capillary; (iii) neither solutes nor sample matrix elements interact or bind to the capillary wall; and (iv) the conductivity of the buffer substantially exceeds the total conductivity of the sample components.

3.2 Mobility

The fundamental parameter governing CZE is electrophoretic mobility (Eq. (2.3)). Since mobility depends on a solute's charge/size ratio, the buffer pH is the most important experimental variable. The impact of pH on mobility, corrected for EOF, is illustrated in Fig. 3.2 (1).

Among the solutes described in Fig. 3.2 are acids, zwitterions, and an alkali metal, sodium. Within the pH range covered in the figure, the net charge on sodium is constant, as is the mobility. Acetate and glutamate are negatively charged and show electrophoretic mobility toward the positive electrode (negative mobility). Zwitterions such as amino acids, proteins, and peptides show charge reversal at their p*I* concurrent with shifts in the direction of electrophoretic mobility.

A mobility plot is an invaluable tool for methods development. At a glance, the optimal pH for separation is clear, and problem areas can be noted. The p*I* or pK_a of a solute can frequently be found from the mobility plot.

The net charge on a solute at any pH can be calculated using the Henderson–Hasselbalch equation for bases,

$$pH = pK_a + \log(1/a - 1), \tag{3.1}$$

and for acids,

$$pK_a = pH + \log(1/a - 1), \tag{3.2}$$

where a = the fraction ionized.

For monovalent ions, the calculation is straightforward. For zwitterions such as amino acids, the contributions of the acidic and basic portions, and side chains, if any, must be combined to calculate the net charge, as shown in Table 3.1. In Fig. 3.3, the Henderson–Hasselbalch equation is solved for bases between pH 2.5 and 11.0.

The net charge on small peptides is calculated in a similar fashion. The pK_a values for the C and N terminus, along with values for side chains, must be used to determine the contribution to the charge from each moiety. For proteins and large

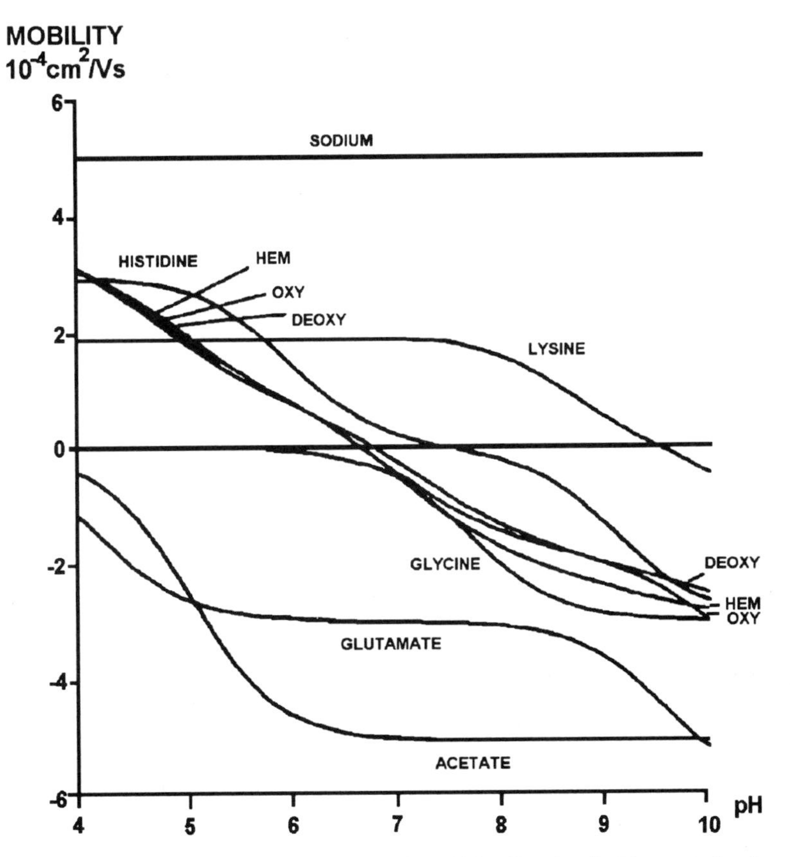

Figure 3.2. Effect of buffer pH on electrophoretic mobility. Reprinted with permission from *J. Chromatogr.* **480,** 35, copyright ©1989 Elsevier Science Publishers.

Table 3.1. Calculation of Net Charge for Lysine at pH 7

	Functional Group	pK_a	Charge
H_2N-CH-COOH	Carboxylate	2.18	−1.000
$(CH_2)_4$	α-Amino group	8.95	+0.989
NH_2	ε-Amino group	10.53	+1.000
		Net Charge	+0.989

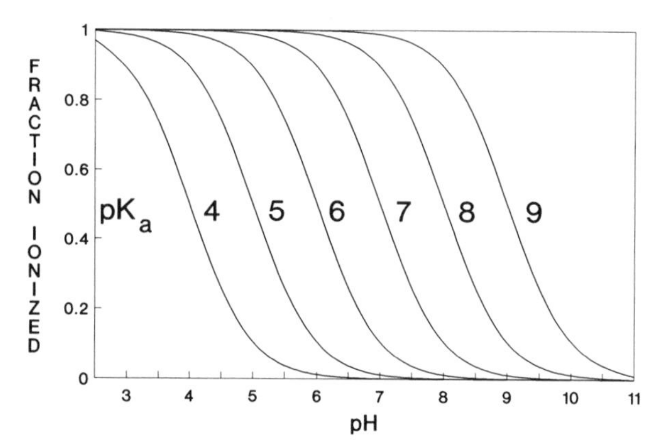

Figure 3.3. Solution for the Henderson–Hasselbalch equation for bases of specified pK from pH 2.5 to 11.

peptides, this simple model does not work because of charge suppression, and more complex calculations are required (2, 3).

These calculations can be useful to correlate mobility with pH. Grossman *et al.* (4) developed an empirical relationship that linked mobility to a complex function, $\ln(q + 1)/n^{0.43}$, where q is the charge and n is the number of amino acids. Deyl *et al.* (5, 6) and Rickard *et al.* (7) found a best-fit correlation for mobility with $q/M^{2/3}$ where M is the molecular weight. This is consistent with Offord's model, which describes mobility of large molecules. For small molecules, $M^{1/3}$ provides a better fit. Grossman's model falls between these two values (2), as might be

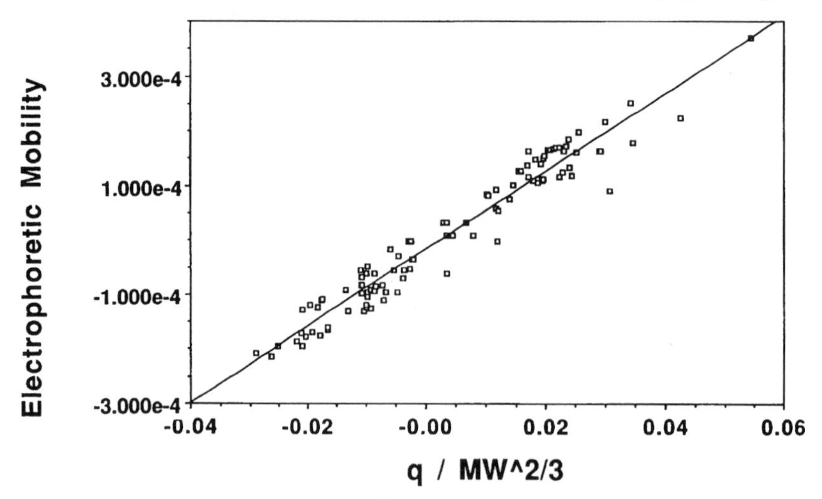

Figure 3.4. Correlation of mobility and $q/M^{2/3}$ for a human growth hormone digest (separated at pH 2.35, 8.0, and 8.15), insulin-like growth factor II digest (separated at pH 2.35 and 8.15). Reprinted with permission from *Anal. Biochem.* **197,** 197, copyright ©1991 Academic Press.

expected when separating peptides containing between 3 and 39 amino acids. The accuracy of the $q/M^{2/3}$ versus mobility model is illustrated in Fig. 3.4. Data from a series of peptides from two separate digests separated by CZE at three different pH values are plotted.

There is a practical side to this story. If the pK_a and molecular weight of a substance are known, the use of mobility calculations to select the initial experimental conditions is a worthwhile undertaking. While optimal separation conditions cannot be predicted using this model, the calculations are effective as a first approximation.

The profound effect of pH on mobility (8,9) is illustrated in Fig. 3.5 for two peptides differing by one amino acid, with sequences AFKAING and AFKADNG (8). At pH 2.5, the calculated charges on these two peptides are 1.41 and 1.36, respectively. At pH 4.0, the calculated charges become 1.02 and 0.46. It is expected and observed that greater resolution is found for the higher-pH buffer, since the mobilities are better distinguished.

Adjustment of pH solves many separation problems. Others may require additional experimental modifications. The next step is to examine the use of buffer additives to address these situations.

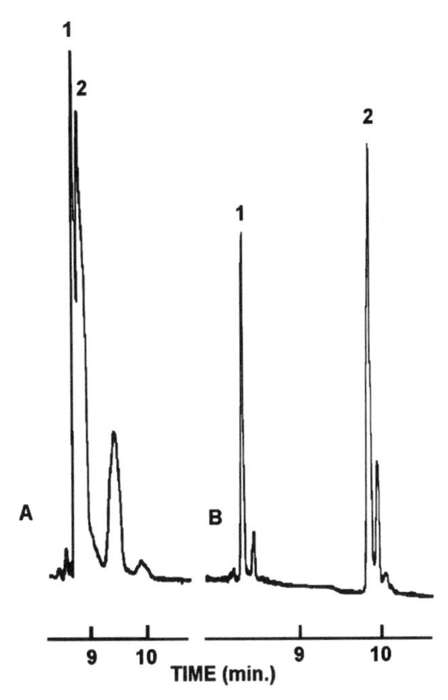

Figure 3.5. Effect of buffer pH on the selectivity of peptide separations by CZE. Capillary length: 45 cm to detector (65 cm total) × 50 μm i.d.; buffer: citric acid, 20 mM, (a) pH 2.5, (b) pH 4.0; field strength: 277 V/cm; current: (a) 24 μA, (b) 12 μA; temperature: 30°C; detection: UV, 200 nm; peptides: (1) AFKAING, (2) AFKADNG. Reprinted with permission from *Anal. Chem.* **61,** 1186, copyright ©1989 Am. Chem. Soc.

3.3 Buffer Additives

Buffer additives are used to address several problems:

1. to modify mobility;
2. to modify electroosmotic flow;
3. to solubilize solutes or sample matrix compounds; or
4. to reduce or eliminate solute-capillary wall interaction.

A listing of various buffer additives and their functions is given in Table 3.2. Other buffer additives can change the mechanism of separation. For example, use of cellulose polymers imparts a sieving mechanism resembling gel electrophoresis (Chapter 5). Addition of surfactants such as sodium dodecyl sulfate shifts the mechanism to electrokinetic chromatography (Chapter 7).

Separation of solutes that comigrate at any pH requires selective modification of mobility. Selection of additives that bind to the analyte can permit separation, provided the appropriate additive can be identified. The additive usually modifies the net charge of the ion.

Metal ions such as Cu(II), Zn(II), Ca(II), and Fe(II) often coordinate to nitrogen, oxygen, and sulfur. Mosher (10) separated histidine-containing dipeptides with Zn(II) and Cu(II) additives in 100 mM phosphate buffer, pH 2.5. Addition of 20–30 mM $ZnSO_4$ gave baseline resolution of DL-His–DL-His diastereomers. Other groups that can interact with metal ions include thiol, indoyl, *N*-terminal amino, imidazole, and β-carboxyglutamate. Phosphorylated proteins interact with Mg(II) and Mn(II). Separations of calcium-binding proteins such as calmodulin, parvalbumin, and thermolysin, along with zinc-binding proteins such as carbonic anhydrase and thermolysin, have been reported (11, 12) using the respective metal ion. Issaq *et al.* (13) found the best conditions for separating small peptides included 50 mM Zn(II) in pH 2.5, 50 mM phosphate buffer.

Table 3.2. Buffer Additives for CZE

Additive	Function
Inorganic salts	Protein conformational changes, reduce wall interactions
Zwitterions	Reduce wall interactions
Metal ions	Modify mobility
Organic solvents	Solubilizer, modify EOF, reduce wall interactions
Urea	Solubilize proteins
Sulfonic acids	Ion pairing agents, surface charge modifiers
Cellulose polymers	Modify EOF
Amine modifiers	Tie up active sites on capillary wall
Cationic surfactants	Charge reversal on capillary wall

Addition of salts to buffers can alter mobility and selectivity for some proteins (14). Both the cation and the anion of the additive can have some impact. It is possible that the added salt modifies the protein's electrical double layer, conformation, or ionization of individual amino acids. Separations are given in Fig. 3.6.

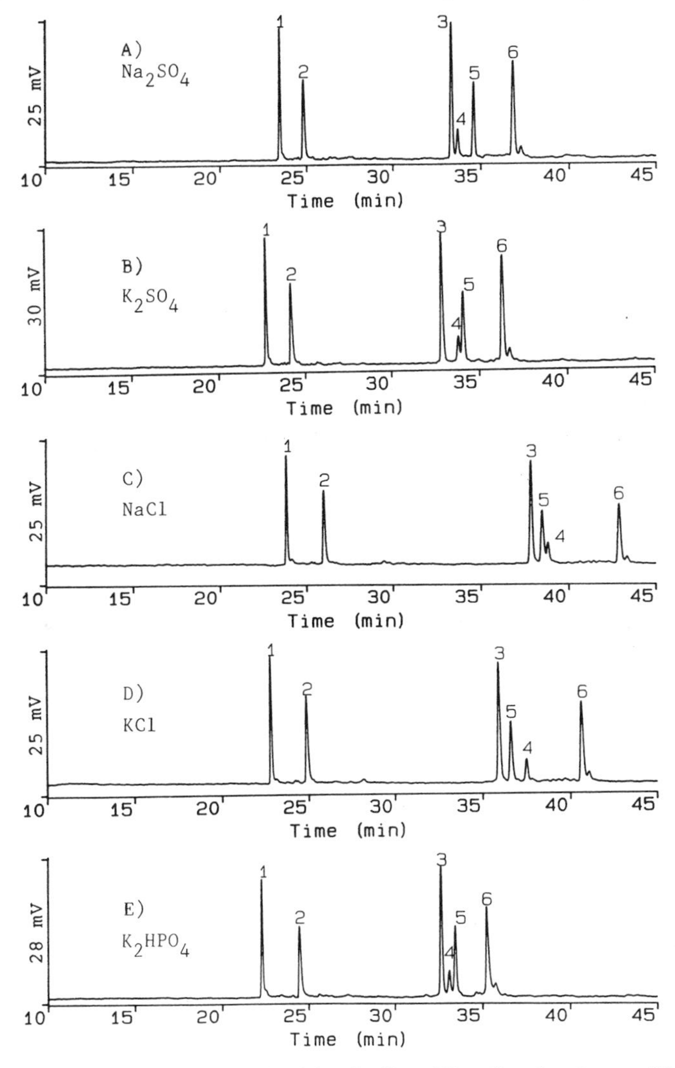

Figure 3.6. Effect of salt additive on selectivity. Capillary: 75 cm (length to detector, 55 cm) × 50 μm i.d.; buffer: 50 mM 1,3-diaminopropane, the specified salt and titrated to pH 7 with phosphoric acid; field strength: 200 V/cm; solutes: (1) lysozyme, (2) cytochrome *c*, (3) ribonuclease A, (4) trypsinogen, (5) α-chymotrypsinogen A, (6) rhu-IL-4; salt concentration: (A–B, E) 40 mM, (C–D) 80 mM. Reprinted with permission from *J. Microcol. Sep.* **3,** 241, copyright ©1991 Microseparations Inc.

Hydrophobic interaction between peptides or proteins and alkyl sulfonic acids increases the net negative charge of the solute (15). Solutes with differing affinity for the sulfonic acid are differentiated. The alkylsulfonic acid's tail binds to a hydrophobic site on the analyte with the anionic head group extending out into the bulk solution. Increasing the negative charge on the solute is helpful in reducing wall effects (Section 3.4). Selectivity can be altered as well by selecting the length of the alkyl chain; pentanesulfonic through decanesulfonic acids are good choices. Increasing the sulfonic acid concentration generally improves selectivity at the expense of Joule heating; 50 mM is sufficient for many separations.

Organic solvents can modify both mobility and EOF (16). Fujiwara and Honda (17) used 1:1 acetonitrile:20 mM phosphate buffer, pH 7, to separate amino- and methylbenzoic acid isomers. Chadwick and Hsieh (18) separated *cis* and *trans*-retinoic acid in 1:1 borate:acetonitrile. The organic solvent is also useful to improve the solubility of solutes (17, 19). Organic modifiers, while useful in CZE, are very important as additives to MECC buffers (Section 7.6).

The addition of 7 M urea to 200 mM borate buffer, pH 9.2 was used to solubilize hydrophobic membrane proteins (20). A 25 μm i.d. capillary was required because of the high ionic strength borate buffer. The use of fresh urea does not increase the ionic strength or the conductivity of the buffer solution.

Hjertén reported that a cellulose polymer such as hydroxypropylmethyl cellulose can reduce the EOF (21). This additive provides EOF reduction without increasing the conductivity of the buffer. Reducing EOF is not critical in CZE unless resolution needs improvement. Reduction of EOF is more important in CIEF (Chapter 4) and CITP (Chapter 6). These same cellulose polymers also serve as physical gels for performing size separations of DNA by CGE (Chapter 5).

3.4 Solute–Wall Interactions

A. The Problem of Wall Effects

A key advantage of HPCE compared with HPLC is the absence of the chromatographic packing. The vast surface area of the packing material is responsible in part for irreversible adsorption of solutes, particularly proteins. However, the composition of the capillary surface still provides opportunities for protein adsorption. Binding of solutes to the capillary wall leads to bandbroadening, tailing, and irreproducibility of separations. If the kinetics of adsorption/desorption is slow, broadened tailed peaks occur. Irreversible adsorption leads to modification of the capillary, altered EOF, and loss of resolution.

Fig. 3.7 illustrates the electrostatic binding of a protein to the capillary wall. At most pH values, the capillary wall has a negative charge because of silanol ionization. Separation of a protein at a pH below its p*I* produces a cationic solute

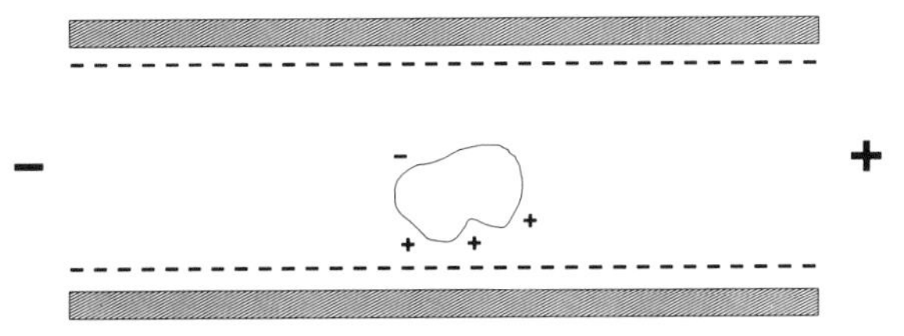

Figure 3.7. Pictoral representation of the electrostatic interaction of a cationic protein with the anionic capillary wall.

that ion-pairs to the capillary wall. Hydrophobic binding occurs as well between the epoxide moiety of fused-silica and a hydrophobic solute. Since separations occur in aqueous media, hydrophobic solutes are not well solvated, further enhancing this potential binding mechanism.

Wall effects on bare silica have proved to be a difficult problem since the early days of HPCE (22).

Since then, several solutions have been proposed, including the use of

1. extreme pH buffers,
2. high-concentration buffers,
3. amine modifiers,
4. dynamically coated capillaries, and
5. treated or functionalized capillaries.

B. Extreme-pH Buffers

In 1986, Lauer and McManigill (23) reasoned that anionic proteins should be repelled from the anionic capillary wall. Selecting a pH that is 1–2 units above the p*I* of a protein produces separations free of wall effects. There are several limitations to this approach: (i) Hydrophobic binding is not eliminated. (ii) Proteins with high p*I*s, such as histones, do not form anions until pH 13 is reached. To reach that pH, 100 mM sodium hydroxide is required, a solution that is too conductive to be a useful electrolyte. (iii) Alkaline pHs may not be optimal for the separation. Despite these limitations, this useful approach was the first to employ charge manipulation to reduce wall interactions.

Two years later, McCormick (24) chose a different approach. At pH 1.5, the silanol groups at the capillary wall are not ionized. While the proteins are cationic at that pH, electrostatic interactions should be eliminated. McCormick proposed that phosphate groups bind to the capillary wall, further decreasing the activity of the silanol groups. Zhu *et al.* (25) found that pH 2.5 phosphate buffer prevented nonspecific adsorption of hemoglobin on both bare and polyacrylamide-coated capillaries. Borate, phosphate (pH 7), and especially acetate gave appreciable protein adsorption. The low pH approach works well but is limited by hydrophobic binding and poor selectivity. Some proteins may precipitate in low-pH buffers as well.

C. High-Concentration Buffers

Green and Jorgenson (26) found that buffers with high concentrations of salts decreased adsorption of proteins on the capillary wall. The added salts probably occupy potential adsorption sites. Potassium sulfate was the additive of choice based on performance and UV absorption background. A buffer comprising 100 mM CHES, pH 9, with 250 mM potassium sulfate and 1 mM EDTA produced 140,000 theoretical plates for trypsinogen. The approach was limited because low field strengths and narrow i.d. capillaries were necessary because of Joule heating problems. The combination of low field strength and low EOF yielded prolonged run times. Sensitivity was also problematic because of background UV absorption by the electrolyte.[1]

To address the limitations of high ionic strength buffers, Bushey and Jorgenson (27) substituted a zwitterionic species such as glycine, betaine, or sarcosine for the ionic salt. When the buffer pH is adjusted to the p*I* of the zwitterion, its net charge, mobility, and conductivity approach zero. The best results were found using 40 mM phosphate–250 mM potassium sulfate, pH 7; 100 mM CHES–250 mM potassium sulfate, pH 9; and 40 mM phosphate–2 M betaine–100 mM potassium sulfate, pH 7.6. In principle, any of the Good's buffers (Table 2.3) could be used at the appropriate pH provided their UV absorption is sufficiently low. A protocol for protein separation and a reagent kit are commercially available.[2]

Even modest increases in the ionic strength can have profound impact on the separation. Increasing the buffer concentration from 10 mM tricine, pH 8.1, to 100 mM tricine at the same pH gave dramatic improvements for the separation of a human growth hormone tryptic digest (28).

D. Amine Modifiers

Amine derivatives such as triethylamine have been used for years to suppress silanol effects in HPLC. Alkyldimethylamines have been shown to be most effective (29). Lauer and McManigill (23) added 5 mM putrescine (1,4-diaminobutane) to improve the resolution and peak shape of myoglobin. The additive was not effective for lysozyme and cytochrome *c*, and the EOF was reduced by 60%. Stover *et al.* (30) used putrescine to reduce wall effects for histidine containing heptapeptides. Tsuda *et al.* (31) found that 1% ethylenediamine reduced tailing during the MECC separation of polyamines without EOF reduction. Rohlicek and Deyl (32) used 5 mM 1,5-diaminopentane to improve protein separations. Bullock and Yuan (14) used 30–60 mM 1,3-diaminopropane to improve resolution of basic proteins such as lysozyme, cytochrome *c* and ribonuclease. Monovalent amines such as triethylamine or *n*-propylamine were not as effective. The use of 1,3-diaminopropane

[1] High ionic strength buffers are useful for micropreparative separations (Section 10.11).

[2] Waters Division of Millipore, Milford, MA.

allows separations of basic proteins at pHs below their isoelectric points. Since the pH can be independently controlled over a wide range, a buffer can be optimized without considering wall effects. Separations are shown in Fig. 3.8.

E. Dynamically Coated Capillaries

Wiktorowicz and Colburn (33), Cunico *et al.* (34), Tsuji and Little (35), and Emmer *et al.* (36, 37) use an approach that reverses the charge of the capillary wall to reduce protein adsorption. Charge reversal is accomplished using a cationic surfactant such as cetyltrimethylammonium bromide (CTAB) or hexadimethrine bromide (polybrene).

The mechanism of charge reversal is illustrated in Fig. 3.9. Ion-pair formation between the cationic head group of the surfactant and the anionic silanol group

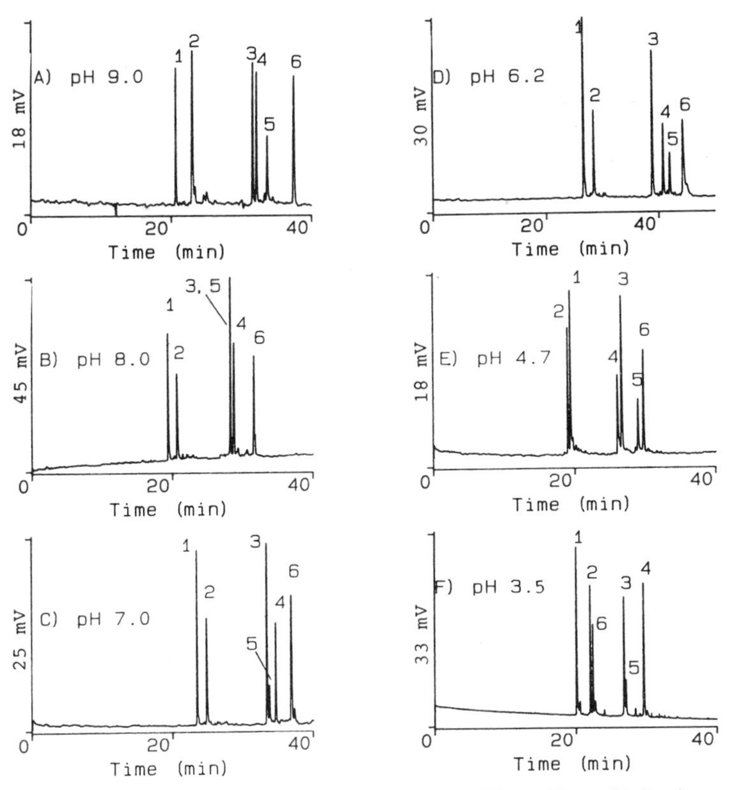

Figure 3.8. Separation of basic proteins on an untreated fused-silica capillary with diaminopropane as a buffer additive. Capillary: 75 cm (55 cm to detector) × 50 μm i.d.; buffers: pHs as noted in the figure with 30–60 mM diaminopropane as an additive; field strength: 200–240 V/cm; solutes: (1) lysozyme; (2) cytochrome *c*; (3) ribonuclease; (4) α-chymotrypsin; (5) trypsinogen; (6) rhuIL-4. Reprinted with permission from *J. Microcol. Sep.* **3**, 241, copyright ©1991 Microseparations Inc.

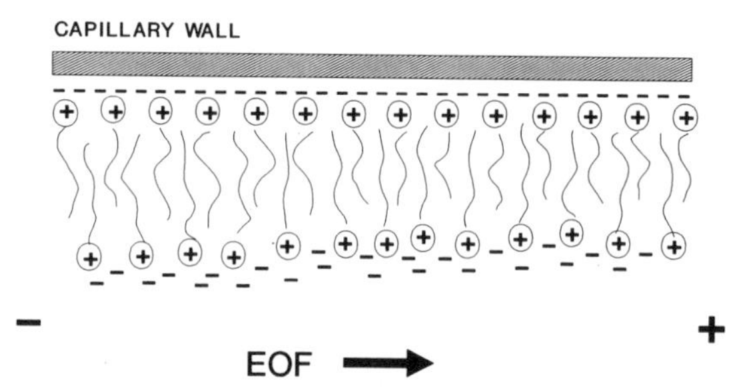

Figure 3.9. Pictoral representation of cationic surfactant mediated charge reversal at the capillary wall with concomitant reversal of the electroosmotic flow.

occurs. The hydrophobic surfactant tail extending into the bulk solution cannot be solvated by water. Its solvation need is satisfied by binding to the tail of another surfactant molecule. As a result, the cationic head group of the second surfactant molecule is in contact with the bulk solution. The capillary wall behaves with cationic character because of this treatment, and the EOF is directed toward the positive electrode. For most separations, it is necessary to operate the HPCE instrument with reverse (sample-side negative) polarity. Using this approach, a buffer pH is selected that is below the pI of the target protein. The cationic protein is now repelled from the cationic wall.

In the Wiktorowicz approach,[3] the cationic surfactant is not present in the run buffer. The capillary is coated before the separation step. Excess surfactant is flushed from the capillary and replaced with run buffer. This coating is stable for many runs, after which the capillary is simply recoated. The coating is sufficiently stable to permit CE/MS of proteins without detection of the surfactant (38).

In Emmer's method, a fluorinated surfactant[4] is added to the run buffer at a concentration of 100 μg/mL. Over 300,000 theoretical plates was reported for lysozyme, a protein with a p*I* of 11. This approach, along with Wiktorowicz's, does not require a covalently treated capillary, extreme pH, amine additives, or high ionic strength buffers.

F. Functionalized Capillaries

Traditional silane chemistry used to prepare HPLC packings has been successfully transferred to HPCE. The goal is to prepare a stable and inert surface that provides no retention. Even a minor amount of retention will dramatically reduce efficiency because of mass-transfer effects.

[3] A protocol and kit tradenamed μ-Coat, is available from Applied Biosystems, Inc., Foster City, CA.

[4] Fluorad FC 134 from 3M Company, St. Paul, MN.

An ideal surface is very hydrophilic; a buffer additive may be used to maintain hydrophilic character. Since HPCE is performed in aqueous media, any hydrophobic character in the surface treatment will result in undesirable retention. Jorgenson and Lukacs (39) prepared a silylated capillary to reduce the EOF and improve the resolution of dansyl amino acids. In 1985, Hjertén (40) first produced treated capillaries using polyacrylamide or methylcellulose as the coating (40). Solute adsorption was reduced, but the polyacrylamide coating was not very stable, particularly at high buffer pH. Since then, a variety of coatings, listed in Table 3.3, have been synthesized.

Very little data have been published on most of these coatings. Polyacrylamide-treated capillaries received the most attention because of polyacrylamide's history of use in slab-gel electrophoresis. These surface-treated capillaries are commercially available. The coating has a limited lifetime, particularly in alkaline buffers. One manufacturer is reportedly working on a stabilized version. Vinyl-bound polacrylamide is reportedly more stable (42) than those prepared by the method of Hjertén (40). While vinyl-bound polyacrylamide yielded over 200,000 theoretical plates, this was less than predicted by theory (42). That is taken as evidence for modest interaction between proteins and the capillary wall. Such interaction must

Table 3.3. Coated Capillaries for HPCE

Coating	Reference
Polymethylglutamate	(41)
Polyacrylamide[a]	(40, 42–45)
Polyvinylpyrrolidone	(24)
Polyethylene glycol	(46–48)
Polyethyleneimine	(48, 49)
Maltose	(50)
Methylcellulose	(40)
Epoxydiol	(50, 51)
Arylpentafluoro	(52)
Trimethylsilane	(39, 53)
Octyl[b]	(54)
Octadecyl[c]	(54)
"Polar phase"[d]	(54)
Hydroxylated polyether	(55)
Alkylsilane, Brij 35[e]	(56)

[a]Polyacrylamide-coated capillaries are available from Bio-Rad, Richmond, CA.
[b]CElect-H1, available from Supelco, Bellefonte, PA.
[c]CElect-H2, available from Supelco, Bellafonte, PA.
[d]CElect-P1, available from Supelco, Bellefonte, PA.
[e]CE-100-C18, available from Isco, Lincoln, NE.

have rapid kinetics, since the peak skew or tailing values are small. Other capillaries that are commercially available are noted in Table 3.3.

Without going into exhaustive detail, it can be noted that problems with stability, efficiency, need for additional additives, and reproducibility plague many of these new surface treatments. The hydrophobic surfaces require the addition of surfactants to tie up binding sites and generate a hydrophilic coating. The need for additional additives is not necessarily bad. It's just another factor to be considered.

Lowering of the EOF occurs in most coated capillaries. Under these circumstances, it may be necessary to reverse the voltage polarity to run the separation (57). Other positively charged surface treatments reverse the direction of the EOF (49). Towns and Regnier found that a surface coating of 30 Å is required to suppress the underlying silanol effects. The dependence of EOF on buffer pH can also be reduced (56).

Nashabeh and El Rassi (55) produced fuzzy and interlocked polyether coatings for CZE of proteins. The fuzzy capillary (Fig. 3.10) yielded 250,000 theoretical plates with run-to-run, day-to-day, and capillary-to-capillary migration time precision of 0.2–1.8, 0.6–2.3 and 4.2–6.7%, respectively. The capillaries lasted for several weeks when used below pH 6. At pH 6.5–7.0, lifetimes of more than 80 hours were reported.

The use of hydrophobic coatings with surfactant additives can produce a well-coated hydrophilic surface (56). Non-ionic surfactants such as Brij-35 or

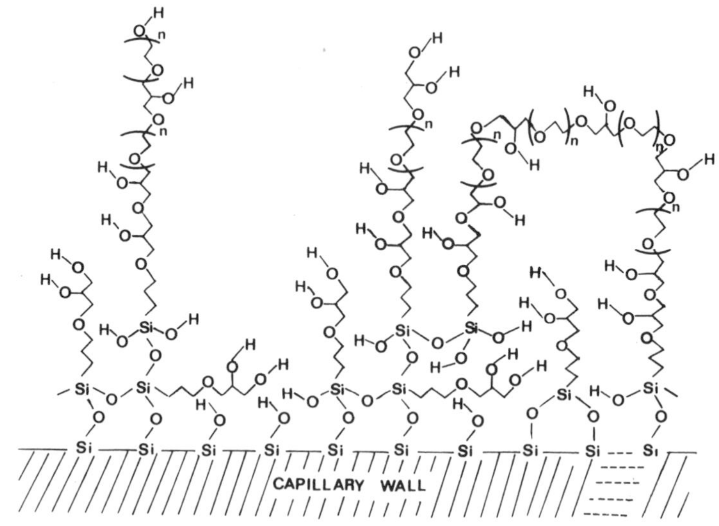

Figure 3.10. Illustration of the idealized structure of fuzzy polyether coating. Reprinted with permission from *J. Chromatogr.* **559**, 367, copyright ©1991 Elsevier Science Publishers.

Tween 20 at levels of 0.01% in the buffer yield good separations of acidic (Fig. 3.11a) or basic (Fig. 3.11b) proteins. Approximately 240,000 theoretical plates have been obtained for myoglobin.

G. What About GC Capillaries?

Hydrophobic GC capillaries can probably be substituted for hydrophobic coated capillaries as described above for a very low cost per capillary. DB-1 (dimethylpolysiloxane), DB-17 (50% methyl/50% phenylpolysiloxane), and others are available pre-cut, including detection windows.[5] These capillaries can also be purchased in bulk as GC capillaries, but detection windows must be cut using either hot sulfuric or nitric acid. The use of these capillaries has not been extensively studied with HPCE.

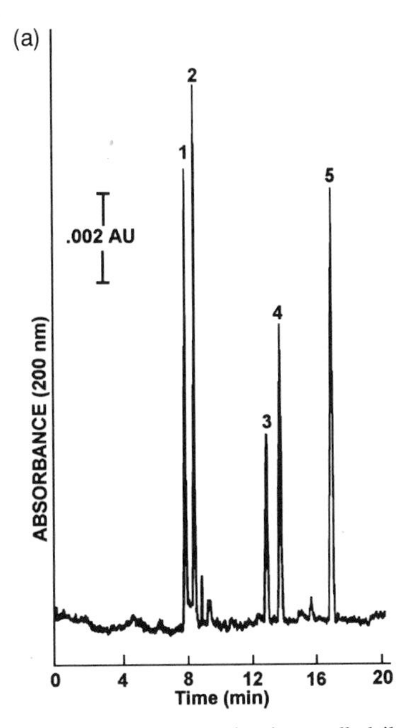

Figure 3.11. (a) Electropherogram of basic proteins in an alkylsilane-treated capillary with a non-ionic surfactant buffer additive. Capillary: 50 cm × 75 μm i.d. alkylsilane-treated capillary; buffer: 10 mM phosphate, pH 7 with 0.001% Brij-35; field strength: 300 V/cm; detection: 200 nm; solutes: (1) lysozyme; (2) cytochrome *c*; (3) ribonuclease A; (4) α-chymotrypsinogen; (5) myoglobin. Reprinted with permission from *Anal. Chem.* **63,** 1126, copyright ©1991 Am. Chem. Soc.

[5] J&W Scientific, Folsom, CA.

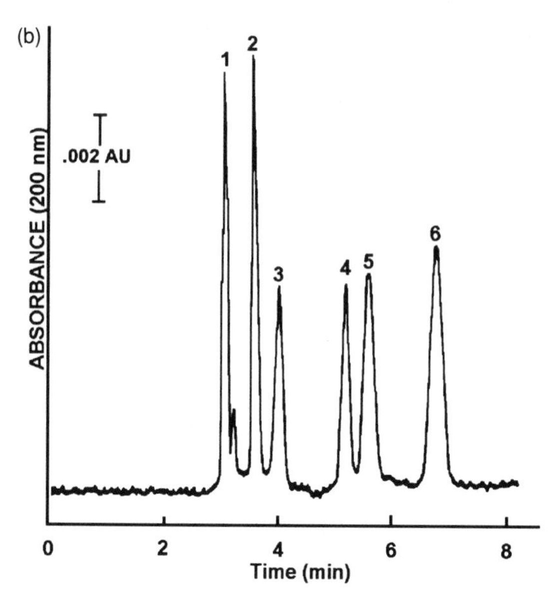

Figure 3.11 *(continued).* (b) Electropherogram of acidic proteins in an alkylsilane-treated capillary with a non-ionic surfactant buffer additive. Capillary: 30 cm × 75 mm i.d. alkylsilane-treated capillary; conditions as in Fig. 3.14; solutes: (1) myoglobin; (2) conalbumin; (3) transferrin; (4) B-lactoglobulin B; (5) B-lactoglobulin A; (6) ovalbumin. Reprinted with permission from *Anal. Chem.* **63,** 1126, copyright ©1991 Am. Chem. Soc.

H. Treated Capillaries Versus Buffer Additives

Good separations of proteins are possible using either buffer additives or treated capillaries. It is impossible to predict if one method will predominate over the other. From the standpoint of cost, buffer additives are clearly preferable, since bare fused-silica capillaries are disposable items. On the other hand, buffer designs are more complex using additives, and the additives may interfere with mass spectrometry. For micropreparative work, removal of the additive may also be necessary. When separating ions or small molecules, treated capillaries are occasionally useful.

3.5 Bandbroadening

In Section 2.4, Jorgenson and Lukacs's model describing the efficiency of HPCE is presented. This model assumes that diffusion is the only cause of bandbroadening. At low voltages, molecular diffusion is a leading cause of bandbroadening. At higher voltages, the Joule heating problem causes parabolic flow. Adsorption of solutes at the wall also leads to dispersion, as does solute overload (Section 11.4).

Extracolumn processes such as injection (Chapter 9) and detection (Chapter 10) are also problematic.

The peak variance or dispersion can be expressed by

$$\sigma^2 = \sigma^2_{inj} + \sigma^2_{col} + \sigma^2_{det}\,, \tag{3.3}$$

where σ^2_{inj}, σ^2_{col}, and σ^2_{det} are the respective variances due to injection, the column, and the detector. The causes and control of injector- and detector-related bandbroadening are covered in Chapters 9 and 10, respectively. This section will deal with bandbroadening on the capillary.

On-capillary dispersion comprises contributions from injection overload, hydrostatic flow, diffusion, adsorption, and Joule heating. Injection overload (Section 11.4) and hydrostatic flow due to fluid imbalances are easily avoided. The contribution from diffusion, expressed by the Einstein equation (2.11), is directly proportional to the separation time.

The variance contribution from Joule heating has been extensively studied by Hjertén (21), Foret *et al.* (58), Grushka *et al.* (59), Knox (60), and Jones and Grushka (61). Grushka *et al.* (59) concluded that temperature effects are negligible in narrow-bore capillaries. Use of wide-bore capillaries is possible when low-conductivity buffers are used. The problem with low-conductivity buffers is decreased loading capacity (Section 11.4) and increased wall effects. Using Ohm's Law plots (Section 2.7) helps ensure that Joule heating is not a significant cause of dispersion. Operation at the voltage prescribed by the Ohms's Law plot minimizes diffusion-related dispersion because the field strength is optimized as well.

Adsorption or retention in HPCE is determined by a solute's adsorption/desorption kinetics with the capillary wall. These rate constants are not easily evaluated. A first approximation of the impact of wall effects can be understood using a random-walk model from chromatographic theory (29). Random-walk theory considers a solute moving down the capillary in discrete steps. The peak variance is expressed as

$$\sigma_s^2 = 2\left(\frac{k'}{1+k'}\right)^2 v_{ep} t_a L\,, \tag{3.4}$$

where t_a = the time for desorption.

Retention occurs whenever t_d, the time for desorption, is greater than > t_a. Tailing occurs because a desorbed solute does not return at once to the buffer solution. Retained solutes have a migration velocity of zero. Solutes in the buffer move at a rate determined by their migration velocity, and thus move ahead of retained material. Evaluation of the function k' versus $(k' / (1 + k'))^2$ (Fig. 3.12) describes a sigmoidal relationship approaching 1 at large k'. To achieve the theoretical efficiency of CZE, $k' = 0$. As the figure illustrates, even modest retention will lead to severe bandbroadening. The appropriate buffer additives or capillary coatings are required to minimize this form of bandbroadening.

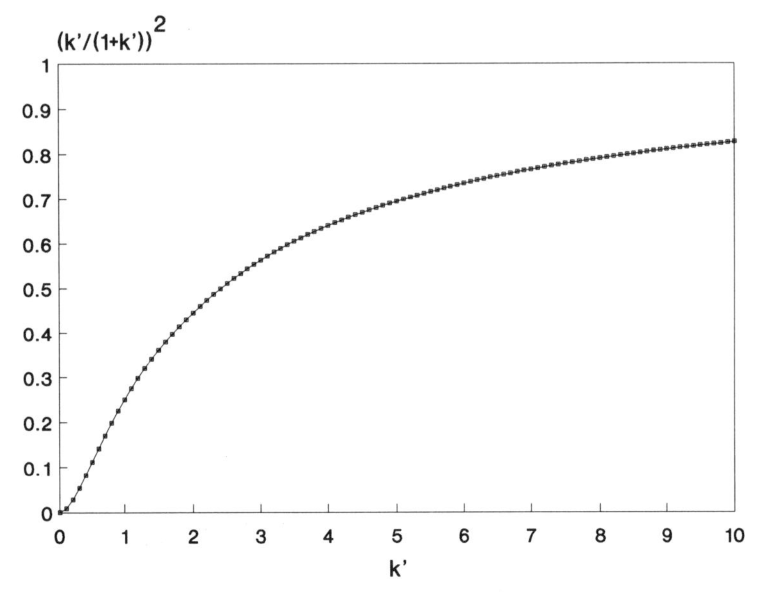

Figure 3.12. Solution of the exponential portion of the random walk model describing band-broadening in open tubular chromatography due to slow adsorption/desorption kinetics.

Capillary coatings do not always reduce wall interactions. Use of a 50 μm i.d. coated capillary instead of a 25 μm i.d. coated tube has been reported to improve resolution of histones (62). This is contrary to the expected results because of Joule heating. It's probable that the decrease in relative surface area of the wall in the larger-i.d. capillary reduces the impact of retention due to wall interaction.

3.6 Applications

A. Capillary Ion Electrophoresis

Small ions such as Na^+, K^+, Cl^-, Br^-, NO_2^-, transition metal ions, and lanthanide ions present unique separation and detection problems. Many of these species are completely ionized between pH 2 and 12. Other ions such as borate and carbonate have pK_as of 9.24 and 10.25, respectively; ammonia has a pK_b of 4.75. Their mobilities are substantially affected by pH. Even for fully ionized species, control of pH is still important since the EOF and resolution are affected.

The ionic equivalent conductance (IEC) is useful to predict separation, since this parameter correlates with mobility (63–65). Metal ions including alkali, alkaline earth, transition metals, and lanthanides may have similar IECs, so complexing reagents such as α-hydroxyisobutyric acid (HIBA) or citrate are required (64). Most anions have sufficiently differing IECs so separations are possible without special additives. Buffer ionic strength adjustment provides only subtle selectivity changes (63). Detection is problematic, since most of these ions do not have appreciable UV absorption.

Despite these problems, high-speed separations of small ions and metal–ion complexes are readily accomplished by CZE using absorbance (66), indirect absorbance (63, 64, 67–77), fluorescence (78), indirect fluorescence (79–82), and conductivity detection (83, 84). The subject was recently reviewed by Jandik *et al.* (70), tracing the history of the technique back to 1967 when Hjérten separated bismuth and copper in a 3 mm i.d. rotating capillary.

Inorganic ions are usually separated by ion chromatography (IC) with conductivity detection. The advantages of CZE compared to IC are speed, resolution, and the lack of a need for gradient elution. Operating costs are considerably lower, since ion-exchange columns are usually expensive. Reagent kits and protocols for separating small ions are commercially available.[6] Some comparative CZE separations that illustrate the speed and resolution advantages compared to IC are shown in Fig. 3.13.

Jandik *et al.* (70) outlined the important principles that govern small ion separations. These principles are general for many CZE applications.

1. The direction of EOF. To perform separations rapidly, the EOF and electrophoretic flow must be in the same direction (63). For anions, a buffer additive to reverse the EOF is used. For the highest resolution, a slower or countermigrating EOF is useful but the separation times are longer. This technique may be necessary when separating two ions with similar mobilities when their concentrations are substantially different.
2. Mobility or conductivity matching. When the conductivity of a sample zone approaches or exceeds that of the background electrolyte, bandbroadening will occur (Section 11.4). This becomes a problem at high solute concentrations. When the analyte is dilute compared with the buffer, stacking or peak compression occurs (Section 9.6). At high solute concentration, "anti-stacking" or bandbroadening is observed because of voltage-drop variations over the injection and run buffer zones. In other words, the linear dynamic range can be affected by the solute's ionic strength (Section 11.4). At high solute concentrations, the use of several co-ions such as chromate (high mobility), phthalate (intermediate mobility), and hydroxybenzoate (low mobility) produces good peak symmetry. These effects are amplified in small ion separations because of the wide variation in solute mobility.

[6] Waters Division of Millipore, Milford, MA.

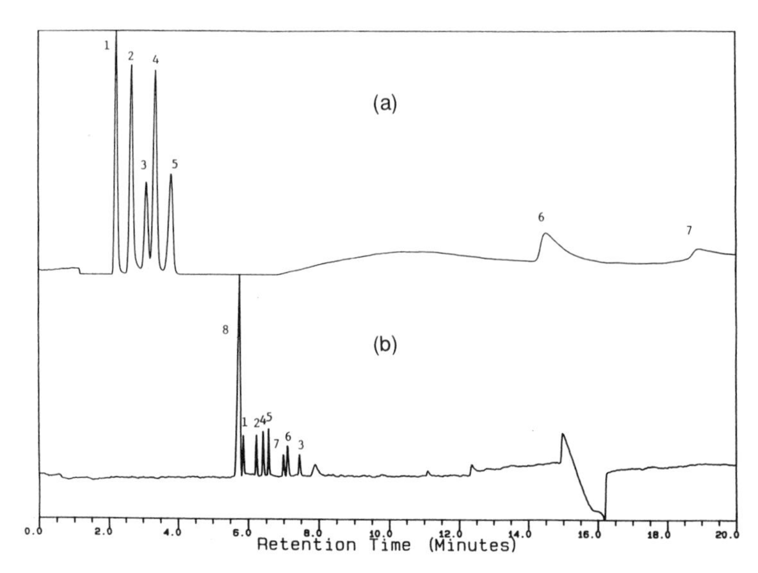

Figure 3.13. Separation of an anion standard by (a) ion chromatography and (b) capillary zone electrophoresis. (a) Column: Vydac 302IC4.6; eluent: isophthalic acid, 250 μg/mL, pH 4.6; flow rate: 2.5 mL/min.; injection size: 25 μL; detection: 280 nm. (b) Capillary: 65 cm × 75 μm i.d.; buffer: 2 mM borate, 40 mM boric acid, 1.8 mM chromate titrated to pH 7.8 with diethylene tetramine; voltage: 20 kV, reverse polarity; detection: indirect UV, 280 nm. Key: (1) chloride, (2) nitrate, (3) chlorate, (4) nitrate, (5) sulfate, (6) thiocyanate, (7) perchlorate, (8) bromide. Reprinted with permission from *J. Chromatogr.* **602,** 241, copyright ©1992 Elsevier Science Publishers.

3. UV absorption spectrum of the additive. For indirect detection (Section 10.8), the UV absorption of the additive should be ±100 nm away from that of the analyte. This is necessary to maximize the differences in absorption between the buffer and solute. Sensitivity will suffer whenever this condition is not fulfilled. Additives such as *p*-anisate, benzoate, hydroxybenzoate, phthalate, salicylate, and chromate are useful visualization reagents for indirect detection. Reagents with high molar absorptivity (69) used at low concentration (85) produce the best sensitivity. Mobility matching also improves sensitivity, since peak dispersion is reduced (85). Sensitivities frequently reach the parts-per-billion range (64, 69). The visualization reagent must have the same charge as the ions being separated (Section 10.8).
4. Stacking. Trace enrichment is required to achieve satisfactory limits of detection (Section 9.6).
5. Isotachophoresis (Chapter 6). A low-mobility ion can be added to the sample solution to permit on-line isotachophoretic trace enrichment (Section 9.7) when electrokinetic injection is employed (69).

A variety of applications are listed by sample matrix or solute in Table 3.4. Both organic acids and inorganic ions are separated.

B. Small Molecules

A variety of small synthetic pharmaceuticals, pesticides and herbicides can be separated by CZE as shown in Table 3.5. Fig. 3.14 illustrates the separation of nonsteroidal anti-inflammatory drugs in coated and uncoated fused-silica capillaries (88). Several important features can be seen in this figure.

1. The peak efficiencies are higher in the untreated capillary. That indicates these drugs are interacting with the polyacrylamide coating. Unless specific wall interactions occur, small molecules are best separated in bare-silica capillaries.
2. The order of elution is opposite on the treated and untreated capillaries. Since the EOF is suppressed on the treated capillaries, the anionic solutes are migrating toward the positive electrode. On the untreated capillary, the EOF sweeps the solutes toward the negative electrode, so the least mobile species now migrates most rapidly. The field strength polarity must be switched both for injection and separation when using the treated capillary.
3. Increasing the buffer pH speeds the separation for both treated and untreated capillaries. In the treated capillary, migration is toward the positive electrode (reverse polarity). At high pH, the compounds that are carboxylic acid derivatives develop greater charge, and thus higher mobility. In the untreated capillary, this effect would lengthen the separation time if not for the increased EOF due to higher pH.

Table 3.4. Applications for Small-Ion Separations by CZE

Application	Solutes	Reference
Urine	Anions	(76)
Kraft black liquor	Small anions, acids	(86)
Dental plaque	Small anions, acids	(86)
Saliva	Small anions, acids	(86)
Shampoo	Alkylsulfonates	(86)
Juice and wine	Acids	(65)
Electroplating solutions	Metal–ion complexes	(66)
Water	Anions	(72)
Parenteral solutions	Alkali and alkaline earth cations	(74)
Explosive residues	Small anions	(77)
Lanthanides		(64, 87)

Table 3.5. Separation of Small Molecules by CZE

Solute	Reference	Comments
Aminobenzoic acids	(89)	
Anthracyclines	(90)	Blood plasma
Benzylpenicillin	(91)	
Catecholamines	(92)	Borate complexes
Cefixime	(93)	Digestive organ extract
Cimetidine	(94)	
Ephedrine alkaloids	(95)	
Flavonoids	(96)	Plant extracts
Glyphosate	(97)	Serum extract
Hypoxanthine	(98)	Fish extract
Methotrexate	(99)	Laser fluorescence
Paralytic shellfish poison	(100)	Mass spectrometry
Paraquat, diquat	(101)	
	(102)	Serum extract
Polyamines	(103)	Rat intestine extract
	(31)	Rat tissues
Sulfonamides	(104)	Pork meat extracts
	(105)	Optimization method
Thiols	(106)	Compared to μ-LC
	(107)	Blood extracts
	(108)	
Triazine herbicides	(109)	On-line preconcentration
	(110)	
Tricyclic antidepressants	(111)	Optimization method
Uric acid	(112)	
	(113)	Urine and serum
Vitamin B_2	(114)	
Xanthines	(112)	

C. Carbohydrates

Carbohydrate separations fall into two broad categories: simple sugars and oligosaccharides. Large oligosaccharides are sometimes separated by CGE (Chapter 5). Derivatization is frequently required to obtain sufficient sensitivity (115–119). Borate complexes facilitate separation of even neutral carbohydrates (115, 116, 120). Indirect fluorescence (121) and absorption (122) detection have been reported as well. Other reported applications include disaccharide compositional analysis of heparin (123), separation of chrondroitin disaccharides (124, 125), separation of oligosaccharides formed by chitanase acting on *N*-acetylchitooligosaccharide–fluorescent conjugates (119), separation of monoterpene glycosides of *Paeonia radix* (126), analysis of oligosaccharides in ovalbumin (127), and separation of glycolipid gangliosides (128).

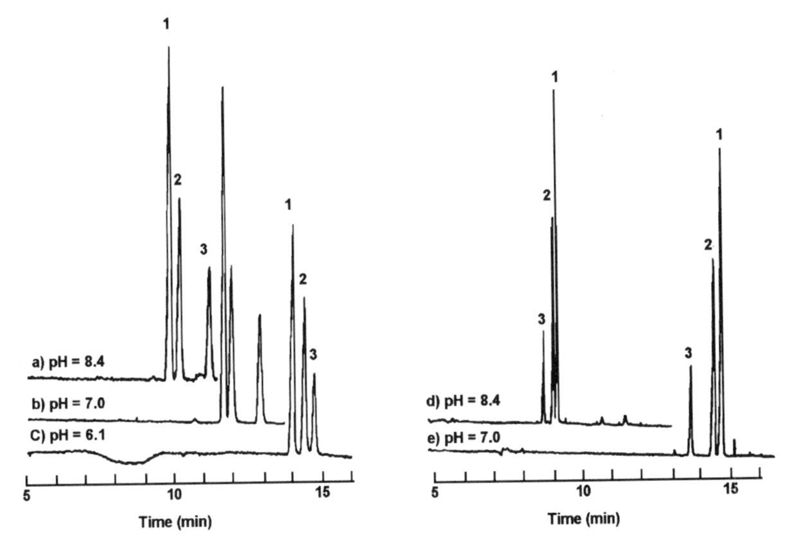

Figure 3.14. Separation of anti-inflammatories by CZE at various pHs in a 40 cm polyacrylamide-coated (left) and a 70 cm uncoated (right) capillary. Field strength: 275 V/cm; detection: UV, 215 nm; injection: electromigration, 2 s at cathode (–) for the coated capillary and anode (+) for the uncoated capillary; buffers: 20 mM borate–100 mM boric acid, pH 8.4 (46 μA); 30 mM phosphate–9 mM borate, pH 7.0 (70 μA); 80 mM MES-30 mM tris, pH 6.1 (20 μA); solutes: (1) naproxen; (2) ibuprofen; (3) tolmetin. Reprinted with permission from *J. Microcol. Sep.* **2,** 166, copyright ©1990 Microseparations Inc.

D. Proteins, Peptides, and Amino Acids

Most of the reported CZE applications are in this broad field. Amino acids (AAs) are best separated by MECC, particularly when all of the common protein hydrolysate AAs must be resolved. CZE is appropriate for AA separations when only several need to be determined. Jellum *et al.* (129) separated sulfur-containing AAs as monobromobimane derivatives to diagnose homocystinuria. Other AA separations were used to diagnose various metabolic disorders. The advantage of CZE is speed. Separations are complete in under 10 min.

Capillary zone electrophoresis is complementary to HPLC for peptide mapping studies (28, 130). Microheterogeneity not detected by HPLC can be resolved by CZE, even for glycoprotein fragments (131). Tryptic digests are usually run by gradient elution reverse-phase liquid chromatography (Fig. 3.15a). Run times of several hours are commonplace. By CZE, a run can be completed in 12 min (Fig. 3.15b) with modest resolution (132). Lengthening the run time to 60 min further improves the resolution (data not shown). Speed is again the compelling advantage. This feature is conducive to screening large numbers of samples searching for variants, decomposition products or post-translational modifications. The CZE

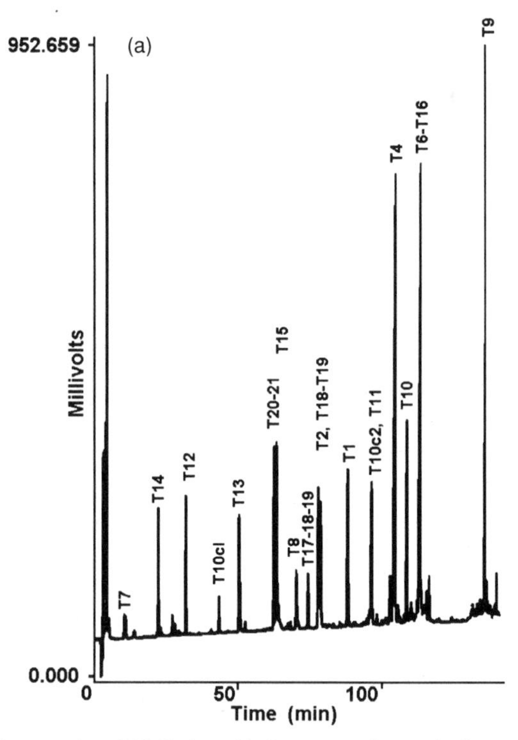

Figure 3.15. (a) Reverse-phase HPLC of peptide fragments of a tryptic digest of rhGH. Tryptic fragments are numbered sequentially from the *N*-terminus of the protein, and the suffix "c" indicates a peptide fragment resulting from a chymotrypsin-like cleavage. Column: 150 × 4.6 mm i.d. Nucleosil C_{18}; temperature: 35°C; mobile phase: A, 0.1% TFA in water; B, 100% acetonitrile; gradient: 100% A, hold 5 min, linear ramp to 37% B over 120 min, linear ramp to 57% B over 10 min; flow rate: 1 mL/min; detection: UV, 214 nm; injection size: 200 μL. Reprinted with permission from *J. Chromatogr.* **480,** 379, copyright ©1989 Elsevier Science Publishers.

mechanism of separation bears no relationship to that of reverse-phase LC. A scatter plot comparing LC retention time to CZE migration time (not shown) for the individual peaks in Figs. 3.15a and 3.15b yields a random distribution.

CZE has some significant disadvantages compared with HPLC concerning tryptic mapping:

1. The overall peak capacity of CZE can be lower than that of gradient elution LC.
2. The difficulty of fraction collection (Section 10.11) means that peak identification through protein sequencing can be a problem.
3. The reproducibility of CZE is not as robust as that of gradient elution LC. There have been reports of poor precision both for migration time and

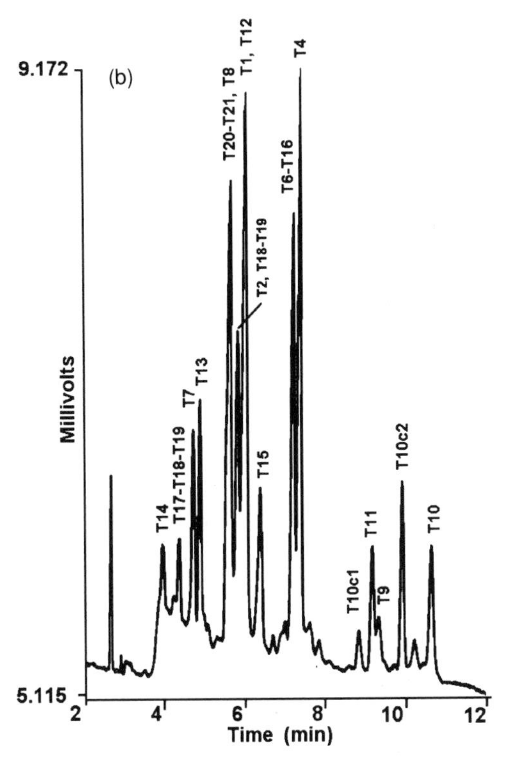

Figure 3.15 *(continued)*. (b) CZE of peptide fragments of a tryptic digest of rhGH. Capillary: polyacrylamide-coated fused silica, 20 cm × 25 μm i.d.; buffer: phosphate buffer, pH 2.5; injection: 5 s at 8 kV; detection: UV, 200 nm. Peaks were identified by injection of individual fractions from the HPLC separation. Reprinted with permission from *J. Chromatogr.* **480,** 379, copyright ©1989 Elsevier Science Publishers.

peak area. Many of these substandard results can be related to deficiencies in the experimental or instrumental design (Section 11.2).

High-sensitivity peptide maps using solid-phase digestion methods, derivatization, and laser fluorescence detection have been reported (133–135). CZE/MS has also been used for separating and monitoring peptide maps (136).

Serum proteins are traditionally separated in the clinical lab by agarose gel electrophoresis. The patterns obtained are diagnostic for various disease states. Use of an alkaline, high ionic strength buffer produces high-speed separations of important serum proteins in an untreated fused-silica capillary (137, 138).[7]

[7] The complete buffer composition has not been disclosed. Beckman Instruments is introducing a multicapillary instrument for clinical use. Serum protein analysis is one of the first dedicated applications.

A 25 μm i.d. capillary is employed to reduce the heating effects of the high field strength. The separations (Fig. 3.16a) are reproducible and are similar to the agarose gel patterns. The run time can be as short as 90 s. Increasing the buffer's ionic strength lowers the EOF and improves the resolution (Fig. 3.16b).

Additional applications for proteins and peptides are listed in Table 3.6.

E. Nucleic Acids, Nucleosides, Nucleotides

Mono-, di-, and triphosphates of adenosine, guanosine, cytidine, and uridine are separable on a polyacrylamide coated capillary at pH 4 (161). Polycytidines up to 10-mer could be separated as well. The higher homologues require a gel. Ribonucleoside triphosphates (Fig. 3.17) separate on a coated capillary at pH 2.7 (162). Ribonucleotide (163) and nucleotide (164–167) separations appear straightforward.

Table 3.6. CZE of Proteins and Peptides

Application	Reference	Comment
Calmodulin	(139)	
Collagens	(6)	Peptide maps
Concanavalin A	(140)	
Erythropoietin	(141)	Glycoform microheterogeneity
Growth hormone–releasing factor	(142)	Degradation products
	(143)	
Hepatitis-B vaccine	(144)	Process control
Hirudins	(145)	
Histones	(146)	
	(147)	Coated capillary
	(148)	
Human growth hormone	(132)	
Immune complexes	(149)	
Insulin	(150)	Quantitation in parenteral solutions
	(151)	
Interleukin-3	(152)	Micropreparative
Leucinostatins	(153)	Coated capillary
Luteinizing hormone–releasing hormone	(154)	From ovine hypothalamus
Membrane proteins	(20)	Uses 7 M urea
Metal-binding proteins	(12)	Impact of Ca^{++} and EDTA
Monoclonal antibodies	(2)	Mobility modeling
	(155)	Quantitative analysis
	(156)	Micropreparative
Motilin peptides	(157)	Structure/mobility studies
Ribonuclease-B	(158)	Glycoform population
Tissue plasminogen activator (TPA)	(159)	Also isoelectric focusing
Transferrins	(160)	Glycoforms and also CIEF

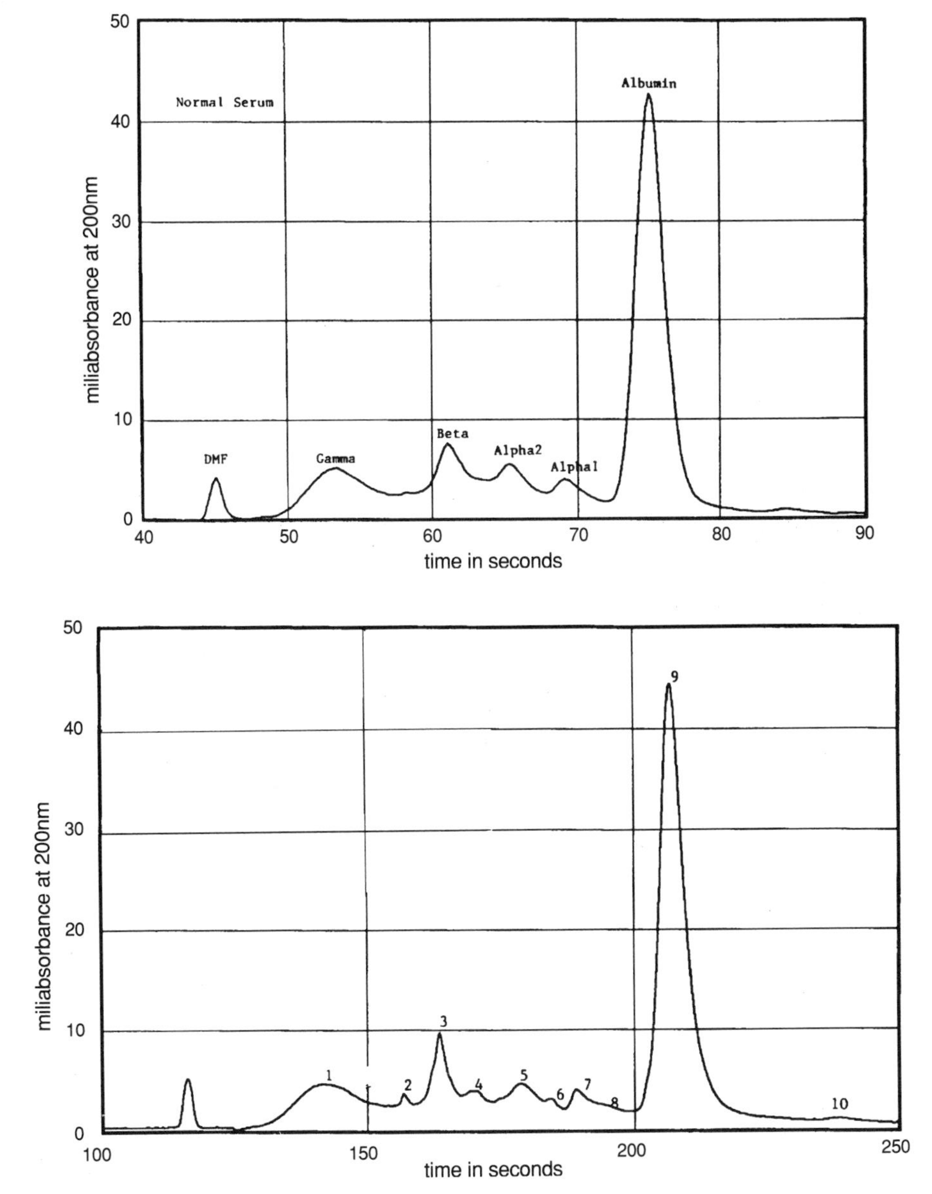

Figure 3.16. CZE protein profile of a normal control serum. Capillary: 25 cm × 25 μm i.d.; voltage: 20 kV; buffer: (a) proprietary, pH 10.0; (b) high ionic strength, pH 10; temperature: 22°C; detection: UV, 200 nm. Key: (1) DMF, (2) γ-globulin, (2′) complements, (3) transferrin, (4) β-lipoproteins, (5) haptoglobin, (6) α_2-macroglobulin, (7) α_1-antitrypsin, (8) α_1-lipoproteins, (9) albumin, (10) prealbumin. Reprinted with permission from *J. Chromatogr.* **559,** 445, copyright ©1991 Elsevier Science Publishers.

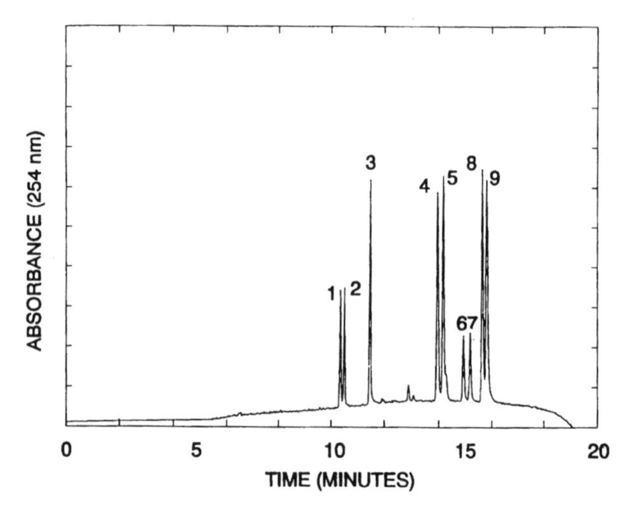

Figure 3.17. Separation of a ribonucleotide triphosphate and deoxyribonucleotide mixture. Capillary: 69.9 cm (62.8 cm to detector) × 75 μm i.d., polyacrylamide-coated; buffer: 50 mM phosphate, pH 2.7, 2 mM EDTA; voltage: 20 kV; injection: pressure, 5 (20 nL). Key: (1) UTP, (2) dTTP, (3) ITP, (4) GTP, (5) dGTP, (6) dCTP, (7) CTP, (8) dATP, (9) ATP. Reprinted with permission from *J. Chromatogr.* **559,** 247, copyright ©1991 Elsevier Science Publishers.

Various adducts such as benzopyrene deoxyguanosyl-5-monophosphate (168) and cytosine-*b*-arabinoside (169) have also been separated.

F. Other Applications

Supramolecular species such as viruses (170, 171), bacteria (170), red blood cells (172), and small particles (173–176) can be separated by CZE. The degree of orientation of particles such as the rod-shaped tobacco mosaic virus (TMV) with the electric field can influence the mobility (171). While migration velocity is expected to increase with field strength, the mobility should remain constant unless Joule heating effects are significant. For TMV, mobility increases with field strength, while 0.364 μm spherical latex particles show no such effect in the identical run buffer. It appears that the rod-shaped virus is better aligned with the field at higher field strengths, thereby reducing frictional drag through the buffer.

3.7 Methods Development

Developing a method by CZE can be complex because multiple experimental variables may need optimization. Selecting the appropriate pH through the use of mobility plots does not always give an adequate separation. Prior to optimizing the pH, it may be necessary to use buffer additives to control selectivity or suppress wall effects. If the buffer ionic strength is too low, loading effects (Section 11.4) will mask even the best attempts at optimization.

Nielsen and Rickard (28) reported an empirical method for optimizing a complex separation of human growth hormone (hGH) tryptic digest fragments. The sample consisted of peptide fragments whose structures and chemical properties are given in Table 3.7. This complex sample contains fragments ranging in size from one to 32 amino acid residues. Fragments 6–16 and 20–21 consist of two chains connected by a disulfide bond. Fragments 1 and 3 are basic peptides likely to exhibit wall interactions. Both hydrophobic peptides (fragments 4, 6–16, 9, and 10) and hydrophilic peptides (fragments 3, 5, 7, 14, and 17) are present in the sample.

First, it is necessary to ensure adequate buffer ionic strength. Separations in 10 mM and 100 mM tricine, pH 8.1, are shown in Figs. 3.18a and 3.18b. Since pH 8.1 is close to the p*I* of tricine, the higher buffer concentration did not draw a very high current. The impact of buffer concentration is substantial because of loading effects (Section 11.4), reduction of wall effects, and reduction of the EOF. Adding sodium chloride to further increase the ionic strength was not useful.

Table 3.7. Structure and Properties of Peptide Fragments from the Enzymatic Digestion of hGH

Fragment Number	Isoelectric Point	Hydro-phobicity	Molecular Weight	Amino Acid Residues	Amino Acid Sequence
1	10.1	18.1	930	8	FPTIPLSR
2	5.8	17.5	978	8	LFDNAMLR
3	10.4	−12.3	382	3	AHR
4	4.2	45.5	2,343	19	LHQLAFDTYQEFEEAYIPK
5	6.4	−15.7	404	3	EQK
6–16	7.3	74.8	3,763	32	6: YSFLQNPQTSLCFSESIPTPSNR 16: NYGLLYCFR
7	4.5	−19.0	762	6	EETQQK
8	5.9	9.4	844	7	SNLELLR
9	6.4	72.1	2,056	17	ISLLLIQSWLEPVQFLR
10	3.5	33.3	2,263	21	SVFANSLVYGASDSNVYDLLK
11	4.0	13.0	1,362	12	DLEEGIQTLMGR
12	4.0	−3.4	773	7	LEDGSPR
13	9.2	0.4	693	6	TGQIFK
14	9.0	−14.2	626	5	QTYSK
15	3.8	0.9	1,490	13	FDTNSHNDDALLK
17	9.0	−13.9	146	1	K
18	6.1	10.5	1,382	11	KDMDKVETFLR
19	4.0	12.1	1,253	10	DMDKVETFLR
20–21	5.9	19.9	1,401	13	20: IVQCR 21: SVEGSCGF

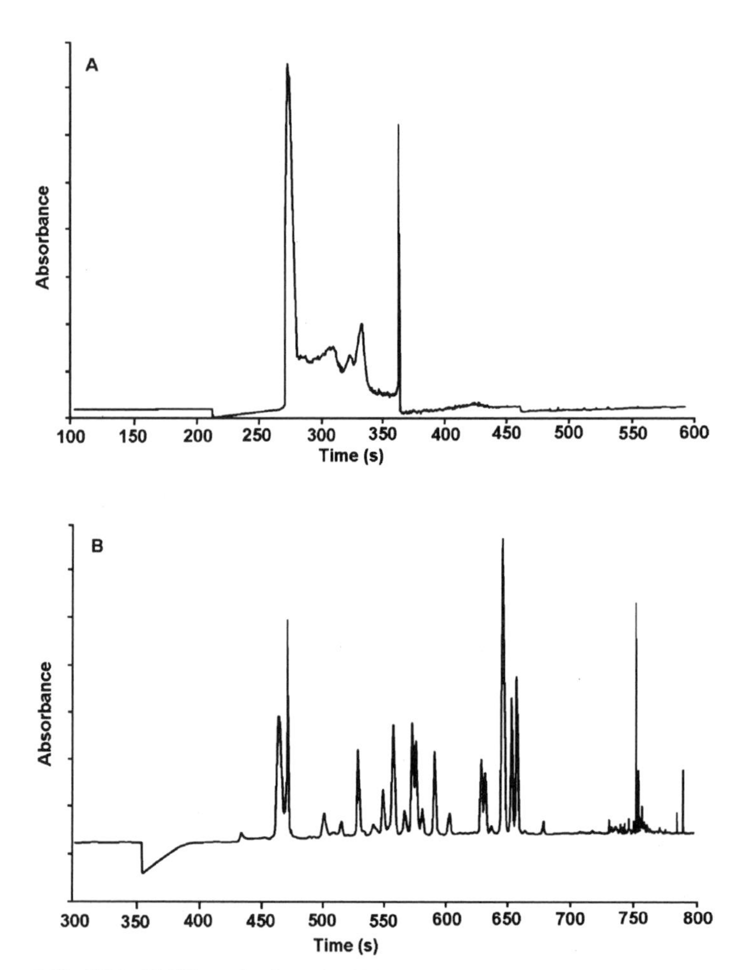

Figure 3.18. CZE of hGH tryptic digest in pH 8.1 tricine buffer: (a) 10 mM and (b) 100 mM. Capillary: 100 cm (80 cm to detector) × 50 μm i.d.; voltage: 30 kV; temperature: 30°C; detection: UV, 200 nm; injection: 3 s vacuum (10 nL); sample concentration: 90 mM for each fragment (2 mg/mL). Reprinted with permission from *J. Chromatogr.* **516,** 99, copyright ©1990 Elsevier Science Publishers.

Four different buffer pHs (2.4, 6.1, 8.1, and 10.4) were initially studied. The best separations were found at pH 2.4 (data not shown) and 8.1, though in both cases a number of fragments were found to overlap. The separation at pH 8.1 was 2.5 times faster than that at pH 2.4, so the higher pH was considered a better choice.

Addition of an amine modifier, in this case morpholine further improved the resolution (Fig. 3.19), presumably by reducing wall interactions between the peptide fragments and free silanol groups.

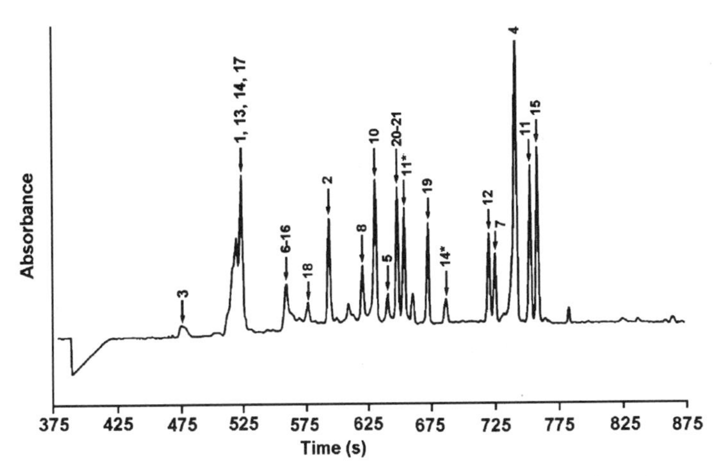

Figure 3.19. CZE of hGH tryptic digest in pH 8.1 tricine with 30 mM morpholine. Other conditions as in Fig. 3.18. The peaks marked (*) are degradation products. Reprinted with permission from *J. Chromatogr.* **516,** 99, copyright ©1990 Elsevier Science Publishers.

It should be noted that no single set of conditions resolved all of the tryptic fragments. The same was true for gradient elution LC, even with a two-hour long gradient run. At pH 2.4, overlapping fragments 1, 13, 14, and 17 are all resolved.

This basis for methods development stresses an important concept. Before a separation can be fine-tuned, freedom from wall effects and sufficient buffer capacity are a prerequisite. It makes little sense to proceed without these features compensated for. The insufficient resolution at the beginning of the separation calls for more experimental work. Perhaps with the addition of cyclodextrins (Section 7.9), surfactants (Section 7.3), metal ions (Section 3.3), or sulfonic acids (Section 3.3), complete resolution in a single run may possible. It may also be possible to calculate a better pH based on the arguments given in Section 3.2.

References

1. G. O. Roberts, P. H. Rhodes, and R. S. Snyder, *J. Chromatogr.* **480,** 35 (1989).
2. B. J. Compton, *J. Chromatogr.* **559,** 357 (1991).
3. B. J. Compton and E. A. O'Grady, *Anal. Chem.* **63,** 2597 (1991).
4. P. Grossman D., J. C. Colburn, and H. H. Lauer, *Anal. Biochem.* **179,** 28 (1989).
5. Z. Deyl, V. Rohlicek, and R. Struzinsky, *J. Liq. Chromatogr.* **12/13,** 2515 (1989).
6. Z. Deyl, V. Rohlicek, and M. Adam, *J. Chromatogr.* **480,** 371 (1989).
7. E. C. Rickard, M. M. Strohl, and R. G. Nielsen, *Anal. Biochem.* **197,** 197 (1991).
8. P. D. Grossman J. C. Colburn, H. H. Lauer, R. G. Nielsen, G. S. Sittampalam, and E. C. Rickard, *Anal. Chem.* **61,** 1186 (1989).

9. P. D. Grossman, K. J. Wilson, G. Petrie, and H. H. Lauer, *Anal. Biochem.* **173,** 265 (1988).
10. R. A. Mosher, *Electrophoresis* **11,** 765 (1990).
11. H. Kajiwara, H. Hirano, and K. Oono, *J. Biochem. Biophys. Meth.* **22,** 263 (1991).
12. H. Kajiwara, *J. Chromatogr.* **559,** 345 (1991).
13. H. J. Issaq, G. M. Janini, I. Z. Atamna, G. M. Muschik, and J. Lukszo, *J. Liq. Chromatogr.* **15,** 1129 (1992).
14. J. A. Bullock and L.-C. Yuan, *J. Microcol. Sep.* **3,** 241 (1991).
15. G. M. McLaughlin, J. A. Nolan, J. L. Lindahl, R. H. Palmieri, K. W. Anderdson, S. C. Morris, J. A. Morrison, and T. J. Bronzert, *J. Liq. Chromatogr.* **15,** 961 (1992).
16. C. Schwer and E. Kenndler, *Anal. Chem.* **63,** 1801 (1991).
17. S. Fujiwara and S. Honda, *Anal. Chem.* **59,** 487 (1987).
18. R. R. Chadwick and J. C. Hsieh, *Anal. Chem.* **63,** 2377 (1991).
19. R. Weinberger, E. Sapp, and S. Moring, *J. Chromatogr.* **516,** 271 (1990).
20. D. Josić, K. Zeilinger, W. Reutter, A. Bottcher, and G. Schmitz, *J. Chromatogr.* **516,** 89 (1990).
21. S. Hjertén, *Chromatogr. Rev.* **9,** 122 (1967).
22. J. W. Jorgenson and K. D. Lukacs, *Science* **222,** 266 (1983).
23. H. H. Lauer and D. McManigill, *Anal. Chem.* **58,** 166 (1986).
24. R. M. McCormick, *Anal. Chem.* **60,** 2322 (1988).
25. M. Zhu, R. Rodriguez, D. Hansen, and T. Wehr, *J. Chromatogr.* **516,** 123 (1990).
26. J. S. Green and J. W. Jorgenson, *J. Chromatogr.* **478,** 63 (1989).
27. M. M. Bushey and J. W. Jorgenson, *J. Chromatogr.* **480,** 301 (1989).
28. R. G. Nielsen and E. C. Rickard, *J. Chromatogr.* **516,** 99 (1990).
29. J. W. Dolan and L. R. Snyder, *Troubleshooting LC Systems* (Humana Press, 1989).
30. F. S. Stover, B. L. Haymore, and R. J. McBeath, *J. Chromatogr.* **470,** 241 (1989).
31. T. Tsuda, Y. Kobayashi, A. Hori, T. Matsumoto, and O. Suzuki, *J. Microcol. Sep.* **2,** 21 (1990).
32. V. Rohlicek and Z. Deyl, *J. Chromatogr.* **494,** 87 (1989).
33. J. E. Wiktorowicz and J. C. Colburn, *Electrophoresis* **11,** 769 (1990).
34. R. L. Cunico, V. Gruhn, L. Kresin, D. E. Nitecki, and J. E. Wiktorowicz, *J. Chromatogr.* **559,** 467 (1991).
35. K. Tsuji and R. J. Little, *J. Chromatogr.* **594,** 317 (1992).
36. A. Emmer, M. Jansson, and J. Roeraade, *J. Chromatogr.* **547,** 544 (1991).
37. A. Emmer, M. Jansson, and J. Roeraade, *HRC & CC* **14,** 738 (1991).
38. P. Thibault, C. Paris, and S. Pleasance, *Rapid Commun. Mass Spectrom.* **5,** 484 (1991).
39. J. W. Jorgenson and K. D. Lukacs, *Anal. Chem* **53,** 1298 (1981).
40. S. Hjertén, *J. Chromatogr.* **347,** 191 (1985).
41. D. Bentrop, J. Kohr, and H. Engelhardt, *Chromatographia* **32,** 171 (1991).
42. K. A. Cobb, V. Dolník, and M. Novotny, *Anal. Chem.* **62,** 2478 (1990).
43. M. V. Novotny, K. A. Cobb, and J. Liu, *Electrophoresis* **11,** 735 (1990).
44. M. Huang, W. P. Vorkink, and M. L. Lee, *J. Microcol. Sep.* **4,** 233 (1992).
45. J. Kohr and H. Engelhardt, *J. Microcol. Sep.* **3,** 491 (1991).
46. G. J. M. Bruin, J. P. Chang, R. H. Kuhlman, K. Zegers, J. C. Kraak, and H. Poppe, *J. Chromatogr.* **471,** 429 (1989).

47. J. A. Lux, H. Yin, and G. Schomburg, *HRC & CC* **13,** 145 (1990).
48. M. Huang, W. P. Vorkink, and M. L. Lee, *J. Microcol. Sep.* **4,** 135 (1992).
49. J. K. Towns and F. E. Regnier, *J. Chromatogr.* **516,** 69 (1990).
50. G. J. M. Bruin, R. Huisden, J. C. Kraak, and H. Poppe, *J. Chromatogr.* **480,** 339 (1989).
51. J. K. Towns, J. Bao, and F. E. Regnier, *J. Chromatogr.* **599,** 227 (1992).
52. S. A. Swedberg, *Anal. Biochem.* **185,** 51 (1990).
53. A. T. Balchunas and M. J. Sepaniak, *Anal. Chem.* **59,** 1466 (1987).
54. A. M. Dougherty, C. L. Woolley, D. L. Williams, D. F. Swaile, R. O. Cole, and M. J. Sepaniak, *J. Liq. Chromatogr.* **14,** 907 (1991).
55. W. Nashabeh and Z. E. Rassi, *J. Chromatogr.* **559,** 367 (1991).
56. J. K. Towns and F. E. Regnier, *Anal. Chem.* **63,** 1126 (1991).
57. C. A. Bolger, M. Zhu, R. Rodriguez, and T. Wehr, *J. Liq. Chromatogr.* **14,** 895 (1991).
58. F. Foret, M. Deml, and P. Boček, *J. Chromatogr.* **452,** 601 (1988).
59. E. Grushka, R. M. McCormick, and J. J. Kirkland, *Anal. Chem.* **61,** 241 (1989).
60. J. H. Knox, *Chromatographia* **26,** 329 (1988).
61. A. E. Jones and E. Grushka, *J. Chromatogr.* **466,** 219 (1989).
62. L. R. Gurley, J. E. London, and J. G. Valdez, *J. Chromatogr.* **559,** 431 (1991).
63. W. R. Jones and P. Jandik, *J. Chromatogr.* **546,** 445 (1991).
64. A. Weston, P. R. Brown, P. Jandik, W. R. Jones, and A. L. Heckenberg, *J. Chromatogr.* **593,** 289 (1992).
65. B. F. Kenney, *J. Chromatogr.* **546,** 423 (1991).
66. M. Aguilar, X. Huang, and R. N. Zare, *J. Chromatogr.* **480,** 427 (1989).
67. W. R. Jones, P. Jandik, and R. Pfeifer, *American Lab.* **May,** 40 (1991).
68. W. R. Jones and P. Jandik, *American Lab.* June, 51 (1990).
69. P. Jandik and W. R. Jones, *J. Chromatogr.* **546,** 431 (1991).
70. P. Jandik, W. R. Jones, A. Weston, and P. R. Brown, *LCGC* **9,** 634 (1991).
71. G. M. Janini and H. J. Issaq, *J. Liq. Chromatogr.* **15,** 927 (1992).
72. G. Bondoux, P. Jandik, and W. R. Jones, *J. Chromatogr.* **602,** 79 (1992).
73. F. Foret, S. Fanali, A. Nardi, and P. Boček, *Electrophoresis* **11,** 780 (1990).
74. M. Koberda, M. Konkowski, P. Youngberg, W. R. Jones, and A. Weston, *J. Chromatogr.* **602,** 235 (1992).
75. A. Weston, P. R. Brown, A. L. Heckenberg, P. Jandik, and W. R. Jones, *J. Chromatogr.* **602,** 249 (1992).
76. B. J. Wildman, P. E. Jackson, W. R. Jones, and P. G. Alden, *J. Chromatogr.* **546,** 459 (1991).
77. K. A. Hargadon and B. R. McCord, *J. Chromatogr.* **602,** 241 (1992).
78. D. F. Swaile and M. J. Sepaniak, *Anal. Chem.* **63,** 179 (1991).
79. L. Gross and E. S. Yeung, *J. Chromatogr.* **480,** 169 (1989).
80. L. Gross and E. S. Yeung, *Anal. Chem.* **62,** 427 (1990).
81. E. S. Yeung, *Acct. Chem. Res.* **22,** 125 (1989).
82. E. S. Yeung and W. G. Kuhr, *Anal. Chem.* **63,** 275 (1991).
83. X. Huang, T. K. J. Pang, M. J. Gordon, and R. N. Zare, *Anal. Chem.* **59,** 2747 (1987).
84. X. Huang, M. J. Gordon, and R. N. Zare, *J. Chromatogr.* **425,** 385 (1988).

85. F. Foret, S. Fanali, L. Ossicini, and P. Boček, *J. Chromatogr.* **470,** 299 (1989).
86. J. Romano, P. Jandik, W. R. Jones, and P. E. Jackson, *J. Chromatogr.* **546,** 411 (1991).
87. M. Chen and R. M. Cassidy, *J. Chromatogr.* **602,** 227 (1992).
88. A. Wainright, *J. Microcol. Sep.* **2,** 166 (1990).
89. M. W. F. Nielen, *J. Chromatogr.* **542,** 173 (1991).
90. N. J. Reinhoud, U. R. Tjaden, H. Irth, and J. van der Greef, *J. Chromatogr.* **574,** 327 (1992).
91. A. M. Hoyt and M. J. Sepaniak, *Anal. Lett.* **22,** 861 (1989).
92. T. Kaneta, S. Tanaka, and H. Yoshida, *J. Chromatogr.* **538,** 385 (1991).
93. S. Honda, A. Taga, K. Kakehi, S. Koda, and Y. Okamoto, *J. Chromatogr.* **590,** 364 (1992).
94. S. Arrowood and A. M. Hoyt, *J. Chromatogr.* **586,** 177 (1991).
95. Y.-M. Liu and S.-J. Sheu, *J. Chromatogr.* **600,** 370 (1992).
96. U. Seitz, P. J. Oefner, S. Nathakarnkitool, M. Popp, and G. K. Bonn, *Electrophoresis* **13,** 35 (1992).
97. M. Tomita, T. Okuyama, Y. Nigo, B. Uno, and S. Kawai, *J. Chromatogr.* **571,** 324 (1991).
98. J. H. T. Luong, K. B. Male, C. Masson, and A. L. Nguyen, *J. Food Sci.* **57,** 77 (1992).
99. M. C. Roach, P. Gozel, and R. N. Zare, *J. Chromatogr.* **426,** 129 (1988).
100. P. Thibault, S. Pleasance, and M. V. Laycock, *J. Chromatogr.* **542,** 483 (1991).
101. J. Cai and Z. E. Rassi, *J. Liq. Chromatogr.* **15,** 1193 (1992).
102. M. Tomita, T. Okuyama, and Y. Nigo, *Biomedical Chromatogr.* **6,** 91 (1992).
103. T. Matsumoto, T. Tsuda, and O. Suzuki, *Trends in Anal. Chem.* **9,** 292 (1990).
104. M. T. Ackermans, J. L. Beckers, F. M. Everaerts, H. Hoogland, and M. J. H. Tomassen, *J. Chromatogr.* **596,** 101 (1992).
105. C. L. Ng, H. K. Lee, and S. F. Y. Li, *J. Chromatogr.* **598,** 133 (1992).
106. B. L. Ling, W. R. G. Baeyens, and C. Dewaele, *HRC & CC* **14,** 169 (1991).
107. B. L. Ling, W. R. G. Baeyens, and C. Dewaele, *Anal. Chim. Acta* **255,** 283 (1991).
108. J. S. Stamler and J. Loscalzo, *Anal. Chem.* **64,** 779 (1992).
109. J. Cai and Z. El Rassi, *J. Liq. Chromatogr.* **15,** 1179 (1992).
110. F. Foret, V. Sustacek, and P. Boček, *Electrophoresis* **11,** 95 (1990).
111. K. Saloman, D. S. Burgi, and J. C. Helmer, *J. Chromatogr.* **549,** 375 (1991).
112. I. Z. Atamna, G. M. Janini, G. M. Muschik, and H. J. Issaq, *J. Liq. Chromatogr.* **14,** 427 (1991).
113. C. Masson, J. H. T. Luong, and A.-L. Nguyen, *Anal. Lett.* **24,** 377 (1991).
114. E. Kenndler, C. Schwer, and D. Kaniansky, *J. Chromatogr.* **508,** 203 (1990).
115. S. Honda, S. Iwase, A. Makino, and S. Fujiwara, *Anal. Biochem.* **176,** 72 (1989).
116. S. Honda, S. Suzuki, A. Nose, K. Yamamoto, and K. Kakehi, *Carbohydrate Research* **215,** 193 (1991).
117. J. Liu, O. Shirota, and M. Novotny, *Anal. Chem.* **63,** 413 (1991).
118. W. Nashabeh and Z. El Rassi, *J. Chromatogr.* **514,** 57 (1990).
119. K.-B. Lee, Y.-S. Kim, and R. J. Linhardt, *Electrophoresis* **12,** 636 (1991).
120. S. Hoffstetter-Kuhn, A. Paulus, E. Gassmann, and H. M. Widmer, *Anal. Chem.* **63,** 1541 (1991).
121. T. W. Garner and E. S. Yeung, *J. Chromatogr.* **515,** 639 (1990).
122. A. E. Vorndran, P. J. Oefner, H. Scherz, and G. K. Bonn, *Chromatographia* **33,** 163 (1992).
123. S. A. Ampofo, H. M. Wang, and R. J. Linhardt, *Anal. Biochem.* **199,** 249 (1991).

124. A. Al-Hakim and R. J. Linhardt, *Anal. Biochem.* **195,** 66 (1991).
125. S. L. Carney, D. J. Osborne, *Anal. Biochem.* **195,** 132 (1991).
126. S. Honda, K. Suzuki, M. Kataoka, A. Makino, and K. Kakehi, *J. Chromatogr.* **515,** 653 (1990).
127. S. Honda, A. Makino, S. Suzuki, and K. Kakehi, *Anal. Biochem.* **191,** 228 (1990).
128. Y. Liu and K.-F. Chan, *Electrophoresis* **12,** 402 (1991).
129. E. Jellum, A. K. Thorsrud, and E. Time, *J. Chromatogr.* **559,** 455 (1991).
130. R. G. Nielsen, R. M. Riggin, and E. C. Rickard, *J. Chromatogr.* **480,** 393 (1989).
131. W. Nashabeh and Z. El Rassi, *J. Chromatogr.* **536,** 31 (1991).
132. J. Frenz, S.-L. Wu, and W. S. Hancock, *J. Chromatogr.* **480,** 379 (1989).
133. K. A. Cobb and M. V. Novotny, *Anal. Chem.* **64,** 879 (1992).
134. L. Amankwa and W. G. Kuhr, *Anal. Chem.* **64,** 1610 (1992).
135. K. A. Cobb and M. Novotny, *Anal. Chem.* **61,** 2226 (1989).
136. P. Ferranti, A. Malorni, P. Pucci, S. Fanoli, A. Nordi, and L. Ossicini, *Anal. Biochem.* **194,** 1 (1991).
137. F.-T. A. Chen, C.-M. Liu, Y.-Z. Hsieh, and J. C. Sternberg, *Clin. Chem.* **37,** 14 (1991).
138. F.-T. A. Chen, *J. Chromatogr.* **559,** 445 (1991).
139. K. F. J. Chan and W. H. Chen, *Electrophoresis* **11,** 15 (1990).
140. K. Hettarachchi and A. P. Cheung, *J. Pharm. & Biomed. Anal.* **9,** 835 (1991).
141. A. D. Tran, S. Park, P. J. Lisi, O. T. Huynh, R. R. Ryall, and P. A. Lane, *J. Chromatogr.* **542,** 459 (1991).
142. J. Bongers, T. Lambros, A. M. Felix, and E. P. Heimer, *J. Liq. Chromatogr.* **15,** 1115 (1992).
143. Z. Prusik, V. Kasika, P. Mudra, J. Stepanek, O. Smekel, and J. Hlavacek, *Electrophoresis* **11,** 932 (1990).
144. W. M. Hurni and W. J. Miller, *J. Chromatogr.* **559,** 337 (1991).
145. V. Steiner, R. Knecht, O. Bornsen, E. Gassman, S. R. Stone, R. Raschdorf, J.-M. Schlaeppi, and R. Maschler, *Biochemistry* **31,** 2294 (1992).
146. H. Lindner, W. Helliger, A. Dirschlmayer, M. Jaquemar, and B. Puschendorf, *Biochem. J.* **283,** 467 (1992).
147. L. R. Gurley, J. S. Buchanon, J. E. London, D. M. Stavert, and B. E. Lehnert, *J. Chromatogr.* **559,** 411 (1991).
148. C. Miller, J.-F. Hernandez, A. G. Craig, J. Dykert, and J. Rivier, *Anal. Chim. Acta* **249,** 215 (1991).
149. R. G. Nielsen, E. C. Rickard, P. F. Santa, D. A. Sharknas, and G. S. Sittampalam, *J. Chromatogr.* **539,** 177 (1991).
150. M. Lookabaugh, M. Biswas, and I. S. Krull, *J. Chromatogr.* **549,** 357 (1991).
151. R. G. Nielsen, G. S. Sittampalam, and E. C. Rickard, *Anal. Biochem.* **177,** 20 (1989).
152. R. I. Hecht, J. F. Coleman, J. C. Morris, F. S. Stover, and C. D. Demarest, *Prep. Biochem.* **19,** 363 (1989).
153. M. G. Quaglia, S. Fanali, A. Nardi, C. Rossi, and M. Ricci, *J. Chromatogr.* **593,** 259 (1991).
154. J. P. Advis, L. Hernandez, and N. Guzman, *Peptide Research* **2,** 389 (1989).
155. N. A. Guzman and L. Hernandez, *A Rapid Procedure for the Quantitative Analysis of Monoclonal Antibodies by HighPerformance Capillary Electrophoresis* (T. E. Hugli, ed.), Techniques in Protein Chemistry (Academic Press, 1989).
156. N. A. Guzman, M. A. Trebilcock, and J. P. Advis, *Anal. Chim. Acta* **249,** 247 (1991).

157. J. R. Florance, Z. D. Konteatis, M. J. Macielag, R. A. Lessor, and A. Galdes, *J. Chromatogr.* **559,** 391 (1991).

158. P. M. Rudd, I. G. Scragg, E. Coghill, and R. A. Dwek, *Glycoconjugate J.* **9,** 86 (1992).

159. K. W. Yim, *J. Chromatogr.* **559,** 401 (1991).

160. F. Kilár and S. Hjertén, *J. Chromatogr.* **480,** 351 (1989).

161. V. Dolník, J. Liu, J. F. Banks, M. V. Novotny, and P. Boček, *J. Chromatogr.* **480,** 321 (1989).

162. R. Takigiku and R. E. Schneider, *J. Chromatogr.* **559,** 247 (1991).

163. X. Huang, J. B. Shear, and R. N. Zare, *Anal. Chem.* **62,** 2049 (1990).

164. T. Tsuda, G. Nakagawa, M. Sato, and K. Yagi, *J. Appl. Biochem.* **5,** 330 (1983).

165. T. Tsuda, K. Takagi, T. Watanabe, and T. Satake, *HRC & CC* **11,** 721 (1988).

166. A. L. Nguyen, J. H. T. Luong, and C. Masson, *Anal. Chem.* **62,** 2490 (1990).

167. L. Hernandez, B. G. Hoebel, and N. A. Guzman, *ACS Symp. Ser.* **434,** 50 (1990).

168. E. Jackim and C. Norwood, *HRC & CC* **13,** 195 (1990).

169. D. K. Lloyd, A. M. Cypess, and I. W. Wainer, *J. Chromatogr.* **568,** 117 (1991).

170. S. Hjertén, K. Elenbring, F. Kilár, J. L. Liao, A. J. C. Chen, C. J. Siebert, and M. D. Zhu, *J. Chromatogr.* **403,** 47 (1987).

171. P. D. Grossman and D. S. Soane, *Anal. Chem.* **62,** 1592 (1990).

172. A. Zhu and Y. Chen, *J. Chromatogr.* **470,** 251 (1989).

173. B. B. VanOrman and G. L. McIntire, *J. Microcol. Sep.* **1,** 289 (1989).

174. B. J. Herren, S. G. Shafer, J. Van Alstine, J. M.Harris, and R. S. Snyder, *J. Colloid & Interfacial Sci.* **115,** 46 (1987).

175. H. K. Jones and N. E. Ballou, *Anal. Chem.* **62,** 2484 (1990).

176. R. M. McCormick, *J. Liq. Chromatogr.* **14,** 939 (1991).

Chapter 4

Capillary Isoelectric Focusing

4.1 Introduction

A. Basic Concepts

The basic parameter of electrophoresis, mobility, is defined by the charge/mass ratio of a solute. If this ratio, specifically the net charge, changes during a run, the electrophoretic properties will change as well. Should a solute become neutral during the run, migration will cease. Modification of mobility can be performed on-line with a pH gradient generated by a series of reagents known as carrier ampholytes. Therein lies the fundamental premise of isoelectric focusing (IEF).

Conventional IEF is performed in an anticonvective medium such as the slab-gel. The advantages and disadvantages of gels, discussed in Section 1.1, apply equally to IEF. As opposed to the usual gel format, the pore size of an IEF gel should be sufficiently large to reduce the impact of molecular sieving that lengthens the run time. Agarose gels have an extremely large pore size, 500 nm for a 0.16% gel (1). Polyacrylamide gels with large pore sizes can be formulated as well.

Detection in the slab-gel is time consuming and semiquantitative. The carrier ampholytes must be washed out of the gel to avoid reaction with the staining reagent. Small peptides are not detectable because they are lost during the wash step (2). Gels are unnecessary in capillary isoelectric focusing (CIEF) because the ampholyte separation medium is effectively supported and contained by the capillary walls. CIEF is performed in free solution.

Mobilization represents an additional difference between capillary and conventional CIEF. Since the focusing process produces electrically neutral solutes, some means of eluting the bands is required for detection. In the slab-gel, mobilization is unnecessary because detection is performed by staining. Hjertén and Zhu's original work (3) described a scheme that forces the proteins to move past the detector window. Such a technique is known as mobilization.

B. Separation Mechanism

Figure 4.1 illustrates the mechanism of CIEF. A sample is mixed with a series of reagents known as *carrier ampholytes.* The capillary is filled with the sample–reagent blend, and a voltage is applied. Carrier ampholytes have the capacity to generate a pH gradient along the length of the capillary under the influence of the applied electric field. Without considering at this stage how the gradient is formed, let us examine the behavior of a zwitterionic solute toward the pH gradient.

There are two possible behaviors. At a pH below the solute's p*I*, the zwitterion is positively charged and migrates toward the cathode. As the solute migrates through the pH gradient, it encounters progressively higher pHs. At some point, the solute enters a region along the gradient where the ampholyte pH is equal to its own p*I*. At this point, the solute's net charge becomes zero, and migration ceases.

The second possible behavior occurs when the solute is negatively charged. This occurs along the pH gradient when the pH is greater than the solute's p*I*. The solute migrates toward the anode encountering progressively lower pHs. While the direction of migration through the pH gradient is opposite to the case described above, the net result is the same. Solute migration ceases whatever the direction of approach to the pH of neutrality or p*I*. When a capillary is filled with a mixture of ampholytes and solutes, both behaviors occur, as shown in Fig. 4.1. CIEF is a true focusing technique. If a solute diffuses into a buffer region where it becomes charged, it will migrate back to the region of zero charge.

The CIEF system is substantially different from zone electrophoresis. In CZE, the buffer system is homogeneous throughout the length of the capillary and throughout the duration of the run. In CIEF, a heterogeneous pH gradient is created inside the capillary when a voltage is applied with the carrier ampholytes. The breadth of the pH gradient is dependent on which series of ampholytes are selected. Ampholytes are commercially available to cover both wide and narrow

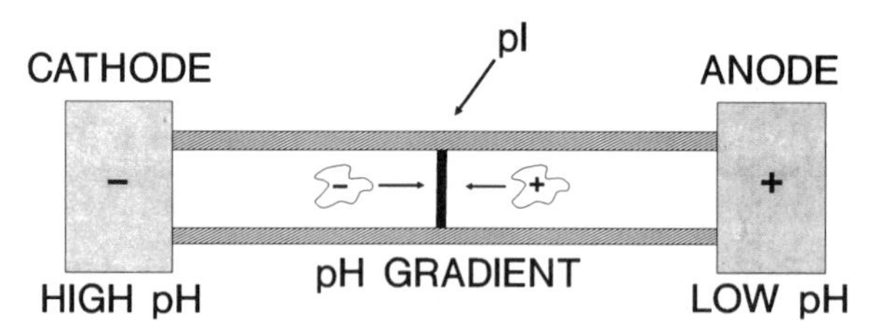

Figure 4.1. Isoelectric focusing. Migration of a protein through a pH gradient to its isoelectric pH.

pH ranges, as shown in Fig. 4.2.[1] The Ampholines (LKB-Pharmacia) are a series of polyamino, polycarboxylic acids. Servalytes (Serva) are polyamine-polysulfonic acids. Pharmalytes (LKB-Pharmacia) are branched polyamino, polycarboxylic acids. Ampholyte concentrations of 0.5–2.0% are used for most applications.

Isoelectric focusing provides a mechanism that permits the separation of solutes, primarily zwitterionic proteins, based on their isoelectric points. Using CIEF, there is no need to develop a buffer system to separate solutes based on mobility. If the solutes have sufficiently different p*I*s, they will separate. This mechanism of separation contrasts sharply with CZE (Chapter 3) and CGE (Chapter 5), where the basis for separation is the charge/mass ratio and molecular size, respectively. Righetti's textbook (1) is recommended for further background on conventional IEF methodology.

C. pH Gradient Formation

Before the field is started, the individual ampholytes are uniformly distributed throughout the capillary. The pH is uniform as well and represents the average pH of the ampholyte blend (Fig. 4.3a). Individual ampholytes may be cationic or anionic at the start of the run, depending on their p*I*s.

When voltage is applied, the ampholyte mixture begins to separate into individual components. Complete separation of adjacent ampholytes is never attained, since this would cause a discontinuity in the pH gradient. Positively charged (high p*I*) ampholytes migrate toward the negative electrode, and vice-versa.

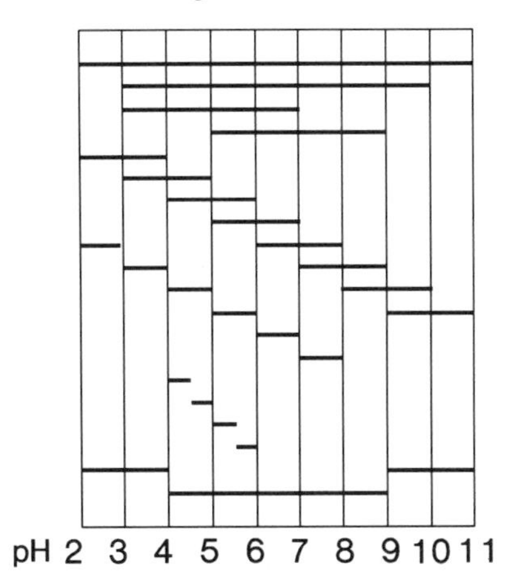

Figure 4.2. Carrier ampholyte pH ranges. Each horizontal line represents the pH range covered by an ampholyte blend.

[1] Carrier ampholytes are available from Bio-Rad (Richmond, CA), Pharmacia LKB (Piscataway, NJ) and Serva Biochemicals (Paramus, NJ).

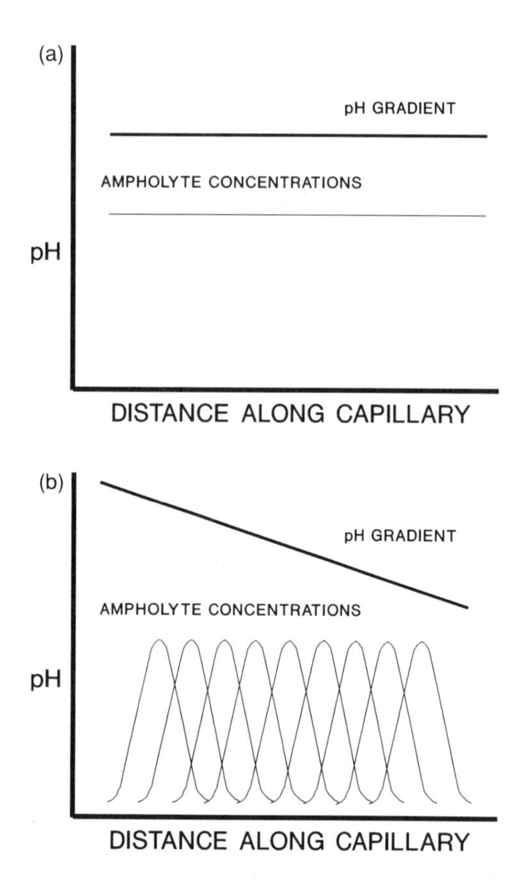

Figure 4.3. Representation of a pH gradient and individual ampholyte concentrations (a) before focusing and (b) after focusing.

The migration of ampholytes causes the electrolyte pH to increase as the cathode is approached. Conversely, the pH in the anodic region begins to decline as the low-p*I* (negatively charged) ampholytes migrate in that direction.

Eventually, migration will cease as each ampholyte encounters a pH where its net charge becomes zero. This occurs at the individual isoelectric point or p*I* of each ampholyte. If a solute such as a protein is present in the carrier ampholyte blend, it, too, migrates while charged, until it encounters a pH equivalent to its p*I*. The system does not distinguish between a protein and a carrier ampholyte; both behave as zwitterions and migrate according to their respective p*I*s. At steady state, the ampholyte distribution and resulting pH gradient are illustrated in Fig. 4.3b. The measured current should approach zero, because no further movement from either ampholytes or solutes is expected, except diffusion.

D. Electrode Buffer Solutions

The pHs of the respective electrode buffer solutions are critical in CIEF. The anodic buffer or *anolyte* must have a pH that is lower than the pI of the most acidic ampholyte. Similarly, the cathodic buffer or *catholyte* must have a higher pH than the most basic ampholyte. If these conditions are not met, ampholytes will bleed into an electrode buffer reservoir. Diluted versions of electrode solutions used in conventional IEF are appropriate for CIEF.

Sodium hydroxide (20 mM) and phosphoric acid (10–20 mM) solutions are frequently used as catholyte and anolyte solutions, particularly with the pH 3–10 gradient. For narrow gradients such as pH 6–8, weaker acids and bases such as 40 mM glutamic acid (anolyte) and 40 mM arginine (catholyte) are used (4). While not yet studied for CIEF, ampholytes themselves can serve as anolyte and catholyte (1). The same holds true for Good's buffer solutions of the appropriate p*I*. HEPES (p*I* 7.51) is used as the anolyte for narrow-range alkaline gradients (1).

E. Resolving Power

The resolving power, ΔpI, of CIEF is described by (1)

$$\Delta pI = 3\sqrt{\frac{D(dpH/dx)}{E(d\mu/dpH)}}, \tag{4.1}$$

where D is the diffusion coefficient, E is the field strength, μ is the mobility of the protein, and $d\mu/dpH$ describes the mobility–pH relationship. The term dpH/dx represents the change in the buffer pH per unit of capillary length. This adjustable parameter is controlled by selecting an appropriate ampholyte pH range. For highest resolution, a narrow pH ampholyte range should be selected. To separate a wide-p*I* range of proteins, although at lower resolution, select a broad pH range ampholyte blend. Under optimal conditions, resolution of 0.02 pH units can be achieved (1).

4.2 Capillary Coatings and Electroosmotic Flow

The use of coated capillaries[2] is advocated in CIEF to suppress the EOF and minimize protein adsorption (3, 5, 6).

The reduction of EOF is desired in CIEF for several reasons:

1. The EOF may sweep the solutes past the detector before focusing.

[2] A polyacrylamide coated capillary is available from Bio-Rad (Richmond, CA).

2. Silanol ionization at the capillary wall is pH-dependent. In the presence of a pH gradient, the degree of ionization will vary along with the pH. As a result, the measured EOF is the average value derived from the integrated contributions along the entire length of the capillary. This can generate an osmotic pressure throughout the capillary, causing a hydrodynamic-like flow profile that leads to bandbroadening. However, the focusing effect from CIEF probably overcomes this form of bandbroadening.
3. In the presence of EOF, the low-pH anolyte enters the capillary because the net flow is directed toward the cathode. As more of the anolyte fills the capillary, the EOF will continually decline as ionized silanol groups are protonated. This leads to nonlinearity of the pH gradient. Under nonlinear conditions, the migration time is not proportional to the p*I*. Calibration is difficult when this occurs, because multiple standards are required.

Bolger *et al.* (6) found that the EOF interfered with CIEF. Using short capillaries, some bands pass the detector window before the completion of focusing. The authors incorrectly concluded that coated capillaries were essential for CIEF. Mazzeo and Krull (4, 7) controlled the EOF with cellulose polymers and ran CIEF in an untreated fused-silica capillary. The combination of cellulose polymers and carrier ampholytes probably suppressed protein adsorption on the capillary wall. With the controlled EOF, the so-called mobilization step (described later) is unnecessary.

In another approach, Chen and Wiktorowicz (2) run CIEF with dissolved methylcellulose in a DB-1-coated capillary.[3] Mobilization is required and accomplished hydrodynamically with the voltage on. This combination reduces bandbroadening because the bands remain focused as they are drawn past the detector.

4.3 Performing a Run

The three discrete steps of capillary CIEF in the method developed by Hjertén (3, 5, 6, 9–13) are loading, focusing, and mobilization. Until 1991, this was the only working protocol for CIEF. Alternative methods developed by Mazzeo and Krull (4, 7), Thormann *et al.* (8), and Chen and Wiktorowicz (2) will be discussed as well. The conditions for four separate protocols are summarized in Table 4.1.

A. Loading

Using a modification of Hjertén's method, desalted sample is mixed with a 1–2% solution of carrier ampholytes to give a final concentration of approximately

[3] J & W Scientific, Folsom, CA.

Table 4.1. Comparison of Protocols for CIEF[a]

Method	Capillary (Coating)	Anolyte	Catholyte	Injection	Focusing Voltage	Mobilization
Mazzeo	60 cm (40 cm to detector) × 50–75 μm i.d.	10 mM H_3PO_4	20 mM NaOH	full	20 kV	EOF
Thormann	90 cm (70 cm to detector) × 75 μm i.d.	10 mM H_3PO_4	20 mM NaOH	partial	20 kV	EOF
Chen	72 cm (50 cm to detector) × 50 μm i.d.(DB-1)	100 mM H_3PO_4	20 mM NaOH	partial	30 kV	Hydrodynamic, 30 kV
Hjertén[b]	14 cm × 25 μm i.d. (polyacrylamide)	10 mM H_3PO_4	20 mM NaOH	full	12 kV	Salt, 8 kV

[a]For ampholyte blends covering pH 3–10
[b]The conditions specified in Ref.15 are cited.

250 μg/mL of protein. The entire mixture is loaded into a 14–20 cm × 25 μm i.d. polyacrylamide-coated capillary.

Removing the salt (Section 11.5) is extremely important. Otherwise, localized Joule heating and ionic strength effects on pH may preclude the formation of a stable and reproducible pH gradient. Since protein aggregation and precipitation can occur during the subsequent focusing step, dispersing agents such as 10–50% ethylene glycol, 2% Triton X-100, or Brij-35 are frequent additives (14). These additives are quite gentle and seldom alter the p*I* of the individual proteins. Hydrophobic solutes such as calcium-binding proteins are difficult to separate by CIEF because of aggregation and subsequent precipitation. While urea can solubilize these proteins, they will be denatured under this condition. All solutions used in CIEF should be carbonate-free, or else pH gradient drift may occur.

Alternative Injection Schemes Filling the entire capillary with sample requires the use of a relatively large amount of protein, predissolved in ampholytes. Thormann *et al.* (8) and Chen and Wiktorowicz (2) use an aliquot of sample dissolved in ampholytes (8) or water (2) to partially fill the capillary.

The method of Thorman *et al.* (8), run in 75 μm i.d. bare-silica capillaries, begins with capillary conditioning using 0.1 M sodium hydroxide containing 0.3% hydroxypropylmethyl cellulose (HPMC). The capillary is then filled with catholyte (20 mM sodium hydroxide, 0.06–0.5% HPMC). Samples are dissolved in 2.5–5% (w/v) Ampholine at a concentration of 0.06–1 mg/mL. Sample application by hydrodynamic injection (Chapter 9) gave a zone length of 5 cm.

Chen and Wiktorowicz (2) start with a DB-1 coated capillary, 72 cm × 50 μm i.d., with a film thickness of 50 μm. The capillary is filled with 20 mM sodium hydroxide containing 0.4% methylcellulose. The ampholytes (0.5% Servalyte with 0.4% methylcellulose) are then loaded into the capillary. Samples and markers, dissolved in water, are then injected, followed by a small plug of ampholyte solution. The ampholyte plug insulates the sample from the anolyte (100 mM phosphoric acid). The catholyte is 20 mM sodium hydroxide. Of the three injection schemes discussed, this method is advantageous because the sample need not be prepared in ampholytes. On activation of the voltage, the proteins migrate throughout the capillary toward the pH corresponding to their p*I*. Sensitivity is probably reduced since less material is injected, but this has not been studied.

B. Focusing

The second step in CIEF is focusing (Fig. 4.4). The buffer reservoirs are filled with the appropriate solutions as specified in Table 4.1. When the voltage is activated, the current is relatively high, since both ampholytes and solutes are highly mobile at the start of the run. As the focusing proceeds, the pH gradient forms, along with solute migration to the appropriate position along the gradient. When the ampholytes and solutes approach their respective p*I*s, their mobilities begin to slow, and as a result, the current declines.

Focusing takes only a few minutes at high field strength. Monitoring the current is a useful indicator to optimize the focusing time. Typically, the current will decline to and level off at 0.5–1.0 μA with the 25 μm i.d. capillary. Overfocusing results in protein precipitation, as shown by spikes on the electropherogram (8, 10).

C. Mobilization

With focusing complete, it is necessary to mobilize or elute the contents of the capillary past the detector to record the electropherogram.

There are three ways to accomplish this:

1. Electrophoretic (salt) mobilization;
2. hydrodynamic mobilization;
3. electroosmotic mobilization.

Points 1 and 2 are separate and discrete processes, whereas point 3 occurs simultaneously during the focusing step.

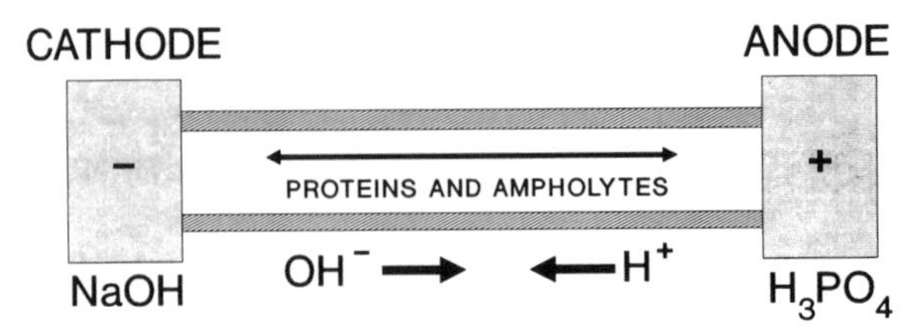

Figure 4.4. Diagram of the isoelectric focusing process.

1) Electrophoretic (Salt) Mobilization[4] This form of mobilization uses salts (3) or zwitterions (14) as additives to a buffer reservoir to effect pH changes in the capillary when the voltage is applied. The direction of mobilization can be anodic or cathodic, as indicated in Figs. 4.5a and 4.5b, depending upon which reservoir the salt is added to. Cathodic mobilization is usually selected unless very acidic proteins at the far end of the capillary need to be seen. As mobilization proceeds, the current always increases.

Electrophoretic mobilization occurs with salt addition to a buffer reservoir because of the requirement for electroneutrality (14). At steady state this condition is satisfied by

$$C_{H^+} + \Sigma\, C_{NH_3^+} = C_{OH^-} + \Sigma\, C_{COO^-}\,, \tag{4.2}$$

where C_{H^+}, C_{OH^-}, C_{NH3^+} and C_{COO^-} are the concentrations in equivalents per liter of all charged species in the ampholyte–solute blend. If a salt (anion) is added to the catholyte, this expression becomes

$$C_{H^+} + \Sigma\, C_{NH_3^+} = C_{OH^-} + \Sigma\, C_{COO^-} + C_{Y^{m-}}, \tag{4.3}$$

where $C_{Y^{m-}}$ represents an anion (m is the charge on that anion). The added salt, now the anion, competes with OH^- for electromigration into the capillary. Since fewer

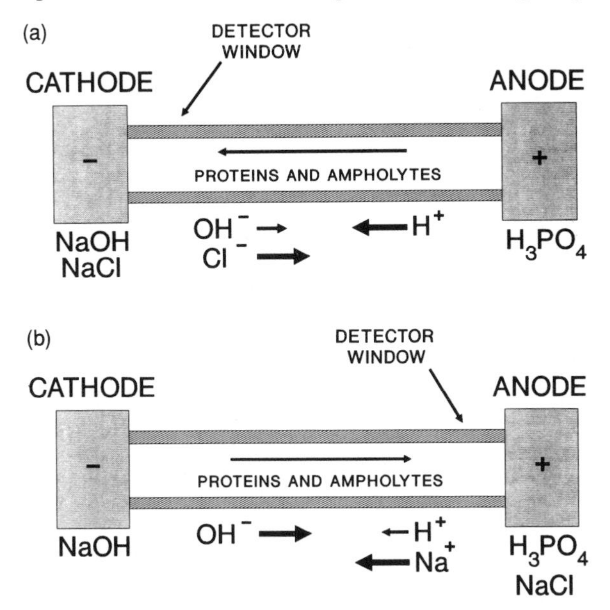

Figure 4.5. Diagrams of (a) cathodic and (b) anodic mobilization.

[4] Coated Capillaries and methodology are available from Bio-Rad, Richmond, CA.

hydroxyls enter the capillary, the pH declines. Proteins previously at their p*I*s become cationic and begin to migrate toward the cathode. For anodic mobilization, an expression equivalent to Eq. (4.3) is

$$C_{X^{n+}} + C_{H^+} + \Sigma C_{NH_3^+} = C_{OH^-} + \Sigma C_{COO^-}, \quad (4.4)$$

where $C_{X^{n+}}$ represents the cation added to the anolyte. The added anion is of no consequence because it never migrates into the capillary. The same holds true for the cation when cathodic mobilization is used. Increasing the concentration of salt speeds the mobilization process at the expense of thermal deformation due to Joule heating.

Examples of cathodic and anodic mobilization for human hemoglobin and transferrin are shown in Fig. 4.6. This early work employed 3 mm i.d. capillaries that were rotated at 40 rpm to reduce convection. While the applied field strength is low and speed of separation relatively slow by today's standards, the fundamental aspects of CIEF are well illustrated. As expected, the electropherograms are approximate mirror images of each other. Better resolution is always seen during the early portions of the electropherograms. This effect is due to the nature of the alteration of the pH gradient by added salt.

The relationship of the change in pH versus the distance from the end of a slab-gel is shown in Fig. 4.7 (14) for anodic mobilization. The pH changes are substantial along the first 4 cm of the gel and then converge to no measurable differences. This means that the relationship between the migration time and p*I* is not linear. The loss of linearity is a consequence of mobilization, not focusing. This is confirmed in Fig.4.8 using data published by Wehr *et al.* (12) and Yim (15).

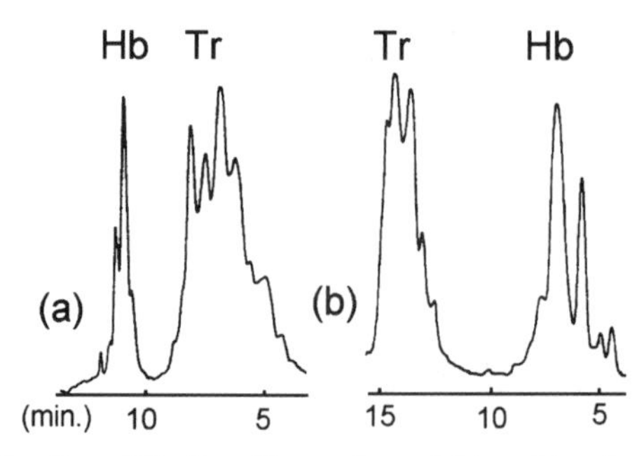

Figure 4.6. Electrophoretic mobilization of human hemoglobin and transferrin by capillary CIEF. Capillary: agarose-plugged methylcellulose-coated 38 cm × 3 mm i.d., rotated at 40 rpm; ampholytes: 1% solution of Pharmalyte (pH 3–10); anolyte: phosphoric acid, 20 mM; catholyte: sodium hydroxide, 20 mM; focusing: 1,200 V for 20 min; mobilization: (a) anolyte 20 mM sodium hydroxide catholyte 20 mM sodium hydroxide; (b) anolyte 20 mM phosphoric acid catholyte 20 mM phosphoric acid; mobilization voltage: 3,000 V; detection: UV, 280 nm. Redrawn with permission from *J. Chromatogr.* **346,** 265, copyright ©1985 Elsevier Science Publishers.

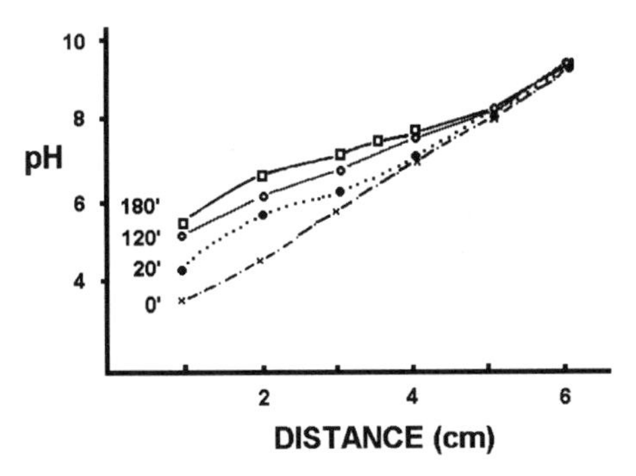

Figure 4.7. The pH gradient at different mobilization times on a 2 mm slab gel (3%C, 6%T) cast in a 2.5% solution of Bio-Lyte (pH 3–10) on a water-cooled microscope slide. The pH measurements were made with a surface electrode. Focusing: with 20 mM sodium hydroxide as catholyte and 20 mM phosphoric acid as anolyte; mobilization: anodic, with 20 mM phosphoric acid with 80 mM sodium chloride as the anolyte. Redrawn with permission from *J. Chromatogr.* **387,** 127, copyright ©1987 Elsevier Science Publishers.

2) Hydrodynamic Mobilization[5] The problem of mobilization of proteins at the far end of the capillary is addressed by hydrodynamic mobilization (3). After focusing, the capillary contents may be pumped or aspirated past the detector. Hydrodynamic bandbroadening is reduced by applying the focusing voltage during elution.

Chen and Wiktorowicz (2) adapted the approach with modern instrumentation. Instead of pumping out the capillary contents, a high-precision vacuum is used.

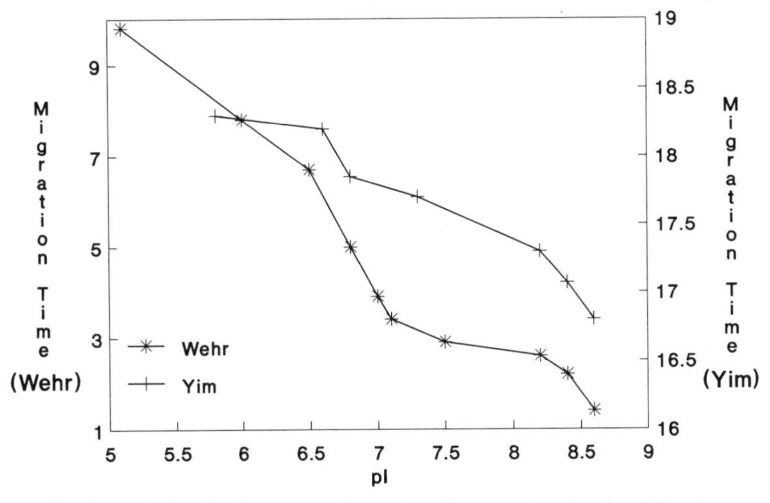

Figure 4.8. Graph of migration time versus p*I* for a series of proteins using the salt mobilization process.

[5] A protocol is available from Applied Biosystems Inc., Foster City, CA.

The full focusing voltage, 30 kV, is applied during the mobilization step. A linear relationship between mobility and p*I* occurs over the full pH range. The disadvantage of this method is the inability to add selectivity to the mobilization step (Section 4.4.C). This can be overcome in part through the use of narrow-range ampholytes.

3) Electroosmotic Mobilization Electroosmotic mobilization occurs together with the focusing step (4,7,8). The EOF is reduced by adding 0.1 % methylcellulose to the ampholyte blend. This is sufficient to ensure that focusing occurs before solute migration past the detection window. The advantage here is that buffers do not need changing, nor does the voltage have to be turned off and on. Since a bare silica capillary can be employed, the potential for instability of a coating, particularly at high pH is avoided.

The disadvantage is nonlinearity in the pH gradient. This is only a problem when protein identification by p*I* is important. Quantitative analysis is not affected and the issue of linearity has not been thoroughly studied. Published data (4) covering a narrow pH range are insufficient to measure the linearity of the system.

D. Detection

Carrier ampholytes absorb UV light below 250 nm. For most modes of HPCE, buffer absorption is not a problem because the optical pathlength is short (Chapter 10) and the buffer is homogeneous. The reagent background is simply zeroed out. In CIEF, the electrolyte composition is continually changing during focusing and subsequent mobilization. Monitoring the electropherogram at 200 nm produces a substantial background, masking all but the most concentrated proteins in the sample (8).

This problem precludes the use of 200 nm for detection, the most sensitive wavelength for proteins. Sufficient selectivity toward proteins is obtained at 280 nm. The loss of sensitivity at this wavelength compared to 200 nm is dependent on the aromatic amino acid composition of the protein. Proteins with few aromatic residues produce poor sensitivity. Limits of detection (LOD) have not been reported in the CIEF literature. Through examination of published electropherograms, an LOD of several μg/mL is expected for most proteins.

Derivatization has not been reported in CIEF. It is possible that specific pre- or post-capillary labeling can solve some of these detection problems (Section 10.7).

4.4 Focusing and Mobilization Problems

A. Basic Proteins

The use of 0.5–1% TEMED (*N*,*N*,*N*,*N*-tetramethylethyldiamine) as an additive to the ampholyte blend extends the range of the pH gradient (4, 10) by preventing migration of basic proteins past the detector window. During focusing, TEMED migrates to the cathodic end of the capillary, effectively blocking basic proteins such as cytochrome *c* (p*I* 9.6). Alternatively, anodic mobilization can be employed. A high-p*I* zwitterion[6] could probably be used as well, but this has not been reported.

B. Acidic Proteins

The failure to effectively mobilize solutes at the far end of the capillary is addressed in part with a zwitterionic mobilization reagent. When cathodic mobilization is employed, the problem occurs with acidic (low-p*I*) proteins. Adding a zwitterion to the catholyte (instead of salt) with a p*I* lower than the most acidic protein is effective in mobilizing the far end of the capillary (10). During mobilization, the low-p*I* zwitterion migrates near the anode until it becomes neutral. Ampholytes and solutes with greater p*I*s all acquire a positive charge and mobilize toward the cathode.

C. Selective Mobilization

Zwitterionic mobilizers can eliminate unwanted peaks from the extremes of electropherograms. For example, performing cathodic mobilization with p*I* 6.9 zwitterion eliminates all bands with p*I*s below that value (10). For anodic mobilization, all bands with p*I*s above the zwitterion will not be detected.

Another approach is to select the appropriate ampholyte pH range. Proteins outside that range will not be detected because they will mobilize into the respective electrode buffer reservoir. In any event, the use of narrow pH range ampholytes often requires zwitterionic anolyte and/or catholyte.

D. Impact of the Injection Mode

Most work published in CIEF designates loading the entire capillary with sample–ampholyte blend. Another less obvious advantage of partial capillary injection occurs during focusing. If the whole capillary is filled with sample, basic proteins will always focus near the cathode. If the focusing point is beyond the detector, the proteins will not be seen using cathodic mobilization unless measures discussed earlier are implemented. If only part of the capillary is filled, the capillary length, voltage, or focusing time can be adjusted so the most basic proteins never pass the detection window before mobilization.

4.5 Applications

There are hundreds of variations of human hemoglobin that result from single-point amino acid mutations. Abnormal hemoglobin is found in one of 10,000 individuals with electrophoresis as the diagnostic test. CIEF is useful for performing a

[6] Select a zwitterion based on p*I* from Table 2.3.

high-speed separation separating some important variants such as hemoglobin S, the sickle-cell variant that results from a 6 Glu → Val replacement. Zhu *et al.* (10) resolved hemoglobins S, C, F, and A in 6 min using salt mobilization (Fig. 4.9). The separation of hemoglobin A from F is remarkable, since the two proteins differ by only 0.05 p*I* units. As the separation was run in pH 3–10 ampholytes, even greater resolution is possible using a narrower-range ampholyte.

Chen and Wiktorowicz (2) separated, in part, RNase T_1 (p*I* 2.9), RNase ba (p*I* 9.0), and site directed mutants (p*I* 3.1, 3.1, 3.3) of RNase T_1 using the hydrodynamic mobilization method. The two mutants of p*I* 3.1 were not separable. The separation shown in Fig. 4.10 includes the marker proteins RNase A (p*I* 9.5),

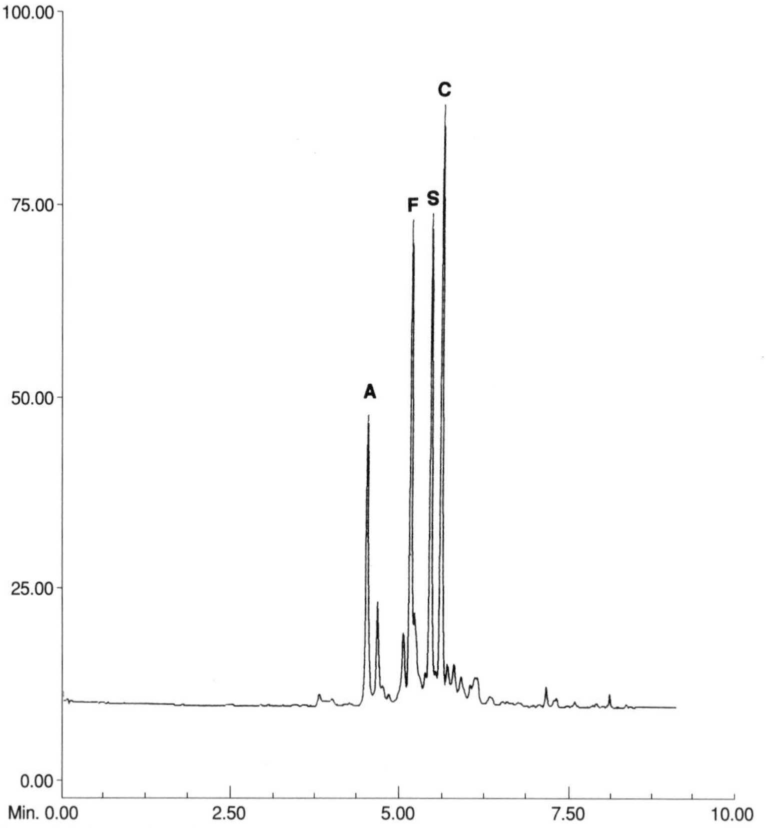

Figure 4.9 Separation of hemoglobin variants by capillary CIEF in a 12 cm × 25 μm i.d. coated capillary using pH 3–10 ampholytes. Focusing and mobilization were carried out at 8 kV. Protein concentration: 250 μg/mL of each protein; detection: 280 nm; isoelectric points are: hemoglobin A, p*I* 7.1; F, p*I* 7.15; S, p*I* 7.25; and C, p*I*, 7.5. Reprinted with permission from *J. Chromatogr.* **559,** 479, copyright ©1991 Elsevier Science Publishers.

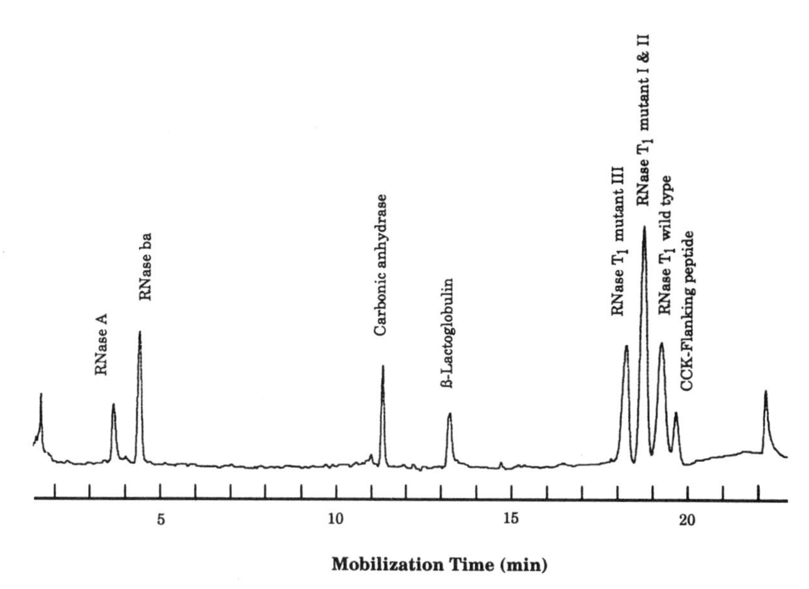

Figure 4.10. Capillary CIEF of RNase ab, RNase T_1, and RNase T_1 site-directed mutants. Capillary: DB-1-coated 72 cm (50 cm to detector) × 50 μm i.d.; ampholytes: Servalyt 3-10, 0.5% with 0.4% methylcellulose; injection: 30 s vacuum (200 nL); focusing: 6 min at 30 kV; mobilization: vacuum (5 in. Hg) with 30 kV voltage; detection: UV, 280 nm; solutes: 200 μg/mL each (4 ng injected). Reprinted with permission from *Anal. Biochem.* **206,** 84, copyright ©1992 Academic Press.

carbonic anhydrase (p*I* 5.9), β-lactoglobulin (p*I* 5.1), and CCK-Flanking peptide (p*I* 2.75). The authors reported the DB-1 capillary was stable for at least 420 runs over a two-month period. When marker proteins were used for internal standardization, the p*I* measurement precision ranged from 0.5 to 3.3% relative standard deviation. A peak area calibration curve was linear versus injection time (volume) from 10 to 80 s (50–400 nL). Studies varying solute concentration were not performed. This can be important because solute concentration affects the migration time when salt mobilization is employed (15).

Production and purification of recombinant proteins are performed to yield a homogeneous product with regard to the polypeptide chain. Despite this lengthy purification, numerous peaks are seen by CIEF (Fig. 4.11). Microheterogeneity of human recombinant tissue plasminogen activator (rtPA) is expected based on the number of available glycosylation sites and the degree of glycosylation. The degree and type of glycosylation can impact on the activity, clearance rate, and immunogenicity of a protein (15).

Type I and II rtPA are two glycosylation variants of rtPA. After fractionation on a Sepharose column, concentration, and dialysis, the fractions were further separated by CIEF (Fig. 4.11) and CZE (not shown). There are four potential

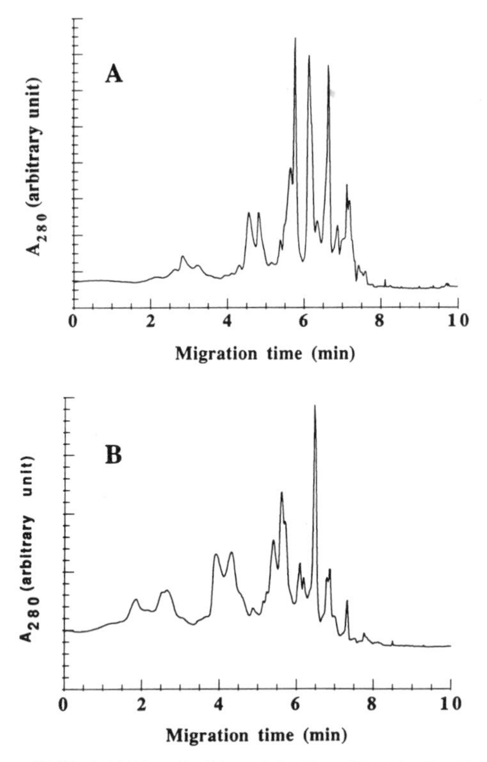

Figure 4.11 Capillary CIEF of (a) Type I rtPA and (b) Type II rtPA. Capillary: polyacrylamide-coated 14 cm × 25 μm i.d.; ampholytes: 2% solution of ampholyte pH 6–8, 2% CHAPS, 6 M urea; anolyte: 10 mM phosphoric acid; catholyte: 20 mM sodium hydroxide; cathodic mobilizer: 20 mM sodium hydroxide + 80 mM sodium chloride; focusing voltage: 2 min at 12 kV; mobilizing voltage: 8 kV; detection: 280 nm. Reprinted with permission from *J. Chromatogr.* **559,** 401, copyright ©1991 Elsevier Science Publishers.

glycosylation sites on rtPA, of which three are actually occupied. Two variants form during the recombinant synthesis. Type I rtPA is glycosylated at asparagine residues 117, 184, and 448, whereas Type II rtPA is glycosylated only at Asn-117 and 448. The carbohydrate at Asn-117 is a high-mannose oligosaccharide. At Asn-448, a complex branched oligosaccharide that may contain sialic acid is attached to galactose. The structural complexity and variation are reflected in the complex electropherograms for each rtPA type.

In other applications, Kilár and Hjertén (9, 13, 16) separated human serum transferrin isoforms, including the di-, tri-, tetra-, penta-, and hexasialo moieties. Nielsen *et al.* (17) examined separation of antibody–antigen complexes, and Silverman *et al.* (18) separated the isoforms of a monoclonal antibody.

References

1. P. G. Righetti, *Isoelectric focusing: Theory, Methodology and Applications* (T. S. Work and R. H. Burdon, eds.), Laboratory Techniques in Biochemistry and Molecular Biology (Elsevier Biomedical Press, Amsterdam, 1983).
2. S.-M. Chen and J. E. Wiktorowicz, *Anal. Biochem.* **206,** 84 (1992).
3. S. Hjertén and M.-D. Zhu, *J. Chromatogr.* **346,** 265 (1985).
4. J. R. Mazzeo and I. S. Krull, *Anal. Chem.* **63,** 2852 (1991).
5. S. Hjertén, K. Elenbring, F. Kilár, J. L. Lias, A. J. C. Chen, C. J. Liebert, and M. Zhu, *J. Chromatogr.* **403,** 47 (1987).
6. C. A. Bolger, M. Zhu, R. Rodriguez, and T. Wehr, *J. Liq. Chromatogr.* **14,** 895 (1991).
7. J. R. Mazzeo and I. S. Krull, *J. Microcol. Sep.* **4,** 29 (1992).
8. W. Thormann, J. Caslavska, S. Molteni, and J. Chmelik, *J. Chromatogr.* **589,** 321 (1992).
9. F. Kilár and S. Hjertén, *J. Chromatogr.* **480,** 351 (1989).
10. M. Zhu, R. Rodriguez, and T. Wehr, *J. Chromatogr.* **559,** 479 (1991).
11. M. Zhu, D. L. Hansen, S. Burd, and F. Gannon, *J. Chromatogr.* **480,** 311 (1989).
12. T. Wehr, M. Zhu, R. Rodriguez, D. Burke, and K. Duncan, *Amer. Biotech. Lab.* , 22 (1990).
13. F. Kilár and S. Hjertén, *Electrophoresis* **10,** 23 (1989).
14. S. Hjertén, J.-L. Liao, and K. Yao, *J. Chromatogr.* **387,** 127 (1987).
15. K. W. Yim, *J. Chromatogr.* **559,** 401 (1991).
16. F. Kilár, *J. Chromatogr.* **545,** 403 (1991).
17. R. G. Nielsen, E. C. Rickard, P. F. Santa, D. A. Sharknas, and G. S. Sittampalam, *J. Chromatogr.* **539,** 177 (1991).
18. C. Silverman, N. Komar, K. Shields, G. Diegnan, and J. Adamovics, *J. Liq. Chromatogr.* **15,** 207 (1992).

Chapter 5

Size Separations in Capillary Gels and Polymer Networks

5.1 Introduction

Slab-gel electrophoresis is the predominant technique for the separation of peptides, proteins, and polynucleotides. The slab-gel format provides mechanical stability for the separation, reduces solute dispersion from convection and diffusion, and permits handling for detection, scanning, storage, etc., as described in Section 1.1. Unlike IEF gels, which exist solely for these points, the slab-gel also provides the mechanism for separation.

Gels are porous matrices comprising polymeric materials dissolved in a solvent, usually an aqueous buffer. The pore size of the gel is determined by the concentration of the polymeric reagent and the three-dimensional structure of the matrix. Chemical cross-linkers further influence structural rigidity and pore size of the gel when incorporated during the polymerization process. The porous structure of the gel provides effective separations of macromolecules based on molecular size.

True gels are composed of lyophilic (solvent-loving) colloidal particles, the best known of which is polyacrylamide. Lyophobic (solvent-hating) colloidal materials such as agarose are known as sols. Agarose solutions must be prepared in boiling aqueous buffer to dissolve a sufficient amount of material to form the gel matrix.

Other polymeric materials such as methylcellulose derivatives and dextrans can also define molecular pores. These systems are best described as polymer networks, entangled polymers, or physical gels. Since the mechanism of separation in gels and entangled polymers is molecular sieving, they are both described in this chapter.

The transition from the slab-gel to the capillary format began in 1983 (1), when Hjertén filled a 150 μm i.d. tube with polyacrylamide. He later reported size separations of proteins by capillary sodium dodecyl sulfate–polyacrylamide gel electrophoresis (SDS-PAGE)(2). Since then, applications for proteins (3–5), oligonucleotides (6–14), DNA restriction digests (15–20), polymerase chain reaction products (18, 20–23), DNA sequencing reaction products (22, 24–36), oligosaccharides (37, 38), and poly(styrenesulfonates) (39) have appeared.

In slab-gel electrophoresis, a gel is poured and polymerized just before use.[1] The gel must be sufficiently viscous to eliminate flow and permit handling. Tackiness of the material must also be avoided. These constraints are not necessary in CGE though gel viscosity is an important consideration. Low-viscosity gels can be pumped in and out of the capillary. For each run or several runs, a fresh gel matrix can be employed. High-viscosity gels are either polymerized *in situ* or pumped in under high pressure. These gels must be stable enough to tolerate multiple runs, because the entire capillary must be replaced if the gel deteriorates. Clearly, pumpable gels are advantageous for CGE, provided adequate speed and resolution are obtained.

5.2 Separation Mechanism

The separation mechanism for CGE is illustrated in Fig. 5.1. Driven by the electric field, solutes migrate toward the appropriate electrode through the gel matrix. Small molecules pass through the pores unimpeded. Larger biopolymers travel a tortuous path, moving through the pores of the gel in a snakelike reptilian fashion. Under properly controlled conditions, the solute's mobility is inversely proportional to its size.

Grossman (40) reviewed the mechanisms of migration through polymer networks. The Ogston model treats a molecule as a point source where the migration velocity is determined by a solute's mobility modified by the probability of an encounter with a restricting pore. Large biopolymers do not necessarily behave following this model. The molecule can deform during its transit through the gel.

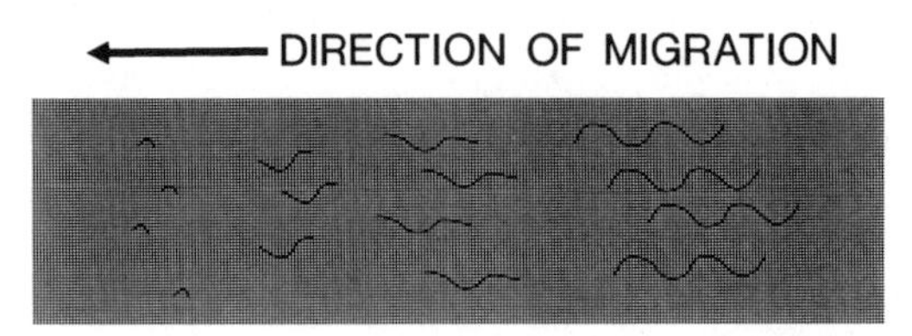

Figure 5.1. Pictorial description of the mechanism of size separation by HPCE.

[1] Rehydratable gels are also commercially available, but these have no role in HPCE.

Envision a long strand of DNA wriggling through the gel matrix in a snakelike manner. Such a process is known as reptation. It is also observed that fragment resolution decreases with the size of the molecules. Molecules such as DNA align with the electric field in a size-dependent manner that is biased towards the larger strands. The dependence of mobility on the molecular size is obscured by this process, which is known as *biased reptation.* This effect limits the size of DNA molecules that can be separated using conventional slab-gel techniques. Beyond 20,000 base pairs (bp), pulsed-field techniques (Section 5.5) must be employed.

Separations in gels are based on molecular size. When molecular weight measurements are required, the macromolecule must be denatured. For proteins, this entails reduction of disulfide bonds and unfolding by heating with a solution composed of sodium dodecyl sulfate (SDS) and a reducing agent, β-mercaptoethanol or dithiothreitol (DTT). After these processes, most proteins have roughly the same shape and charge density. SDS binds to and denatures proteins via electrostatic and hydrophobic interactions; about 1.4 g SDS is bound per gram of protein. Also, SDS is anionic, so all proteins become negatively charged and migrate toward the anode. Under these conditions, a protein's mobility is proportional to its molecular weight and the gel can be calibrated with a sizing standard. SDS-PAGE (polyacrylamide gel electrophoresis) is the standard method for separating proteins based on molecular weight.

Some proteins exhibit nonideal behavior during SDS-PAGE. Glycoproteins and lipoproteins do not bind the same amounts of SDS per gram of protein. The correlation of mobility to molecular weight falls apart under those circumstances. Aggressive denaturing conditions may be required to alleviate this problem (41).

DNA and RNA are usually denatured to disrupt aggregates and hydrogen-bonded structures by heating in buffer to 60–90°C for a few minutes. Heated denaturing gels containing 8 M urea, formamide, or methylmercuric hydroxide may be required to prevent renaturation (42).

5.3 Materials for CGE

A. Classification of Materials

A variety of gelling agents are compatible with CGE. Classical reagents such as polyacrylamide and agarose have been adapted to the capillary format. Other low-viscosity solutions not suitable for slab-gels work well in capillaries. Both crosslinked and linear polymers can define pores in solution (Fig. 5.2) through either covalent bonding or polymeric entanglement. These materials are classified as chemical and physical gels, respectively.

A better classification of gels and molecular sieves for CGE is based on viscosity. Low-viscosity gels are pumpable and easily filled into a capillary.

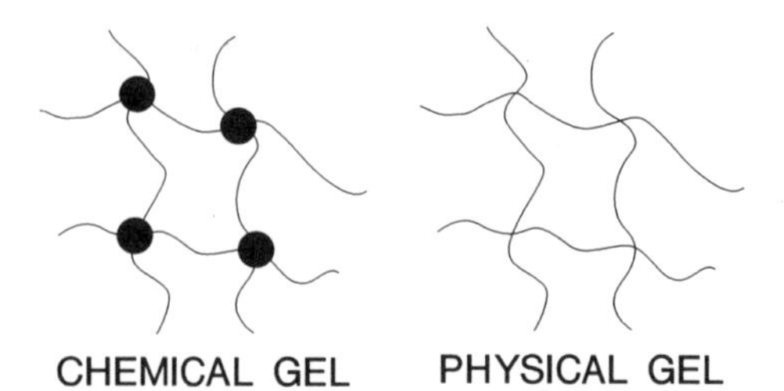

Figure 5.2. Illustration of the differences between chemical and physical gels. Pores in chemical gels are defined via covalent bonds. In physical gels, polymeric entanglement defines the porous network.

Regeneration after several runs is simple as well. High-viscosity gels are either prepared *in situ* or pumped into the capillary under pressure. Upon gel failure, the entire capillary must be replaced. Pumpable gels are advantageous because they are user-prepared, easy to regenerate, and less costly. Excepting linear polyacrylamide, materials such as agarose, dextrans, starches, and methylcellulose derivatives do not even require user polymerization, another substantial advantage.

When low-viscosity gels are used, a coated capillary is frequently required to suppress the EOF. Otherwise, solutes may elute in a reverse order, i.e., large molecules elute before small ones. Migration-time imprecision and wall effects may also occur when untreated capillaries are used.

B. Polyacrylamide

Raymond and Weintraub (43) first used polyacrylamide as a matrix for electrophoresis in 1959. Work of Ornstein (44) and Davis (45) firmly established the use of the material for gel electrophoresis.

The gel composition of this electrically neutral material is defined by %T, the total amount of acrylamide, and %C, the amount of crosslinker. Percent T is calculated by

$$\%T = \text{acrylamide(g)} + \text{bisacrylamide(g)} / 100\ \text{mL}\,. \tag{5.1}$$

Percent C is

$$\%C = \frac{\text{bisacrylamide(g)}}{\text{bisacrylamide(g)} + \text{acrylamide(g)}} \times 100\,. \tag{5.2}$$

Bisacrylamide is frequently the crosslinker, though there are many alternatives.

The pore size of the gel is controlled by both %T and %C. Highly crosslinked gels are usually employed to increase the pore size. A 30% C gel has a pore size of 200–250 Å (46). Liu *et al.* (47) prepared capillaries with up to 40% T using sequential polymerization techniques (48), although these would be expected to be far too rigid to survive the high field strengths of CGE. The pore size can also be increased by using low %T gels. These dilute gels are not useful in the slab-gel format because of insufficient mechanical stability. However, they can be used in CGE where low-viscosity materials are advantageous, since they are pumpable. Large pore size gels (lower %T) are used for separating DNA sequencing reaction products, whereas the narrow-pore media are best for proteins and small oligonucleotides. Selection of the %T is based on the molecular size to be separated (Table 5.1). Lower %T yields higher speeds, while higher %T gives enhanced resolution; however, the range of molecular sizes separable in a single run declines with high %T gels.

The preparation of a stable, bubble-free gel capillary is a complex undertaking. It is anticipated that few users will attempt to manufacture their own gels, but rather that most will opt for a commercial version. Bubble production and gel shrinkage during polymerization are often confounding problems that do not occur in the open environment of the slab-gel.

Table 5.1. Applications with Polyacrylamide Gels in CGE

Application	Gel Composition	Buffer	Reference
Polyadenylates	5%T, 5%C	0.1 M Tris, 0.1 M boric acid, 7 M urea, pH 8.8	(56)
Polydeoxycytidines	3, 5, 10 %T	20 mM boric acid, 7 M urea, 2 mM EDTA, pH 8.6	(14)
Polydeoxycytidines	6%T, 3%C	0.1 M Tris, 0.25 M borate, 7 M urea, pH 7.5	(55)
Oligonucleotides	6%T, 5%C	0.1 M Tris, 0.25 M boric acid, 7 M urea, pH 7.6	(54)
Oligonucleotides	7.5% T, 3.3% C	50 mM Tris, 50 mM boric acid, 7 M urea	(9)
Oligothymidylic acids	2.5%T, 3.3%C	89 mM Tris, 69 mM boric acid, 7 M urea, 2 mM EDTA, pH 8.6	(10)
Restriction fragments	6, 9, 12% T	0.1 M Tris, 0.1 M boric acid, 2 mM EDTA, pH 8.3	(17)
DNA sequencing	3% T, 5% C	0.1 M Tris-borate, 2.5 mM EDTA, 7 M urea, pH 7.6	(27)
DNA Sequencing	4% T, 5% C	90 mM Tris, 90 mM boric acid, 2.5 mM EDTA, 1.3 mM TEMED, pH 8.3	(29)
Hybridization products	9% T	25 mM Tris-borate, 25 mM EDTA, pH 8.0	(57)
Proteins (10–100 kD)	5%T	375 mM Tris, 0.1% SDS, pH 8.8, 1.8–2.7 M ethylene glycol	(4)
Proteins (40–200 kD)	3%T	375 mM Tris, 0.1% SDS, pH 8.8, 1.8–2.7 M ethylene glycol	(4)
Proteins	10-12.5%T, 3.3%C	90 mM Tris-phosphate, 0.1% SDS, 8 M urea, pH 8.61	(3)
Oligosaccharides	25.5% T, 3% C	0.1 M Tris, 0.25 M borate, 2 mM EDTA, pH 8.48	(47)

Many schemes for producing semistable gels have appeared in the literature (3, 4, 25, 48–54). These include bonding the gel to the capillary wall with a bifunctional silane reagent (3), using ethylene glycol to reduce bubble formation (4), high-pressure polymerization (25), various schemes for sequential polymerization such as isotachophoretic polymerization (48), programmed-temperature polymerization (48), and photopolymerization (48, 53). Surface-treating the capillary before polymerization may also play an important role (54). Gamma radiation could serve as an initiator in place of ammonium persulfate and *N*, *N*, *N*, *N*-tetramethylethylenediamine (TEMED) (55). It is unclear at this time which approach is superior. The method of preparation, field strength, buffer composition, and temperature all play a role in determining the longevity of gel-filled capillaries.

The integrity of the polymerized gel often fails because of the high field strengths that are employed in CGE. Gel failure is manifested by unstable currents and bubble formation, followed by the loss of electrical continuity. Gel failure can occur without warning, a significant problem during automated unattended runs. Injection of a single high ionic strength sample can ruin a gel. On the other hand, spectacular separations of oligonucleotides such as those illustrated in Figs. 5.3 and 5.4 have been reported. These separations often yield millions of theoretical plates. With low ionic strength samples, gels can last for 20 h at 200–300 V/cm — time enough for approximately 50 runs(9).

Both cross-linked and linear polyacrylamide gels (Table 5.1) have been used in slab-gel electrophoresis and CGE. In slab-gels, the lowest polyacrylamide concentration forming a mechanically stable gel is 2% T, 2.2% C (46) for a cross-linked gel and 10% T (17) for a linear gel. In CGE, mechanical stability is less important, and linear polyacrylamide gels as low as 3% T have been reported (14). However, resolution of oligodeoxycytidylic acids is poor at such a low %T. A 10% T gel is required for unit base resolution. Linear polyarylamide gels of up to 10% T can be loaded into the capillary with a specially modified syringe (14).

1) Field Strength Increasing the field strength speeds the separation up to the point that Joule heating becomes significant or gel breakdown occurs. Running rigid gels beyond 400 V/cm leads to premature breakdown, though the impact of Joule heating can be compensated in part with effective cooling systems.

A more fundamental limitation occurs with 15–20 kbp of DNA. The fragments line up with the electric field, leading to size-independent migration. Solutes exhibiting this behavior will show a positive deviation from linearity in a plot of velocity versus field strength. The problem occurs in CGE with smaller fragments compared to the slab-gel (17). This is a consequence of the high field strengths employed in CGE. Only pulsed-field techniques (Section 5.5) can reduce this problem, but these instruments are not commercially available.

The magnitude of the field strength also influences the resolution of the smaller fragments (17). The greatest resolution is observed at field strengths as low as 100 V/cm, though other workers found better resolution at 200 V/cm (58). Variations

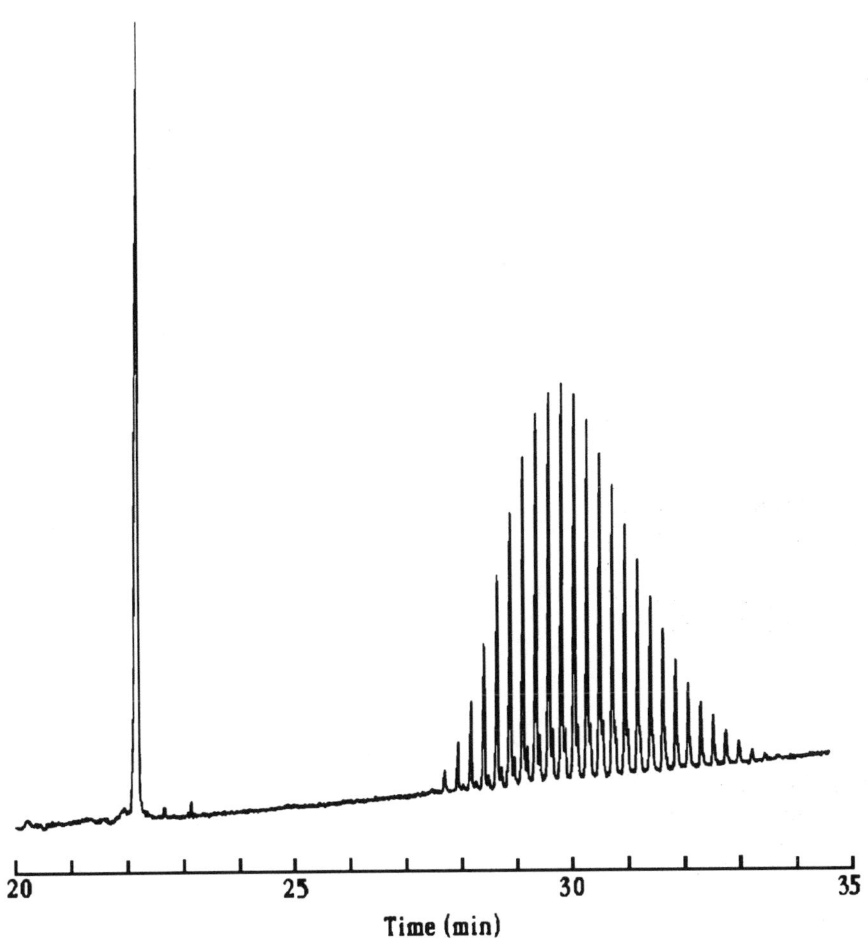

Figure 5.3. CGE of poly $(dA)_{20}$ and poly $(dA)_{40-60}$ on a 9%T linear polyacrylamide gel. Capillary: 60 cm (45 cm to detector) × 75 μm i.d.; field strength: 308 V/cm; current: 8.8 μA; buffer: 100 mM Tris-borate, pH 8.3, 2 mM EDTA, 7 M urea; injection: electrokinetic, 10 kV for 0.5 s; detection: UV, 260 nm. Reprinted with permission from *J. Chromatogr.* **516,** 33, copyright ©1990 Elsevier Science Publishers.

in gel composition may account for these differences. In any case, separations become lengthy at low field strength, so the operating voltage should be selected based on the required resolution.

2) Capillary Length Longer capillaries provide greater resolution at the expense of run time. For oligonucleotide separations of greater than 80 mer, a 40 cm capillary is recommended[2] with a Beckman eCAP U100P polyacrylamide gel-filled

[2] eCAP U100P instruction sheet, Beckman Instruments.

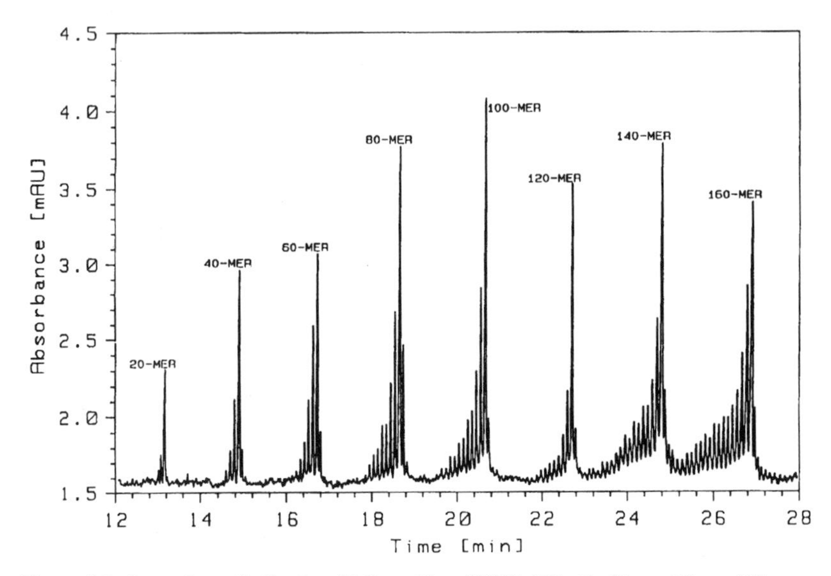

Figure 5.4. Separation of oligothymidylic acids pd(T)20–160. Capillary: 75 cm (50 cm to detector) × 100 μm i.d. filled with 2.5%T, 3.3%C; buffer: 89 mM Tris, 69 mM boric acid, 7 M urea, 2 mM EDTA, pH 8.6; voltage: 400 V/cm; current: 40 μA; detection: UV, 260 nm; injection: 10 kV for 10 s. Reprinted with permission from *Electrophoresis* **11,** 702, copyright ©1990 VCH.

capillary. The number of theoretical plates is linearly dependent of the capillary length in a 6%T, 5%C gel (58).

3) Temperature Guttman and Cooke (16) studied the impact of temperature on the separation of restriction fragments. In the constant-voltage mode, increasing the temperature decreased the migration times as expected. Loss of resolution accompanies the rise in temperature, particularly for the larger fragments. Conformation changes in the fragments are suspected as the root cause. Peak asymmetry also appears above 30°C.

4) Injection Electrokinetic injection (Section 9.3) must be employed, or the rigid gel may be extruded from the capillary. Furthermore, the ionic strength of the injection solution is critical in determining the amount of material injected into the capillary (Sections 5.4 and 9.6).

5) Operating Hints with Rigid Gell-Filled Capillaries Gel-filled capillaries are fragile. The following guidelines will help maximize capillary lifetime.

Consult the manufacturer's package insert for specific instructions on capillary operation.

1. After installation in an instrument, immediately immerse the capillary in buffer to prevent gel dehydration. The run buffer is generally the identical buffer in which the gel was originally polymerized. When a capillary is

removed from the instrument, the ends must be promptly immersed in a buffer solution to prevent drying.

2. Run a performance standard to ensure that both the capillary and instrument are performing to specifications.
3. Use electrokinetic injection, or else gel may be extruded from the capillary.
4. If the operating current fluctuates rapidly, a bubble has probably formed. If the bubble is confined near the ends of the capillary, the capillary may be saved by removing a few centimeters from the end. On the detector side, this may not be possible because there is little excess length.
5. Running at higher temperatures is possible, but once a temperature of 50°C is exceeded, it should never be lowered during the lifetime of the capillary.
6. Field strengths over 400 V/cm will lead to premature breakdown of the gel.
7. The sample solution should be of low ionic strength to maximize the amount of injected material. Ionic strength is critical for quantitative analysis and separation efficiency. Differences in ionic strengths will result in substantial variation of the response factors.
8. The use of a reference marker and the reporting of relative migration times will offset in part run-to-run and capillary-to-capillary variation.
9. Electrode buffer solutions should be replaced frequently.

C. Agarose

Agarose is a complex group of polysaccharides extracted from the agarocytes of *Rhodophyceae*, a marine alga found predominately in the Pacific and Indian Oceans. Neutral, pyruvated, and sulfated fractions have been isolated, though all fractions contain some charged groups. The material is insoluble in cold water and slowly soluble in hot water. A 1% solution forms a stiff gel upon cooling. In 1961, Hjertén first employed agarose as a support for zone electrophoresis.

Righetti (46) summarized the properties of this remarkable material. The polysaccharide is a double helix with a pore structure more rigid than a polyacrylamide strand. Even in dilute concentrations, the agarose structure has high mechanical strength. Pores with diameters of 500–800 nm have been reported. It is not surprising that agarose is useful in separating large segments of DNA. Even more remarkable is the low viscosity of the gelled material; it is pumpable, a significant advantage in CGE.

Boček and Chrambach (59) separated DNA fragments ranging in size from 72 to 1,353 base pairs (bp) at 40°C using 0.3–2.0% solutions of agarose[3] in a homemade polyacrylamide-coated capillary. Agarose solutions are prepared by weight in boiling buffer (89 mM Tris, 89 mM boric acid, 2.5 mM disodium EDTA [TBE buffer]). Additional water is added to compensate for losses due to evaporation.

[3] Seaplaque GTG (Cat. No. 50112), FMC Corporation, Rockland, MD 04841.

The capillary, 27 cm (20 cm to detector) × 150 μm i.d., is filled with hot (50°C) agarose using the high-pressure mode of a Beckman P/ACE 2000 HPCE instrument set at 40°C. The electrode buffer reservoirs contain only buffer without agarose, since the reservoirs could only be kept at ambient temperature, which is below the gel point of the polymer (26°C). The field strength is −185 V/cm (reverse polarity), so negatively charged DNA migrates toward the anode. Electrokinetic injection, −1 kV for 15 s, is used with detection at 260 nm. The range of the separation can be increased to 12 kB with 1.7% Seaprep[4] agarose at 40°C in TBE buffer (60), as shown in Fig. 5.5. Approximately 30 runs can be made before refilling the capillary with gel (61). Peak resolution of 3 bp is obtained in the early portions of the electropherogram at the specified agarose concentration. The expanded electropherogram (not shown) separates as high as 12,216 bp.

A polyacrylamide or similarly coated capillary is required to nullify the EOF. At 40°C, this coating is stable through 100 injections over a one month period (60). Slight decreases in mobility are noted over that injection span. This is caused by increased EOF due to slow deterioration in the coating. Coatings such as DB–1[5] may be useful here, provided there are no wall effects.

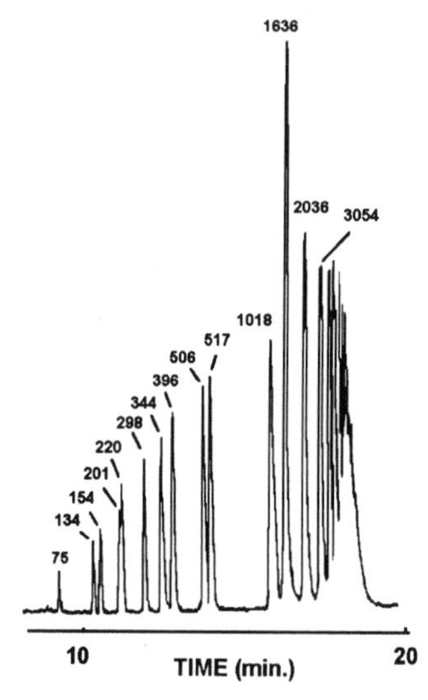

Figure 5.5. Separation of a 1 kb ladder of DNA fragments in 1.7% SeaPrep agarose at 40°C. Capillary: 33 cm (27 cm to detector) × 150 μm i.d.; buffer: 89 mM Tris-borate, 2.5 mM EDTA; field strength: 185 V/cm; injection: 1 kV for 16 s; detection: UV, 260 nm. Reprinted with permission from *Electrophoresis* **13,** 31, copyright ©1992 VCH.

[4] Seaprep agarose (catalog no. 50302), FMC Corporation, Rockland, MD 04841.

[5] J & W Scientific, Folsom, CA.

The need to maintain the agarose temperature above its gel point is a distinct disadvantage. This can be remedied by using substituted agaroses that have a low gel temperature. A vinyl-substituted agarose[6] used at room temperature is equivalent in resolving power to SeaPlaque (61).

Motsch *et al.*(62) fill bare-silica capillaries with 0.5–0.7% agarose,[7] heated to 90°C. Nitrogen pressure at 20 bar is required to fill a 100 cm × 75 μm i.d. capillary. The gel sets after 1–2 h at room temperature, but the capillaries are stable for only a few days. A buffer of 10 mM disodium hydrogen phosphate and 10 mM sodium hydrogen phosphate is used for gel preparation and separation. Run times were 5–10 times longer compared to those of Boček and Chrambach (59, 60).

D. Methylcelluloses

1) Introduction to Polymer Networks Methylcellulose and its derivatives are the prototypical polymer networks. Following the first reports by Hjertén *et al.* (63) and Zhu *et al.* (64), much of the work with physical gels has been accomplished using this material. Grossman (65) found that the pore size in these polymer networks depends on the polymer concentration and the radius of the mesh-forming polymer chain. At low concentration, the polymers are isolated from one another and exhibit no overlap. The overlap threshold (Φ^*) can be estimated from

$$\Phi^* = N^{-4/5}, \tag{5.3}$$

where N is the number of segments in the polymer chain. The overlap threshold can be evaluated experimentally by monitoring viscosity of the polymeric solution. A plot of viscosity versus Φ^* becomes nonlinear at a concentration of 0.29% hydroxyethylcellulose (65), indicating the onset of polymer overlap. The pore size, $\xi(\Phi)$, can be estimated from

$$\xi(\Phi) \approx a\Phi^{-3/4}, \tag{5.4}$$

where a is the length of one repeat unit along the polymer chain. Refer to Ref. 65 for instructions to estimate the statistical segment length. Pore sizes of 223–350 Å are obtained with polymeric solutions of 0.4% hydroxyethylcellulose. Pores of this size are sufficient for separating DNA restriction fragments and polymerase chain reaction products (18), but are too large for the separation of all but the largest proteins. For comparison, the pore size of an 8% T linear polyacrylamide gel is in the range of 100–200 Å. The broad distribution of pore sizes may contribute to the wide range of size selectivity of the polymer network (66).

Theory predicts that polymer networks can provide narrow pores using short-chain polymers at a concentration near the overlap threshold (65). The polymer

[6] AcrylAide (Catalog no. 51013), FMC Corporation, Rockland, ME 04841.

[7] Agarose Wide Range (Catalog No. A-2790), Sigma Chemical.

network systems described here can yield resolution of 3 bp of DNA. If unit base resolution is required as in DNA sequencing, polyacrylamide gel-filled capillaries are the best choice at the present time. However, if small pore sizes are forthcoming, polymer networks will displace the use of rigid gels in CGE.

2) Polymer Concentration Optimizing the methylcellulose concentration is simple. Figure 5.6 shows the variation in migration time versus the number of base pairs for a 1 kb DNA sizing ladder. Since the viscosity of the solution, and thus the electrolyte loading time, increases with methylcellulose concentration, the lowest concentration necessary for the separation should be selected (66). Depending on the capillary length and diameter, it can take up to 30 min to fill a 50 cm × 75 μm i.d. capillary with a 0.6% methyl cellulose solution. Separation between 1,000 and 10,000 bp can be achieved with 0.2% methylcellulose. Approximately 4–5 runs can be performed before the capillary must be refilled (67). Using 0.4% methylcellulose, the refill time is less than 5 min.

3) Injection Schwartz *et al.* (18), separated DNA restriction fragments on an OV-17 (50% phenyl, 50% methylsiloxane) coated capillary with 0.5% hydroxypropylmethylcellulose (HPMC)-4000 as a buffer additive. Electrokinetic injection (Section 9.3) with a stacking buffer (Section 9.6) provided higher-efficiency separations than hydrodynamic injection (Fig. 5.7). Electrokinetic injection is particularly useful for DNA separations because the mobility of all fragments is identical in the absence of the gel matrix. This eliminates the bias that is normally

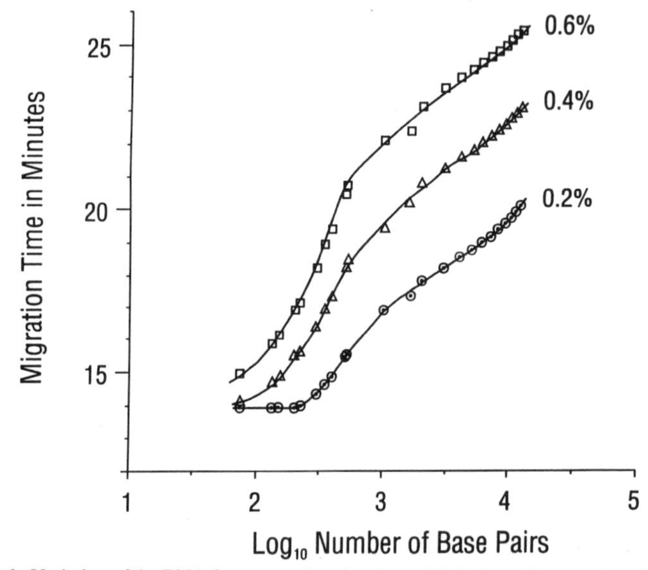

Figure 5.6. Variation of the DNA fragment migration time with the log of the number of base pairs of a 1 kb ladder using 0.2, 0.4 and 0.6% methylcellulose in 100 mM Tris-borate, 2 mM EDTA, pH 8 on a polyacrylamide-coated capillary. Capillary: 100 cm (80 cm to detector) × 75 μm i.d.; voltage: –30 kV; detection: UV, 260 nm. Reprinted with permission from *J. Liq. Chromatogr.* **15,** 1063, copyright ©1992 Marcel Dekker.

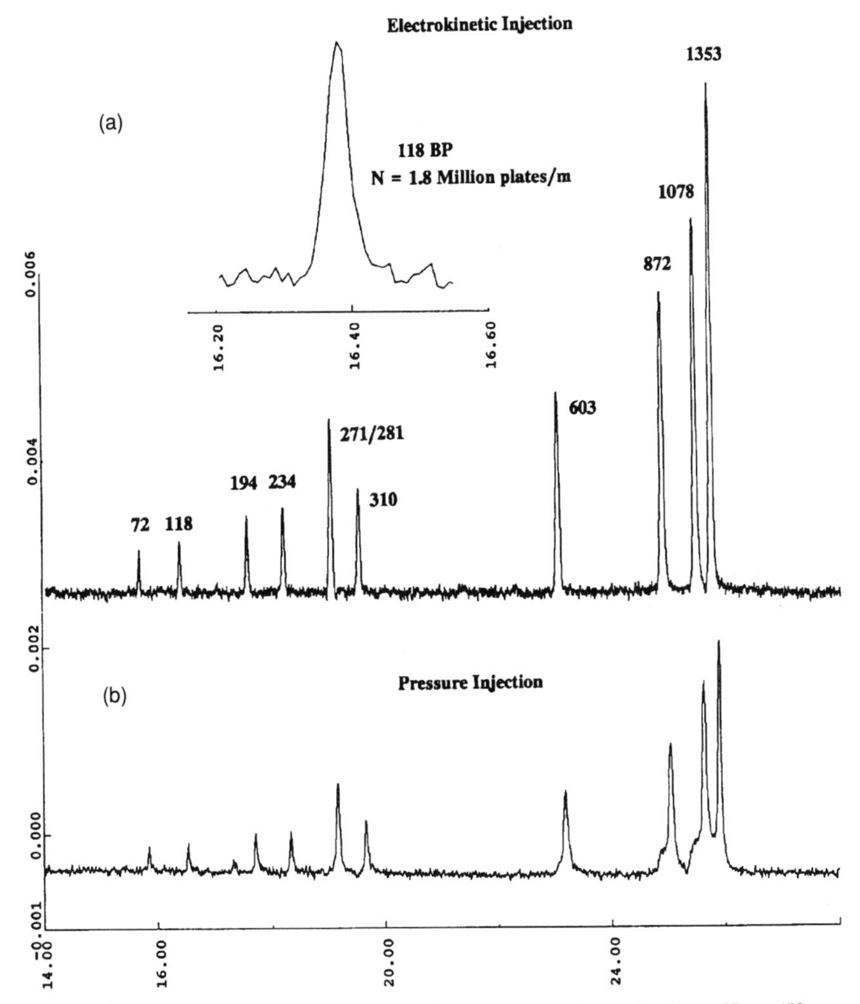

Figure 5.7. Separation of a *Hae* III restriction digest of ϕX 174 DNA. Capillary: 57 cm (50 cm length to detector) × 100 μm i.d. coated with 0.1 μm thick OV-17; buffer: 89 mM tris borate, 2 mM EDTA (pH 8.5) and 0.5% HPMC-4000; voltage: 10 kV; temperature: 25 °C; detection: UV, 260 nm; sample concentration: 25 μg/mL. Injection: (a) electrokinetic, 5 s at 2 kV; (b) pressure, 60 s at 3.44 MPa. Reprinted with permission from *J. Chromatogr.* **559,** 267 (1991), copyright ©1991 Elsevier Science Publishers.

caused by this type of injection. In other reports, hydrodynamic injection has been used successfully with methylcellulose polymer networks (67, 68).

4) Field Strength Optimization of the field strength can be performed via an Ohm's law plot (Section 2.7); however, as with polyacrylamide gels, beware of biased reptation effects for DNA fragments above 15 kbp (68). Temperature effects affecting DNA structure and thus mobility may occur as well. Also, since DNA is negatively charged at pH 8, the voltage must be set to negative polarity. MacCrehan *et al.* (66) found optimal resolution of the 506/517 bp fragments in 0.4% methylcellulose at 250 V/cm. The optimal efficiency was at 350 V/cm; however, the mobility differential ($\Delta\mu$) was superior at the lower field strength. As described by Eq. (2.15), $\Delta\mu$ has more significance in controlling resolution. Nathakarnkitkool *et al.* (68) reported optimal resolution at a field strength of 200 V/cm.

5) Use with Coated Capillaries Suppression of the EOF is important when physical gels are used. While the cellulose additives do this to some extent, the EOF may still be sufficiently strong that the solute's electrophoretic flow is overcome and larger fragments are detected first. Nathakarnkitkool *et al.* (68) used both bare silica and a phenylmethylsiloxane[8]-treated capillary for separating restriction fragments and PCR products. The use of coated capillaries together with 0.4–0.5% methylcellulose (Fig. 5.8) further reduces the EOF and restores the order of migration to smaller fragments followed by larger ones (19, 66).

The need to control the EOF in bare silica increased the complexity of the buffers and experimental conditions. In particular, the bare silica capillary requires careful washing with hydroxide followed by equilibration with buffer. Phenylmethyl-coated capillaries require only flushing with distilled water.

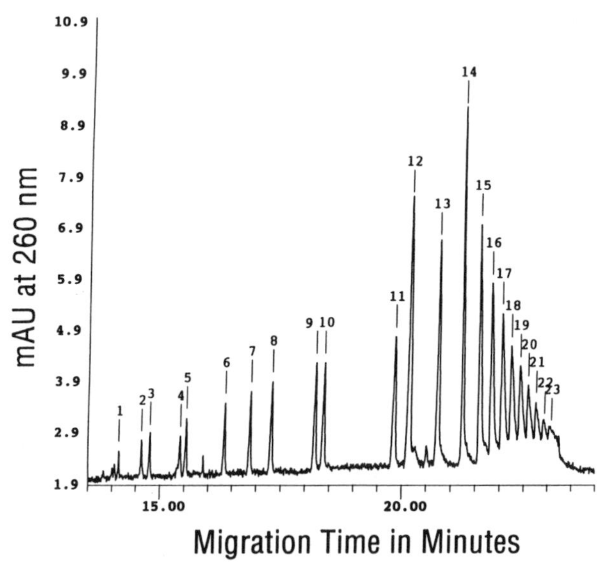

Figure 5.8. Separation of a 1 kb DNA ladder using 0.4% methylcellulose. Conditions described in Fig. 5.6. Key (bp): (1) 75; (2) 134; (3) 154; (4) 201; (5) 220; (6) 298; (7) 344; (8) 396; (9) 506; (10) 517; (11) 1,018; (12) 1,636; (13) 2,036; (14) 3,054; (15) 4,072; (16) 5,090; (17) 6,108; (18) 7,126; (19) 8,144; (20) 9,126; (21) 10,180; (22) 11,198; (23) 12,216. Reprinted with permission from *J. Liq. Chromatogr.* **15,** 1063, copyright ©1992 Marcel Dekker.

[8] Catalog No. 10041, Restek Corporation, Bellefonte, PA.

6) Precision Migration time precision (18), expressed as percent relative standard deviation, ranged from 0.09 to 0.24%. Peak height precision was 1–7%, and peak area precision ranged from 2 to 9%, all at DNA concentrations of 10–25 μg/mL. At 5 μg/mL, precision began to deteriorate as the limit of detection was approached (18). With polyacrylamide-coated capillaries (19), migration time precision was run-to-run 0.4%, day-to-day 0.5%, and capillary-to-capillary 0.9%. The polyacrylamide-treated capillaries[9] lasted for 50 injections before the coating deteriorated with the pH 8.0 buffer. Reasons for the discrepancies in precision between these two reports are not obvious.

7) Polymer Size and Type There have been few studies exploring on a comparative basis selection of polymer network type and concentration. Schwartz *et al.* (18) compared the use of HPMC-4000,[10] HPMC-100, polyethyleneglycol (PEG), and polyacrylamide for restriction fragment separations. Mobilities of the larger fragments tend to be higher in the short-chain, low-viscosity HPMC-100. This is inconsistent with Grossman's theory (65) that predicts smaller pore sizes (lower mobility) for the short-chain polymer. Below 100 bp, mobilities are higher in HPMC-4000. Restriction fragments were not separable in 5% PEG.

Hydroxylated celluloses such as hydroxypropylmethylcellulose (HPMC) are easier to get into solution. Methylcellulose solutions are best prepared in hot (70°C) water, allowed to stand overnight, and filtered through a fractional micron filter. The filtration step is lengthy because of the viscosity of the solution.

E. Dextrans

Dextrans are branched polysaccharides produced by bacteria growing on a sucrose substrate. They form viscous slimy solutions when dissolved in water. Unlike agarose, which is helical in structure, dextran pores are defined by polymeric entanglement.

This material overcomes many limitations of polyacrylamide for the separations of proteins. Besides being pumpable and replaceable, the solution has a low absorbance in the low UV, permitting sensitive detection (69). Polyacrylamide gels are usually monitored at higher wavelengths—260 nm for DNA and oligonucleotides, and 280 nm for proteins. While these wavelengths are suitable for oligonucleotides, the molar absorbtivities are poor for proteins at 280 nm.

Proteins can be size-separated using an electrolyte containing 0.1% SDS, 10% dextran 2 M (molecular weight 2 mD) in 60 mM 2-amino-2-methyl-1,3-propane diol (AMPD)-cacodylic acid (CACO), pH 8.8 (69) (Fig. 5.9). There have been no reported studies of oligonucleotides or DNA using this material.

[9] Bio-Rad, a manufacturer of polyacrylamide capillaries, claims to have stabilized the surface. Hundreds of injections have reportedly been obtained on a single capillary with an alkaline buffer.

[10] The numerical designator refers to the viscosity at 25°C.

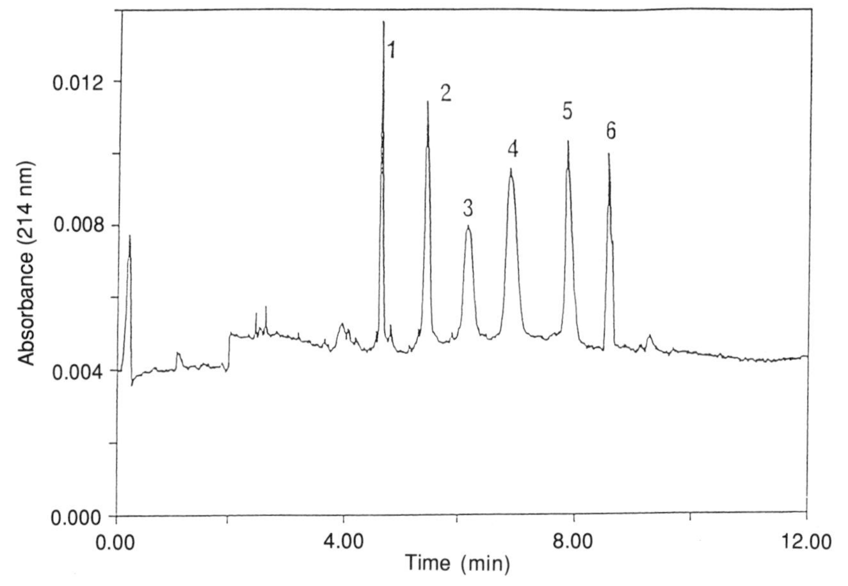

Figure 5.9. Separation of standard SDS–protein complexes in a dextran polymer network. Capillary: dextran-grafted, polyacrylamide-coated 23 cm (18 cm length to detector) × 75 μm i.d.; buffer: 60 mM AMPD-CACO, pH 8.8, 1% SDS with 10% w/V dextran 2 M; field strength: 400 V/cm; current: 30 μA; injection: 2 s at 300 V/cm; detection: UV, 214 nm. Key: (1) myoglobin; (2) carbonic anhydrase; (3) ovalbumin; (4) bovine serum albumin; (5) β-galactosidase; (6) myosin. Reprinted with permission from *Anal. Chem.* **64,** 2665, copyright ©1992 Am. Chem. Soc.

F. Polyethylene Glycol

Polyethylene glycol (PEG), a linear polymer, can also be employed as a polymer network. A 3% solution of PEG (MW 100,000) can separate SDS-proteins from 14 to 94 kD with detection at 214 nm (69). Protein separations are comparable with PEG, but the run times are somewhat longer compared to dextrans (Fig. 5.10). Migration time RSDs of 0.4–0.5% are found when the PEG solution is replaced after each run. Ferguson plots (Section 5.4A,B) are linear and intercept the *y*-axis at the CZE mobility values. This indicates a true size separation. More than 300 injections were performed without degradation in performance.

G. Intercalating Reagents

Intercalating reagents are additives to polymer-network or gel systems that form complexes with specific solutes. These additives are employed to alter selectivity and improve detection. For example, the bands comprising 271 and 281 base pairs (bp) are not resolved using the buffer recipe given in Fig. 5.7. If 10 μM

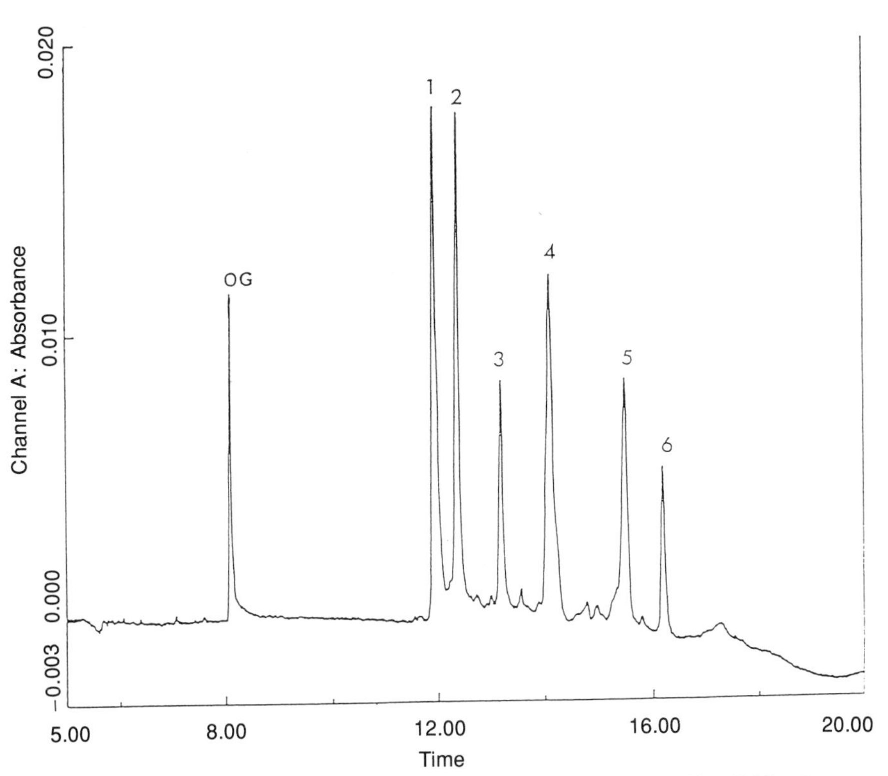

Figure 5.10. Separation of low molecular weight standard SDS-proteins with a PEG polymer network. Capillary: dextran-grafted, polyacrylamide-coated 46 cm (40 cm length to detector) × 100 μm i.d.; buffer: 100 mM Tris-CHES, 0.1% SDS, pH 8.8 with 3% w/V PEG 100,000; field strength: 300 V/cm; injection: pressure (0.5 psig) for 20 s; detection: UV, 214 nm. Key: OG (internal standard, Orange G); (1) α-lactalbumin; (2) ovalbumin; (3) carbonic anhydrase; (4) ovalbumin; (5) bovine serum albumin; (6) phosphorylase B. Reprinted with permission from *Anal. Chem.* **64,** 2665, copyright ©1992 Am. Chem. Soc.

ethidium bromide (EtBr) is added to the buffer, the bands are completely resolved as shown in Fig. 5.11, though the run time increases by 40% (18).

Ethidium bromide is an intercalating reagent that inserts between base pairs of the DNA double helix. Since the reagent is cationic, the mobility of the DNA–EtBr complex decreases because of the reduction of the ionic charge. When bare silica capillaries are used instead of the phenylmethyl (OV-17) coating, 1–5 μM EtBr in the buffer decreases the run time slightly because the direction of migration is determined by the EOF (68). However, the advantage of this approach is undermined by the other difficulties of using bare silica for this application.

Incorporation of 1 μg/mL of EtBr in a linear polyacrylamide gel matrix improves not only the separation, but also the detector sensitivity (70). The UV

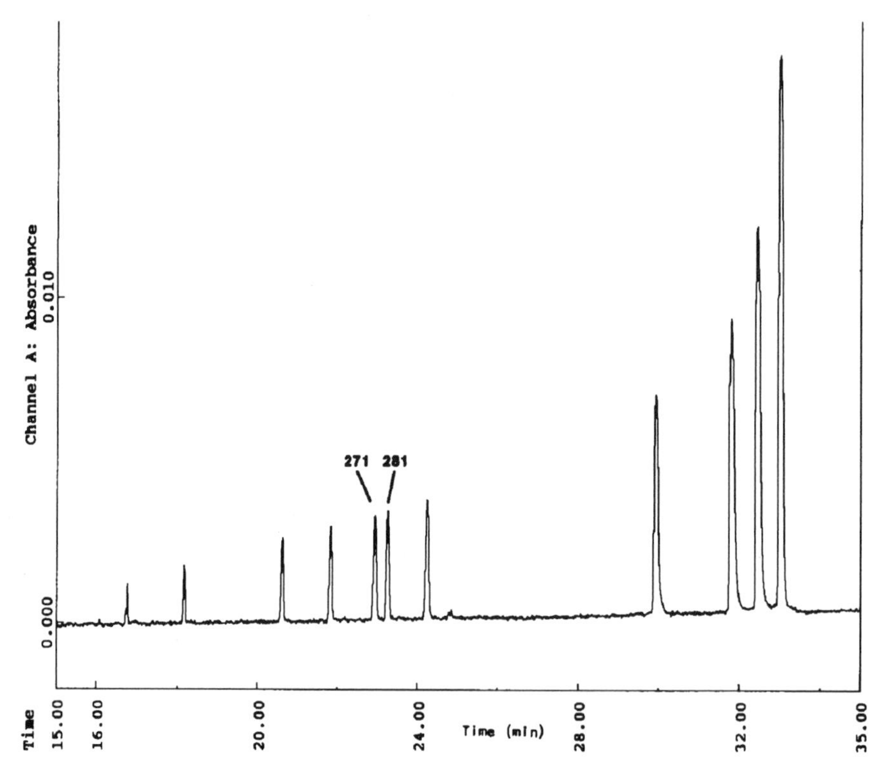

Figure 5.11. Effect of ethidium bromide on DNA restriction fragment resolution. Conditions: see Fig. 5.8 except: sample concentration: 10 μg/mL; injection: 2 kV for 10 s; buffer: as in Fig. 5.8 with the addition of 10 μg/mL EtBr. Reprinted with permission from *J. Chromatogr.* **559,** 267, copyright ©1991 Elsevier Science Publishers.

absorbance values in the presence of EtBr are enhanced two- to three-fold because of the stronger absorptivity of the DNA–EtBr complex. Since EtBr intercalates between the base pairs of double-stranded DNA, there is no improvement for the separation of oligonucleotides. While restriction digests are well separated on linear polyacrylamide, the use of the methylcelluloses is advantageous because the polymerization steps are not required.

Ethidium bromide fluoresces strongly when intercalated in the DNA matrix. In the bulk aqueous buffer, EtBr fluorescence is quenched through collisions with solvent molecules. In HPCE, this feature cannot be effectively exploited because the absorption spectrum of EtBr does not match the emission lines of low-cost lasers. Schwartz and Ulfelder (21) separated restriction fragments and PCR products using thiazole orange as the intercalator with 0.5% HPMC in a buffer consisting of 89 mM Tris-borate, 2 mM EDTA, pH 8.5, and 0.1–1 μg/mL thiazole orange. Since the dye absorbs strongly at 488 nm, argon-ion laser fluorescence could be used to measure the intercalated product. The limit of detection is improved by a factor of 400 compared to UV detection. Unlike EtBr complexes, the DNA–thiazole orange complex is sensitive to the DNA–dye concentration ratio. Good peak shapes are found when a 9:1 molar ratio of DNA:dye is employed.

In a related development, Swaile and Sepaniak (71) used hydrophobic probes to monitor proteins. While the reported separations were by CZE, it is likely that such a scheme can be employed in CGE. Dyes such as 1-anilinonaphthalene-8-sulfonate (ANS) and 2-*p*-toluidinonaphthalene-6-sulfonate (TNS) fluoresce only when bound to a protein. Incorporation of 200 μM TNS in the run buffer optimized the sensitivity for conalbumin with helium–cadmium laser-induced fluorescence detection. A limit of detection of 360 nM was reported.

H. Commercially Available Gels and Polymer Network Reagents

Both gel-filled capillaries and reagents for polymer network separations are commercially available. Table 5.2 contains a compilation of available material. Unfortunately, because of the competitive marketplace, many manufacturers choose not to reveal their specific recipes. It is likely, though, that most of these formulas are composed of reagents that have been described in this chapter.

Table 5.2. Commercial Gels and Polymer Network Reagents

Application	Sieving Material	Buffer	Trade Name	Source	Notes
Oligonucleotides	Linear non-polyacrylamide gel-filled capillary	75 mM Tris-phosphate, 10% methanol, pH 7.6	MICRO-GEL_{100}	ABI	
	Denaturing linear polyacrylamide gel-filled capillary	100 mM borate, 250 mM Tris, 7 M urea, pH 7.45	eCAP U100P	Beckman	
	5% T, 5% C polyacrylamide gel-filled capillary	100 mM Tris-borate, 7 M urea, pH 8.3	μPAGE-5	J & W	
PCR products	Linear polymer	100 mM Tris-borate, 2 mM EDTA, pH 8.45	PCR kit	Bio-Rad	Coated capillary
	3% T, 3% C polyacrylamide gel-filled capillary	100 mM Tris-borate, 7 M urea, pH 8.3	μPAGE-3	J & W	
Proteins	Linear polymer	Contains 0.1% SDS	eCAP SDS200	Beckman	Coated capillary
	Linear polymer	Contains 0.2% SDS	ProSort	ABI	Bare silica capillary

5.4 Applications

A. Proteins

Cohen and Karger (3) reported that the separation of myoglobin and several of its fragments was performed by capillary SDS-PAGE on a 12.5%T, 3.3%C gel. The relationship between the mobility of each fragment and its log molecular weight was found to be linear. Larger proteins such as pepsin (MW 34,700) are better separated on a more porous 10%T, 3.3%C gel. Proteins always migrate faster on more porous (lower %T) gels.

Validation of the size separation is assessed with a Ferguson plot. These plots are infrequently performed by CGE with rigid gels since at least three different %T gels in separate capillaries are required. When properly denatured, all proteins should have identical charge/mass ratios. This is confirmed because the Ferguson plot (data not shown) indicates all of the myoglobin proteins converging to the same point on the y-axis at 0%T.

Tsuji (4) applied CGE to separate protein standards and recombinant material. With a high concentration of ethylene glycol in the polymerization solution and run buffer, stable bubble-free capillaries were used for as many as 200 runs. A pumpable version of this gel might be more suitable for routine use.

The limitations of rigid gels are overcome by using UV-transparent and pumpable denaturing polymer networks. Proprietary formulations are being marketed by ABI and Beckman. The Beckman process has been correlated with conventional slab-gel electrophoreis for more than 50 different proteins. The linear dynamic range is from a few micrograms to one milligram per milliliter, while the molecular weight linear range is from 29 to 205 kD. In addition, Beckman has shown migration time RSDs of 0.28–0.38% and peak area RSDs of 0.87–7.0%, externally standardized. The ABI material has been shown linear from 14.2 to 205 kD with reported migration time RSDs of less than 1%.

The applicability of polymer networks for performing size separations is illustrated in Fig. 5.12. Data generated from the separation of a crude catalyse preparation are compared for runs by HPCE and a densitometer scan of a slab-gel. With the x-axis approximately normalized for molecular weight, a remarkable correlation between these two techniques is observed. Using dextrans or PEG networks, Ganzler *et al.* (69) separated rat plasma proteins with direct plasma injection, light and heavy chains of human IgG, an *E. coli* crude extract, and the reduced F_{ab} fragments of a monoclonal antibody. Identification of protein dimers, trimers, etc., deglycosylation, and other size-modifying chemistries are additional applications for this technique.

B. Oligonucleotides and Restriction Digests

Spectacular separations of synthetic oligonucleotides yielding millions of theoretical plates have been reported on linear polyacrylamide as well as cross-linked

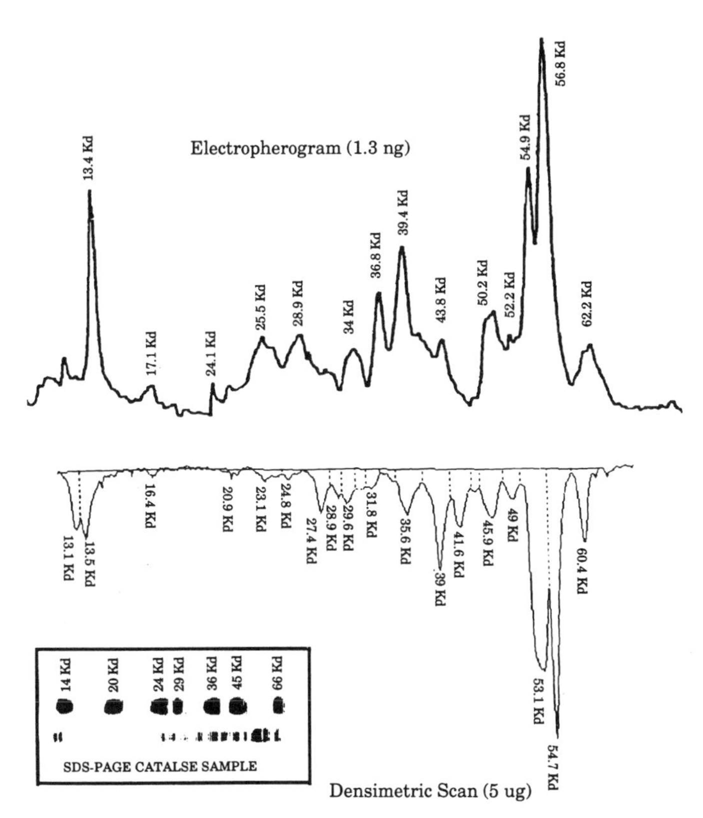

Figure 5.12. Size separation of a crude denatured (boiled for 15 min in 1% SDS and 1% mercaptoethanol) catalyse preparation (1 mg/mL) by HPCE (upper trace) and slab-gel electrophoresis (lower trace). The lower trace is a densimetric scan of a 10%T, 2.6%C gel. Courtesy of Applied Biosystems.

gels (6–14). The separation shown in Fig. 5.3 of poly(dA)$_{20}$ and poly(dA)$_{40-60}$ run on a 9%T gel yields unit base resolution and shows partial separation of the phosphorylated and dephosphorylated oligos. These physical gels are prepared *in situ* but were reported to be stable for weeks. CGE is particularly useful for identifying failure sequences that occur during the synthesis of oligonucleotides.

A comparison between HPLC and CGE for poly (dA) standards showed compelling advantages for capillary electrophoresis (56). By HPLC, on a mixed mode Neosorb-LC-N-7R column, 1–70 mer were separated with unit base resolution with a gradient run of 150 min generating 10,000 theoretical plates. By CGE, on a 5%T, 5%C gel at 200 V/cm, 6–255 mer were separated in 62 min generating 2,300,000 plates (7 million plates/meter). For small oligos containing fewer than 30 bases, HPLC generally provides adequate speed and resolution. For larger oligos, CGE provides substantial separation advantages. The data in Table 5.3 provide an effective means of selecting either HPLC or CGE as the separation tool.

Table 5.3. Comparative Performance of HPLC and CGE[a]

Method	Separated Polynucleotides	Analysis Time (min)
Ion-Exchange HPLC		
Partisil SAX	1–30 mer	50
Nucleogen-DMA-60	1–37 mer	110
MonoQ	1–27 mer	15
Gen-Pak FAX	40–60 mer	39
TSK Gel DEAE-NPR	20–70 mer	17
Reversed-Phase LC		
Zorbax ODS	2–10 mer	25
μBondapak C_{18}	1–19 mer	60
Mixed-Mode LC		
Neosorb-LC-N-7R	1–75 mer	95
CGE		
	20–160 mer	25
	19–330 mer	70
	19–300 mer	115
	19–340 mer	70
	1–430 mer	130

[a]Reprinted in part from *J. Chromatogr.* **558**, 273 (1991).

Schwartz *et al.* (18) employed a HPMC polymer network system to monitor a PCR-amplified human immunodifficiency virus (HIV) infected cell line (U1.1). The cell line contained one copy of HIV-1 provirus and two copies of HLA-DQ-α, which is normally present in all healthy cells. Using specific primers, a 115 bp region of HIV-1 and a 242 bp region of HLA-DQ-α were co-amplified by 35 cycles of PCR. Separations with and without ultracentrifugation are shown in Fig. 5.13. The ultracentrifuge simultaneously desalts and concentrates the DNA in the sample.

Since the concentration of the polymeric additive is easily varied, generation of a Ferguson plot (log μ vs. %T) is simple. Plotting the HPMC-4000 concentration versus mobility for a ϕX174 Hae III restriction digest (Fig. 5.14) shows imperfect convergence at 0% polymer (18). Since all lines do not intersect the *y*-axis at the same point, there may be differences in free-solution mobility between some fragments. This may lead to errors in assigning the proper molecular size to each fragment. The addition of 7 M urea to the buffer might correct this problem.

Guttman *et al.* (6) studied both native and denaturing gels for the separation of oligonucleotides. In native gels, the relative migration order is not constant for homooligomers, but is dependent on the base number. For base numbers less than 14, the migration order is A>C>G>T; for bases larger than 18, the order becomes G>A>C>T. The authors suspect this discrepancy is caused by molecular bending

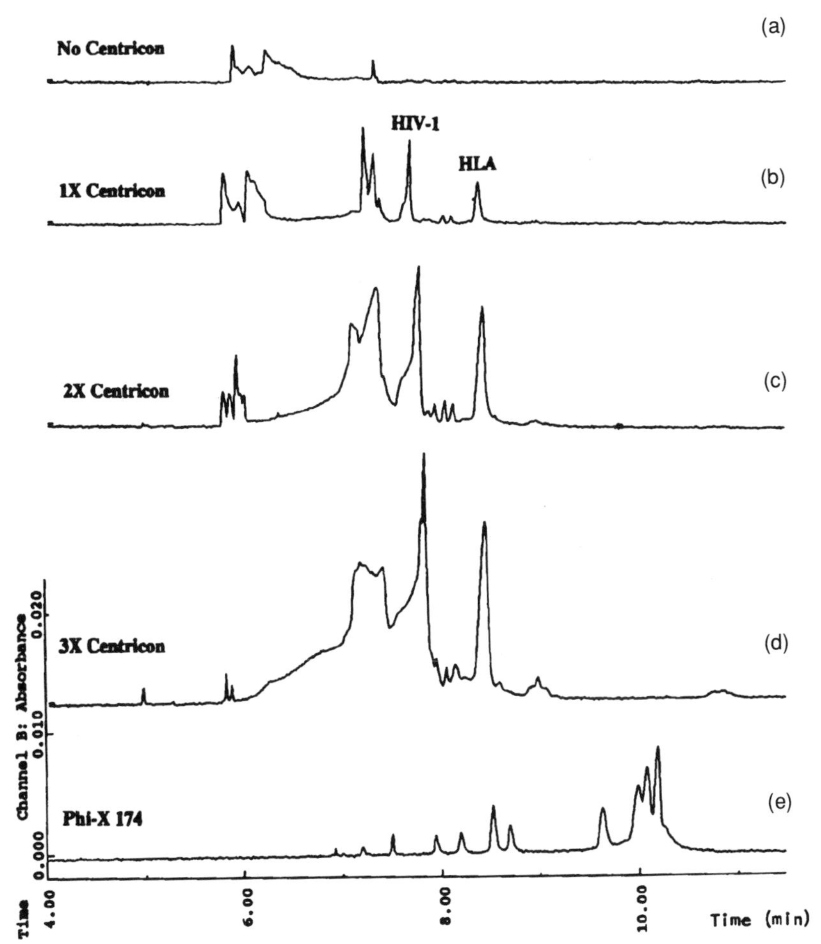

Figure 5.13. Separation of a PCR-amplified cell line containing HIV-1 provirus. Voltage: 20 kV; injection: electrokinetic, 10 kV for 10 s. Other conditions as in Fig.5.7. (a) No ultracentrifugation; (b)1× ultracentrifugation; (c) 2× ultracentrifugation; (d) 3× ultracentrifugation; (e) ϕX174 DNA standard. Reprinted with permission from *J. Chromatogr.* **559,** 267, copyright ©1991 Elsevier Science Publishers.

due to self-association of guanosine. In denaturing gels, the order of migration is the same regardless of the chain length.

C. DNA Sequencing

The human genome initiative (HGI) is a long-term project designed to ultimately sequence the entire human genome. HGI will, according to Leroy Hood, "profoundly change the study of biology and the treatment of disease" (72). HGI

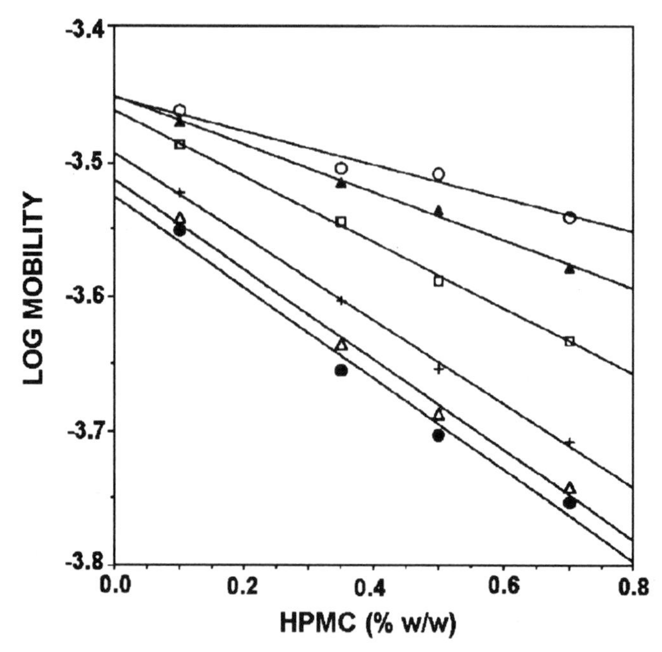

Figure 5.14. Ferguson plot (log μ vs. % w/w HPMC) for a buffer containing HPMC-4000 as the polymer network. Mobilities of selected ϕX174-Hae III digest fragments were used to generate the plot. Key: ○ = 118 bp; ▲ = 194 bp; □ = 310 bp; + = 603 bp; △ = 872 bp; ● = 1,353 bp. Reprinted with permission from *J. Chromatogr.* **559,** 267, copyright ©1991 Elsevier Science Publishers.

proposes to map and sequence the 24 different human chromosomes. To achieve this goal, which is presently funded at $150 M/year, new developments in mapping, sequencing, and data handling will be required. To date, only 5 million of the 3 billion bases in the human genome have been sequenced and entered in a data bank. About 300 gigabytes of computer storage will ultimately be needed.

Capillary electrophoresis may play a role in the development of a genomic-scale instrument for the separation of DNA-sequencing reaction products. This class of instrument must be capable of separating 1,000,000 bases/day. High-speed and sensitive separations of the sequencing reaction products are possible by CGE with laser fluorescence detection (24–26, 28–36).

Figure 5.15 shows a comparison of separations by CGE on a 4%T, 5%C gel compared to that obtained on an automated DNA sequencer. While CGE is sixfold faster, only a single sample can be run at a time since HPCE is a serial technique; samples are processed individually. Slab-gel electrophoresis is analogous to thin layer chromatography; it's a parallel technique capable of processing many samples simultaneously. To recapture the inherent speed of the slab-gel, a multicapillary instrument is required. Such design was first reported by Zagursky and McCormick (29) in 1990. More elegant designs were recently developed by Huang *et al.* (36).

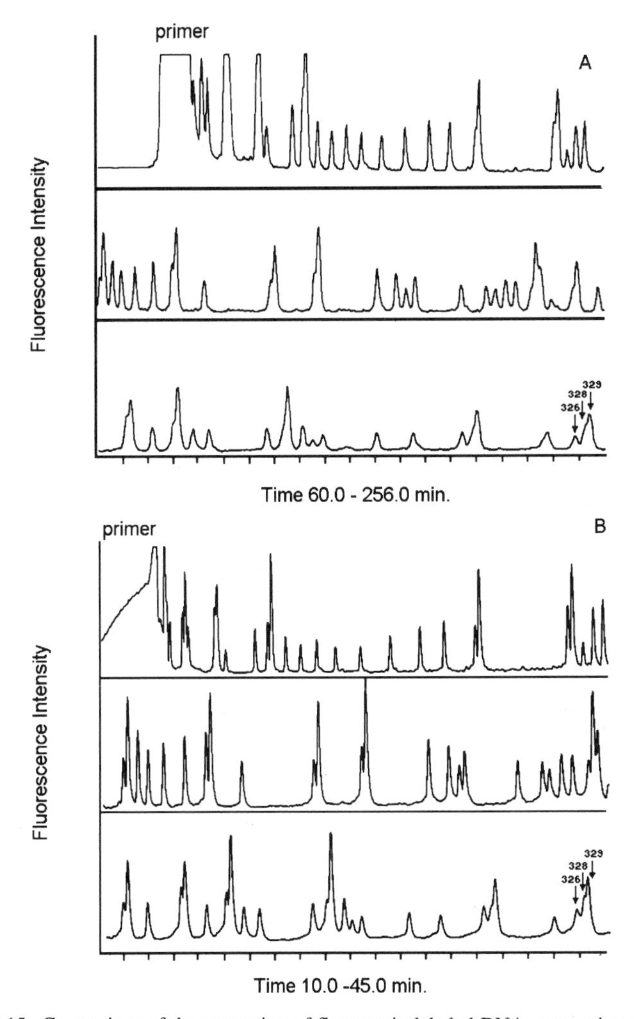

Figure 5.15. Comparison of the separation of fluorescein-labeled DNA sequencing reactions by conventional slab gel electrophoresis and capillary electrophoresis: (a) conventional electrophoresis on the Applied Biosystems 370A automated DNA sequencer; (b) capillary gel electrophoresis on the same sample. Reprinted with permission from *Anal. Chem.* **62,** 900, copyright ©1990 Am. Chem. Soc.

The viability of these designs will be dictated by the stability of the gels, not instrumental factors. Since maintaining a single capillary is problematic, compounding the problem with multiple capillaries can lead to substantial instrument downtime. If polymer networks can be applied in place of rigid gels, the multiplex instrument approach may be practical.

There are other technologies that may be more suitable for a genomic-scale instrument (73). In the electrophoresis domain, the use of ultrathin gels may combine the advantages of high field strength with those of a planar separation technique. Other techniques include single molecule detection, mass spectrometry, hybridization reactions, and even scanning tunneling microscopy. However, these novel technologies still require new invention before they can be reduced to practice.

D. Blotting

Blotting involves the electrophoretic transfer of material from the gel matrix onto a membrane. Hybridization reactions using DNA probes can be used to identify the blotted material. As an alternative to traditional blotting, it is possible to perform a precapillary hybridization with subsequent separation of the reaction products by CGE (57).

Figures 5.16 a–c show electropherograms of Joe-labeled 17-mer sequencing primer (GTAAAACGACGGCCAGT), complementary pC2 34-mer (TCGAATT-CACTGGCCGTCGTTTTACAACGTCGTC), and a mixture of the two annealed

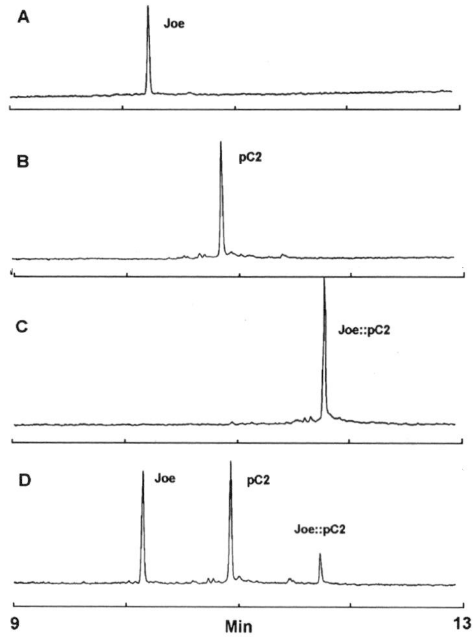

Figure 5.16. Identification of a DNA molecule by hybridization with a fluorescence-tagged oligonucleotide probe using CGE. (a) Joe-labeled 17-mer alone (5 μg/mL in 10 mM Tris-borate–EDTA (TBE)); (b) pC2 alone (5μg/mL in 10 mM TBE); (c) equal amounts of Joe-labeled primer and pC2 in 10 mM TBE driven to complete hybridization by incubation in dry ice; (d) mixture in (c) heated to 65°C for 5 min. Capillary: 45 cm (25 cm to detector) × 75 μm i.d. filled with a 9%T nondenaturing linear polyacrylamide gel; buffer: 25 mM Tris-borate–EDTA, pH 8.0; injection: electrokinetic, 13.5 kV for 5 s; field strength: 300 V/cm; detection: UV, 260 nm. Reprinted with permission from *J. Chromatogr.* **559,** 295, copyright ©1991 Elsevier Science Publishers.

at 65°C for 10 min, then slowly cooled to room temperature in 30 min. The third peak in Fig. 5.16c was identified as a hybrid of the two reagents by laser fluorescence as well as thermal dissociation (Fig. 5.16d).

Faster and more complete hybridization is possible with incubation in dry ice, possibly because of the lack of salt in the annealing process. Salt was eliminated from the annealing process because of the deleterious impact on injection response and reproducibility. The profound impact of the effect of salt is given in Table 5.4.[11] Failure to control the salt (ionic strength) concentration causes serious quantitative problems. Refer to Chapters 9 and 11 for more information regarding this most critical aspect of HPCE.

This work opens possibilities for the simultaneous separation of multiple species that would be advantageous in the clinical setting. The procedure works well with single-stranded oligonucleotides, but yields inconsistent results with double-stranded DNA molecules.

5.5 Pulsed-Field Capillary Electrophoresis

A consequence of biased reptation is the limitation of DNA fragment size that can be separated using a continuous electric field. The behavior of DNA in an electric field is illustrated in Fig. 5.17. Without the field, the molecule exists as a tightly coiled moiety. Upon application of the voltage, the molecule lengthens as it aligns with the field. Beyond 20 kbp, fragments are no longer resolved by

Table 5.4. Effect of Buffer Concentration on Electrokinetic Injection[a,b]

DNA Concentration (μg/mL)	Tris-HCl Concentration (mM)	Relative Peak Height		
		Fragment 234 bp	Fragment 271 bp	Fragment 603 bp
1,000	10	1	1	1
500	5	5	6	5
100	1	33	27	24
20	0.2	139	139	119
5	0.05	436	442	364

[a]Data from *J. Chromatogr.* **559,** 295 (1991).

[b]Restriction digest, ϕX 174 *Hae* III, was separated on a 9%T nondenaturing gel. The peak heights corresponding to 234, 271, and 603 bp fragments were compared. More DNA was injected when the sample was diluted with distilled water. A nearly 500-fold increase in peak height was observed when the sample was diluted 1,000-fold. Injection was done electrokinetically at 15 kV for 5 s.

[11] Methods for reducing the salt concentration are given in Section 11.4.

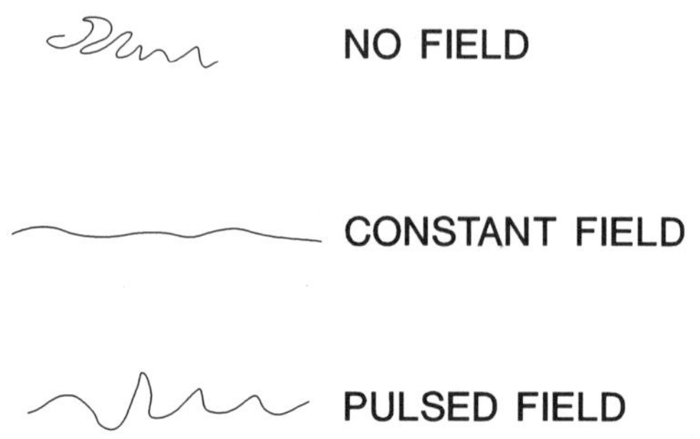

NO FIELD

CONSTANT FIELD

PULSED FIELD

Figure 5.17. Illustration of the configuration of a DNA molecule in the absence of an applied field, in the presence of a constant field, and with a pulsed field.

constant-field electrophoresis. To extend the range of resolution, pulsed-field electrophoresis must be employed (74–76). In its simplest format, the electric field driving the separation is pulsed. During the off cycle, the DNA molecules relax. Upon return of the field, molecules begin migration in a partially deformed state because alignment with the field is a kinetic process. This mechanism provides additional selectivity for the separation of large fragments.

Alternatively, multiple electrode systems that provide two electric fields can be employed. For example, electrodes can be positioned at angles of 90 or 120°. The applied voltage can be alternated between the electrode sets at frequencies designed to enhance selectivity for various size ranges. Upon application of the pulse, DNA molecules reorient in the gel. Longer fragments reorient slowly, and their migration through the gel is hindered. Isolation and mapping of large segments of genomic DNA is an important application of this technique. Separation of several million base pairs is possible using pulsed-field techniques.

Adaptation of this technique to CGE is in its infancy. Heiger *et al.* (17, 77) developed an instrument to deliver the pulsed field in either the unidirectional (single polarity) or field-inversion (polarity reversed pulse) formats, which position the electrodes at a relative angle of 180°. Other angular placements that are used in slab-gel electrophoresis are not compatible in capillaries. Waveform distortion because the high resistance of the gel–buffer system limited the pulse frequency to less than 100 Hz. While this problem may be solved by increasing ionic strength and using active cooling to remove heat, commercialization of this technique is probably years away.

5.6 The Future of CGE

Because it is the least mature of all HPCE techniques, substantial advances are expected in this field. Clearly, size separations will not be performed in a manner analogous to the slab-gel. The problems with rigid gels in capillary electrophoresis suggests a limited future for this approach. Fragility of the gels to the electric field, the sample matrix, dehydration, and unpredictable failure during unattended automated runs are among the problems. The use of pumpable, user-replaceable polymer networks is a more suitable and rugged procedure. Replacement of the separation medium every run or several runs will ensure optimal repeatability of each separation. This problem is particularly severe in multiple capillary instruments designed for DNA sequencing (36).

It appears that polymer networks can be formulated to cover the range of pore sizes necessary for the separation of many proteins and DNA fragments. Extension of and better control of the pore size is expected in the future.

References

1. S. Hjertén, *J. Chromatogr.* **270,** 1 (1983).
2. S. Hjertén, K. Elenbring, F. Kilár, J. L. Liao, A. J. C. Chen, C. J. Liebert, and M. D. Zhu, *J. Chromatogr.* **403,** 47 (1987).
3. A. S. Cohen and B. L. Karger, *J. Chromatogr.* **397,** 409 (1987).
4. K. Tsuji, *J. Chromatogr.* **550,** 823 (1991).
5. A. Widhalm, C. Schwer, D. Blaas, and E. Kenndler, *J. Chromatogr.* **549,** 446 (1991).
6. A. Guttman, R. J. Nelson, and N. Cooke, *J. Chromatogr.* **593,** 297 (1992).
7. A. S. Cohen, D. R. Najarian, A. Paulus, A. Guttman, J. A. Smith, and B. L. Kayer, *Proc. Natl. Acad. Sci.* **85,** 9660 (1988).
8. A. Guttman, A. S. Cohen, D. N. Heiger, and B. L. Karger, *Anal. Chem.* **62,** 137 (1990).
9. A. Paulus and J. L. Ohms, *J. Chromatogr.* **507,** 113 (1990).
10. A. Paulus , E. Gassman, and M. J. Field, *Electrophoresis* **11,** 702 (1990).
11. R. S. Dubrow, *American Lab.*, **March,** p. 64 (1991).
12. H.-F. Yin, M. H. Kleemis, J. A. Lux, and G. Schomburg, *J. Microcol. Sep.* **3,** 331 (1991).
13. D. Demorest and R. Dubrow, *J. Chromatogr.* **559,** 43 (1991).
14. J. Sudor, F. Foret, and P. Boček, *Electrophoresis* **12,** 1056 (1991).
15. A. S. Cohen, D. Najarian, J. A. Smith, and B. L. Karger, *J. Chromatogr.* **458,** 323 (1988).
16. A. Guttman and N. Cooke, *J. Chromatogr.* **559,** 285 (1991).
17. D. N. Heiger, A. S. Cohen, and B. L. Karger, *J. Chromatogr.* **516,** 33 (1990).
18. H. E. Schwartz, K. Ulfelder, F. J. Sunzeri, M. P. Busch, and R. G. Brownlee, *J. Chromatogr.* **559,** 267 (1991).
19. M. Strege and A. Lagu, *Anal. Chem.* **63,** 1233 (1991).
20. K. J. Ulfelder, H. E. Schwartz, J. M. Hall, and F. J. Sunzeri, *Anal. Biochem.* **200,** 260 (1992).
21. H. E. Schwartz and K. J. Ulfelder, *Anal. Chem.* **64,** 1737 (1992).
22. F. J. Sunzeri, T.-H. Lee, R. G. Brownlee, and M. P. Busch, *Blood* **77,** 879 (1991).
23. A. Mayer, F. Sunzeri, T.-H. Lee, and M. P. Busch, *Arch. Pathol. Lab. Med.* **115,** 1228 (1991).
24. H. Swerdlow and R. Gesteland, *Nucl. Acids Res.* **18,** 1415 (1990).

25. H. Drossman, J. A. Luckey, A. J. Kostichka, J. D'Cunha, and L. M. Smith, *Anal. Chem.* **62,** 900 (1990).
26. J. A. Luckey, H. Drossman, A. J. Kostichka, D. A. Mead, J. D'Cunha, T. B. Norris, and L. M. Smith, *Nucl. Acids Res.* **18,** 4417 (1990).
27. A. S. Cohen, D. R. Najarian, and B. L. Karger, *J. Chromatogr.* **516,** 49 (1990).
28. H. Swerdlow, S. Wu, H. Harke, and N. J. Dovichi, *J. Chromatogr.* **516,** 61 (1990).
29. R. J. Zagursky and R. M. McCormick, *BioTechniques* **9,** 74 (1990).
30. L. M. Smith, *Nature* **349,** 812 (1991).
31. J. Z. Zhang, D. Y. Chen, S. Wu, H. R. Harke, and N. J. Dovichi, *Clin. Chem.* **37,** 1492 (1991).
32. A. E. Karger, J. M. Harris, and F. Gesteland, *Nucl. Acids Res.* **19,** 4955 (1991).
33. D. Y. Chen, H. P. Swerdlow, H. R. Harke, J. Z. Zhang, and N. J. Dovichi, *J. Chromatogr.* **559,** 237 (1991).
34. H. Swerdlow, J. Z. Zhang, D. Y. Chen, H. R. Harke, R. Grey, S. Wu, N. J. Dovichi, and C. Fuller, *Anal. Chem.* **63,** 2835 (1991).
35. D.Y. Chen, H. P. Swerdlow, H. R. Harke, J. Z. Zhang, and N. J. Dovichi, *SPIE* **1435,** 161 (1991).
36. X. C. Huang, M. A. Quesada, and R. A. Mathies, *Anal. Chem.* **64,** 967 (1992).
37. J. Liu, O. Shirota, and M. Novotny, *J. Chromatogr.* **559,** 223 (1991).
38. J. Liu, O. Shirota, and M. V. Novotny, *Anal. Chem.* **64,** 973 (1992).
39. J. B. Poli and M. R. Schure, *Anal. Chem.* **64,** 896 (1992).
40. P. D. Grossman, in *Capillary Electrophoresis: Theory and Practice* (P. D. Grossman and J. C. Colburn, eds.) (Academic Press, Boston, 1992), p. 215.
41. M. J. Dunn, in *New Directions in Electrophoretic Methods* (J. W. Jorgenson and M. Phillips, eds.) (American Chemical Society, Washington DC, 1987), p. 20.
42. D. Grierson, in *Gel Electrophoresis of Nucleic Acids: A Practical Approach* (D. Rickwood and B. D. Hames, eds.) (IRL Press, Oxford, 1990), p. 1.
43. S. Raymond and L. Weintraub, *Science* **130,** 711 (1959).
44. L. Ornstein, *Ann. N.Y. Acad. Sci.* **121,** 321 (1964).
45. B. J. Davis, *Ann. N.Y. Acad. Sci.* **121,** 404 (1964).
46. P. G. Righetti, *Isoelectric Focusing: Theory, Methodology and Applications* (T. S. Work and R. H. Burdon, eds.), Laboratory Techniques in Biochemistry and Molecular Biology (Elsevier Biomedical Press, Amsterdam, 1983).
47. J. Liu, V. Dolnik, Y.Z. Hsieh, and M. Novotny, *Anal. Chem.* **64,** 1328 (1992).
48. V. Dolník, K. A. Cobb, and M. Novotny, *J. Microcol. Sep.* **3,** 155 (1991).
49. Y. Baba, T. Matsuura, K. Wakamoto, and M. Tsuhako, *Chem. Lett.*, p. 371 (1991).
50. Y. Baba, T. Matsuura, K. Wakamoto, Y. Morita, Y. Nishitsu, and M. Tsuhako, *Anal. Chem.* **64,** 1221 (1992).
51. M. Chiari, C. Micheletti, P. G. Righetti, and G. Poli, *J. Chromatogr.* **598,** 287 (1992).
52. M. Huang, W. P. Vorkink, and M. L. Lee, *J. Microcol. Sep.* **4,** 233 (1992).
53. T. Wang, G. J. Bruin, J. C. Kraak, and H. Poppe, *Anal. Chem.* **63,** 2207 (1991).
54. H. F. Yin, J. A. Lux, and G. Schomburg, *HRC & CC* **13,** 624 (1990).
55. J. A. Lux, H. F. Yin, and G. Schomberg, *HRC & CC*, **13,** 436 (1990).
56. Y. Baba, T. Matsuura, K. Wakamoto, and M. Tsuhako, *J. Chromatogr.* **558,** 273 (1991).
57. J. W. Chen, A. S. Cohen, and B. L. Karger, *J. Chromatogr.* **559,** 295 (1991).

58. J. Macek, U. R. Tjaden, and J. van der Greef, *J. Chromatogr.* **545,** 177 (1991).
59. P. Boček and A. Chrambach, *Electrophoresis* **12,** 1051 (1991).
60. P. Boček and A. Chrambach, *Electrophoresis* **13,** 31 (1992).
61. Personal communication, A. Chrambach, August 1992.
62. S. R. Motsch, M.-H. Kleemis, and G. Schomberg, *HRC & CC* **14,** 629 (1991).
63. S. Hjertén, L. Valtcheva, K. Elenbring, and D. Eaker, *J. Liq. Chromatogr.* **12,** 2471 (1989).
64. M. Zhu, D. L. Hansen, S. Burd, and F. Gannon, *J. Chromatogr.* **480,** 311 (1989).
65. P. D. Grossman and D. S. Soane, *J. Chromatogr.* **559,** 257 (1991).
66. W. A. MacCrehan, H. T. Rasmussen, and D. M. Morthrop, *J. Liq. Chromatogr.* **15,** 1063 (1992).
67. Personal communication, H. Rasmussen, August 1992.
68. S. Nathakarnkitool, P. J. Oefner, G. Bartsch, M. A. Chin, and G. K. Bonn, *Electrophoresis* **13,** 18 (1992).
69. K. Ganzler, K. Greve, A. S. Cohen, B. L. Kayer, A. Guttman, and N. C. Cooke, *Anal. Chem.* **64,** 2665 (1992).
70. A. Guttman and N. Cooke, *Anal. Chem.* **63,** 2038 (1991).
71. D. F. Swaile and M. J. Sepaniak, *J. Liq. Chromatogr.* **14,** 869 (1991).
72. A. R. Neuman, *Anal. Chem.* **62,** 1260 (1990).
73. G. L. Trainor, *Anal. Chem.* **62,** 418 (1990).
74. M. V. Olson, *J. Chromatogr.* **470,** 377 (1989).
75. R. Anand and E. M. Southern, in *Gel Electrophoresis of Nucleic Acids: A Practical Approach* (D. Rickwook and B. D. Hames, eds.) (IRL Press, Oxford, 1990), p. 101.
76. D. C. Schwartz, in *New Directions in Electrophoretic Methods* (J. W. Jorgenson and M. Phillips, eds.) (American Chemical Society, Washington, DC, 1987), p. 167.
77. D. N. Heiger, S. M. Carson, A. S. Cohen, and B. L. Karger, *Anal. Chem.* **64,** 192 (1992).

Chapter 6

Capillary Isotachophoresis

6.1 Introduction

Commercial instrumentation for capillary isotachophoresis (CITP), also known as displacement electrophoresis, has been available since 1974 with the introduction of the LKB Tachophor, the Shimadzu IP-1B, and later, the Spišská Nová Ves (Czechoslovakia) CS ZKI 001. Several textbooks (1, 2), extensive reviews (3,4) and compilations from international symposia (5–7) have also appeared on this subject. Despite a wealth of commercial instrumentation along with extensive theory and applications, CITP did not become a routine separation tool in most analytical laboratories. This is particularly true in the United States and Japan. In parts of Europe, CITP has become an important technique, especially in Czechoslovakia. Separations of drugs and endogenous biological substances in body fluids, drugs and their production by-products, food analysis, and environmental separations have all been shown to be practical by CITP (3). It is important within the context of this book to try to understand the enigma of CITP and make some generalizations where the technique fits into the total analytical scheme.

Most of this chapter will be devoted to users of conventional capillary electrophoresis instrumentation that may wish to occasionally perform CITP. It is assumed that users of a dedicated CITP instrument have already availed themselves of the aforementioned texts in the field or participated in manufacturer training courses.

6.2 Separation Mechanism

Isotachophoresis (CITP) literally translates to "electrophoresis at uniform speeds." This means that the transit time of a solute through the capillary under isotachophoretic conditions is independent of mobility. To understand this concept and its practical implications, some arguments described in Section 2.1 (Electrical Conduction in Fluid Solution) will be employed.

The separation scheme of CITP is illustrated in Fig. 6.1. Unlike CZE, the buffer system for CITP is heterogeneous. Before the beginning of a run, both the capillary- and detector-side buffer reservoirs are filled with a *leading electrolyte,* or leader. The leader is selected to have a mobility greater than that of any of the components to be separated. For example, when anions are separated by anionic CITP, the highly mobile chloride ion is most frequently selected as the leader. The sample-side buffer reservoir is filled with a *terminating electrolyte,* the mobility of which is less than that of any of the components in the sample mixture. These conditions, with the sample introduced between the two electrolytes, are shown at the top of Fig. 6.1.

The electrophoretic velocity is

$$v_{ep} = \mu_{ep} E \,. \tag{6.1}$$

Since the buffer system and sample are heterogeneous, the field strength developed over each zone will be inversely proportional to the conductivity of the individual zone. The conductivity depends on the mobility and concentration of each solute. Ohm's law must be followed for all electrophoretic systems, and under these conditions,

$$I \propto E_L \kappa_L + E_S \kappa_S + E_T \kappa_T \,, \tag{6.2}$$

where κ = the respective conductivities of leader, solute, and terminator and E = the field strength over each isotachophoretic zone. Unlike CZE, CITP bands are always in contact with each adjacent zone. This feature is necessary to maintain electrical continuity throughout the system, since there is no supporting electrolyte.

Initially, all sample components are blended. As the run proceeds, individual solutes begin to migrate as determined by their individual mobilities. As the bands begin to sort out, the field strength over each sample zone begins to change. Highly

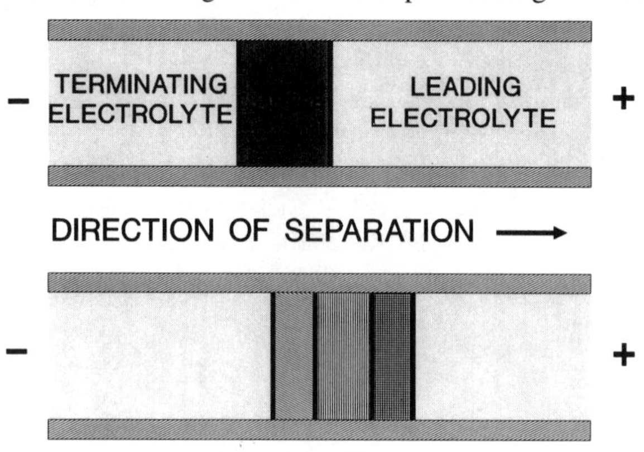

Figure 6.1. Schematic representation of a three-component separation by anionic CITP at the moment of injection (top) and after separation (bottom).

mobile sample components are also highly conductive. As a result, E and thus ν_{ep} are reduced for highly mobile bands. Conversely, sample components with low mobility generate greater field strengths, thereby increasing ν_{ep}. When the system reaches equilibrium, each solute's electrophoretic velocity becomes normalized to the same value. The isotachophoretic velocity at equilibrium for a single component separation can be expressed as

$$\nu_{ITP} = E_L \, \mu_L = E_A \, \mu_A = E_T \, \mu_T \,, \tag{6.3}$$

where μ = the respective mobilities of leader, solute, and terminator, and E = the field strength over each isotachophoretic zone. At the steady state, all zones move at the same speed and exhibit stable, well-defined boundaries. This feature is illustrated in the bottom section of Fig. 6.1.

Focusing is also a consequence of velocity normalization. Should a band diffuse into a neighboring zone, its velocity will automatically adjust based on the encountered field strength. As a result, the band will either speed up or slow down, thereby rejoining its own zone. Should a component be dilute, the CITP zone will be compressed because the solute concentration at equilibrum must be normalized as well (Eq. (6.2)). Dilute sample zones generate higher field strengths, resulting in a correction of the zonal concentration. This form of peak compression can result in substantial trace enrichment, perhaps the most important characteristic of CITP. This unusual feature results in the solute's quantitative information being expressed as the length of the zone rather than the zone height or area. The concentration of all CITP bands is related to the concentration of the leading electrolyte as described by the Kohlrausch regulating function

$$C_i = C_L \frac{\mu_i(\mu_L + \mu_C)}{\mu_L(\mu_i + \mu_C)} \,, \tag{6.4}$$

where C_i = analyte concentration at steady state; C_L = concentration of the leader; and the mobility terms are for the analyte, the leader, and the leader's counterion. Since the concentration of the leader is selected to be much higher than the solute concentration, band compression always occurs because of the high field strength over the solute zones.

Like CZE, CITP can be used to determine both cations and anions. Unlike CZE, separate CITP runs are usually required to determine each ionic form. The two modes of CITP are aptly named *cationic* and *anionic CITP*.

6.3 Instrumental Aspects

CITP has been generally performed under conditions of zero EOF though this is not a fundamental constraint. Many separations have been reported using relatively wide 500–800 μm i.d. Teflon capillaries. It is possible to perform CITP on 50–75 μm i.d. fused-silica capillaries with or without HPMC added as an EOF modifier. The use of fused-silica is particularly important if UV detection is to be facilitated. Otherwise, a flow cell is required. Even with microfabrication techniques, flow-cell design is generally not practical when using small-bore capillaries.

Conventional CITP instrumentation has fewer constraints compared to capillary electrophoresis because the capillary diameters are an order of magnitude larger. Injection can be performed with a syringe through a septum at the head of the capillary. With CE instruments, injection must be electrokinetic or hydrodynamic (Chapter 9).

Detection for CITP is usually based on a bulk property of the electrolytic solution. Conductivity detection measures the specific resistance of the zone as it exits the capillary. Once calibrated, the step height is directly proportional to the mobility of the individual solute. Both the quantitative and qualitative features of CITP are illustrated in Fig. 6.2. At the start of the run, the conductivity is high because the presence of the highly mobile leader. As leader and each successive solute exit the capillary, the conductivity decreases because less mobile species are migrating past the electrodes. Finally, the conductivity reaches its minimal value when the capillary is completely filled with the terminating electrolyte, the least mobile species. Other detection techniques such as thermal (measures the Joule heat produced within each zone) or the potential gradient (measures the voltage drop across each zone) are employed less frequently. When commercial CE instrumentation is used, detection is invariably UV absorbance or fluorescence because these are the only modes that are commercially available.

Runs are generally performed in the constant-current mode. Under this condition, as less mobile species enter the capillary, the voltage increases proportionately. This feature permits higher-speed separations without self-heating problems. Alternatively, constant voltage can be employed, but at the expense of run time.

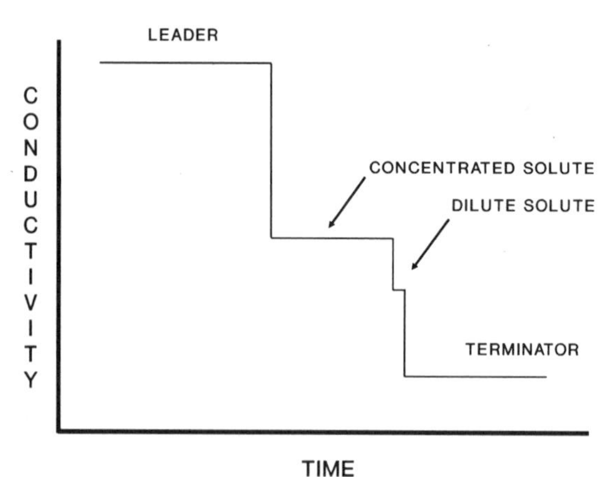

Figure 6.2. Rendition of the output of a two-component CITP separation with conductivity detection.

Operation on coated and uncoated fused-silica capillaries using UV detection is practical (8–11). The isotachopherograms may show differences in the order of migration in the presence or absence of the EOF. This is shown graphically in Fig. 6.3 for both cationic and anionic CITP (8).

For cationic CITP in an uncoated capillary, the order of migration is identical to that found in the absence of EOF. The capillary is first filled with the leader as in most forms of CITP, and the voltage polarity is positive at the sample side. Under these conditions, both the electroosmotic and isotachophoretic (electrophoretic) flows are directed toward the negative electrode. Such is not the case for anionic CITP. The EOF remains directed toward the cathode, but the isotachophoretic migration is now directed toward the anode. In this case, the capillary is filled with terminator and the voltage polarity is sample-side positive. This differs from the usual polarity when anionic CITP is performed in the absence of the EOF. Since the EOF is sufficient to overcome the isotachophoretic migration, the net flow is toward the cathode. In other words, the CITP separation occurs opposite to the EOF. This explains why the capillary must be filled with the terminator solution. Under these conditions, the order of migration is reversed compared to the example where the EOF is suppressed by a capillary coating. These features are shown in Fig. 6.4 with separations in coated and uncoated fused-silica capillaries.

Figure 6.4 illustrates some other important features of CITP. To further the separation between two components, a spacer can be utilized. The spacer is selected to have a mobility intermediate between that of the solutes to be separated. While

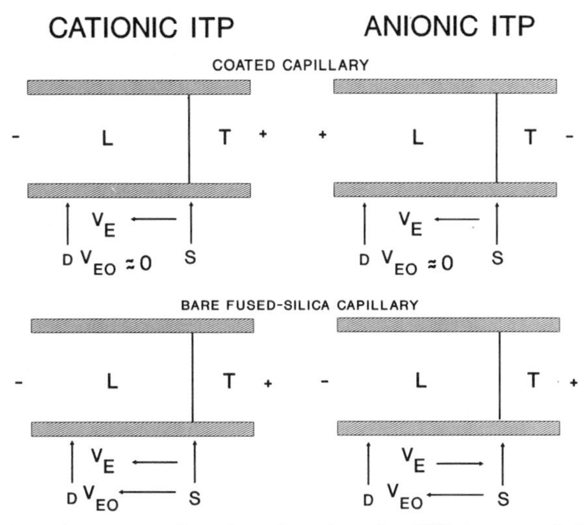

Figure 6.3. Schematic representation of configurations for CITP in open-tubular fused-silica capillaries. L = leading electrolyte; T = terminating electrolyte; S= sample inlet; D = detector position; V_{eo} = electroosmotic velocity; V_e = electrophoretic velocity. Redrawn with permission from *J. Chromatogr.* **516,** 211, copyright ©1990 Elsevier Science Publishers.

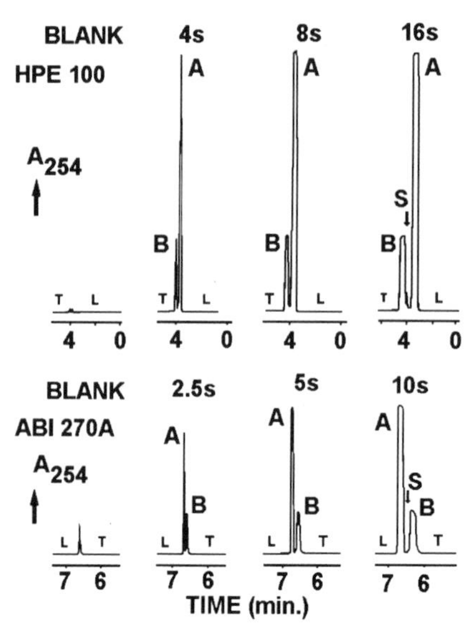

Figure 6.4. Anionic capillary CITP of amarath red (A, 1 mM) acetate (spacer, S 2 mM) and bromophenol blue (B, 1 mM). Leader: 10 mM HCl adjusted with histidine to pH 6; terminator: MES-histidine, pH 6; capillaries: (upper panel), 20 cm × 25 μm i.d. polyacrylamide-coated capillary from Bio-Rad; (lower panel), untreated 60 cm × 50 μm i.d. fused-silica capillary. Conditions: **(upper panel)** injection: 8 kV for 4, 8, 16 s; voltage: 8 kV; current: 2 decreasing to 0.7 μA; temperature: ambient; **(lower panel)** injection: vacuum injection (about 5 nL/s) for 2.5, 5, and 10 s; voltage: 30 kV; current: 6 decreasing to 4 μA rising to 6 μA during detection of the zones; temperature: 35°C. Detection: UV, 254 nm for both instruments. Reprinted with permission from *J. Chromatogr.* **516,** 211, copyright ©1990 Elsevier Science Publishers.

the spacer does not improve the actual separation, it can facilitate detection and/or fraction collection. If UV detection is used, the spacer should not absorb at the monitoring wavelength. Another feature shown in the figure is the impact of the size of the injection zone. In CZE, large-volume injections tend to reduce the number of theoretical plates and the resolution between adjacent bands (data not shown, refer to Section 9.1). In CITP, the resolution between adjacent bands can improve if a spacer is utilized. Note that an increase in the injection volume is expressed in the zone width as opposed to the peak height.

6.4 Buffer Selection

First, a decision must be made whether to use the anionic or cationic mode of CITP. For basic solutes having high pK_a, select the cationic mode; use the anionic mode for acidic solutes.

A. Leading Electrolyte

As in most electrophoretic methods, the pH is the first variable to be considered. Choose a pH such that the analytes are ionized, have differing mobilities, are soluble, and are stable. If solubility is a problem, buffer additives such as nonionic surfactants (Triton X-100, Brij 35), urea, and zwitterionic surfactants (sulfobetains) can be added both to the leader and the sample (2). Osmotic flow modifiers such as HPMC, PEG, polyvinyl alcohol, and Triton X-100 may be used, particularly if fused-silica tubing is employed (2).

For anionic CITP, chloride is generally selected as the leading ion. Potassium or NH_3^+ is frequently selected for cationic CITP. Other less mobile ions may be selected to simplify the isotachopherogram. Solutes with a mobility greater than that of the leader do not separate by CITP. These components will migrate faster than the leader and follow a CZE mechanism.

The counterion is selected based on the requisite pH. Zwitterions such as amino acids or zwitterionic buffers are generally used. Mixed counterions can be used to expand the overall pH range of the buffer. Refer to Section 2.10 for a listing of buffers and their useful pH ranges.

The concentration of the leading ion will control the length of the zones. For narrow zones, use higher-concentration electrolytes; for broad zones, use a lower concentration. These concentrations are all relative to the sample concentration. For general usage, 10 mM buffers are often sufficient.

B. Terminating Electrolyte

The terminating electrolyte must be selected to have a lower mobility than any of the sample components. This can be accomplished by selecting a material that is weakly charged at the operating pH. A better choice is to use a completely charged terminating ion with a relatively low intrinsic mobility. Such a selection will yield improved precision because the charge will not be dependent on small changes in pH. Large molecules such as peptides or proteins have relatively low mobilities, so the first option is frequently the only solution. Overall, if the differences in mobility between the leader and terminator are maximized, the highest-resolution separations will be obtained at the expense of separation time. The use of a high-mobility terminator removes low-mobility components from the isotachopherogram in much the same fashion as selection of a less-mobile leader. Polymeric species such as carboxylate-derivatized PEG 6000 have exceptionally low mobility.

Table 6.1 lists some buffer recipes for both cationic and anionic CITP at a series of pHs. Table 6.2 on page 139 provides application and leader/terminator recipe advice.

Table 6.1. Composition of Common CITP Buffers[a]

Anionic CITP	pH 3.3	pH 6.0	pH 8.8
Leading ion	10 mM HCl	10 mM HCl	10 mM HCl
Leading counterion	ß-alanine	L-histidine	ammediol
Leading additive	0.2% HPME	0.2% HPMC	0.2% HPMC
Terminating ion	10 mM caproic acid	10 mM MES	10 mM ß-alanine
Terminating counterion		Tris	$Ba(OH)_2$
Terminating pH		6.0	9.0
Cationic CITP		**pH 2.0**	**pH 4.5**
Leading ion		10 mM HCl	10 mM KOAc
Leading counterion			HOAc
Terminating ion		10 mM Tris	10 mM HOAc
Terminating counterion		HCl	
Terminating pH		8.5	

[a]Key: Ammediol, 2-amino-2-methyl-1,3-propanediol; HPMC, hydroxypropylmethylcellulose; MES, 2-(*N*-morpholino)-ethane sulfonic acid; Tris, tris(hydroxymethyl)aminomethane. Data from Ref. 16.

6.5 Advantages and Problems with CITP

The compelling advantage of CITP is the ability to trace-enrich a low ionic strength and dilute sample. It is possible to inject several hundred nanoliters of sample, provided the capillary is of sufficient length to permit stable zone formation. Limits of detection by mass spectrometry (Section 10.9) can be improved 100-fold over CZE (12). In micropreparative work, the potential for collecting significant amounts of material appears promising.

The peak capacity of CITP can be very high. This is a consequence of the fundamental requirement that adjacent bands be in contact with each other to maintain electrical continuity. A properly developed method has little unused separation space.

On the other hand, methods development is more complex compared to CZE. In the latter technique, it is possible to obtain separations rapidly with little knowledge about the sample characteristics. The unusual presentation of the data is confusing to analysts who are used to chromatography. Since step height is analogous to retention or migration time, the relative ordering of bands along the *x*-axis is no longer related to the qualitative content of the zone. When components that are not present in standard blends are introduced into the separation, the relative ordering of zones along the *x*-axis may be shifted.

The advantages of CITP are likely to be exploited in two areas: mass spectrometry (12–14) and on-line trace enrichment (15). Both of these important subjects will be covered in Sections 10.9 and 9.7, respectively.

Table 6.2. Applications[a]

Application	Leader	Terminator	Reference
Cationic Systems			
Small cations	10 mM KOH, bicine, 0.2% HEC, pH 8.35	5 mM Tris acetate, pH 4.9	17
Creatinine	10 mM KOAc, HOAc, pH 5.3, 0.2% HPMC	5 mM HCl	18
Minoxidil	5 mM KOH, PICA, pH 5.4, 0.3% Triton X-100	5 mM formic acid	19
Na^+, Ba^{2+}, Sr^{2+}	20 mM $KHCO_3$, citric acid, pH 5.0	10 mM $MgCl_2$	20
ß-Phenylalkylamines, amphetamines, dopamines	10 mM KOAc, pH 4.5–5.4, 0.5% PVA	20 mM 6-ACA, 20 mM ß-ALA or 10 mM HOAc	21
Triethanolamine	5 mM ethanolamine, HOAc, pH 5.3, 0.2% HPMC	4 mM HIS, HOAc, pH 5.3	18
Anionic Systems			
Small anions	10 mM HCl, ß-ALA, 2.5 mM Mg^{2+}, 0.2% HEC, pH 3.5	5 mM citrate	17
Serum carboxylic acids	10 mM HCl, ß-ALA, pH 4.2, 0.2% HPMC	10 mM nicotinic acid	22
Cysteinyl leukotrienes	5 mM HCl, Tris, pH 7, 0.25% HPMC	10 mM phenol, pH 10	23
Fibrinogen	5 mM MES, 10 mM AMM, pH 9.1, 0.5% HPMC	10 mM 6-AHA-AMM, pH 10.8	24
Food colorants	10 mM HCl, ß-ALA, pH 3.5, 0.1% HMEC	5 mM acetic acid	25
	10 mM HCl, HIS, pH 6.0, 0.1% HPEC	5 mM caproic acid	
HA, MHA, MA, PGA	5 mM HCl, 20 mM ß-ALA, pH 3.75, 0.4% HPMC	5 mM caproic acid	26
Membrane proteins	5 mM H_3PO_4, 20 mM AMM, pH 9.2, 0.25% HPMC	not specified	27
Pyrethroids	10 mM HCl, creatinine, 0.05% PVA, pH 4.8	5 mM MES	28
Sialic acid	5 mM HCl, ß-ALA, pH 3.8, 0.3% HPMC	10 mM caproic acid	29
TFA	10 mM HCl, ß-ALA, pH 3.6, 0.1% Triton X-100	10 mM caproic acid	18

[a]ACA, amino caproic acid; AHA, aminohexanoic acid; ALA, alanine; AMM, ammediol; HEC, hydroxyethylcellulose; HIS, histidine; HPMC, hydroxypropylmethylcellulose; PICA, picolinic acid; PVA, polyvinyl alcohol.

6.6 Applications

The vast majority of CITP work has been performed in 200–500 μm i.d. Teflon capillaries (2, and references therein). This section will focus on separations employing narrow i.d. fused-silica capillaries with capillary electrophoresis instrumentation using UV and scanning UV detection. Since there are few reports (8–14) meeting these criteria, selected applications and buffer recipes for CITP in Teflon capillaries are given in Table 6.2. It is likely that many of these applications can be transferred onto fused-silica capillaries, especially when coatings and EOF modifiers are employed.

A. Proteins

The suppression of EOF is advantageous for the separation of proteins by CITP. In the same fashion as CZE or CIEF, the use of HPMC (0.3–0.5%) as a leader additive is useful in limiting the EOF. HPMC can also be useful in suppressing wall effects (9). In untreated fused-silica, substantial wall interaction can be noted.

Proteins can be separated as anions or cations. In the cationic mode, the detector-side electrode serves as the cathode (negative electrode); thus, both the electrophoretic and electroosmotic flows are in the same direction. In the anionic mode, the detector-side electrode is the anode (positive electrode). Since the analytes are negatively charged, the electrophoretic migration is opposite the EOF.

Multiwavelength detection can also be useful as an aid to peak identification. Fig. 6.5 shows three-dimensional cationic CITP of a blank (A) and a mixture of lysozyme (LYSO), conalbumin (CAL), and ovalbumin (OVA), along with the spacers, creatinine (CREAT) and γ-amino-*n*-butyric acid (GABA). Without the addition of 0.3% HPMC, the proteins had poorly defined zones with relatively long elution times (not shown). Even in the presence of HPMC, conalbumin is partially adsorbed (data not shown). Presumably, the proteins were adsorbed onto the capillary wall. With the addition of the additive, sharp zones are observed, even for the blank where impurities indicate the boundary between the leader and terminator. Using multiwavelength detection, both spacers and proteins are properly identified in a single experiment.

The selection of spacer molecules can be troublesome because each spacer must have the proper mobility and absorption characteristics. The use of carrier ampholytes permits the addition of a variety of molecules with spacer potential, as shown in Fig. 6.6 for anionic CITP. In Fig. 6.6a, an RNAase sample shows impurities at the edge of the band. Figure 6.6b shows a blank separation and absorption spectra of a series of pH 3.5–10 ampholytes (1%). Figure 6.6c illustrates the admixture of ampholytes and sample yielding greater resolution of the RNAase impurities.

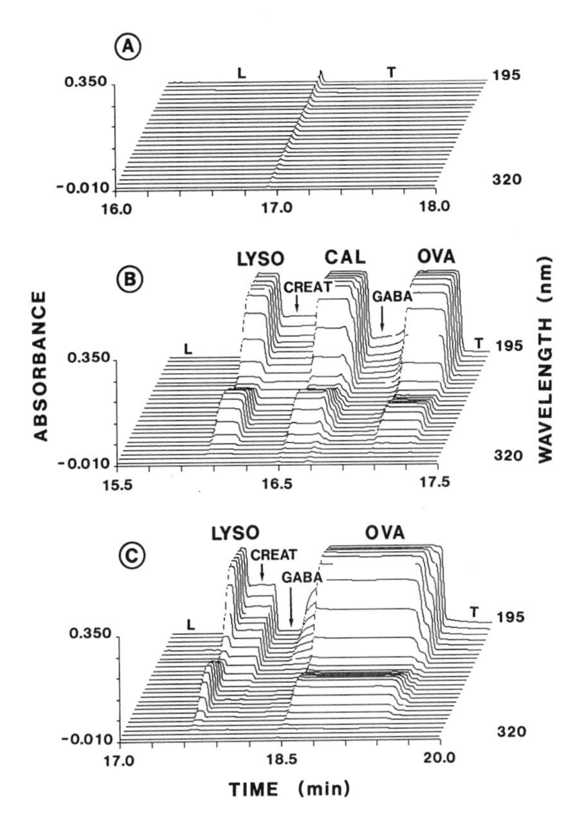

Figure 6.5. Three-dimensional cationic CITP of (A) blank, (B) lysozyme (LYSO), creatinine (CREAT), conalbumin (CAL), γ-amino-*n*-butyric acid (GABA), and ovalbumin (OVA), and (C) OVA spiked with CREAT and GABA. Capillary: 90 cm (length to detector, 70 cm) x 75 μm i.d.; leader: 10 mM potassium acetate and acetic acid with 0.3% HPMC, pH 4.75; terminator: 10 mM acetic acid; sample: proteins, 10–30 mg/mL dissolved in leader without HPMC; voltage: 20 kV; injection: gravity by raising end of capillary 34 cm; detection: multiwavelength UV; current: 12 μA declining to 2 μA. Reprinted with permission from *J. Chromatogr.* **558,** 423, copyright ©1991 Elsevier Science Publishers.

It is likely that a variety of capillary coatings (DB-1, DB-50, or alkylsilane) and buffer additives (polyvinyl alcohol, HPMC, or urea) similar to what is used for CZE and CIEF can also be employed in CITP. Ideally, the buffer additives will be neutral compounds that do not contribute to the conductivity or buffering capacity of the system.

B. Serum Proteins

The necessity for using capillary coatings (certain additives may provide similar protection) to reduce wall binding is also found in CITP. The problem with

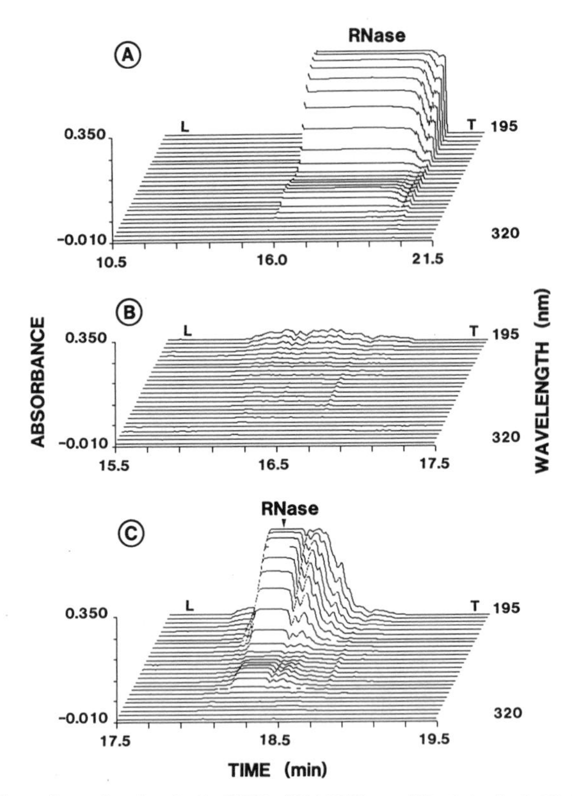

Figure 6.6. Three-dimensional cationic CITP of (A) RNAase (90 s injection), B) 1% solution of pH 3.5–10 ampholine (30 s injection), (C) RNAase and ampholine (30 s injection). Other conditions are described in Fig. 6.5. Reprinted with permission from *J. Chromatogr.* **558,** 423, copyright ©1991 Elsevier Science Publishers.

wall binding of proteins on uncoated fused-silica is exacerbated by the CITP buffering system. Unlike CZE, CITP zones are not contained within a carrier electrolyte. As a result, the ionic strength within an CITP zone can be low. In low ionic strength media, the risk of electrostatic binding between proteins (or any solute) and the wall increases. Figure 6.7 illustrates the separation of serum proteins in both coated (a) and uncoated (b) fused-silica capillaries. Substantial evidence of wall interactions is found for uncoated capillaries.

In the uncoated capillary, the tube is filled with terminator because CITP occurs in the direction away from the detector, but the EOF determines the direction of migration, as in Fig. 6.3. The polarity of the voltage is sample-side positive. For the coated capillary, the tube is filled with leader and run under normal polarity because the EOF is effectively suppressed.

The use of ampholyte spacers provides further fractionation of serum proteins. Admixture of a solution of Bio-Lyte 3/10 buffered with ß-alanine, pH 9.2, with

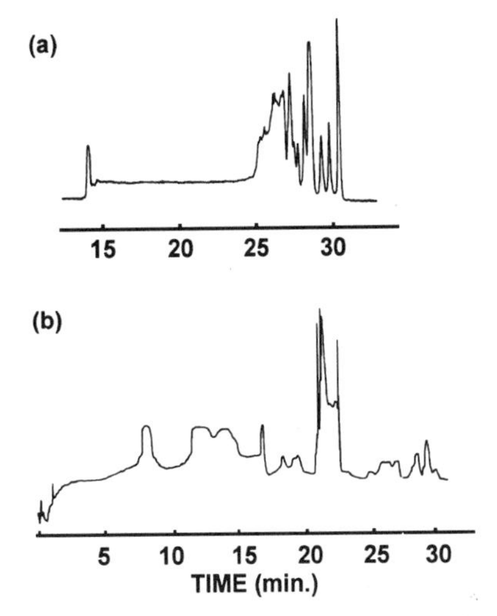

Figure 6.7. Capillary anionic CITP of human serum in polyacrylamide-coated (a) and non-coated (b) fused-silica capillaries. Capillary: 34 cm x 25 μm i.d.; injection: 15 nL (30 mm) voltage: 10 kV; detection: UV, 280 nm; leader: 10 mM HCl adjusted to pH 8.3 with Tris; terminator: 0.1 M ß-alanine adjusted to pH 9.2 with barium hydroxide. Capillary filled with (a) leader and (b) terminator. Reprinted with permission from *J. Chromatogr.* **550,** 811, copyright ©1991 Elsevier Science Publishers.

human serum gives the separation shown in Fig. 6.8, which was run under the conditions for Fig. 6.7. The limitation here is the use of 280 nm for detection. Most proteins have their greatest UV absorbance in the low UV. The use of such a low wavelength in CITP is generally not useful because of a shifting baseline due to the UV adsorption characteristics of the buffers. This is particularly troublesome when carrier ampholytes are used as spacers, since these substances absorb as well at lower wavelengths.

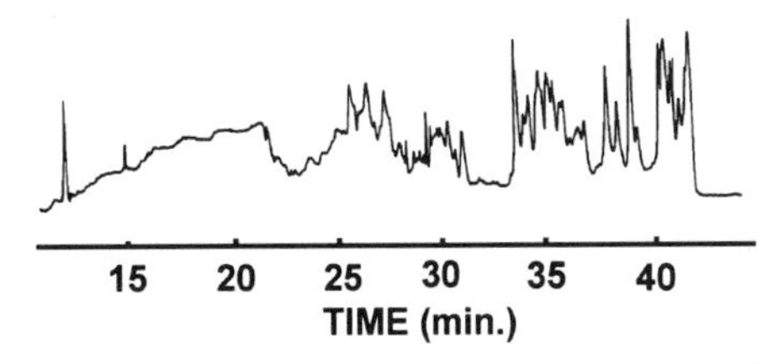

Figure 6.8. Capillary CITP of human serum with carrier ampholytes as discrete spacers. Conditions: see Fig. 6.7a. Sample: 80 μL human serum mixed with 15 μL of a solution consisting of 168 μL ß-alanine, pH 9.2, and 60 μL Bio-Lyte 3/10, adjusted to pH 9.0 with 7 μL of 1 M sodium hydroxide. Reprinted with permission from *J. Chromatogr.* **550,** 811, copyright ©1991 Elsevier Science Publishers.

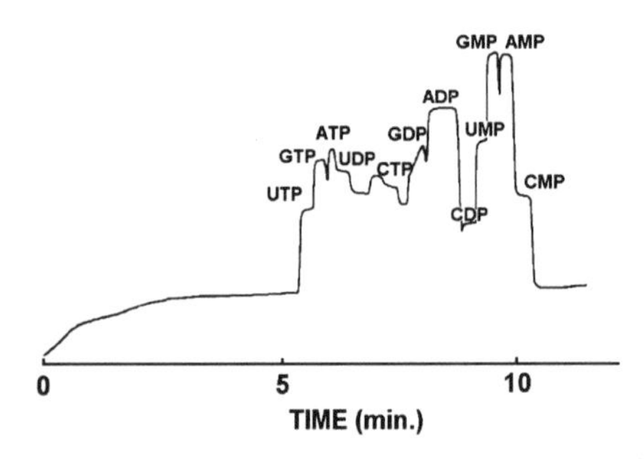

Figure 6.9. Cationic CITP of nucleotides in a polyacrylamide-coated capillary. Sample volume: 4 nL (2 mm); capillary: 29 cm x 50 μm i.d.; detection: 254 nm; leader: 10 mM HCl adjusted to pH 3.9 with solid ß-alanine; terminator: 10 mM caproic acid. Reprinted with permission from *J. Chromatogr.* **550,** 811, copyright ©1991 Elsevier Science Publishers.

C. Nucleosides

Nucleosides (Fig. 6.9) are separated in a 29 cm × 50 μm i.d. polyacrylamide-coated capillary employing UV detection at 254 nm (11). The time of separation is four-fold faster than that obtained with a 200 μm i.d. capillary. Despite the focusing effect of CITP, the use of smaller i.d. capillaries is still required for high-speed separations because of the Joule heating problem. The use of a 25 μm i.d. capillary gives 160-fold less thermal broadening than found for 200 μm i.d. capillary at a field strength of 860 V/cm (11).

References

1. F. M. Everaerts, J. L. Beckers, and Th. P. E. M Verheggen, *Isotachophoresis Theory, Instrumentation and Applications* (Elsevier Scientific Publishing, 1976).
2. P. Boček, M. Deml, P. Gebauer, and V. Dolník, *Analytical Isotachophoresis* (VCH, 1988).
3. P. Boček, P. Gebauer, V. Dolník, and F. Foret, *J. Chromatogr.* **334**, 157 (1985).
4. V. Kašička and Z. Prusík, *J. Chromatogr.* **569,** 123 (1991).
5. *J. Chromatogr.* **320,** 1–268 (1985).
6. *J. Chromatogr.* **390,** 1–197 (1987).
7. *J. Chromatogr.* **545,** 223–484 (1990).
8. W. Thormann, *J. Chromatogr.* **516,** 211 (1990).
9. P. Gebauer and W. Thormann, *J. Chromatogr.* **558,** 423, (1991).
10. S. Hjertén, K. Elenbring, F. Kilár, J. L. Liao, A. J. C. Chen, C. J. Liebert, and M. D. Zhu, *J. Chromatogr.* **403,** 47 (1987).
11. S. Hjertén and M. K-Johansson, *J. Chromatogr.* **550,** 811 (1991).
12. H. R. Udeseth, J. A. Loo, and R. D. Smith, *Anal. Chem.* **61,** 228 (1989).
13. R. D. Smith, J. Loo, C. J. Bariniga, C. G. Edmonds, and H. R. Udseth, *J. Chromatogr.* **480,** 211 (1989).

14. R. D. Smith, S. M. Fields, J. A. Loo, C. J. Bariniga, H. R. Udseth, and C. G. Edmonds, *Electrophoresis* **11,** 709 (1990).
15. D. S. Stegehuis, H. Irth, U. R. Tjaden, and J. Van der Greef, *J. Chromatogr.* **538,** 393 (1991).
16. F. S. Stover, *Electrophoresis* **11,** 750 (1990).
17. I. Matejovič and J. Polonský, *J. Chromatogr.* **438,** 454 (1988).
18. J. Sollenberg, *J. Chromatogr.* **545,** 369 (1991).
19. S. Fanali, M. Cristalli, and P. Catellani, *J. Chromatogr.* **405,** 385 (1987).
20. R. G. Trieling, J. C. Reijenga, and H. D. Jonker, *J. Chromatogr.* **545,** 475 (1991).
21. D. Walterová and V. Šimánek, *J. Chromatogr.* **405,** 389 (1987).
22. L. Kriváнková and P. Boček, *J. Microcol. Sep.* **2,** 80 (1990).
23. D. Tsikas, J. Fauler, G. Brunner, and J. C. Frölich, *J. Chromatogr.* **545,** 375 (1991).
24. D. Del Principe, C. Colestra, A. M. Fiorentino, G. Spagnolo, G. Mancuso, and A. J. Menichelli, *J. Chromatogr.* **419,** 329 (1987).
25. J. Karovičová, J. Polonský, A. Príbela, and P. Šimko, *J. Chromatogr.* **545,** 413 (1991).
26. J. Sollenberg and A. Baldesten, *J. Chromatogr.* **132,** 469 (1977).
27. Dj. Josić, A. Böttcher and G. Schmitz, *Chromatographia* **30,** 703 (1990).
28. V. Dombek, *J. Chromatogr.* **545,** 427 (1991).
29. E. Weiland, W. Thorn, and F. Bläker, *J. Chromatogr.* **214,** 156 (1981).

Chapter 7

Electrokinetic Capillary Chromatography

7.1 Introduction

Retention in liquid chromatography is based on the distribution of a solute between two discrete phases, the stationary phase and the mobile phase. A separation between two or more solutes can be achieved whenever the equilibrium distribution between the phases is distinct for each component in the mixture. Under this condition, solute will differentially migrate through a chromatographic column. The separation factor, α, for solutes A and B can be expressed as follows (1):

$$\alpha = \frac{[A]_x[B]_y}{[A]_y[B]_x}, \tag{7.1}$$

where [A] = the concentration of A in phase x or y and [B] = the concentration of B in either of the chromatographic phases.

Nowhere in Eq. (7.1) nor in any of the other fundamental expressions for retention in chromatography is there an absolute stipulation for the velocity of either of the phases. It is generally assumed that one phase is mobile while the other is stationary. If that assumption of a stationary phase is disregarded, it is easy to imagine a chromatographic separation taking place through the equilibrium distribution of a solute between two phases that are moving at differing velocities. This concept forms the basis for many forms of electrokinetic separations.

The semistationary or slow-moving phase in electrokinetic chromatography (EKC) is composed of molecular aggregates or discrete molecules that are dissolved as additives in the background electrolyte. This formulated buffer system contains, on a molecular level, a heterogeneous environment or "pseudophase" that can compete with the background electrolyte in interacting with the solute. The driving forces that control the speeds of the bulk solution and the heterogeneous pseudophase are electroosmotic and/or electrophoretic migration factors.

The creation of the pseudophase can be accomplished with a variety of buffer additives. Surfactants that generate aggregates known as micelles are the most common additives. This form of EKC is known as micellar electrokinetic (capillary) chromatography (MEKC or MECC) (2, 3). Emulsified aggregates have been reported as well (4). Cyclodextrins represent another class of additives known to form a pseudophase without the need to form aggregates. Cyclodextrins can be used together with micellar media for the separation of very hydrophobic solutes (5). Polymer ions (6, 7) and exotic species such as the "Starburst Dendrimer" (8) can also function as EKC additives.

Electrokinetic chromatographic separations are used primarily for the separation of small molecules. Unique applications include chiral recognition that can be accomplished directly with micelles, with cyclodextrins, or by the MECC separation of diastereomers that were prepared by precapillary derivatization. Each of these features will be covered as part of this chapter. A textbook devoted to MECC has recently been published (9).

7.2 Micelles

Surfactants are molecules comprising long hydrophobic "tails" and polar "head groups." Above a certain concentration known as the critical micelle concentration (CMC), surfactant molecules spontaneously organize into roughly spherical to ellipsoidal aggregates known as micelles. This form of molecular organization occurs to lower the free energy of the system.

In aqueous solution, the surfactant's hydrophobic tail cannot be solvated by water molecules. As the concentration of surfactant is increased in the bulk solution, it becomes probable that the molecules begin to find each other. Since the polar head groups are solvated in aqueous solution, the surfactant molecules orient toward each other's tail forming first a dimer, and later trimers and tetramers, etc. These aggregates are known as premicellar assemblies. Finally, at the CMC, the full micellar organization takes shape, a cartoon of which is shown in Fig. 7.1. The aggregate forms with a hydrophobic core, the result of tail-in orientation. Shape and stability of micelles are further determined by electrostatic repulsion of the polar head groups and van der Waals attraction of the lipoid chains (10). Materials that are insoluble in water are frequently dissolved through hydrophobic interaction with the surfactant. With polar head groups at the periphery, electrostatic interaction with external solutes can occur provided ionic surfactants are employed.

The micellar model provides for four zones (11):

1. a hydrocarbonlike core with a diameter of 10–28 Å;

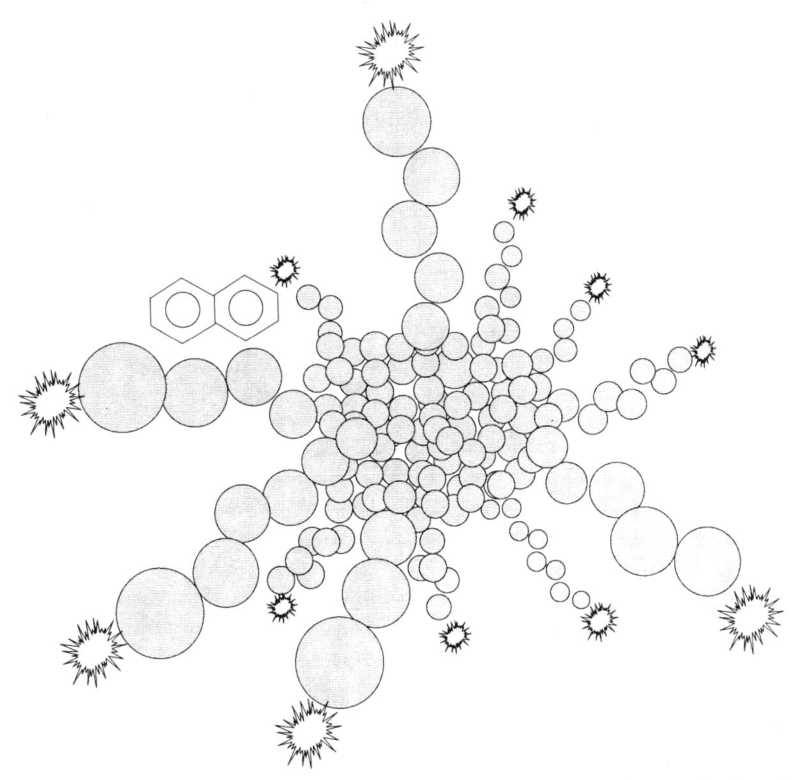

Figure 7.1. Artist's conception of an anionic micelle associated with a solute, naphthalene. The asterisks indicate the anionic head group.

2. the Stern layer that contains the polar head groups and counterions;
3. the Gouy–Chapman layer, an electric double layer that is hundreds of angstroms thick; and
4. the bulk surrounding water.

At the CMC, the bulk properties of the micellar solution are dramatically altered, including surface tension, conductivity, solubilizing power, and the ability to scatter light. Micelles are dynamic entities in equilibrium with the surrounding environment. Surfactant molecules are free to exchange between the micelle and the external media. Solutes dissolved in surfactant solutions are free to exchange within the micelle as well. For example, typical entrance rate constants for arenes dissolved in SDS are $10^9\ s^{-1}$, while the exit rates are $10^4\ s^{-1}$ (12).

Another class of surfactants forms micelles in nonaqueous solvents. Known as inverted micelles, these aggregates have an aqueous core with the hydrophobic portion of the surfactant in contact with the bulk organic solvent (13). There have been no reports to date of the use of inverted micelles in electrokinetic separations.

Micellar solutions play an important role in many phases of analytical and organic chemistry, including catalysis, electrochemistry, spectroscopy, chromatography, and now capillary electrophoresis. The use of these intriguing solutions for analytical chemistry was recently reviewed (14). Of particular relevance to this chapter is the use of micellar mobile phases in liquid chromatography (15). Surfactant solutions above the CMC can serve as mobile phase modifiers and function in a similar fashion to conventional organic solvents in reverse phase liquid chromatography. The phase distribution is more complicated because of the presence of a chromatographic stationary phase, the micellar aggregate, and the bulk aqueous solution. The phase equilibria of a solute among these phases are shown in Fig. 7.2. In MECC, the absence of the stationary phase simplifies the phase distribution.

Sodium dodecyl sulfate (SDS) is the most widely used surfactant for both electrophoresis and chromatography. This surfactant has the proper hydrophilic–lipophilic balance (HLB) for its intended use. In other words, SDS is very water-soluble and has a high degree of lipid solubilizing power. Because of its widespread applicability, the surfactant is available in highly purified form and is very inexpensive. The CMC for SDS is 8 mM, and its aggregation number is 63 (11). While many surfactants can be employed in electrokinetic separations, most of this chapter, as reflected by the scientific literature, will be devoted to the use of SDS.

The pioneering works of Jorgensen and Lukacs (16) for CZE and Armstrong and Nome (15) for micellar liquid chromatography, both appearing in 1981, provided pieces of a puzzle, the solution of which led to the discovery and development of micellar electrokinetic capillary chromatography (MECC). The first reports of this work by Terabe *et al.* appeared in 1984–1985 (2, 3).

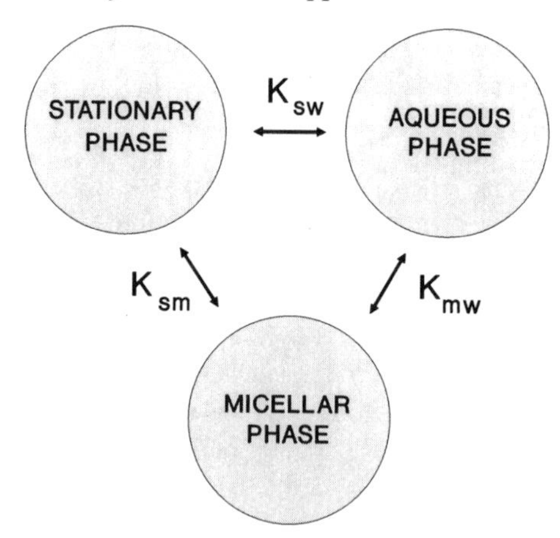

Figure 7.2. Partition coefficients for a solute in micellar liquid chromatography. K_{sw} = stationary phase–water, K_{mw} = micelle–water, and K_{sm} = stationary phase–micellar phase partition coefficients.

7.3 Separation Mechanism

A. Basic Concepts

In untreated fused-silica, the EOF, which is directed toward the cathode, is substantial at pH values ranging from mildly acidic through alkaline. On the other hand, SDS micelles are anionic and electrophorese toward the anode. As a result, the overall micellar velocity is reduced compared with the bulk flow. Electro-osmotic flow overcomes the micellar electrophoretic velocity at this pH range, resulting in a net fluid flow toward the cathode.

Since a solute may partition into and out of the micellar aggregate, its own migration velocity can be affected as well. When the solute is partitioned into the micelle, solute velocity is retarded. When present in the bulk phase or interstitial space between micelles, the solute, if neutral, is simply swept through the capillary by the EOF. MECC has the capability of separating within a single run anionic, neutral, and cationic species.

An illustration of a single-component separation is shown in Fig. 7.3. The term t_0 is analogous to the chromatographic description of the void volume of the column. Similarly, t_R describes the retention of a solute. The term t_{mc} distinguishes MECC from chromatography because it describes the velocity of the pseudophase. Under most separation conditions, all solutes must elute between t_0 and t_{mc}.

The fundamental equation for $\tilde{k}'$[1] accounts for the presence of the mobile pseudophase. As the velocity of the pseudophase approaches zero (a true stationary

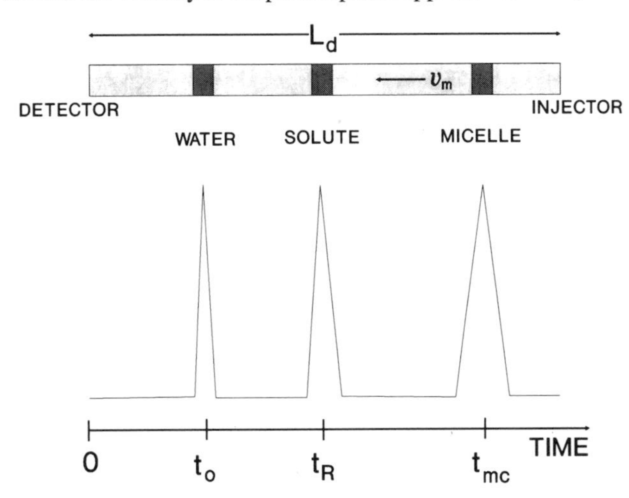

Figure 7.3. Illustration of the zones separated in a capillary (upper trace) along with the detected electropherogram (lower trace) for a hypothetical mixture of water, solute, and micelle. The broadening of the slowly migrating peaks is due to on-capillary detection (Section 10.1).

[1] $\tilde{k}'$ is used to differentiate from the classical term k′ which assumes a true stationary phase.

phase), t_{mc} approaches infinity and Eq. (7.2) reduces to the classical chromatographic expression for k′. A consequence of Eq. (7.2) is that as t_{mc} is approached, the peaks elute at more closely spaced intervals. Terabe *et al.* (3) recognized this effect is similar to that obtained with concave gradient elution LC for solutes with $\tilde{k}' < 150$.

$$\tilde{k}' = \frac{t_R - t_0}{t_0 (1 - t_R/t_{mc})} \tag{7.2}$$

Equation (7.3) describes the resolution between two solutes by MECC. As in Eq. (7.2), when the micellar velocity approaches zero, the equation reduces to the classical expression for chromatographic resolution.

$$R_s = \frac{\sqrt{N}}{4}\left(\frac{\alpha - 1}{\alpha}\right)\left(\frac{\tilde{k}_2'}{1 + \tilde{k}_2'}\right)\left(\frac{1 - t_0/t_{mc}}{1 + (t_0/t_{mc})\,\tilde{k}_1'}\right) \tag{7.3}$$

B. Elution Order

Prediction of the elution order can be straightforward provided a homologous series of compounds are being separated. Fig. 7.4 shows the separation of a series of peptides that differ only by a single amino acid. Peptide 15 has a net charge of –2 and is strongly repelled from the anionic SDS micelle. As a result, the peptide spends much of its time in the bulk phase thereby migrating the fastest of the group. Peptide 2 has a charge of –1 and is repelled less strongly so it spends more time attached to the micelle and exhibits a longer migration time

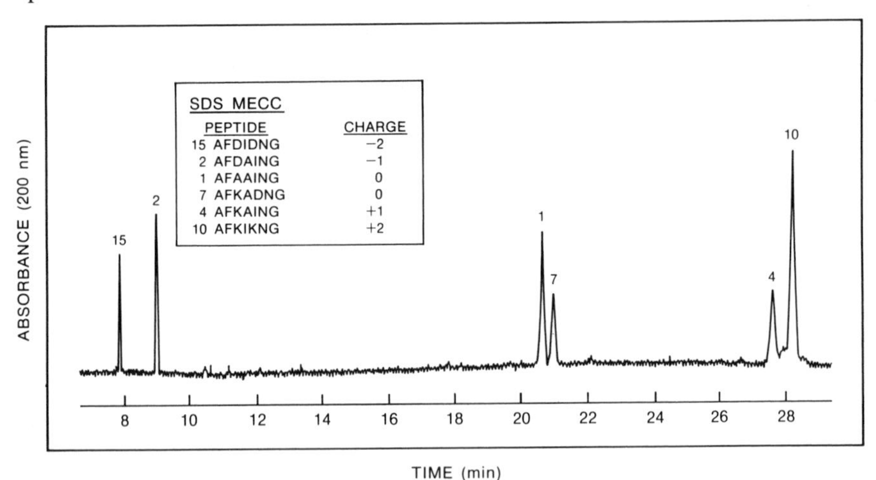

Figure 7.4. MECC of cationic, anionic, and neutral peptides. Capillary: 65 cm (45 cm to detector) × 50 μm i.d.; buffer: 10 mM phosphate, 100 mM SDS, pH 7.0; voltage: 20 kV; injection: vacuum, 2 s; detection: UV, 200 nm. Courtesy of Applied Biosystems, Inc.

compared to peptide 15. Peptides 1 and 7 are neutral and are separated based on hydrophobic effects.[2] Since these peptides are not repelled from the micelle, the migration times are lengthened relative to the anionic peptides. Finally, the cationic peptides migrate through the capillary. These peptides exhibit strong electrostatic interaction with the micelle, and as a result, both have lengthy migration times compared with the other peptides. It is interesting that the migration order of the charged peptides is the reverse of what is expected by CZE. Without the micelle, the cationic peptides migrate most rapidly, since both the EOF and electrophoretic mobility are directed toward the negative electrode. The anionic peptides, under CZE conditions, electrophorese toward the positive electrode but are swept by the EOF toward the negative electrode. The result is longer migration times for these anions.

For solutes that are not part of a homologous series, prediction of the elution order is a daunting task. Both electrostatic and hydrophobic interaction with micelles are in force. If the solutes are charged, they too will experience electrophoresis, at least when they are contained in the bulk solution.

The structures of a series of nonsteroidal anti-inflammatory drugs are shown in Fig. 7.5. These compounds are all aromatic and have carboxylic acid groups as well. Otherwise, phenyl, biphenyl, naphthalene, and other moieties form the structural features of these diverse compounds. Separations by CZE and MECC are shown in Figs. 7.6a and 7.6b (17). There is no apparent rationale for the comparative order of migration for these compounds by either mode of electrophoresis. With reverse-phase LC, the order of elution is peak numbers 3, 1, 5, 2, 4. If only hydrophobic effects were in operation by MECC, it would be expected that the order of elution by LC should be comparable to that by MECC. Since the factors that contribute to the solutes's migration velocity by MECC are complex, a theoretical approach toward the prediction of retention requires a model that considers solute–micelle hydrophobic and electrostatic interactions as well as the solute's electrophoretic properties.

7.4 Sources of Bandbroadening

The factors that contribute to bandbroadening in MECC are, not surprisingly, more complex than CZE because of the existence of the pseudophase. Terabe *et al.* (18) considered many of these factors starting with instrumental parameters common to all CE methods such as injection size, detector bandwidth, and detector rise time. The total peak variance (σ^2) of a system can be expressed as the sum of the individual contributions from the column, injector, and detector (18):

[2] The calculation of neutrality is based on the pH of the bulk solution. Since the pH is much lower at the micellar surface (because of an electrical double layer of protons), it's probable that these "neutral" peptides are cationic in the micellar domain.

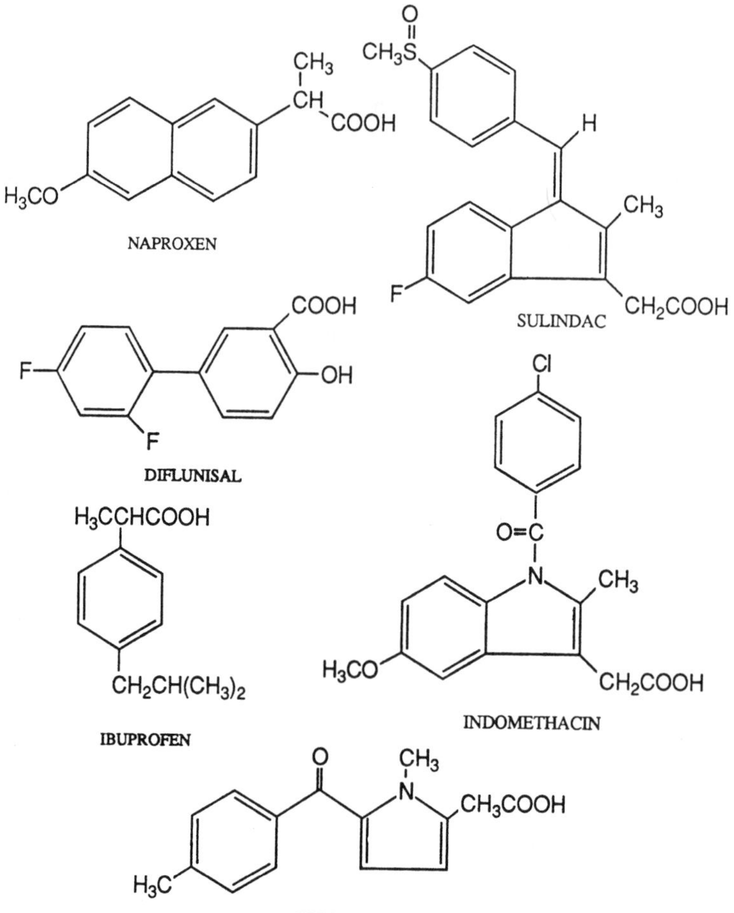

NAPROXEN

SULINDAC

DIFLUNISAL

IBUPROFEN

INDOMETHACIN

TOLMETIN

Figure 7.5. Structures of nonsteroidal anti-inflammatory drugs.

$$\sigma_{\mathrm{tot}}^2 = \sigma_{\mathrm{col}}^2 + \sigma_{\mathrm{inj}}^2 + \sigma_{\mathrm{det}}^2 \,. \qquad (7.4)$$

An injection > 0.8 mm in length causes deterioration in resolution. This holds true only if the injection buffer is identical to the run buffer, a situation that is frequently and often intentionally not followed (Section 9.6). The impact of the detector slit length on variance is small provided the slit is < 0.8 mm.

On-capillary features that contribute to bandbroadening are intrinsic to the method and sometimes unavoidable. According to Terabe (18), these on-capillary factors include longitudinal diffusion, sorption–desorption kinetics (the entry and exit rate of solutes into and out of the micelle), intermicelle mass transfer, radial

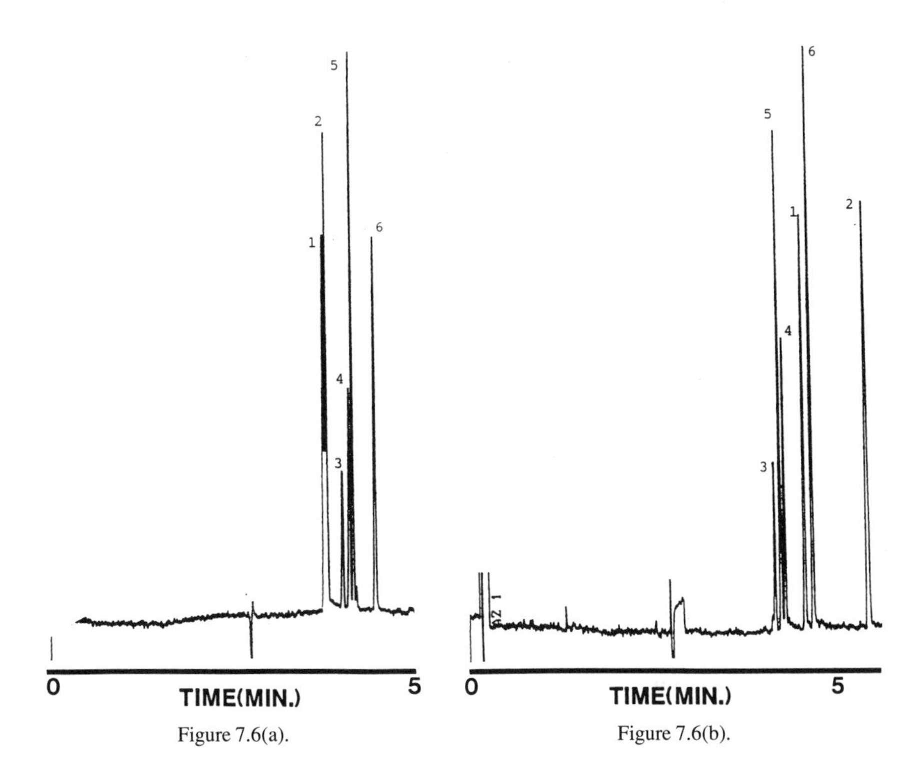

Figure 7.6(a). Figure 7.6(b).

Figure 7.6. (a) CZE and (b) MECC of nonsteroidal anti-inflammatory drugs. Capillary: 64.5 cm (43.5 cm to detector) × 25 μm i.d.; buffer: (a) 20 mM phosphate, pH 7.0; (b) 20 mM phosphate, pH 7.0, 25 mM SDS; voltage: 25 kV; temperature: 30°C; injection: vacuum, 2 s; detection, UV 230 nm. Key: (1) sulindac, 100 μg/mL; (2) indomethacin, 100 μg/mL; (3) tolmetin, 100 μg/mL; (4) ibuprofen, 100 μg/mL; (5) naproxen, 10 μg/mL; (6) diflunisal, 50 μg/mL. Reprinted with permission from *J. Liq. Chromatogr.* **14,** 952, copyright ©1991 Marcel Dekker.

temperature gradient, and micellar heterogeneity (including micelle polydispersity, which is the variance of the aggregation number).

Not all these factors are important. Terabe *et al.* (18) found that longitudinal diffusion was significant for solutes with large $\tilde{k}'$. Sorption/desorption kinetics and heterogeneity were significant at high voltages. Intermicelle mass transfer and temperature gradients were not significant under the conditions studied. The pertinent equations for each of these sources of bandbroadening are given in Table 7.1. Terms are defined in Table 7.2 along with their typical values.

Sepaniak and Cole (19) described results that differed from those of Terabe *et al.* In that work, resistance to mass transfer and thermal gradients were found most important. Modest changes in the CMC of SDS were also reported to be a function of temperature. Operating at high SDS concentrations reportedly increases

Table 7.1. Plate Height Equations for On Capillary Dispersion Processes in MECC

Longitudinal Diffusion	$H_l = \dfrac{2(D_{aq} + \tilde{k}'D_{mc})}{1 + (t_0/t_{mc})\tilde{k}'} \dfrac{1}{\nu_{eo}}$
Mass Transfer	$H_{mc} = \dfrac{2(1 - t_0/t_{mc})^2 \tilde{k}'}{(1 + (t_0/t_{mc})\tilde{k}')(1 + \tilde{k}')^2} \dfrac{\upsilon_{eo}}{k_d}$
Intermicelle Diffusion	$H_{aq} = \left(\dfrac{\tilde{k}'}{1 + \tilde{k}'}\right)^2 \dfrac{(1 - t_0/t_{mc})^2}{1 + (t_0/t_{mc})\tilde{k}'} \dfrac{d^2\upsilon_{eo}}{4D_{aq}}$
Temperature Gradients	$H_t = \dfrac{(1 - t_0/t_{mc})\tilde{k}'}{24(D_{aq} + \tilde{k}'D_{mc})} \dfrac{B^2 I^4}{64\kappa_o^2\pi^4 r_c^2\lambda^2 T_0^4} \upsilon_{eo}$
Polydispersity of the SDS micelle	$H_{ep} = \dfrac{0.026\,(1 - t_0/t_{mc})^2 k'}{1 + (t_0/t_{mc})\tilde{k}'} \dfrac{\nu_{eo}}{k_d}$

efficiency until Joule heating becomes significant. Davis (20) described yet another mode of bandbroadening that is related to changes in the solutes' partition coefficients resulting from Joule heating. Since the temperature near the capillary wall is cooler than at the center, differences in partition coefficients may occur.

This phenomonologic description of variance suggests some practical approaches to minimize bandbroadening in MECC.

Table 7.2. Typical Values for Parameters in Table 7.1

Variable	Definition	Typical Values
B	Viscosity dependence on temperature	2,400
D_{aq}	Diffusion coefficient of a solute in the bulk aqueous phase	3.6 (β–naphthol) – 8.7 (phenol) $\times 10^{-4}$ mm^2/s
D_{mc}	Apparent diffusion coefficient of a solute in the micelle	5.0×10^{-5} mm^2/s
$\tilde{k}'$	Capacity factor	1–50
k_d	Desorption rate constant of a solute from the micelle	4.1×10^2 (perylene) – 4.4×10^6/s (benzene)
κ_o	Electrical conductivity	5.71×10^{-4} S/mm (50 mM SDS)
I	Current	$1.0 \times 10^{-5} - 10 \times 10^{-5}$ A
λ	Thermal conductivity	5.73×10^{-4} W/mm°K
r_c	Capillary radius	0.0125 – 0.05 mm
T_0	Buffer temperature	325°K
t_0	Migration time of an unretained solute	120 – 300 s
t_{mc}	Migration time of a solute totally bound to the micelle	600 – 1,200 s
υ_{eo}	Electroosmotic velocity	0 – 3 mm/s

1. To minimize longitudinal diffusion, maintain a linear velocity of >1 mm/s by adjustment of pH, EOF, and field strength. This is particularly useful for solutes with small capacity factors, since diffusion in the bulk phase is far greater than when the solute is attached to the micelle.
2. For solutes with large capacity factors, it is tempting to increase the field strength to speed the separation. In a similar fashion to LC, this makes mass transfer a significant contribution to variance. It is far better to reduce $\tilde{k}'$ to <5. An added benefit is improved resolution, since MECC is superior in the small $\tilde{k}'$ region.
3. Operating close to the CMC will tend to increase dispersion due to micellar polydispersity. Working at surfactant concentrations at least three times greater than the CMC in water will tend to reduce this problem. It is important to remember that the CMC is affected by temperature and buffer additives, particularly organic solvent modifiers.
4. Perform an Ohm's Law plot and adjust the field strength to minimize the contributions of Joule heating.

7.5 Selecting the Electrolyte System

A. Surfactant Concentration

A general recipe for an MECC electrolyte includes the surfactant, usually SDS, a buffer to fix the pH, and occasionally other additives to adjust $\tilde{k}'$ and/or the overall elution range (t_{mc}/t_o). The SDS concentration generally ranges from 25 to 150 mM. Higher SDS concentrations usually result in longer solute migration times because the probability of partitioning into the micelle increases. As Fig. 7.7 indicates, substantial selectivity can be designed into the separation, depending on the degree of interaction between the solutes and the micellar assembly. That interaction can be hydrophobic or electrostatic. For example, vitamins B_1 and B_{12} are cationic and thus form ion pairs with the anionic micelle. On the other hand, anionic species are repelled from anionic micelles. In this case, increasing the surfactant concentration may not affect the migration time unless hydrophobic interactions are significant (Section 7.8).

B. Effect of pH

A suitable buffer is chosen, depending on the required pH. Many papers have reported on the use of a phosphate–borate blend. The advantage of this composition is the maintenance of a common ionic environment over a pH range covering 6–11. For reasons not entirely clear, the borate–phosphate blend provides better peak symmetry than a borate–acetate blend (21). However, a disadvantage of this blend is its high conductivity. A zwitterionic buffer is usually a better choice to minimize

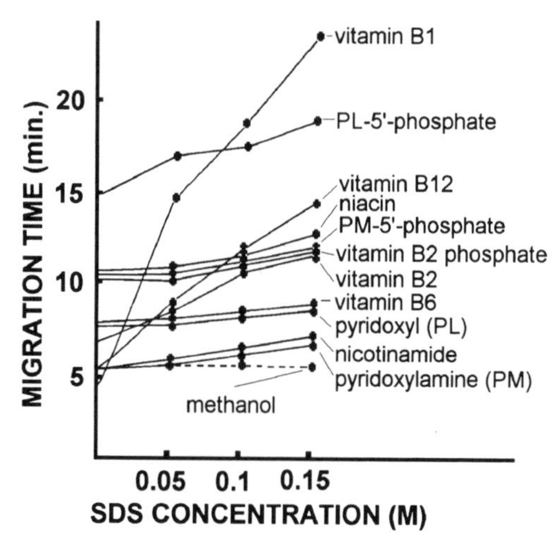

Figure 7.7. Effect of SDS concentration on the retention time of 11 water-soluble vitamins. Buffer: 20 mM phosphate–borate, pH 9.0, plus SDS. Capillary: 65 cm (50 cm to detector) × 50 μm i.d.; volatge: 20 kV; temperature: ambient; detection: UV, 210 nm. Redrawn with permission from *J. Chromatogr.* **465,** 331, copyright ©1989 Elsevier Science Publishers.

the conductivity of the solution, though its selection is less compelling compared to CZE. The surfactant itself is usually the major contributor to the overall ionic strength of the electrolyte. Selecting lithium dodecyl sulfate would be better in this regard, but the cost and purity of that surfactant does not approach that of SDS.

The selection of pH is based on the p*K*s of the solutes and the requisite selectivity. The $\widetilde{k}'$ for neutral compounds is pH-independent. For bases, $\widetilde{k}'$ decreases with increasing pH as the pK_a is approached and the positive charge is reduced. Below the pk_a, bases are strongly retained because of electrostatic interaction with SDS micelles. For acids, $\widetilde{k}'$ decreases as well as the pH of the electrolyte is adjusted above the pK_a. These acidic compounds develop a micelle-repelling negative charge in alkaline systems. Performance of migration-time plots is useful for selecting the optimal pH for the separation of charged solutes. Figure 7.8 illustrates a migration-time plot versus pH for several vitamins.

At low electrolyte pHs, the electrophoretic velocity of the micelles surpasses that of the EOF (22). In these instances, it is necessary to switch the polarity of the power supply to place the positive electrode beyond the detector. The order of migration is switched compared to the high-pH run, since hydrophobic compounds, which spend most of their time attached to the micelle, elute first. These features are illustrated in Figs. 7.9a and 7.9b. While the run time is much greater for the low-pH example, there is room to reduce the capillary length to speed the separation. This mode of separation may be advantageous when hydrophobic compounds have to be resolved. On the other hand, hydrophobic compounds can be rapidly separated with the aid of organic modifiers (Section 7.6), urea (23), or cyclodextrins (Section 7.9).

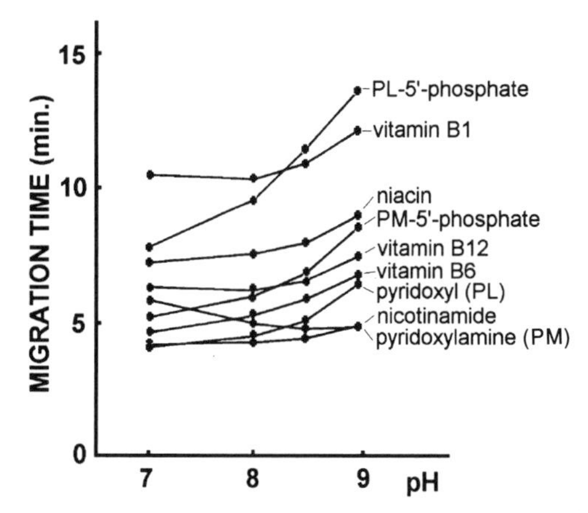

Figure 7.8. Effect of pH on the retention time of 11 water-soluble vitamins. SDS concentration: 50 mM; voltage: 25 kV. Other conditions as in Fig. 7.6. Redrawn with permission from *J. Chromatogr.* **465,** 331, copyright ©1989 Elsevier Science Publishers.

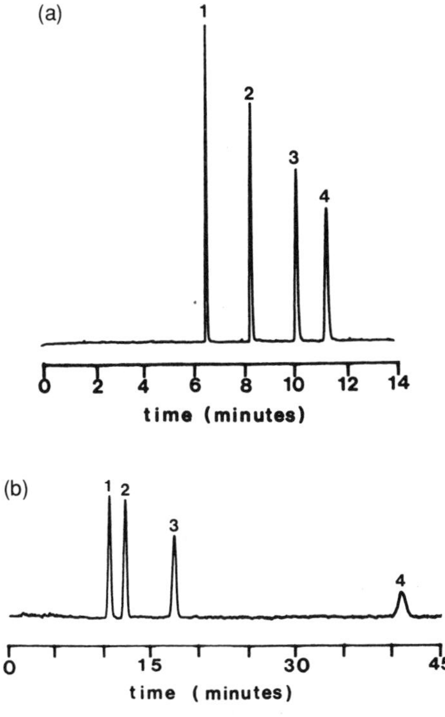

Figure 7.9. Effect of pH on the order of elution of parabens. Capillary: 100 cm (50 cm to detector) × 100 μm i.d.; buffer: 50 mM SDS, 10 mM phosphate, (a) pH 7.0; (b) pH 3.37; voltage: (a) +25 kV, (b) −25 kV; injection: electrokinetic, (a) +5 kV, 5 s; (b) −10 kV, 10 s; detection 254 nm, Key: (1) methyl; (2) ethyl; (3) propyl; and (4) butyl paraben. Reprinted with permission from *J. High Res. Chromatogr.* **12,** 635, copyright ©1989 Dr. Alfred Huethig.

7.6 Elution Range of MECC

Micellar electrokinetic separations have a limited elution range that is defined by the terms t_0 and t_{mc}. After a description of procedures for measuring these terms, the adjustment of the overall elution range and modification of the solute partition coefficients will be discussed.

A. Measurement of t_0

Determining t_0 can be accomplished by measuring the transit time to the detector for a neutral species that has no affinity for the micelle. Methanol, acetone, or formamide is typically selected. The use of the current-monitoring method (Section 2.3) has not been reported for MECC but would probably be sufficient as well. A 5% dilution of the micellar solution should provide sufficient conductivity change without appreciably affecting the EOF. The absorbance-monitoring method (Section 2.3) should work as well with 0.1% acetone added to the sample-side buffer reservoir.

B. Measurement of t_{mc}

The calculation of the capacity factor, $\tilde{k}'$ requires the knowledge of t_{mc}, the micellar migration time. This is determined by employing a probe such as Sudan III, a water-insoluble dye that is bound to micelles (3). When organic solvents are used as additives, the probe method becomes insufficient because the dye can partition into the bulk phase. In this example, the determination of t_{mc} becomes difficult. A homologous series of compounds of increasing hydrophobicity has been employed to determine t_{mc} by an iterative calculation (24).

In Bushey and Jorgenson's method, a series of dansylated aliphatic amines was employed, including C_1, C_6, C_8, and C_{12}, in a buffer system containing 25% methanol and 25 mM SDS. The migration time of dodecylamine differed from octylamine by less than a minute, despite a four-carbon chain length difference, so it was assumed that dodecylamine was migrating at a rate close to the micellar velocity. This was tested by plotting log $\tilde{k}'$ versus the carbon number of all solutes except dodecylamine. A $\tilde{k}'$ for dodecylamine was then extrapolated. The calculated migration time, assumed equal to t_{mc}, was used to calculate a new set of $\tilde{k}'$s using Eq. (7.2). This process was repeated until successive iterations showed no substantial differences in t_{mc}. The process proved that dansylated dodecylamine could be used as a t_{mc} marker with an error of only 0.04%.

C. Increasing the Elution Range

The peak capacity of MECC is directly proportional to $\ln(t_{mc}/t_0)$; therefore, increasing the ratio t_{mc}/t_0 will increase the number of components that can be resolved in a single run (25). Decreasing the EOF with a treated capillary is one means of improving this ratio (25, 26). A C_8 treated capillary is commercially available for this purpose.[3] Hydrophobic sites on the capillary are effectively blocked by SDS, so wall binding should not be a problem here. When using SDS, the net surface charge on the capillary remains anionic, though the charge density appears lower. It is still not clear whether treated capillaries will have advantages in MECC. Adjustment of the elution range can be obtained with organic modifiers and other additives.

D. Decreasing the Elution Range

Brij 35 is a non-ionic surfactant that can form mixed micelles with SDS (27). Increasing the concentration of Brij-35 can decrease the net micellar charge, thereby decreasing t_{mc}. The net effect is to speed up the separation. Changes in resolution can also occur. For example, the benzene/benzaldehyde solute pair is not resolved in an SDS system without the addition of the non-ionic surfactant. This technique can be useful in speeding the separation without increasing the field strength provided the resolution is adequate.

E. Use of Organic Modifiers

Organic modifiers can also be used to modify the elution range (28, 29). However, it is far more useful to consider the use of the modifier in a similar fashion to LC: as a means of adjusting the solute's partition coefficient between the chromatographic phases. The selection of the modifier can increase both t_0 and t_{mc}. The use of methanol or other linear alcohols reduces the EOF, while acetonitrile has a much lower impact.

The full impact of the use of the organic modifier is illustrated in Figs. 7.10a and 7.10b. These separations are for a series of impurities found in heroin seizure samples (30). Many of these impurities are very hydrophobic and elute near t_{mc}. The addition of 15% acetonitrile alters the partition coefficients and dramatically lowers $\tilde{k}'$ for many of these components.

Many organic modifiers are useful in MECC. These include methanol, propanol, acetonitrile, tetrahydrofuran, dimethylformamide, and others. The percent modifier that can be added is limited by the impact of the solvent on the micellar aggregate. Features such as the CMC, aggregation number, and micellar ionization (rate of exchange of surfactant between micelle and bulk solution) are affected by the percent organic modifier. Generally, the use of less than 25% organic modifier does not totally disrupt the micellar aggregate. Higher amounts of modifier may cause sufficient micellar disorder that the separation mechanism shifts over to CZE. This would be indicated on the electropherogram by a loss of selectivity for neutral solutes that are not separable by CZE.

[3] Supelco, Bellefonte, PA (CElect-H150).

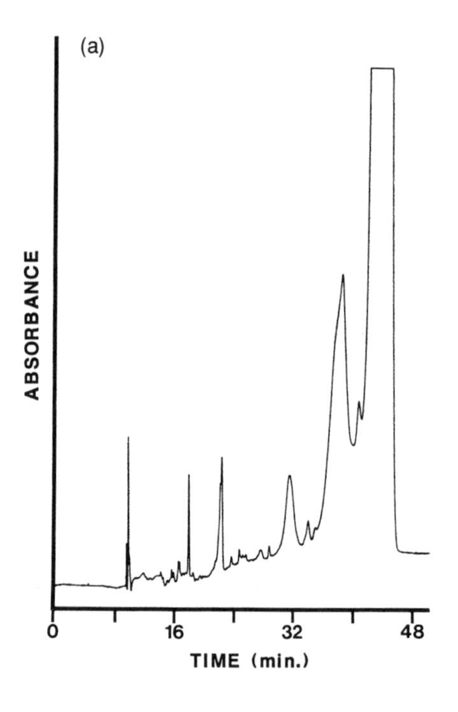

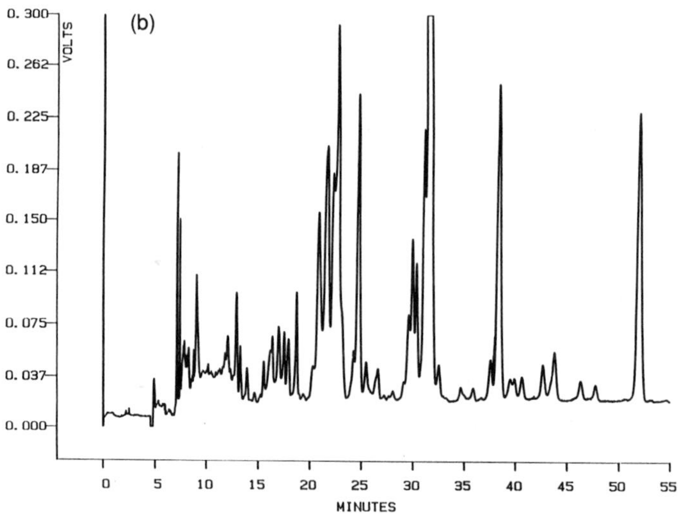

Figure 7.10. Impact of the organic modifier on the MECC separation of heroin impurities. Capillary: (a) 100 cm, (b) 50 cm × 50 μm i.d.; buffer: (a) 100 mM SDS, 10 mM phosphate–borate, pH 8.5; (b) 85 mM SDS, 8.5 mM phosphate–borate, pH 8.5, 15% acetonitrile; temperature: 50°C; detection: UV, 210 nm. Reprinted with permission from *Anal. Chem.* **63,** 823, copyright ©1991 Am. Chem. Soc.

In complex separations, the selection of the appropriate organic modifier or modifier combination may not be straightforward. It is well known in LC that binary and ternary combinations may be required to resolve all peaks. This is also probably true for MECC. While solvent optimization is new in MECC, it is probable that overlapping resolution mapping (31) will provide satisfactory results in relatively few experiments. This technique has been applied to electrokinetic separations using cyclodextrins (32).

F. Effect of Urea

Highly concentrated solutions of urea are frequently employed to solubilize proteins, DNA, hydrocarbons, and amino acids. The mechanism of solubilization probably involves a diminished water structure surrounding the hydrophobic solute (23). In MECC, urea widens the elution range, as indicated in Table 7.3. The ln $\tilde{k}'$ values for hydrophobic solutes such as naphthalene, phenanthrene, and fluoranthene show a linear decrease as the concentration of added urea is increased. The current decreases as well because of an increase in the viscosity (1.66 times more viscous for 8 M urea) of the electrolyte, as well as other changes in ionic mobilities. Separations of 23 PTH amino acids and eight corticosteroids were reported using this technique without the need for organic solvent modifiers (23).

7.7 Alternative Surfactant Systems

A. Surfactant Chain Length

A variety of both anionic and cationic surfactants can be employed in MECC. The alkyl chain can be varied to change the hydrophobicity of the formed micelles. Surfactants with alkyl chains of less than eight carbons are not very useful since their CMCs are far too high; however, they can be used as ion-pairing reagents (33, 34). Alkyl chains of greater that 14 carbons pose solubility problems (35). A few non-ionic surfactants have been employed in MECC either as modifiers of the SDS micelle (36) or alone (37). When used as a lone surfactant, the micelles must

Table 7.3. Migration Times of the Aqueous Phase and Micelle at Different Urea Concentrations[a]

Migration Time	Urea concentration (M)				
	0.0	2.0	4.0	6.0	8.0
t_0 (min)	3.92	3.92	4.65	5.46	6.38
t_{mc} (min)	14.57	16.10	22.76	30.11	36.45
t_0/t_{mc}	0.269	0.243	0.204	0.181	0.175

[a]Buffer: 50 mM SDS in 100 mM borate–50 mM phosphate; voltage: 20 kV. Data from Ref. 23.

gather a surface charge from the adsorption of ions from the bulk solution to generate the requisite electrophoretic mobility. Non-ionic surfactants have only limited utility in MECC.

Table 7.4 provides a listing of surfactants that have been employed in MECC.

B. Bile Salts

Bile salts form micelles with an aggregation number of up to 10 (38). The structure and physical properties of several bile salts are given in Fig. 7.11. The molecular structure of these micellar aggregates differs substantially from the long-chain alkyl variety. The hydroxyl moieties all line up in the same plane; thus, the surfactant possesses both hydrophilic and hydrophobic surfaces. These surfactants tend to be useful for chiral recognition (Section 7.10), for separating cationic solutes that bind strongly to SDS, and for resolving hydrophobic solutes that have a migration time equal to t_{mc}. The interior of the bile salt micelle is less hydrophobic compared to SDS (39). Other operating characteristics, such as pH and organic modifier control, are similar to those for SDS, though bile salts are more tolerant of organic modifiers (39). The CMC of sodium cholate does not change appreciably until the methanol content is above 30%. For SDS, changes in the CMC begin at the 10% methanol level. A separation of corticosteroids is shown in Fig. 7.12.

Table 7.4. Surfactants for MECC

Anionic
Sodium dodecyl sulfate (SDS)
Sodium decyl sulfate (STS)
Sodium taurocholate (STC)
Sodium cholate (SC)
Sodium taurodeoxycholate (STDC)
Sodium deoxycholate
Sodium lauroyl methyltaurate (SLMT)
Cationic
Dodecyltrimethylammonium chloride (DTAC) or bromide (DTAB)
Cetyltrimethylammonium chloride (CTAC) or bromide (CTAB)
Tetradecyltrimethylammonium bromide (TTAB)
Hexyltrimethylammonium bromide (HTAB)
Propyltrimethylammonium bromide (PTAB)
Non-Ionic
Polyoxyethylene-23-lauryl ether (Brij-35)
Octyl glucoside
Triton X-100

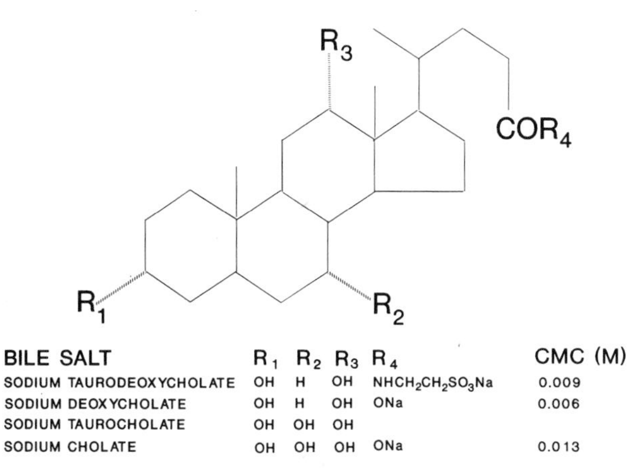

BILE SALT	R_1	R_2	R_3	R_4	CMC (M)
SODIUM TAURODEOXYCHOLATE	OH	H	OH	$NHCH_2CH_2SO_3Na$	0.009
SODIUM DEOXYCHOLATE	OH	H	OH	ONa	0.006
SODIUM TAUROCHOLATE	OH	OH	OH		
SODIUM CHOLATE	OH	OH	OH	ONa	0.013

Figure 7.11. Structure and properties of some bile salt surfactants.

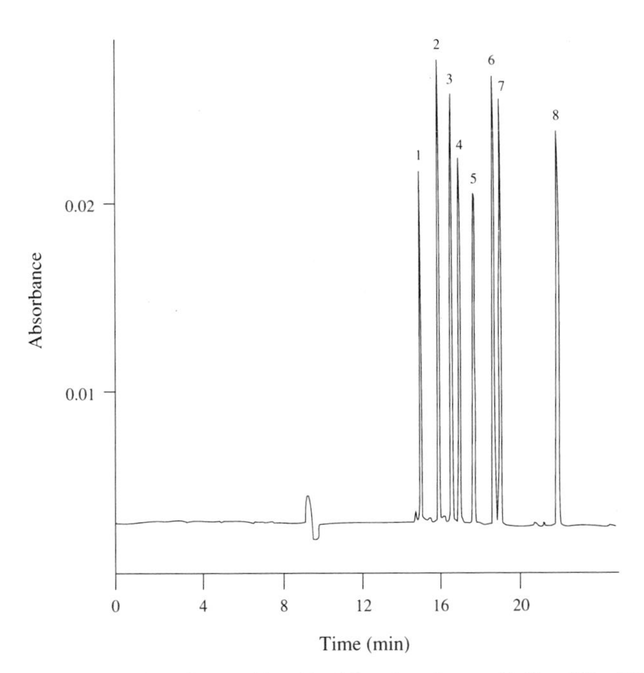

Figure 7.12. Separation of corticosteroids with a bile salt surfactant. Buffer: 100 mM borate, pH 8.45, 100 mM sodium cholate; voltage: 12.5 kV; temperature: 25°C; detection: UV, 254 nm. Key: (1) triamcinalone; (2) hydrocortisone; (3) betamethasone; (4) hydrocortisone acetate; (5) dexamethasone acetate; (6) triamcinalone acetonide; (7) fluocinolone acetionide; (8) fluocinonide. Reprinted with permission from the *Beckman Chromatogram*, August 1990.

C. Cationic Surfactants

Cationic surfactants have the unique capability of reversing the charge of the capillary wall. This reverses the EOF, which requires the power supply polarity to be switched (35, 40). Charge reversal occurs at surfactant concentrations well below the CMC, but without the characteristic selectivity that accompanies MECC.

A high-speed separation of naphthalenedialdehyde (NDA)-derivatized amino acids on a 27 cm × 50 μm i.d. amine-bonded capillary using 200 mM CTAB in 50% acetonitrile is shown in Fig. 7.13 (41). The high acetonitrile content was required to solubilize the hydrophobic NDA–amino acids. The amine coating permits the use of acidic pH without fear of adsorption of cations, because the capillary wall has a positive charge. As the pH of the buffer is lowered, the EOF increases in the direction of the positive electrode. This is opposite to what occurs in untreated fused-silica capillaries. Cationic micelles may be useful to separate acidic (anionic substances) via ion-pairing.

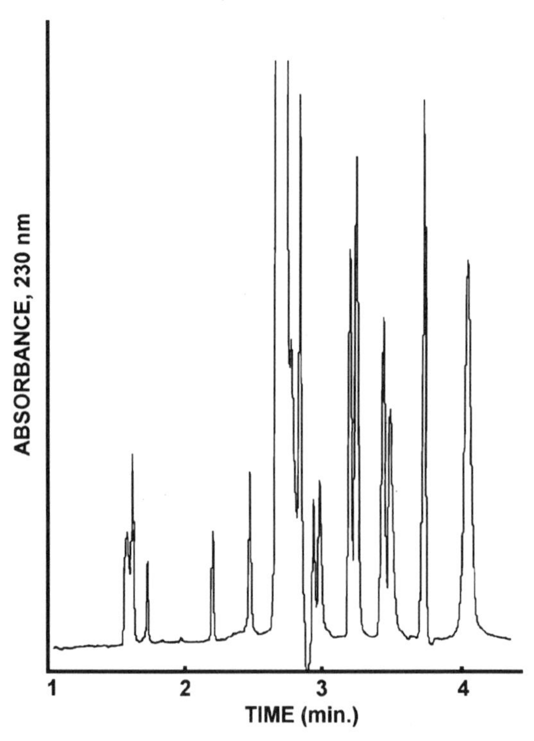

Figure 7.13. MECC of NDA-derivatized amino acids with a cationic surfactant. Capillary: Beckman PC2, 27 cm × 50 μm i.d., a capillary with an amine-treated surface; buffer: 200 mM CTAB in 50% acetonitrile; voltage: 7.5 kV; detection: 230 nm. Courtesy of Beckman Instruments.

7.8 Applications and Methods Development

A summary of applications, beyond those already discussed, is given in Table 7.5. The balance of this section is devoted to the rationale involved in developing a method for the separation of urinary porphyrins (74).

Urinary porphyrins are important precursors in the biosynthetic pathway leading to hemoglobin. Various disease and toxic states interrupt the synthetic cascade, leading to a buildup of porphyrins in urine and other body tissues. These conditions are known as porphyrias. The quantitation of some of the various porphyrins is diagnostic for many of these conditions.

Table 7.5. Applications of Electrokinetic Chromatography

Application	Reference
Aflatoxins	39
Analgesics	42
Anti-inflammatory drugs	17
Aromatic hydrocarbons	5, 39, 43, 44
Aspoxicillin	45
ß-Lactams	33
Barbiturates	46
Bases, nucleosides, oligonucleotides	47
Benzenes, substituted	48, 49
Catechols and catecholamines	50, 52
Cefpiramide	53, 54
Cimetidine	55
Corticosteroids	56, 57
Creatinine	58
Deoxyribonucleosides	59
Deoxynucleosides	60
Dioxins	5
Dns-amino acids	61
Flavonoids	62
Hop bitter acids	63
Illicit drugs	30
Metal chelates	64–67
Methyl-D_3-amine	24
Nucleic acids	68
Nucleosides	69, 70
Organic gunshot and explosive constituents	71
Parabens	22
Peptides	40
Phenols	36, 72
Polychlorinated biphenyls	5

Table 7.5. *continued*

Application	Reference
Phthalates	73
Porphyrins	74
Plant growth regulators (2, 4-D, etc.)	32
PTH-amino acids	75, 76
Sulfonamides	77
Testosterone esters	78
Thiopental	79
Trimetoquinol	80
Uric acid	58
Vitamins	21, 61, 81–83
Xanthines	84

The structure of mesoporphyrin, a synthetic porphyrin not found in nature, is shown in Fig. 7.14. Other species differ in the degree of carboxylation at the perimeter of the porphyrin ring structure. Mesoporphyrin is doubly carboxylated, followed by coproporphyrin (4 COOHs), pentacarboxylporphyrin (5 COOHs), hexacarboxylporphyrin (6 COOHs), heptacarboxylporphyrin (7 COOHs), and uroporphyrin (8 COOHs).

Porphyrins are usually determined by LC via gradient elution. Since the important porphyrins contain between two and eight carboxylic acid substituents, they are good candidates for HPCE as well. A CZE separation is shown in Fig. 7.15a and compared to HPLC (Fig. 7.15b). Isomers of hexacarboxylporphyrin not separated by LC are clearly resolved by CZE. The elution order is reversed for the two techniques. By LC, the more polar carboxylated porphyrins elute first. By CZE, the most highly charged porphyrins migrate toward the positive electrode but are swept toward the negative electrode by the EOF. In this case, the more polar and charged species elute last.

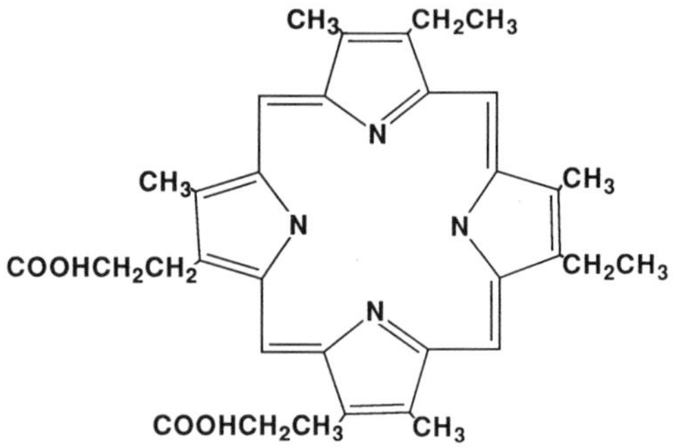

Figure 7.14. Structure of mesoporphyrin.

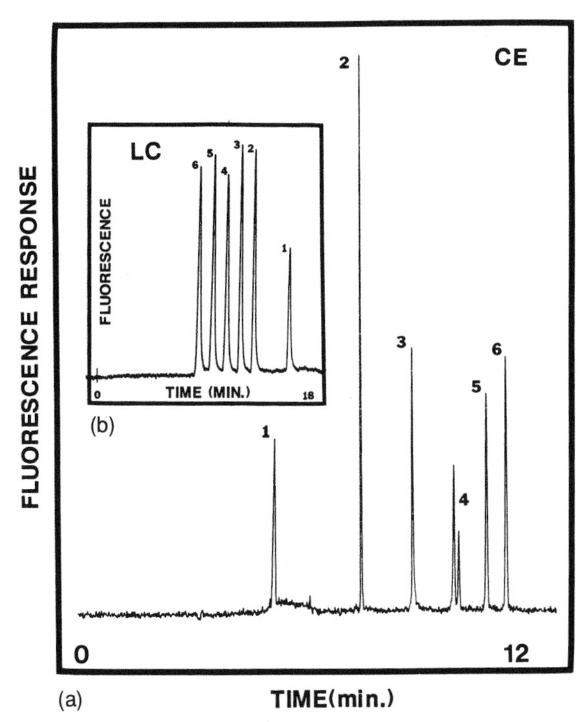

Figure 7.15. (a) CZE of urinary porphyrins. Capillary: 72 cm (50 cm to detector) × 50 μm i.d.; buffer: 20 mM CAPS, pH 11 with 10% methanol; sample porphyrin test mix, 5 nmol/mL in methanol: 20 mM CAPS (50:50); injection: electrokinetic, 12 s at 10 kV; voltage: 30 kV; detection: fluorescence; excitation: 400 nm; emission wavelengths > 595 nm. (b) Gradient elution reverse-phase LC of urinary porphyrins. Key: (1) mesoporphyrin (dicarboxyl); (2) coproporphyrin (tetracarboxyl); (3) pentacarboxyl porphyrin; (4) hexacarboxylporphyrin positional isomers; (5) heptacarboxyl porphyrin; (6) uroporphyrin (octacarboxyl). Reprinted with permission from *J. Chromatogr.* **516,** 271, copyright ©1991 Elsevier Science Publishers.

Although the CZE separation appears adequate, repeated runs show a merging and broadening of peaks, characteristic of solutes binding to the capillary walls. Since the porphyrins are all anionic, mesoporphyrin binds through hydrophobic interaction of the uncharged quadrant of the molecule with the capillary wall.

Because the porphyrins are anionic and hydrophobic, MECC seemed appropriate since SDS is anionic and hydrophobic as well, thereby increasing the likelihood that the active sites on the capillary wall would be saturated. The mechanism of separation would be expected to resemble CZE because the anionic porphyrins should be repelled from the anionic micelle.

In 100 mM SDS, pH 11 (Fig. 7.16a), the elution order is the same as CZE except mesoporphyrin (2-COOH). Mesoporphyrin has its carboxylate groups

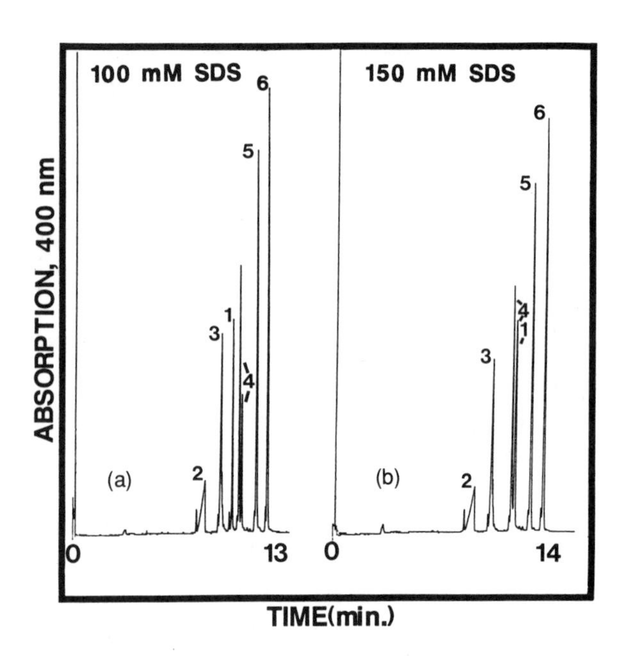

Figure 7.16. MECC of urinary porphyrins: Capillary: 77 cm (55 cm to detector) × 50 μm i.d.; sample: porphyrin test mix, 20 nmol/mL dissolved in 20 mM CAPS, pH 11, 100 mM SDS; (a) run buffer: 20 mM CAPS, pH 11, 100 mM SDS; voltage: 20 kV; temperature 45°C; (b) run buffer: 20 mM CAPS, pH 11, 150 mM SDS; voltage 20 kV; temperature 45°C. Key: see Fig. 7.15. Reprinted with permission from *J. Chromatogr.* **516,** 271, copyright ©1990 Elsevier Science Publishers.

located on one quadrant of the molecule. The other side of the molecule is free to interact hydrophobically with the micelle. At 150 mM SDS (Fig. 7.16b), the mesoporphyrin exhibits a further shift in migration time that is consistent with this argument.

Coproporphyrin (4-COOH), peak 2, exhibits fronting. This is due to a solubility problem —it is poorly soluble in neither the bulk solution nor the micelle. An organic modifier was employed to solve the solubility problem (Fig. 7.17a–c). With 15% methanol, a sharper peak is obtained, but the time of separation is prolonged because of a reduction in the EOF. To speed the separation, both increased temperature and increased voltage were applied. At 30 kV, 45oC, the time of separation is only 13 min and the coproporphyrin peak is now very sharp. Acetonitrile might have been a better choice of modifier, since that solvent does not reduce the EOF.

Next, a real urine sample from a patient suffering from porphyria cutanea tarda, a genetic disorder, was run (Fig. 7.18). Characteristic of this disease is an elevation of peaks 5 and 6. Splitting of these peaks was noted. A photodegraded standard showed the same splitting pattern. The splitting in the sample was due to photodegradation during a 24-hour urine collection.

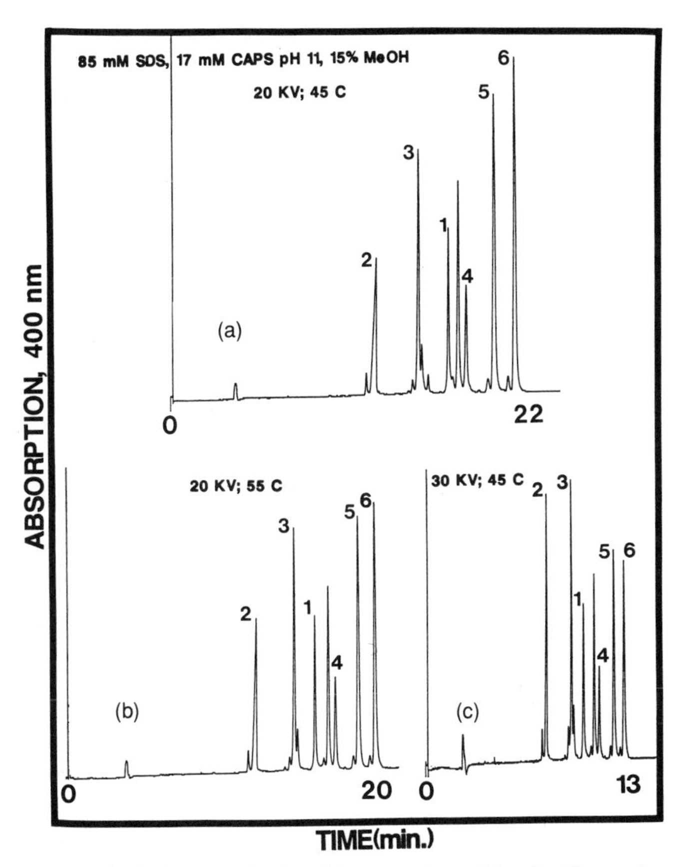

Figure 7.17. MECC of urinary porphyrins with an organic modifier. Capillary and sample: see Fig. 7.16; buffer: 85 mM SDS, 17 mM CAPS, 15% methanol; (a) voltage: 20 kV; temperature: 45°C; (b) voltage: 20 kV; temperature 55°C; (c) voltage: 30 kV; temperature: 45°C. Key: see Fig. 7.15. Reprinted with permission from *J. Chromatogr.* **516,** 271, copyright ©1990 Elsevier Science Publishers.

7.9 Cyclodextrins

Cyclodextrins (CDs) are macrocyclic oligosaccharides that are synthesized by the bacterial enzymatic digestion of starch. The basic structures comprise 6, 7, or 8 glucopyranose units attached by α-1,4 linkages and are referred to as α-, β-, and γ-cyclodextrins (85). In addition, derivatized CDs such as 2-*O*-carboxymethyl-β-CD (48) and heptakis(2,6-di-*O*-methyl)-β-CD (49) have been reported as well. These compounds are shaped like a torus (Fig. 7.19) and have a hydrophobic interior that is optically active. CDs can effectively solubilize water-insoluble solutes by formation of an inclusion complex, provided the size and shape of the compounds conform to the interior dimensions of the torus. The important physicochemical characteristics of CDs are listed in Table 7.6.

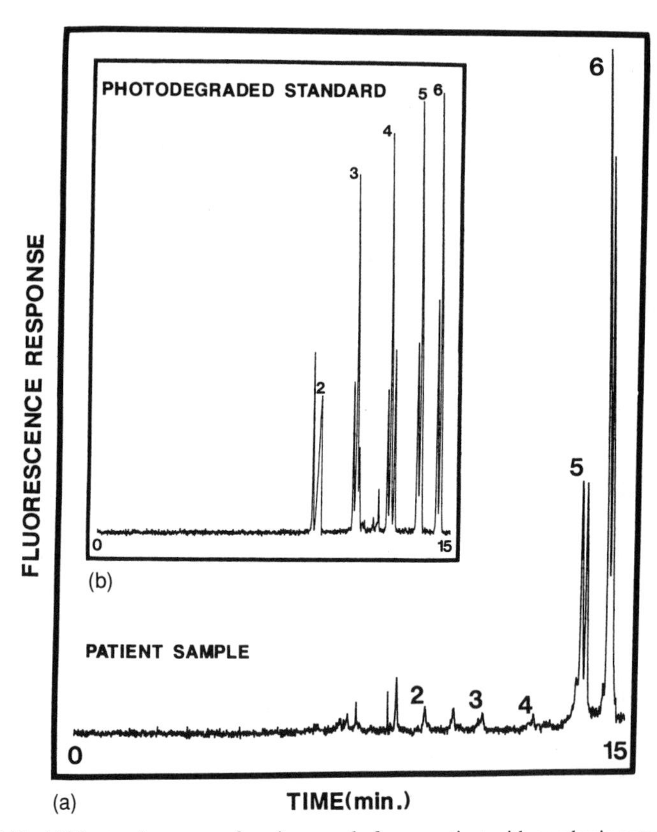

Figure 7.18. (a) Electropherogram of a urine sample from a patient with porphyria cutanea tarda. Sample preparation: centrifugation of urine for 1 min.; injection: 2s vacuum (10 nL); (b) electropherogram of a partially decomposed urine sample. Other conditions given in Fig. 7.16. Key: see Fig. 7.15. Reprinted with permission from *J. Chromatogr.* **516,** 271, copyright ©1990 Elsevier Science Publishers.

Table 7.6. Important Characteristics of Cyclodextrins[a]

Parameter	Type of CD		
	α	β	γ
Molecular weight	972	1,135	1,297
Diameter of cavity (Å)	4.7–6	8	10
Volume of cavity ($Å^3$)	176	346	510
Solubility (g/100 mL, 25°C)	14.5	1.85	23.2
Molecules per unit cell	4	2	6

[a]Data from *Luminescence Applications in Biological, Chemical, and Hydrological Sciences*, ACS Symposium Series 383, p. 169.

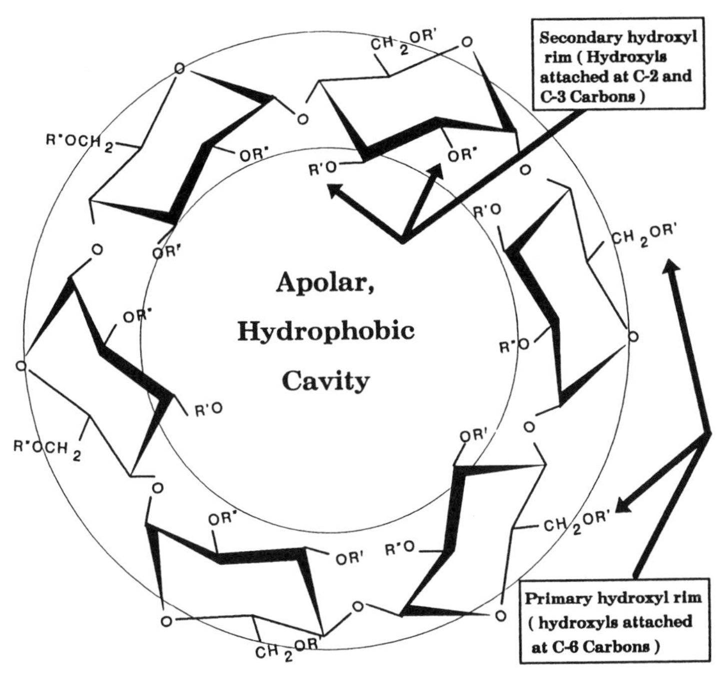

Figure 7.19. Structure of α-cyclodextrin. Reprinted with permission from *J. Liq. Chromatogr.* **15,** 961, copyright ©1992 Marcel Dekker.

Most electrokinetic applications employing CDs are in the field of chiral recognition, which is covered in Section 7.10. There have only been a few nonchiral applications reported for peptides (40), chlorinated benzenes, PCBs and TCDDs (5), corticosteroids (57), and aromatic hydrocarbons (43). Neutral CDs can be used with or without micelles. Without micelles, solutes with normally identical mobilities, e.g., enantiomers, can be separated based on differences in the stability of the formed inclusion complex. When a solute is complexed within a CD, the mobility is substantially affected. Derivatized and charged CDs can form the electrokinetic phase by themselves, thereby generating the potential to separate neutral compounds.

The separation mechanism for CD-MECC is illustrated in Fig. 7.20. Micelles and CDs coexist in aqueous solution with little interaction. Underivatized CDs are neutral and have a hydrophilic outer surface, so there is little driving force for micellar interaction. The CD in this example is simply carried by the EOF toward the negative electrode. SDS micelles electrophorese toward the positive electrode, as usual. Hydrophobic solutes that are normally bound to the micelle can form inclusion complexes with the CDs. The separation mechanism is then based on differences in a solute's partition coefficient between the micelle and the CD.

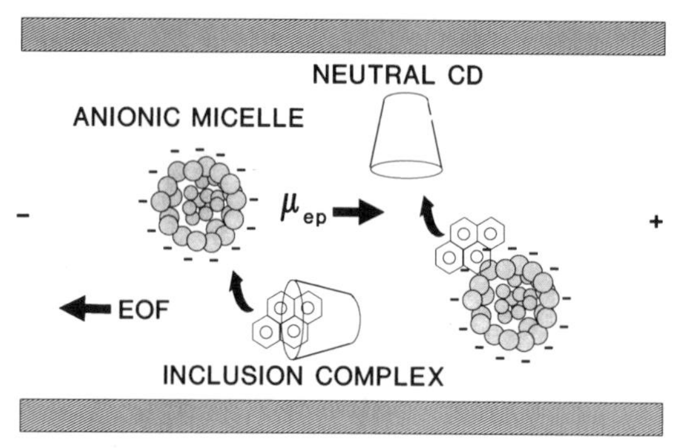

Figure 7.20. Separation mechanism of CD-MECC.

Increasing the CD concentration will decrease $\tilde{k}'$ for compounds that form inclusion complexes (Fig. 7.21). Improvements in the selectivity are obtained when the molecular fit of the solute within the CD cavity is better. Although Fig. 7.21 illustrated a decrease in k' for some corticosteroids with β-CD, there was little change in the selectivity. The use of γ-CD, shown in Fig. 7.22, provides substantial improvements in corticosteroid separations. The larger cavity of the γ-CD better accommodates the bulk of the steroid moiety. Even nonpolar compounds can be well separated by CD-MECC. A separation of polycyclic aromatic hydrocarbon (PAH) priority pollutants is given in Fig. 7.23.

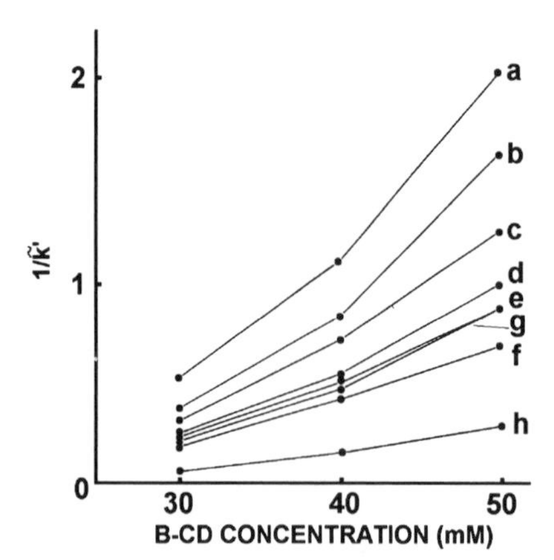

Figure 7.21. Effect of β-CD concentration on $1/k'$ of corticosteroids. Buffer: 50 mM SDS, pH 9.0 borate–phosphate with 4.0 M urea; capillary: 50 cm length to detector × 50 μm i.d.; voltage: 20 kV; temperature: ambient; detection: UV, 220 nm. Key: (a) hydrocortisone; (b) hydrocortisone acetate; (c) betamethasone; (d) cortisone acetate; (e) triamcinolone acetonide; (f) flucinolone acetonide; (g) dexamethasone acetate; (h) flucinonide. Redrawn with permission from *J. Liq. Chromatogr.* **14,** 973, copyright ©1991 Marcel Dekker.

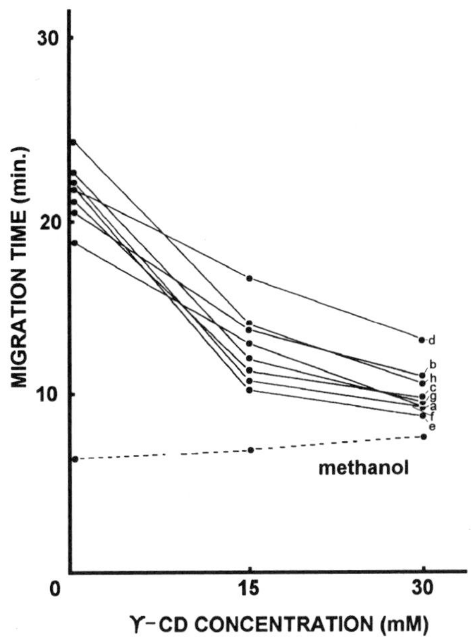

Figure 7.22. Effect of γ-CD concentration on the migration times of eight corticosteroids. The dashed line indicates the migration of methanol, an unretained EOF marker. Other conditions as in Fig. 7.12. Redrawn with permission from *J. Liq. Chromatogr.* **14,** 973, copyright ©1991 Marcel Dekker.

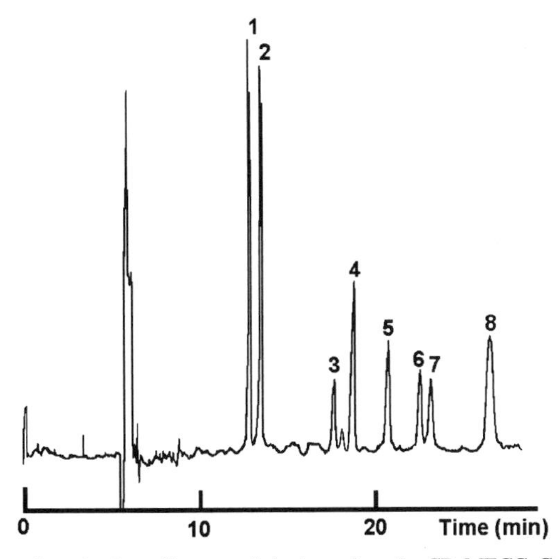

Figure 7.23. Separation of polycyclic aromatic hydrocarbons by CD-MECC. Capillary: 70 cm (51 cm to detector); buffer: 30 mM γ-CD, 100 mM SDS, 5M urea on 100 mM borate buffer, pH 9.0; voltage: 20 kV; current: 41 μA; detection: UV, 210 nm. Key: (1) naphthalene; (2) acenaphthalene; (3) anthracene; (4) fluorene; (5) phenanthrene; (6) chrysene; (7) pyrene; (8) fluoranthene. Redrawn with permission from *J. Chromatogr.* **516,** 23, copyright ©1990 Elsevier Science Publishers.

7.10 Chiral Recognition

A. Basic Concepts

Chiral recognition of racemic mixtures continues to be an active area of research in gas chromatography, liquid chromatography, and, of late, capillary electrophoresis.

Whatever the separation technique employed, chiral recognition is obtained in one of three ways:

1. formation of diastereomers by additives to the mobile phase or carrier electrolyte;
2. formation of diastereomers through interaction with a stationary phase or heterogeneous carrier electrolyte; or
3. precolumn (capillary) derivatization with an optically pure derivatizing reagent.

In the first two cases, diastereomer formation is transient, occurring by electrostatic and/or hydrophobic mechanisms. In the third case, covalently bound and chromatographically separable diastereomers are resolved by MECC. Chiral recognition in CE has been reported using cyclodextrins, bile salts, mixed micellar systems with chiral surfactants, crown ethers, trimolecular peptide–Cu(II)–amino acid complexes, and MECC resolution of preformed diastereomers. An applications summary is given in Table 7.7.

The advantages of HPCE over LC include speed, resolution, and lower operating costs. Chiral columns for LC are relatively expensive.

B. Metal-Ion Complexes

The first examples of chiral recognition in HPCE employed the addition of Cu(II) and *L*-histidine to the buffer solution to resolve dansyl amino acids via a trimolecular complex (95). A Cu(II)–aspartame complex was later shown superior (96). Addition of a surfactant can improve the hydrophobic aspects of the separation, permitting the simultaneous resolution of 14 out of 18 dansyl amino acids (96). The use of 2.5 mM $CuSO_4 \cdot 5H_2O$, 5.0 mM aspartame, and 10 mM ammonium acetate, pH 7.2, gives the best separation. Addition of 20 mM STS provides hydrophobic selectivity via the MECC mechanism. The capillary had to be rigorously conditioned with 100 mM phosphoric acid for several hours to remove any metal hydroxides that may have been precipitated at the capillary wall. After 10-min rinse with KOH, the capillary was conditioned in acetate buffer for 10 hours.

C. Chiral Mixed Micelles

Separations have been reported employing a mixture of SDS and optically active surfactants such as *N*, *N*-dodecyl-*L*-alanine (in the presence of Cu(II)) (107) or sodium *N*-dodecanoyl-*L*-valinate (SDVal)(87, 88). Separations reported to date have been limited to derivatized amino acids.

Table 7.7. Applications of Chiral Recognition

Application	Buffer Additive or Derivative	Reference
Adrenalin	Di-*O*-methyl-β-CD	86
Amino acids	Sodium dodecanoyl valine +SDS	87, 88
	β-CD in a gel	89
	GITC derivatives by MECC	90
	Marfey's reagent by MECC	91
	PTH AA/sodium dodecanoyl valine +SDS	92, 93
	PTH AA/digitonin + SDS	94
	Dansyl AA/bile salts	38
	Dansyl AA/Cu(II)–His	95
	Dansyl AA/copper (II)–Asp	96
Benzetimide	Hydroxypropyl-β-CD	97
Benzoin	Sodium dodecanoyl valine +SDS	98
Binaphthyls	Bile salts	99, 100
Cephalosporins	1% Glucidex, pH 7.5	101
Chloramphenicol	Di-*O*-methyl-β-CD	102
Dopa	[18]-Crown-6-tetracarboxylic acid	103
Diltiazem	Bile salts	104
Epinephrine	Di-*O*-methyl-β-CD	86
Ergots	γ-CD	105
Flurbiprofen	10% mylose	101
Ibuprofen	10% mylose	101
Ketotifen	β-CD	102
Naproxen	10% mylose	101
Propranolol	β-CD	106
Quinagolide	β-CD	103
Terbutaline	β-CD	106
Thioridazine	γ-CD	102
Trimetoquinol	Bile salts	104
Tryptophan	γ-CD	86
Warfarin	Sodium dodecanoyl valine +SDS	98
	2.5% Glucidex	101

Mixtures of SDS with the non-ionic optically active surfactant digitonin (94) have been reported to separate PTH amino acids. A low buffer pH was selected so electrophoresis was much greater than electroendosmosis. The separations times were lengthy and did not adequately resolve the early-eluting solutes.

D. Cyclodextrins

Both native and derivatized cyclodextrins have been successfully employed for chiral recognition of aromatic species. The mechanism of chiral resolution is based on differences in the complex stability between each enantiomer and the CD. Both hydrophobic inclusion within the CD and hydrogen bonding between the analyte and CD hydroxyl groups are in part responsible for resolution. The molecular fit of the solute within the CD cavity is critical if chiral recognition is to be obtained. Neutral CDs can be employed to separate charged species. For basic compounds, a low background electrolyte is required to ensure the development of a positive charge. Since most drug compounds contain an amine functionality, low-pH buffers are most common.

The relatively long history in LC and GC provides a framework for determining when applications will be successful. Armstrong's empirical procedure employs molecular structure to predict enantioselectivity in LC using a functionalized β-CD (108). In this model, sp^2-hybridized carbons connected to the stereogenic center were found to provide enhanced resolution. Conversely, sp^3-hybridized carbons showed diminished stereoselectivity. Amido groups improved selectivity, especially when associated with π-acid (3,5-dinitrobenzoyl) or π-basic (naphthyl) groups. These studies employed 121 compounds to develop the model. It is unknown whether this approach will transfer to CE, considering the dependence on the chromatographic mobile phase. To date, it is the only treatment that provides a model for chiral recognition for β-CD.

Guttman *et al.* (109), recognizing the improvements in resolution at low EOF, immobilized β-CD in a gel. Overcoming the disadvantages of the gel format, Fanali (106) studied the impact of several variables on the chiral selectivity for propranolol and terbutaline using a Bio-Rad polyacrylamide-coated capillary. Heptakis (2,6-di-*O*-methyl)-β-CD was superior to native β-CD, as illustrated in Fig. 7.24. The extra methyl groups of the CD derivative influence the hydrophobicity of the molecule. A separation employing this derivatized CD is shown in Fig. 7.25. α- and γ-CD did not yield any chiral selectivity. In aqueous solution, only heptakis (2,6-di-*O*-methyl)-β-CD resolved the propranolol enantiomers. In

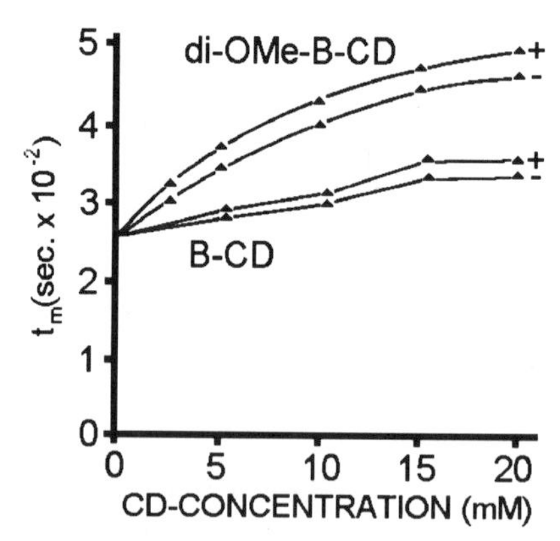

Figure 7.24. Effect of the concentration of CDs on the migration time of (+) and (–) terbutaline. Capillary: polyacrylamide-coated 20 cm × 25 μm i.d.; buffer: 100 mM phosphate, pH 2.5; constant current: 19 μA; voltage: 9 kV; injection: electrokinetic, 8 kV, 10 s of a 10^{-5} M solution. Redrawn with permission from *J. Chromatogr.* **545,** 437, copyright ©1991 Elsevier Science Publishers.

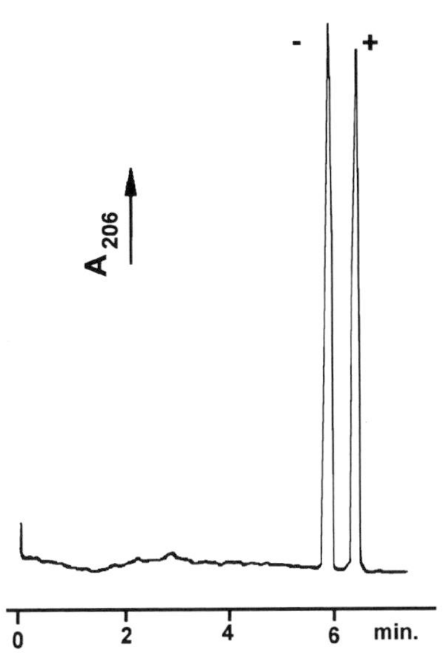

Figure 7.25. Separation of (+) and (–) terbutaline with 5 mM heptakis (2,6-di-*O*-methyl)-β-CD. Conditions as in Fig. 7.24. Reprinted with permission from *J. Chromatogr.* **545,** 437, copyright ©1991 Elsevier Science Publishers.

30% methanol, 50 mM phosphate, pH 2.5, 4 M urea, propranolol enantiomers were resolved with 40 mM β-CD.

Snopek *et al.* (102) used methylhydroxypropylcellulose 15,000 (MHPC) to reduce the EOF and improve the enantiomeric resolution for a variety of drug substances. A separation of chloramphenicol with and without the EOF modifier is shown in Fig. 7.26.

Kuhn *et al.* (103) found both buffer concentration and capillary temperature critical for the separation of quinagolide. These data are shown in Fig. 7.27. The higher buffer concentrations, up to the point where Joule heating plays a role, increase the hydrophobic interaction between the solute and CD through a salting mechanism. A concentration of 50 mM was optimal under the conditions studied. Since Joule heating was a limiting factor, decreasing the capillary diameter and/or the field strength might further improve the resolution by permitting the use of higher-concentration buffers. The buffer type also played a role in chiral resolution. Glycine HCl, pH 2.5, 50 mM gave poorer resolution than phosphate, pH 2.5, and no resolution was found in 50 mM citrate, pH 2.5. They also found a linear dependence of the retention time on the concentration of β-CD in the concentration range of 10–30 mM. No separation occurred below 10 mM CD concentration. These data fit the equation of Gutman *et al.* (109) for chiral recognition:

$$t_r = \frac{L_d}{\mu_f E} + \frac{L_d K}{\mu_f E}[C], \tag{7.5}$$

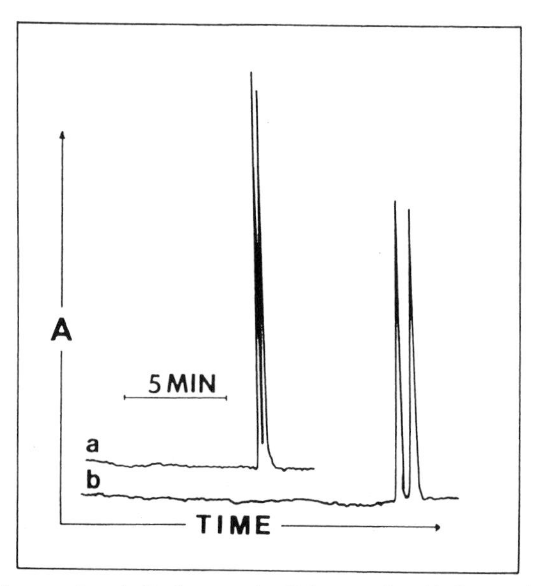

Figure 7.26. Influence of methylhydroxyproplycellulose on the chiral recognition of chloramphenicol enantiomers (a) without MHPC and (b) with 0.1% MHPC. Buffer: 20 mM Tris adjusted to pH 3.5 with citric acid with 10 mM heptakis (2,6-di-*O*-methyl)-β-CD; capillary: 65 cm (45 cm to detector); voltage: 18 kV; current: 6 μA; detection: UV, 254 nm. Reprinted with permission from *J. Chromatogr.* **559,** 215, copyright ©1991 Elsevier Science Publishers.

where t_R is the retention time, L_d is the effective length of the capillary, E the electric field strength, μ_f is the mobility of free solute, K is the complex formation constant, and [C] is CD concentration. The model assumes that the mobility of the complexed solute is much less than that in free solution. The L-isomer usually moves faster than the D-isomer, a reflection of D's greater stability constant.

E. Crown Ethers

Crown ethers, represented in Fig. 7.28 by 18-crown-6, are another class of complexing reagents that have been employed in LC for chiral recognition. Kuhn *et al.* (103) used [18]-crown-6 tetracarboxylic acid to resolve the D,L isomers of phenylalanine, tyrosine, tryptophan, and dopa. The concentration dependence of 18-crown-6 on the resolution factor is shown in Fig. 7.29. A separation of D,L tyrosine and dopa is illustrated in Fig. 7.30.

F. Bile Salts

Bile salts have already been shown to be useful (Section 7.7) for the determination of hydrophobic solutes. This usefulness can be extended to the separation of optical isomers; bile salts are naturally optically active. Unlike CDs, bile salts

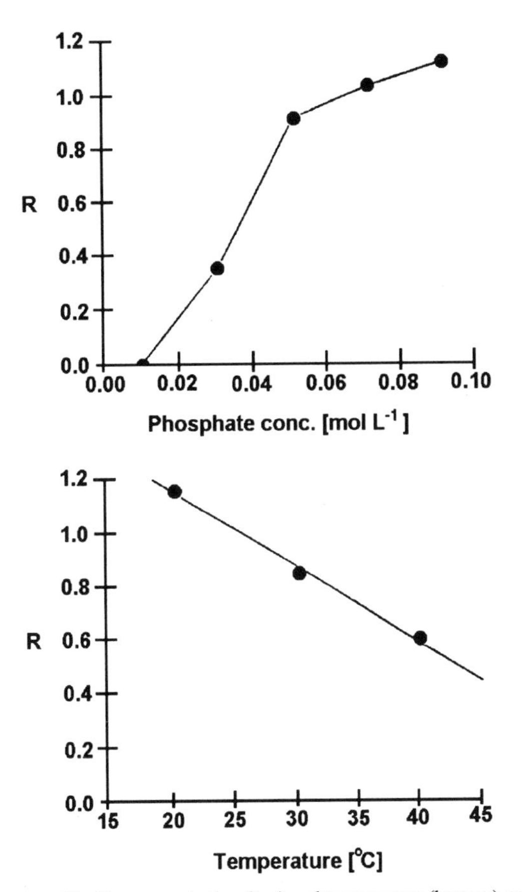

Figure 7.27. Influence of buffer concentration (top) and temperature (bottom) on the resolution of the enantiomers of quinagolide. Capillary: 50 cm × 75 μm i.d.; voltage: 15 kV; buffer: phosphate (top, variable; bottom, 50 mM), pH 2.5, 30 mM β-CD; detection: UV, 214 nm. Reprinted with permission from *Chromatographia* **33,** 32, copyright ©1992 Friedr. Vieweg and Sohn.

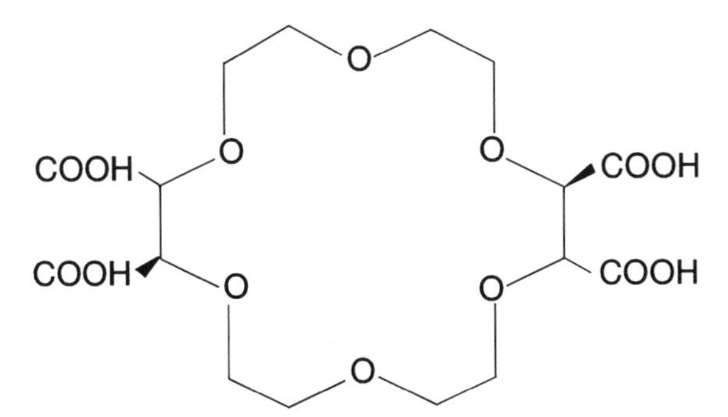

Figure 7.28. Structure of 18-crown-6 tetracarboxylic acid.

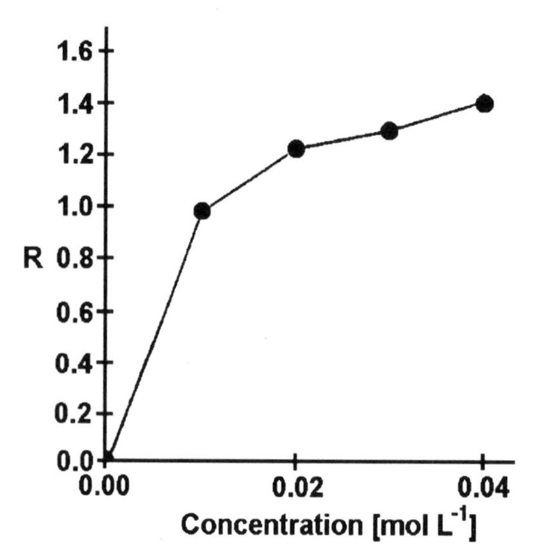

Figure 7.29. Impact of the concentration of 18-crown-6 on the chiral resolution of D,L-dopa. Capillary: 50 cm × 75 μm i.d.; buffer: 18-crown-6 dissolved in water, pH 2.2. Reprinted with permission from *Chromatographia* **33,** 32, copyright ©1992 Friedr. Vieweg and Sohn.

are best utilized under conditions of a strong EOF; this is a consequence of the separation chemistry following the MECC mechanism. Operation at low pH generally results in lengthy separation times.

Bile salts tend to provide enantioselectivity when the structure of the solute is rigid compared to that of the surfactant (100). The selection of the appropriate bile salt has substantial impact on the chiral recognition. There is no way of predicting

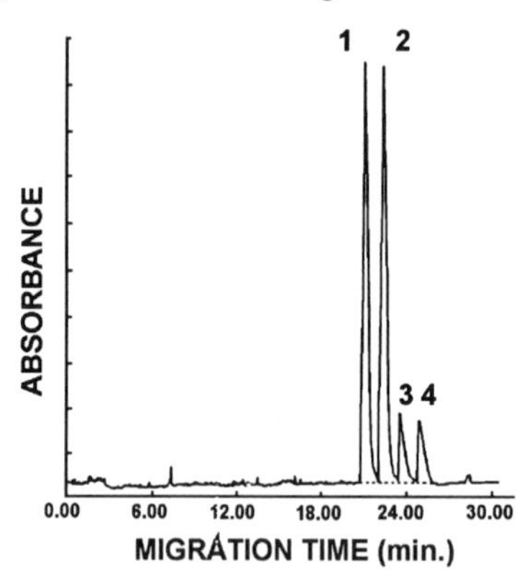

Figure 7.30 Separation of D,L-Tyr and D,L-dopa with 18-crown-6. Capillary: 50 cm × 75 μm i.d.; buffer: 18-crown-6 dissolved in water, pH 2.2; detection: UV, 254 nm; key: (1) L-try, (2) D-try, (3) L-dopa, (4) D-dopa. Reprinted with permission from *Chromatographia* **33,** 32, copyright ©1992 Friedr. Vieweg and Sohn.

which salt will yield the best results at this time. High buffer pHs, where solutes become anionic, reduce chiral recognition, presumably because of repulsion from the anionic micelle (100). Neutral compounds are not affected by pH changes. The use of organic solvents to reduce $\tilde{k}'$ and improve α has been reported (99). A typical separation is shown in Fig. 7.31.

G. Precapillary Derivatization

Separations have been reported by MECC for amino acids derivatized with Marfey's reagent (1-fluoro-2,4-dinitrophenyl-5-*L*-alanine amide)(91) and GITC (2,3,4,6-tetra-*O*-acetyl-β-*D*-glucopyranosyl isothiocyanate)(90).

A variety of reagents have been employed to create synthetic diastereomers that can be separated by LC (110). It is likely that many of these will be applicable to separation by MECC. When a reagent is available in both optically active forms, it is possible to control the order of elution of the enantiomers. Ideally, the

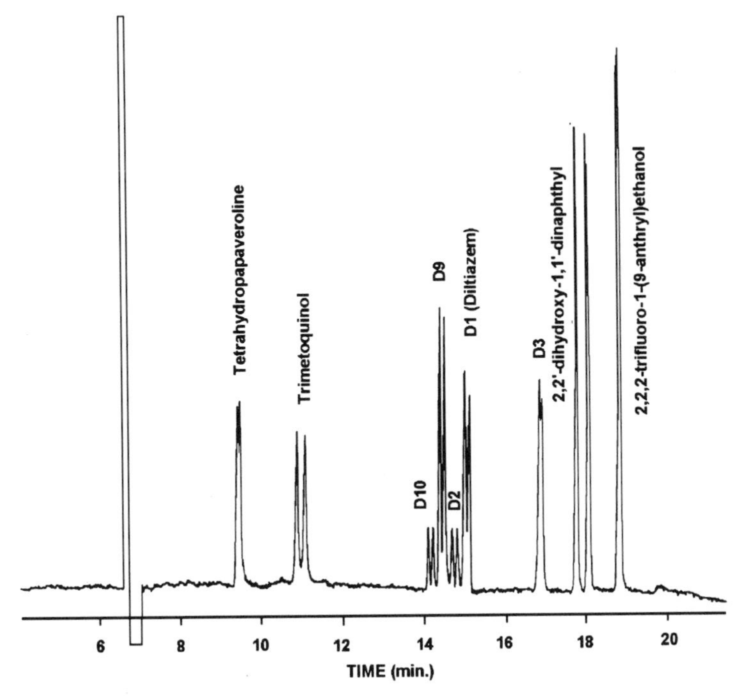

Figure 7.31. Chiral separation of trimetoquinol HCl, tetrahydropapaveroline, five diltiazem-related compounds, 2,2′-dihydroxy-1,1′-dinaphthyl, and 2,2,2-trifluoro-1-(9-anthryl)ethanol. Capillary: 65 cm (50 cm to detector) × 50 μm i.d.; buffer: 50 mM STDC in 20 mM phosphate–borate, pH 7.0; voltage: 20 kV; detection: UV, 210 nm. Reprinted with permission from *J. Chromatogr.* **515,** 233, copyright ©1990 Elsevier Science Publishers.

enantiomer present in the lower concentration should elute first. It is less likely to be lost on the tail of the more concentrated solute.

The ideal reagent will have the following characteristics:

1. Rapid reaction;
2. stable products;
3. no racemization;
4. excess reagent invisible to detector or easily removable;
5. reagent contains or produces a strong chromophore or fluorophore;
6. commercially available in either form of purified enantiomer; and
7. inexpensive.

The advantages of pre-capillary derivatization are:

1. that it provides a highly predictable mechanism for chiral recognition, provided the chiral center is in the proximity of the reaction site; and
2. that it simultaneously produces a good chromophore and chiral selectivity.

The disadvantages of pre-capillary derivatization are that

1. it complicates assay validation;
2. incomplete reactions are possible;
3. excess reagent can complicate separations; and
4. there are extra sample-handling steps.

Another reagent that has been employed in HPCE is FLEC, 1-(9-fluorenyl) ethyl chloroformate (111). This reagent reacts rapidly with primary and secondary amines to form a fluorescent derivative that absorbs at 260 nm. This reagent has been used to separate five of eight enantiomers from a compound with three chiral centers (112). The reagent meets all of the criteria listed under characteristics of an ideal reagent. Its only shortcoming is the need to perform a solvent extraction to remove the excess reagent. This can result in recovery losses for very hydrophobic solutes.

Many analytical chemists are averse to using derivatization procedures to improve separation and detection. This is unfortunate, because the enhanced results often justify the extra work in sample preparation and method validation.

7.11 Optimization

The remarkable ability of EKC to provide a few hundred thousand theoretical plates permits complex separations to be performed, frequently without the need to employ chemometrics-based optimization schemes. Following a scheme such as that illustrated in Fig. 7.32 will usually lead to an adequate separation, often during the course of only a day or so of experimentation.

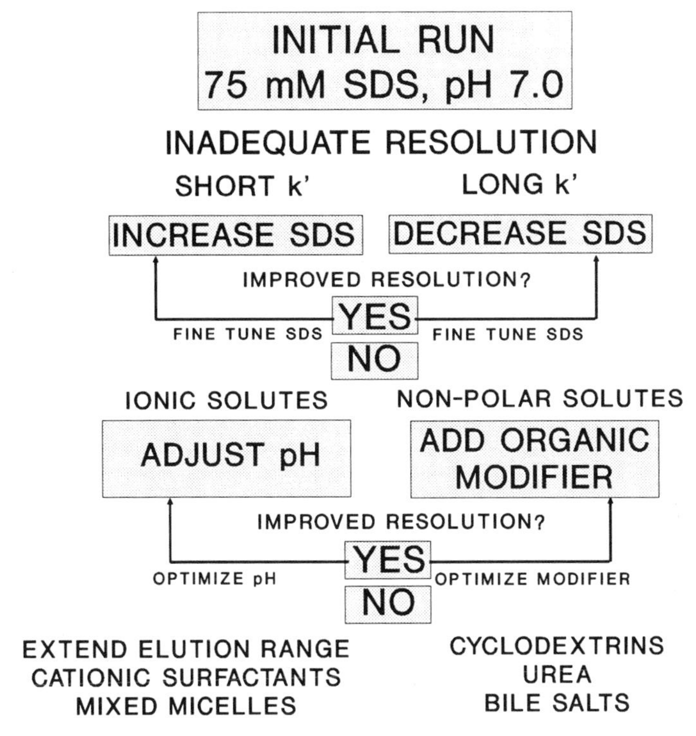

Figure 7.32. An empirical scheme for methods development.

Switching to surfactants other than SDS is normally beneficial for the separation of nonpolar substances and chiral recognition. Even in the former case, the use of cyclodextrins permits the separation of nonpolar species such as aromatic hydrocarbons. Usually, alternative surfactants should be considered only after other experiments covering pH, modifiers, etc. have been performed. Most problems will be solvable using SDS alone, or with additives.

In other instances, statistical tools can be valuable for speeding methods development. This can be particularly true when ternary blends of solvents or cyclodextins are needed to optimize the separation. For example, overlapping resolution maps (ORMs) have been used for years to optimize HPLC separations (31). Recently, Yeo *et al.* (32) applied this technique to optimize the separation of plant growth regulators using mixtures of α-, β-, and γ-CD.

Several groups (113–115) have derived complex expressions to mirror the complex mechanism of EKC, particularly for charged solutes. A pure chemometric approach using Plackett–Burman statistics has also been shown useful (116). Other approaches using simplex optimization and factor analysis may also be fruitful for developing complex separations.

References

1. B. L. Karger, L. R. Snyder, and C. S. Horvath, in *An Introduction to Separation Science* (John Wiley & Sons, 1972), p. 31.
2. S. Terabe, K. Otsuka, K. Ichikawa, A. Tsuchiya, and T. Ando, *Anal. Chem.* **56,** 111 (1984).
3. S. Terabe, K. Otsuka, and T. Ando, *Anal. Chem.* **57,** 834 (1985).
4. H. Watari, *Chem. Lett.*, 391 (1991).
5. S. Terabe, Y. Miyashita, O. Shibata, E. R. Barnhart, L. R. Alexander, D. G. Patterson, B. L. Karger, K. Hosoya, and N. Tanaka, *J. Chromatogr.* **516,** 23 (1990).
6. S. Terabe and T. Isemura, *Anal. Chem.* **62,** 650 (1990).
7. S. Terabe and T. Isemura, *J. Chromatogr.* **515,** 667 (1990).
8. S. Terabe, paper presented at HPCE'92, Amsterdam, Feb. 1992.
9. J. Vindevogel and P. Sandra, *Introduction to Micellar Electrokinetic Chromatography* (Hüthig, 1992).
10. L. R. Fisher and D. G. Oakenfull, *Chem. Soc. Rev.* **6,** 25 (1977).
11. M. Gratzel and J. K. Thomas, in *Modern Fluorescence Spectroscopy* (E. L. Wehry, ed.), Vol. 2, Chap. 4 (Plenum Press, 1976).
12. M. Almegren, F. Grieser, and J. E. Thomas, *J. Am. Chem. Soc.* **12,** 111 (1979).
13. J. B. Peri, *J. Coll. Interface Sci.* **29,** 6 (1969).
14. G. L. McIntire, *Crit. Rev. in Anal. Chem.* **21,** 257 (1990).
15. D. W. Armstrong and F. Nome, *Anal. Chem.* **53,** 1662 (1981).
16. J. W. Jorgenson and K. D. Lukacs, *Anal. Chem.* **53,** 1298 (1981).
17. R. Weinberger and M. Albin, *J. Liq. Chromatogr.* **14,** 953 (1991).
18. S. Terabe, K. Otsuka, and T. Ando, *Anal. Chem.* **61,** 251 (1989).
19. M. J. Sepaniak and R. O. Cole, *Anal. Chem.* **59,** 472 (1987).
20. J. M. Davis, *Anal. Chem.* **61,** 2455 (1989).
21. N. Nishi, N. Tsumagari, T. Kakimoto, and S. Terabe, *J. Chromatogr.* **465,** 331 (1989).
22. H. T. Rasmussen and H. M. McNair, *J. High Res. Chromatogr.* **12,** 635 (1989).
23. S. Terabe, Y. Ishihama, H. Nishi, T. Fukuyama, and K. Otsuka, *J. Chromatogr.* **545,** 359 (1991).
24. M. M. Bushey and J. W. Jorgenson, *Anal. Chem.* **61,** 491 (1989).
25. A. T. Balchunas and M. J. Sepaniak, *Anal. Chem.* **59,** 1466 (1987).
26. J. A. Lux, H. Yin, and G. Schomburg, *HRC & CC* **13,** 145 (1990).
27. H. T. Rasmussen, L. K. Goebel, and H. M. McNair, *J. High Res. Chromatogr.* **14,** 25 (1991)
28. M. J. Sepaniak, D. F. Swaile, A. C. Powell, and R. O. Cole, *HRC & CC* **13,** 679 (1990).
29. J. Gorse, A. T. Balchunas, D. F. Swaile, and M. J. Sepaniak, *J. High Res. Chromatogr.* **11,** 554 (1988).
30. R. Weinberger and I. S. Lurie, *Anal. Chem.* **63,** 823 (1991).
31. J. L. Glajch, J. J. Kirkland, and L. R. Snyder, *J. Chromatogr.* **238,** 269 (1982).
32. S. K. Yeo, C. P. Ong, and S. F. Y. Li, *Anal. Chem.* **63,** 2222 (1991).
33. H. Nishi, N. Tsumagari, T. Kakumoto, and S. Terabe, *J. Chromatogr.* **465,** 331 (1989).
34. H . Nishi, N. Tsumagari, T. Kakumoto, and S. Terabe, *J. Chromatogr.* **477,** 259 (1989).
35. D. E. Burton, M. J. Sepaniak, and M. P. Maskarinec, *J. Chromatogr. Sci.* **25,** 514 (1987).
36. H. T. Rasmussen, L. K. Goebel, and H. M. McNair, *J. Chromatogr.* **517,** 549 (1990).
37. S. A. Swedberg, *J. Chromatogr.* **503,** 449 (1990).

38. S. Terabe, M. Shibata, and Y. Miyashita, *J. Chromatogr.* **480,** 403 (1989).
39. R. O. Cole, M. J. Sepaniak, W. L. Hinze, J. Gorse, and K. Oldiges, *J. Chromatogr.* **557,** 113 (1991).
40. J. Liu, K. A. Cobb, and M. Novotny, *J. Chromatogr.* **519,** 189 (1990).
41. C. H. Shieh, T. Salinas, N. Cooke, S. Terabe, and R. Kerr, paper presented at HPCE'92, Amsterdam (1992).
42. S. Fujiwara and S. Honda, *Anal. Chem.* **59,** 2773 (1987).
43. Y. F. Yik, C. P. Ong, S. B. Khoo, H. K. Lee, and S. F. Y. Li, *J. Chromatogr.* **589,** 333 (1992).
44. Y. F. Yik, C. P. Ong, S. B. Khoo, H. K. Lee, and S. F. Li, *Environ. Monit. Assess.* **19,** 73 (1991).
45. H. Nishi, T. Fukuyama, and M. Matsuo, *J. Chromatogr.* **515,** 245 (1990).
46. W. Thormann, P. Meier, C. Marcolli, and F. Binder, *J. Chromatogr.* **545,** 445 (1991).
47. A. S. Cohen, S. Terabe, J. A. Smith, and B. L. Karger, *Anal. Chem.* **59,** 1021 (1987).
48. S. Terabe, H. Ozaki, K. Otsuka, and T. Ando, *J. Chromatogr.* **332,** 211 (1985).
49. J. Snopek, H. Soini, M. Novotny, E. Smolkova-Kevlemansova, and I. Jellinek, *J. Chromatogr.* **559,** 215 (1991).
50. C. P. Ong, S. F. Pang, S. P. Low, H. K. Lee, and S. F. Y. Li, *J. Chromatogr.* **559,** 529 (1991).
51. R. A. Wallingford and A. G. Ewing, *J. Chromatogr.* **441,** 299 (1988).
52. R. A. Wallingford, P. D. Curry, and A. G. Ewing, *J. Microcol. Sep.* **1,** 23 (1989).
53. T. Nakagawa, Y. Oda, A. Shibukawa, and H. Tanaka, *Chem. Pharm. Bull.* **36,** 1622 (1988).
54. T. Nakagawa, Y. Oda, A. Shibukawa, H. Fukuda, and H. Tanaka, *Chem. Pharm. Bull.* **37,** 707 (1989).
55. H. Soini, T. Tsuda, and M. V. Novotny, *J. Chromatogr.* **559,** 547 (1991).
56. H. Nishi, T. Fukuyama, M. Matsue, and S. Terabe, *J. Chromatogr.* **513,** 279 (1990).
57. H. Nishi and M. Matsuo, *J. Liq. Chromatogr.* **14,** 973 (1991).
58. M. Miyake, A. Shibukawa, and T. Nakagawa, *HRC & CC* **14,** 181 (1991).
59. W. H. Grist, M. P. Maskariinec, and K. H. Row, *Sep. Sci and Tech.* **23,** 1905 (1988).
60. T. Lee, E. S. Yeung, and M. Sharma, *J. Chromatogr.* **565,** 197 (1991).
61. C. P. Ong, C. L. Ng, H. K. Lee, and S. F. Y. Li, *J. Chromatogr.* **559,** 537 (1991).
62. P. G. Pietta, P. L. Mauri, A. Rava, and G. Sabbatini, *J. Chromatogr.* **549,** 367 (1991).
63. J. Vindevogel, P. Sandra, and L. C. Verhagen, *HRC & CC* **13,** 295 (1990).
64. T. Saltoh, H. Hoshino, and T. Yotsuyanagi, *J. Chromatogr.* **469,** 175 (1989).
65. T. Saltoh, C. Kiyohara, and N. Suzuki, *HRC & CC* **14,** 245, (1991).
66. T. Saltoh, H. Hoshino, and T. Yotsuyanagi, *Anal. Sciences* **7,** 494 (1991).
67. K. Saltoh, C. Kiyohara, and N. Suziki, *J. High Res. Chromatogr.* **14,** 245 (1991).
68. K. H. Row, W. H. Griest, and M. P. Maskarinec, *J. Chromatogr.* **409,** 193 (1987).
69. J. Liu, J. F. Banks, and M. Novotny, *J. Microcol. Sep.* **1,** 136 (1989).
70. A. Lahey and R. L. St. Claire III, *Am. Lab.* **22,** 68 (1990).
71. D. M. Northrop, D. E. Martire, and W. A. MacCrehan, *Anal. Chem.* **63,** 1038 (1991).
72. C. P. Ong, C. L. Ng, N. C. Chong, H. K. Lee, and S. F. Y. Li, *J. Chromatogr.* **516,** 263 (1990).
73. C. P. Ong, H. K. Lee, and S. F. Y. Li, *J. Chromatogr.* **542,** 473 (1991).
74. R. Weinberger, E. Sapp, and S. Moring, *J. Chromatogr.* **516,** 271 (1990).
75. K. Otsuka, S. Terabe, and T. Ando, *J. Chromatogr.* **332,** 219 (1985).
76. S. Terabe, Y. Ishihama, H. Nishi, T. Fukuyama, and K. Otsuka, *J. Chromatogr.* **545,** 359 (1991).
77. M. W. F. Nielen and M. J. A. Mensink, *HRC & CC* **14,** 417 (1991).

78. J. Vindevogel and P. Sandra, *Anal. Chem.* **63,** 1530 (1991).
79. P. Meier and W. Thormann, *J. Chromatogr.* **559,** 505 (1991).
80. H. Nishi, T. Fukuyama, M. Matsuo, and S. Terabe, *Anal. Chim. Acta.* **236,** 281 (1990).
81. D. E. Burton, M. J. Sepaniak, and M. P. Maskarinec, *J. Chromatogr. Sci.* **24,** 347 (1986).
82. S. Fujiwara, S. Iwase, and S. Honda, *J. Chromatogr.* **447,** 133 (1988).
83. C. P. Ong, C. L. Ng, N. C. Chong, H. K. Lee, and S. F. Y. Li, *J. Chromatogr.* **547,** 419 (1991).
84. I. Z. Atamna, C. J. Metral, G. M. Muschik, and H. J. Issaq, *J. Liq. Chromatogr.* **14,** 427 (1991).
85. L. A. Blyshak, G. Patonay, and I. M. Warner, in *Luminescence Applications in Biological, Chemical, Environmental, and Hydrological Sciences* (M. C. Goldberg, ed.), ACS Symposium Series 383, Chapter 10 (American Chemical Society, 1989).
86. S. Fanali and P. Boček, *Electrophoresis* **11,** 757 (1990).
87. A. Dobashi, T. Ono, S. Hara, and J. Yamaguchi, *Anal. Chem.* **61,** 1984 (1989).
88. A. Dobashi, T. Ono, S. Hara, and J. Yamaguchi, *J. Chromatogr.* **480,** 413 (1989).
89. A. Guttman, A. Paulus, A. S. Cohen, N. Grinberg, and B. L. Karger, *J. Chromatogr.* **448,** 41 (1988).
90. H. Nishi, T. Fukuyama, and M. Matsuo, *J. Microcol. Sep.* **2,** 234 (1990).
91. A. D. Tran, T. Blanc, and E. J. Leopold, *J. Chromatogr.* **516,** 241 (1990).
92. K. Otsuka, J. Kawahara, K. Tatekawa, and S. Terabe, *J. Chromatogr.* **559,** 209 (1991).
93. K. Otsuka and S. Terabe, *Electrophoresis* **11,** 982 (1990).
94. K. Otsuka and S. Terabe, *J. Chromatogr.* **515,** 221 (1990).
95. E. Gassman, J. E. Kuo, and R. N. Zare, *Science* **230,** 813 (1985)
96. P. Gozel, E. Gassman, H. Michaelson, and R. N. Zare, *Anal. Chem.* **59,** 44 (1987).
97. A. Pluym, W. V. Ael, and M. De Smet, *TRAC* **11,** 27 (1992).
98. K. Otsuka, J. Kawahara, K. Tatekawa, and S. Terabe, *J. Chromatogr.* **559,** 209 (1991).
99. T. O. Cole, M. J. Sepaniak, and W. L. Hinze, *J. High Res. Chromatogr.* **13,** 570 (1990).
100. H. Nishi, T. Fukuyama, M. Matsue, and S. Terabe, *J. Microcol. Sep.* **1,** 234 (1989).
101. A. D'Hulst and N. Verbeke, *J. Chromatogr.* **608,** 275 (1992).
102. J. Snopek, H. Soini, M. Novotny, E. Smolkova-Kevlemansova, and I. Jellinek, *J. Chromatogr.* **559,** 215 (1991).
103. R. Kuhn, F. Stoecklin, and F. Erni, *Chromatographia* **33,** 32 (1992).
104. H. Nishi, T. Fukuyama, M. Matsuo, and S. Terabe, *J. Chromatogr.* **515,** 233 (1990).
105. S. Fanali, paper presented at HPCE '92.
106. S. Fanali, *J. Chromatogr.* **545,** 437 (1991).
107. A. S. Cohen, A. Paulus, and B. L. Karger, *Chromatographia* **24,** 15 (1987).
108. A. Berthod, S.-C. Chang, and D. W. Armstrong, *Anal. Chem.* **64,** 395 (1992).
109. A. Guttman, A. Paulus, A. S. Cohen, N. Grinberg, and B. L. Karger, *J. Chromatogr.* **448,** 41 (1988).
110. M. Ahnoff and S. Einarsson, in *Chiral Liquid Chromatography* (W. J. Lough, ed.), Chapter 4 (Blackie & Sons, 1990).
111. S. Einarsson, B. Josefsson, P. Möller, and D. Sanchez, *Anal. Chem.* **59,** 1191 (1987).
112. R. Weinberger, unpublished results.
113. M. Khaledi, S. C. Smith, and J. K. Strasters, *Anal. Chem.* **63,** 1820 (1991).
114. K. Ghowsi, J. P. Foley, and R. J. Gale, *Anal. Chem.* **62,** 2714 (1990).
115. J. P. Foley, *Anal. Chem.* **62,** 1302 (1990)
116. J. Vindevogel and P. Sandra, *Anal. Chem.* **63,** 1530 (1991).

Chapter 8

Capillary Electrochromatography

8.1 The Electroosmotic Pump

In their seminal work on CZE, Jorgenson and Lukacs also demonstrated an electrically driven separation (1) of 9-methylanthracene from perylene in a 170 μm i.d. Pyrex tube packed with 10 μm Permaphase ODS particles. The separation efficiencies were modest, e.g., 31,000 theoretical plates for perylene, and the authors said that "the performance of these columns appears to offer a modest improvement over conventional (pressure-driven) flow, but may not justify the increased difficulty in working with electroosmotic flow."

Despite these early difficulties, the prospects for employing an electroosmotic pumping system for capillary electroosmotic chromatography (CEC) are compelling. The characteristics of EOF have three potential advantages. First is the facile generation of very low flow rates. The need for such low flows is established in Table 8.1,

Table 8.1. Column Diameters and Nominal Flow Rates for LC and CEC

Category	Column i.d.	Nominal Flow Rate (μL/min)
Analytical LC	4.6 mm	1,000
Small-bore LC	2.0 mm	95
Micro-bore LC	1.0 mm	47
Packed capillary LC	350 μm	5.8
Packed capillary EC	170 μm(1)	1.4
Open-tubular LC	50 μm	0.12
Open-tubular CEC	25 μm (4)	0.029
Open-tubular CEC	10 μm (14)	0.0047
Open-tubular CEC	5 μm (4)	0.0012

for both packed and open-tubular capillaries. Generation of electroosmotically pumped low flow rates is simple and can be controlled by selection of the appropriate additives as well as the applied field (2, 3). Second, the plug flow characteristics of this system might prove advantageous compared to the laminar profile of the pressure-driven system (Section 2.4), particularly for open-tubular separations. Lastly, the electrodriven system operates at low pressure and is pulse-free.

The electroosmotic velocity is described by

$$v_{eo} = \frac{\varepsilon\zeta}{4\pi\eta} E, \tag{8.1}$$

where ε is the dielectric constant of the electrolyte, E is the applied field, η is the viscosity, and ζ is the zeta potential. These factors are identical to the controlling forces in CZE. Changes in temperature, pH, viscosity, organic solvents, and ionic strength all affect the electroosmotic flow. Even in the presence of a coated open-tubular capillary, a robust EOF is generated (4).

The first report of electroosmotic pumping for LC was by Pretorius *et al.* in 1974 (5). Using a 1 mm i.d. packed column, he proved the feasibility of a voltage-driven system. Limitations in injection and detection systems prevented further development.

Stevens and Cortes (3) studied the impact of experimental parameters on the linear velocity and found that a substantial degree of control was possible based on the selection of particle size, solvent, and buffer additives. Linear velocities as high as 3.2 mm/s in packed capillaries were reported. Knox and Grant (6, 7) provided computations of the thermal problems, provided techniques for producing packed columns and drawn capillaries, studied the impact of particle size on the mobile-phase linear velocity, and published comparative separations employing both pressure- and electropumped systems (Fig. 8.1). Bruin *et al.* (4) described optimized results for open-tubular capillaries with internal diameters of 5–25 μm. They found the electropumped system superior to pressure pumping by a factor of two. At a voltage of 600 V/cm, a linear velocity of 1.4 mm/s was achieved. Tsuda (8–10) employed electrochromatography in combination with pressure pumping and found the technique useful for loading large amounts of sample onto a packed capillary (10). Pfeffer and Yeung used a cationic surfactant to enhance the EOF in a bonded-phase polymer-based open-tubular capillary (11). Without the buffer additive, the EOF was found to be too slow to serve as an electropump.

Despite the apparent advantages posed by the electroosmotic pump, it is important to consider its relative merits concerning chromatography and electrophoresis. Both are techniques with substantially different separation mechanisms. The distinguishing feature separating electrophoresis from chromatography is retention. In HPLC, retention is the driving force for separation. In HPCE, retention must be avoided at all costs or the capillary wall will be altered. Such a modification causes a loss of separation efficiency and a change in the EOF. The very nature of retention provides for its own limitations with regard to chromatographic efficiency.

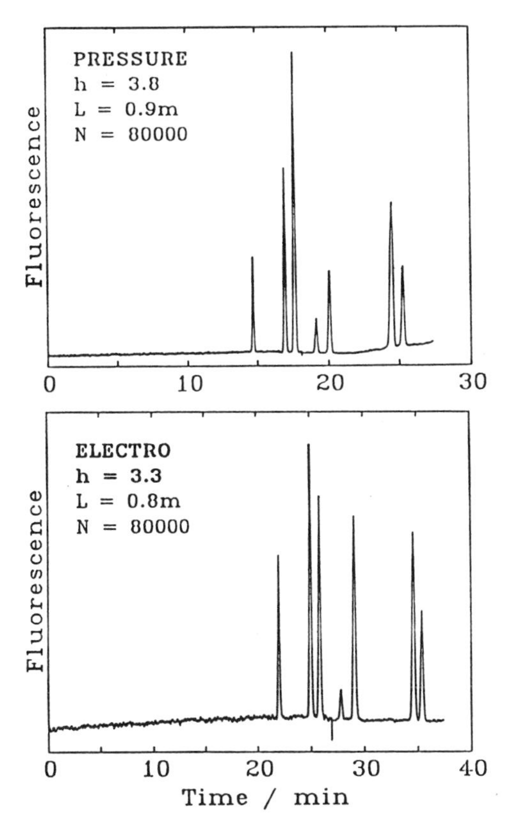

Figure 8.1. Separation of polycyclic aromatic hydrocarbons on a drawn capillary packed with 3 μm Hypersil particles and derivatized *in situ* with octadecylsilane. Capillary: 90 cm (pressure), 80 cm (electro) × 30 μm i.d.; pressure (upper): 25 bar; voltage (lower): 320 V/cm; detection: fluorescence. Order of elution: naphthalene, 2-methylanthracene, fluorene, phenanthrene, anthracene, pyrene, and 9-methylanthracene. Reprinted with permission from *Chromatographia* **32,** 317, copyright ©1991 Friedr. Vieweg and Sohn.

Retention can lead to bandbroadening due to mass-transfer problems in the mobile and stationary phases.

8.2 Bandbroadening

Three processes contribute to band broadening in capillary chromatography: axial dispersion, mass transfer in the mobile phase, and mass transfer in the stationary phase. These contributions (12) can be expressed as

$$H = 2\frac{D_m}{\nu} + C_m\frac{r^2\nu}{D_m} + C_s\nu\,, \tag{8.2}$$

where H = height equivalent to a theoretical plate; ν = mean linear velocity; D_m = solute diffusion coefficient in the mobile phase; r = the capillary radius; and C_m and C_s are the coefficients for resistance to mass transfer for a solute in the mobile and stationary phases, respectively. The first term in the equation simply describes the time-related impact of solute diffusion. This can be considered equivalent for both CZE and HPLC. The third term, restricted mass transport in the stationary phase, does not exist for CZE and is usually insignificant in HPLC. Only the mass transfer term in the mobile phase is affected by the cross-sectional flow profile. In pressure-driven systems, the height of the theoretical plate (H) is given by

$$H = \frac{1 + 6k' + 11k'^2}{96(1 + k')^2}. \tag{8.3}$$

In voltage-driven systems,

$$H = \frac{k'^2}{16(1 + k')^2}. \tag{8.4}$$

Calculated values based on these equations along with their ratio are plotted versus k' in Fig. 8.2. As the ratio indicates, the advantage of electropumped over pressure-pumped systems is realized only at values of very low k'. Rigorous calculations support both these data and conclusions (12, 13). Experimental data given in Fig. 8.1 further support the modest improvements in efficiency with electropumped compared with pressure-pumped systems. An expansion of the fluorene peak (7) is shown in Fig. 8.3. The improvements in efficiency are significant but not dramatic.

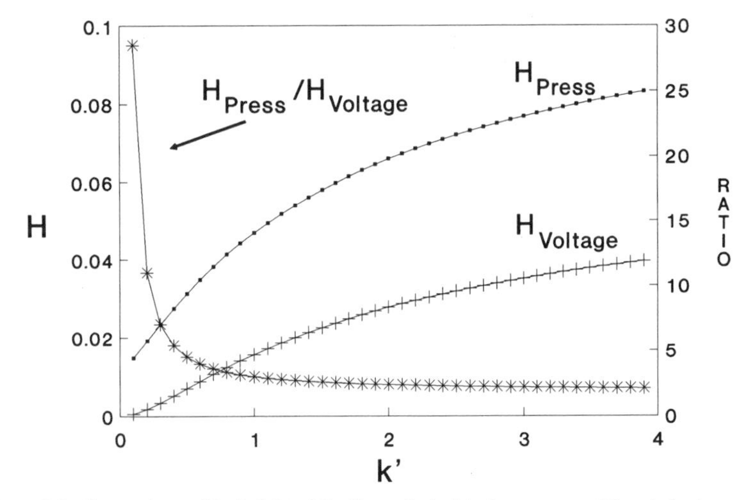

Figure 8.2. Comparison of the height of the theoretical plate for pressure- (P) and electropumped (E) systems versus k'. The graph is based on calculations from Eqs. (8.3) and (8.4).

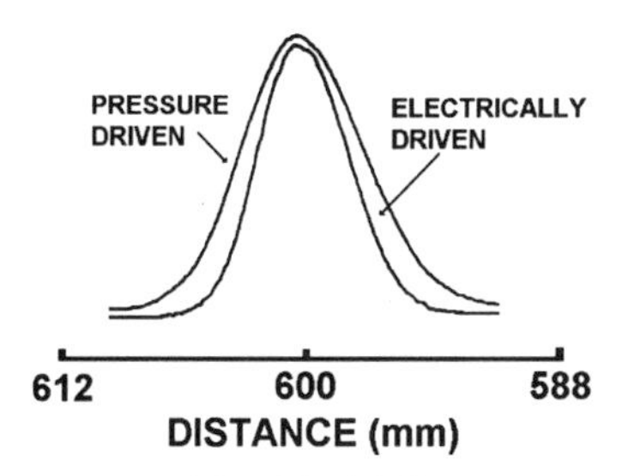

Figure 8.3. Expansion of the fluorene peak from Fig. 8.1. Outer curve: pressure-driven chromatogram; inner curve: electrically driven chromatogram. Reprinted with permission from *Chromatographia* **32,** 317, copyright ©1991 Friedr. Vieweg and Sohn.

8.3 Applications

There have been a few interesting separations reported by CEC. Pfeffer and Yeung (14) showed high-speed separations of some sulfonic acids (Fig. 8.4) on a 10 μm i.d. capillary coated with 0.9% PS-264 or 10% OV-17. An ion-pairing

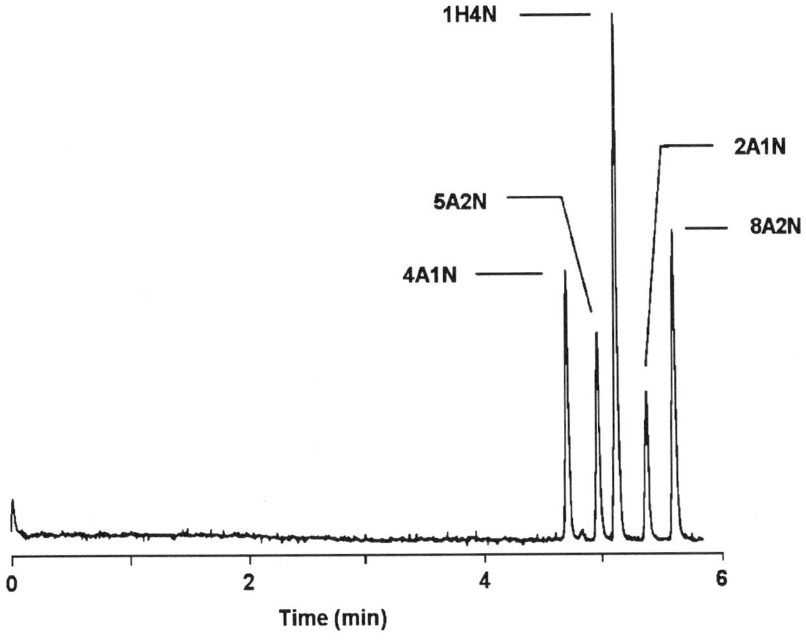

Figure 8.4. Electrochromatography of naphthalene sulfonic acids. Capillary: 50 cm × 10 μm i.d. PS-264 coated; buffer: 10 mM phosphate, pH 7, 1.25 mM tetrabutylammonium hydroxide; voltage: −21 kV; detection: laser fluorescence. Key: 4A1N, 4-amino-1-naphthalene sulfonic acid; 2A1N, 2-amino-1-naphthalene sulfonic acid; 8A2N, 8-amino-2-naphthalene sulfonic acid; 5A2N, 5-amino-2-naphthalene sulfonic acid; 1H4N, 1-naphthol-4-sulfonic acid. Reprinted with permission from *J. Chromatogr.* **557,** 125, copyright ©1991 Elsevier Science Publishers.

reagent, tetrabutylammonium hydroxide, was used as a mobile-phase modifier. The use of a narrow-bore capillary improved the mass transfer problem in accordance with the second term of Eq. (8.2). Improvements over CZE and pressure-driven LC separations were demonstrated with this ion-pair CEC system.

Mayer and Schurig (15) prepared a bonded-phase β-cyclodextrin capillary and separated the enantiomers of 1,1′-binaphthyl-2,2′diylhydrogenphosphate (Fig. 8.5) and 1-phenylethanol. Unfortunately, they did not report comparative separations of these same solutes by EKC with a cyclodextrin buffer additive.

Yamamoto *et al.* (16) reported separations of the drug Isradipin and its by-products (Fig. 8.6) along with a series of benzene derivatives. The excellent separation is consistent with theory, because the k' values are quite small (0.17–0.90). The retention times for these components were reported to range from 1.6 to 2.2% (run/run) and 9% (capillary/capillary). Production of narrow-bore packed capillaries can be problematic.

8.4 Electrochromatography in Perspective

A. Comparisons with Capillary LC

It is clear from Section 8.2 that the performance advantage of electrochromatography, relative to pressure-pumped capillary LC, is only modest. The primary

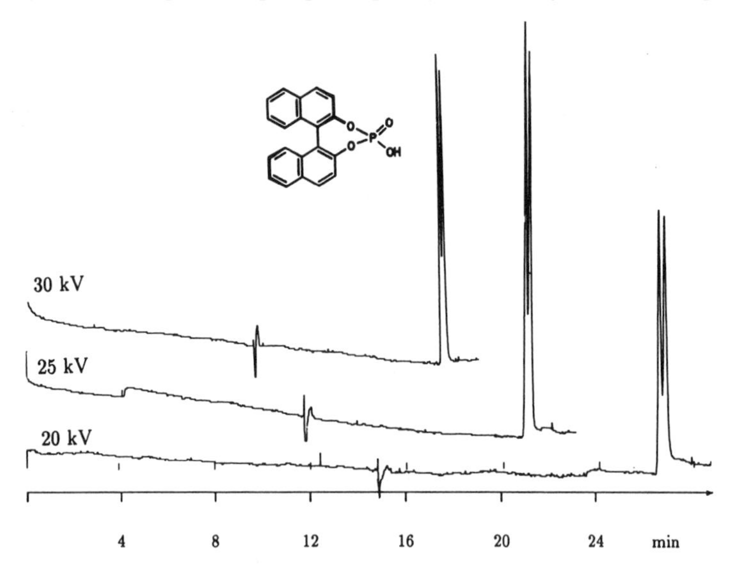

Figure 8.5. Separation of the enantiomers of 1,1′-binaphthyl-2,2′diyl-hydrogen phosphate at different applied voltages. Capillary: Chiralsil-Dex, 80 cm effective length × 50 μm i.d.; buffer: borate–phosphate, pH 7; detection: UV, 220 nm; temperature: 20°C. Reprinted with permission from *J. High Res. Chromatogr.* **15**, 129, copyright ©1992 Dr. Alfred Huethig.

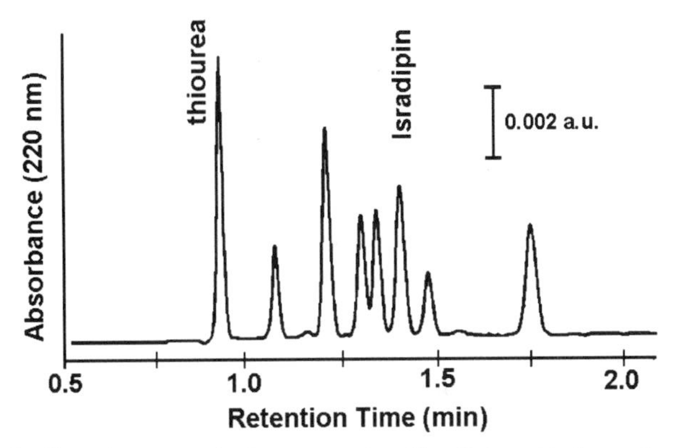

Figure 8.6. Capillary electroosmotic chromatography of Isradipin and its by-products. Capillary: 14.3 cm × 50 μm i.d. packed with Hypersil ODS (3 μm); mobile phase: 2 mM sodium tetraborate (pH 8.7)–80% acetonitrile; voltage: 30 kV; injection: 4 s at 1.5 kV; detection: UV, wavelength not specified. Reprinted with permission from *J. Chromatogr.* **593,** 313, copyright ©1992 Elsevier Science Publishers.

advantage is the simplicity of the pumping system. However, only isocratic separations have been demonstrated to date. While the design of a gradient system is surely possible, the increased complexity may negate the basic advantage.

The use of packed columns can produce a detection problem. The column packing is translucent to UV light, so the sensitivity of absorption detection may be reduced. One might argue that fluorescence detection is required to improve the sensitivity. Absorption detection should work well with the larger-diameter open-tubular capillaries.

On the other hand, 1.6 μm particles can be used to produce packed capillaries (16). High-speed separations (<1 min) become practical with such chromatographic material, as the mobile phase linear velocity is not limited by the back-pressure generated by the packing. It would be necessary to pump at 25,000 psi to obtain a linear velocity of 1 mm/s. With electropumping, a velocity of 3 mm/s is found at 30 kV (16). Using a capillary packed with 3 μm i.d. particles, Yamamoto *et al.* (16) produced a high-speed separation of the drug Isradipin (Fig. 8.6) and its synthetic by-products. Thiourea was added to indicate t_0.

B. Comparisons with Capillary Electrophoresis

No definitive studies have been performed to date. Figure 7.23 shows a separation of eight aromatic hydrocarbons by CD-MECC that can be compared with Fig. 8.1. The selectivities are somewhat different, but the efficiencies are comparable. Since MECC produces separations that resemble gradient-elution LC, the general elution problem is better addressed therein than with an isocratic LC separation. MECC has its own limitations. For example, the interface to the mass spectrometer may prove difficult because of the large quantities of surfactant in the electrolyte. Electrochromatography may prove useful here, because only hydro-organic solvents and small amounts of buffers are used. The buffers can be selected for compatibility with mass spectrometry, e.g., ammonium acetate.

For large-molecule separations performed by CZE, CGE, or CIEF, CEC will probably not be very useful. The nature of the retention process places chromatography at a distinct disadvantage compared to electrophoresis for these applications: (i) gradient elution is generally required for all large-molecule separations; (ii) the lack of a retentive phase in electrophoresis is a primary advantage for large-molecule separations. Proteins are notorious for adsorbing on chromatographic supports.

Electrochromatography is the least mature of the electroseparation techniques described in this book. Pending new and substantive invention, CEC is likely to find only limited applications in the near future.

References

1. J. W. Jorgenson and K. D. Lukacs, *J. Chromatogr.* **218,** 209 (1981).
2. T. Tsuda, K. Nomura, and G. Nakagawa, *J. Chromatogr.* **248,** 241 (1982).
3. T. S. Stevens and H. J. Cortes, *Anal. Chem.* **55,** 1365 (1983).
4. G. J. M. Bruin, P. P. H. Tock, J. C. Kraak, and H. Poppe, *J. Chromatogr.* **517,** 557 (1990).
5. V. Pretorius, B. J. Hopkins, and J. D. Schieke, *J. Chromatogr.* **99,** 23 (1874).
6. J. H. Knox and I. H. Grant, *Chromatographia* **24,** 135 (1987).
7. J. H. Knox and I. H. Grant, *Chromatographia* **32,** 317 (1991).
8. T. Tsuda, *Anal. Chem.* **59,** 521 (1987).
9. T. Tsuda, *Anal. Chem.* **60,** 1677 (1988).
10. T. Tsuda and Y. Muramatsu, *J. Chromatogr.* **515,** 645 (1990).
11. W. D. Pfeffer and E. S. Yeung, *Anal. Chem.* **62,** 2178 (1990).
12. M. Martin and G. Guiochon, *Anal. Chem.* **56,** 614 (1984).
13. M. Martin, G. Guiochon, Y. Walbroehl, and J. W. Jorgenson, *Anal. Chem.* **57,** 559 (1985).
14. W. Pfeffer and E. S. Yeung, *J. Chromatogr.* **557,** 125 (1991).
15. S. Mayer and V. Schurig, *J. High Res. Chromatogr.* **15,** 129 (1992).
16. H. Yamamoto, J. Baumann, and F. Erni, *J. Chromatogr.* **593,** 313 (1992).

Chapter 9

Injection

9.1 Volumetric Constraints on Injection Size

All capillary separation techniques including HPCE have certain constraints on the amount of material that can be injected. Injection in HPCE is designed to allow introduction of sufficient material into the capillary and minimize the extracapillary variance from the process itself. The volumetric problem is expressed in Table 9.1. Because the entire internal volume of a 50 cm × 50 μm i.d. capillary is only 981 nL, the injection volume must be kept quite small.

Since all processes contributing to variance (bandbroadening) are additive (1–3), the total peak variance (σ^2_{tot}) is

$$\sigma^2_{tot} = \sigma^2_{col} + \sigma^2_{inj} + \sigma^2_{det} , \tag{9.1}$$

where σ^2 indicates each variance due to the **col**umn, **inj**ection, and **det**ection, respectively.

The contribution to variance from a plug injection is (1, 2)

$$\sigma^2_{inj} = \frac{l^2_{inj}}{12} \tag{9.2}$$

where l is the length of the injection plug. Since the capillary efficiency is not badly

Table 9.1. Internal Volume Versus Diameter for a 50 cm Capillary

Capillary i.d. (μm)	10	25	50	100	200
Volume/mm (nL)	0.0785	0.491	1.96	7.85	31.4
Total volume (nL)	39.3	245	981	3925	15700

degraded by a 5–10% loss in efficiency, then for a 5% increase in bandbroadening, Otsuka and Terabe reported (2)

$$\sigma_{tot}^2 \leq (1.05\sigma_{col})^2 = 1.103\sigma_{col}^2 , \tag{9.3}$$

and for a 10% increase,

$$\sigma_{tot}^2 \leq (1.10\sigma_{col})^2 = 1.210\sigma_{col}^2 . \tag{9.4}$$

If the injector and detector are the only sources of extracapillary band spreading and $\sigma_{inj}^2 = \sigma_{det}^2$, then allowing for a 5% increase in variance and substituting $\sigma = HL$,

$$l_{inj} \leq \sqrt{12 \cdot 0.103/2}, \ \sigma_{col} = 0.786\sqrt{HL} , \tag{9.5}$$

and for a 10% increase,

$$l_{inj} \leq \sqrt{12 \cdot 0.210/2}, \ \sigma_{col} = 1.120\sqrt{HL} , \tag{9.6}$$

Data based on Eqs. (9.5) and (9.6) are shown in Table 9.2 for a 50 cm × 50 μm i.d. capillary. Peaks exhibiting high efficiencies (high number of theoretical plates) require significantly smaller injection plugs than broader peaks.

Grushka and McCormick's (4) calculations begin with Eq. (9.2). Then using a form of the plate-height equation,

$$H = \frac{2D}{\nu} , \tag{9.7}$$

where D is the solute's diffusion coefficient and υ is the solute's velocity,

$$l_{inj} = \sqrt{24DEt} , \tag{9.8}$$

where t is the elution time, and E, the acceptable increase in H. This model predicts,

Table 9.2. Calculated Maximum Injection Length and Volume for a 50 cm × 50 μm i.d. Capillary[a]

Number of Theoretical Plates (N):	100,000	500,000	1,000,000
Peak width (mm), t_m = 10 min	4.47	2.00	1.41
5% increase			
l_{inj} (mm)	1.24	0.56	0.39
V_{inj} (nL)	2.43	1.09	0.77
10% increase			
l_{inj} (mm)	1.77	0.79	0.56
V_{inj} (nL)	3.48	1.55	1.10

[a]Data from reference (2). Peak widths at half-height calculated from $N = 5.54\left(t_m^2/w_{1/2}^2\right)$

for a protein with $D = 1 \times 10^{-6}$ cm^2/s, a migration velocity of 0.1 cm/s and a migration time of 600 s; the length of the injection plug should be 380 μm, allowing for a 10% decrease in efficiency. For a smaller molecule having $D = 1 \times 10^{-5}$ cm^2/s, the injection plug can be up to 1.2 mm under the same conditions.

The model of Huang *et al.* (5) attributes the width of the injection zone as the most significant determination to the number of theoretical plates obtained. They ended with the following relationship for N:

$$N = 12\frac{L^2}{w_{\text{inj}}^2}. \tag{9.9}$$

All these models assume that the injection electrolyte is identical to the running electrolyte. By proper selection of the injection buffer relative to the run buffer, it is possible to obtain substantial compression of the injection zone. This process is known in electrophoresis as stacking. This family of related processes will be described in detail in Section 9.6.

9.2 Performing an Injection and Run

The steps involved in injecting a sample into a capillary and performing a run are as follows:

1. Equilibrate capillary in run buffer for 1–5 min (longer if capillary is new or separation is sensitive);
2. transfer the capillary–electrode assembly to the sample container;
3. immediately inject the sample via electrokinetic or hydrodynamic injection (1–30 s);
4. if necessary (see Table 9.4), designate a wash station to rinse the outside walls of the capillary;
5. return the capillary–electrode assembly to the sample-side run buffer;
6. promptly engage the high voltage and run the separation;
7. if necessary, run a wash step with 0.1 M sodium hydroxide; and
8. go to Step 1 for the next sample.

9.3 Injection Techniques

A. Hydrodynamic Injection

Hydrodynamic injection is accomplished in one of four ways:

1. by elevating capillary at the sample end, permitting sample introduction by siphoning;

2. by applying pressure on the individual sample vial;
3. by applying vacuum on the detector-side buffer reservoir; or
4. injecting by syringe and employing a splitter to reduce the volume introduced into the capillary.

Until the arrival of commercial instrumentation, nearly all hydrodynamic injections were performed using Method 1. Most commercial systems employ Methods 2 or 3. The volume of material injected per unit time (V_t, nL/s) using Methods 1–3 are determined by the Poiseuille Equation (6):

$$V_t = \frac{\Delta P D^4 \pi}{128 \eta L}, \tag{9.10}$$

where ΔP equals the pressure drop, D is the capillary internal diameter, η is the viscosity, and L is the length of the capillary. For gravity-based injections,

$$\Delta P = \rho g \Delta h, \tag{9.11}$$

where ρ is the density of the sample solution, g is the gravitational constant, and Δh is the height difference between the liquid levels of the sample vial and the detector-side buffer reservoir. Since the flow rate is proportional to the fourth power of the capillary diameter, these values can only be considered approximate. Exact assessment of the injected amount is generally unnecessary, because standards are used to calibrate the system. Table 9.3 contains data calculated for Methods 1–3 for various capillary i.d.s.

Hydrodynamic injection is generally useful for capillaries with i.d.s ranging from 25 to 100 μm. Below that range, it is difficult to inject sufficient material into the capillary. Above the range, hydrodynamic flow rates are far too great to permit a small, reproducible injection.

The impact of the injection zone length on the electropherogram is illustrated in Fig. 9.1. The effect on efficiency and subsequently resolution is obvious. Less

Table 9.3. Injection Volume/s for Pressure (0.5 psi), Vacuum (5 in. Hg) and Gravity (10 cm) Versus Capillary i.d.[a]

	Injection size (nL/s)		
Capillary diameter (μm)	Pressure	Vacuum	Gravity
10	0.000078	0.00038	0.000022
25	0.039	0.19	0.011
50	0.60	2.9	0.17
75	3.0	15	0.87
100	9.6	47	2.7

[a]Capillary length: 100 cm; temperature: 30°C; viscosity: 0.801 gm cm^{-2} s^{-1}; pressure injection: 0.5 psi (Beckman); vacuum injection: 5 in. Hg (Applied Biosystems); gravity injection, Δh = 10 cm (Waters) where g = 980 g cm^{-2} and ρ = 0.997 g mL^{-1}.

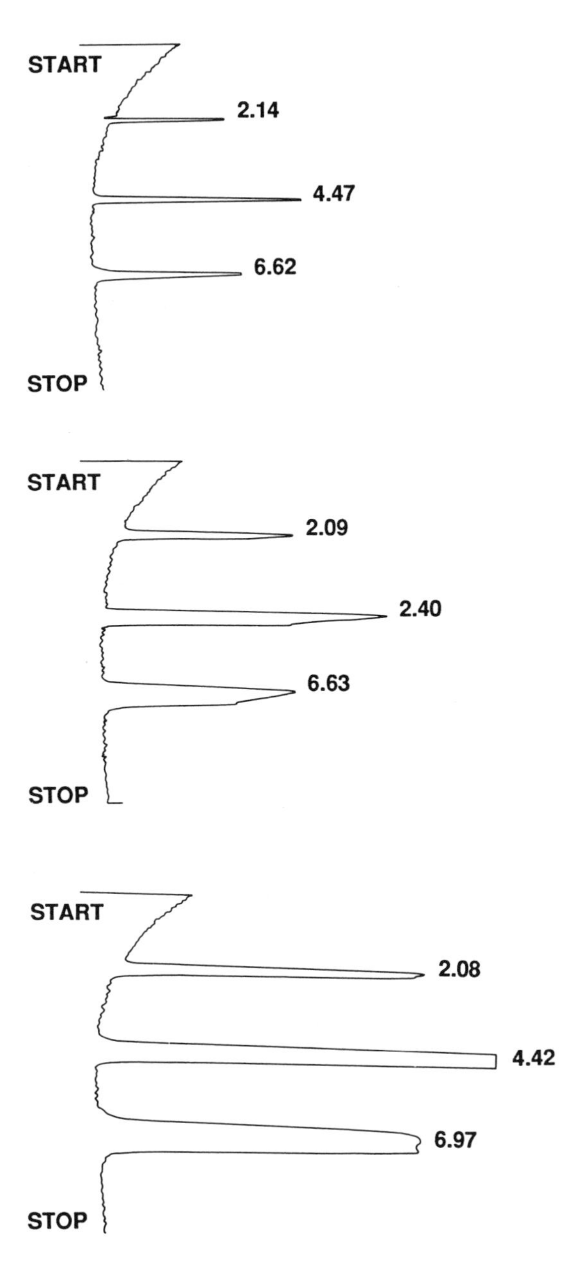

Figure 9.1. Impact of the injection zone width on electrophoretic efficiency. Capillary: polyacrylamide-coated 20 cm × 25 μm i.d.; buffer: pH 2.5 phosphate; solutes: substance P fragments dissolved in run buffer, 0.5 μg/mL each; detection: UV, 200 nm. Injection width: top, 0.6 cm; center, 2.0 cm; bottom, 3.0 cm. Reprinted with permission from *J. Chromatogr.* **480,** 311, copyright ©1989 Elsevier Science Publishers.

obvious is the increase of peak width as the migration time increases (7). This feature is a consequence of the on-capillary detection used in HPCE. Solutes that migrate slowly take more time to transit the detector window than more mobile ones. The result is manifested in apparent peak width. This feature will be covered in more detail in Chapter 10.

Selection of pressure-driven versus vacuum-driven systems is not particularly important unless an interface to a mass spectrometer is required. In this case, the pressure-driven system is preferred choice.

Method 4, syringe injection, is used with a low-cost manual HPCE instrument (8).

B. Electrokinetic Injection

Another injection technique uses electrophoretic and/or electroosmotic migration to inject samples into the capillary (6). By applying a low voltage for a short period, controlled amounts of the sample are simply introduced. As in the case with hydrodynamic injection, large (long) injections result in a substantial loss of resolution (Fig. 9.2), particularly if the injection and run buffers are identical.

The quantity (Q) of a solute injected is given by

$$Q = (\mu_{ep} + \mu_{eo})\, \pi r^2 ECt\,, \tag{9.12}$$

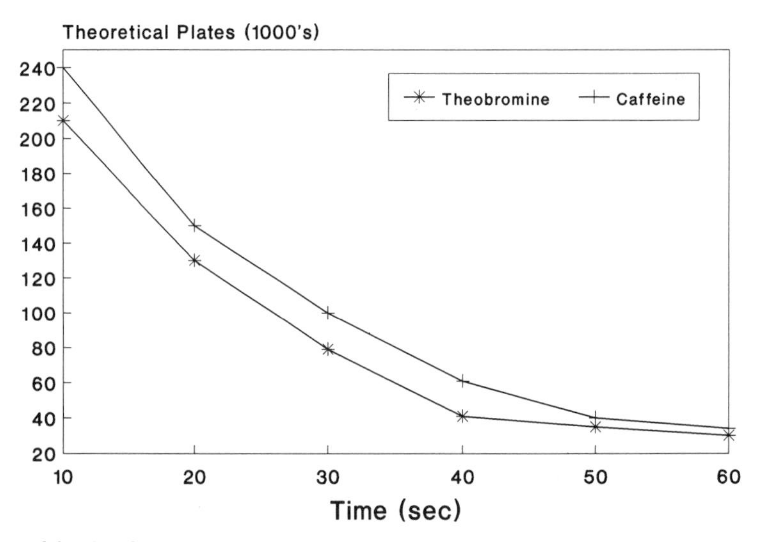

Figure 9.2. Plot of number of theoretical plates versus injection time with electrokinetic injection. Capillary: 100 cm × 75 μm i.d.; buffer: 50 mM SDS, 10 mM disodium phosphate, 6 mM sodium tetraborate, run voltage: 17.5 kV, 30 μA; detection: UV, 280 nm; injection voltage: 5 kV. Data from *Chromatographia* **21,** 583 (1986).

where μ_{ep} and μ_{eo} are the electrophoretic and electroosmotic mobilities respectively; r is the capillary radius; E is the field strength; t is the time of injection; and C is the concentration of each solute. This form of the equation is preferable to the simple volumetric relationship of Eq. (9.11) because solute discrimination may take place with electrokinetic injection (6, 9). High-mobility solutes may be preferentially enriched compared to those with low mobility (Fig. 9.3). Solutes that have identical mobility in free solution show no such bias, e.g., oligonucleotides and DNA fragments (10). This is fortunate, because it is necessary to use electrokinetic injection with gel-filled capillaries. Hydrodynamic injection can result in extrusion of the gel.

Besides their use with high-viscosity gels, electrokinetic injections are generally required when the capillary diameter differs significantly from the specifications imposed by the instrument manufacturers. As Table 9.3 indicates, the problem is substantial, particularly with wide-diameter capillaries. With a 100 μm i.d. capillary, a flow rate of 47 nL/s is achieved with the Applied Biosystems instrument. Since an injection time of <1 s is not recommended because of injection imprecision, electrokinetic injection is required here. Electrokinetic injections are also useful under certain stacking conditions and to provide selective injection of

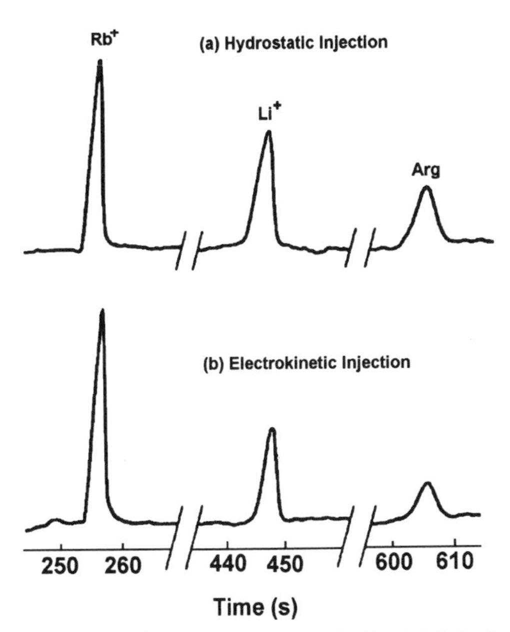

Figure 9.3. Hydrostatic versus electrokinetic injection. Buffer: 20 mM MES adjusted with histidine to pH 6.0; solutes: Rb^+, Li^+, and arginine, 5×10^{-5}M; injection: (top), hydrostatic, $\Delta h = 10$ cm, $t = 10$ s; (bottom), electrokinetic, 1 s at 10 kV; detection: conductivity. Reprinted with permission from *Anal. Chem.* **60,** 375, copyright ©1988 Am. Chem. Soc.

anions or cations (Section 9.6). Excepting these conditions, hydrodynamic injection is generally preferred.

9.4 Injection Problems and Solutions

Unlike HPLC, HPCE cannot use a fixed loop to define the injection volume. The ill-defined injection plug can be affected by extra-injection artifacts. The simple process of inserting and withdrawing a capillary into a solution can cause an extraneous bolus of sample to enter the capillary by (i) hydrostatic pressure; (ii) convective flow; and (iii) diffusion (4). Diffusion is most important for small molecules, short injection zones, or slow injection sequences (14). Avoid very short injection zones and never leave the capillary in the sample for prolonged periods of time. Similarly, once the capillary is returned to the buffer reservoir, begin the run promptly. Be certain that the liquid levels of the sample vial and both buffer reservoirs are at maintained at the same height. Proper internal or external standardization of the separation can minimize, though not totally eliminate, these effects.

The most prominent problems that appear injection-related are actually functions of sample preparation. In HPLC, samples must be prepared in a solvent that has a lower eluting strength than the mobile phase. The situation in capillary electrophoresis is similar, except that ionic strength of the sample solution may be the controlling factor. Since the solute contributes to the total ionic strength, high solute concentration also causes dispersion (Section 11.4).

Table 9.4 contains a description of problems frequently attributed to injection, their cause, and possible remedies. Despite the unique problems imposed by injection, peak area relative standard deviations of less than 1% are common with modern, automated injection systems.

Table 9.4. Injection Problems, Causes, and Solutions

Problem	Cause	Remedy
No injection	Plugged capillary	Replace capillary or remove 1 cm from sample side and retry
Migration time drift	Buffer evaporation	Replace buffer
		Use manufacturer's closure system
	Buffer depletion (11)	Replace buffer
	Siphoning	Balance fluid levels between sample and detector side buffers
		Reduce capillary diameter
Broad peaks	Large injection size	Reduce injection size
	High ionic strength of sample solution (12)	Desalt or dilute sample, prepare in a low ionic strength buffer
	Siphoning	see above

Table 9.4. *conintued*

Problem	Cause	Remedy
Quantitative accuracy	Sample-standard ionic strength differences	Prepare all solutions in constant ionic strength buffer
	Sample-standard viscosity differences	Dilute samples
		Add internal standard
		Prepare samples in a constant viscosity solution
	Sample evaporation	Cool samples if possible
		Use manufacturer's closure system
Peak tailing	Shard at capillary end	Recut capillary
	Capillary end not square	Recut capillary
	Sample residue on outer capillary wall(13)	Dip capillary in a wash solution prior to high-voltage start
Carryover	Sample residue on outer capillary wall	Dip capillary in wash solution between runs

9.5 Alternative Injection Schemes

A variety of alternative injection schemes have been reported. Hartwick (15) described an electrokinetic injection scheme termed "electrosyphon." This multi-electrode setup induces an osmotic flow at the detector end of the capillary, generating a hydrodynamic flow towards the detector. This eliminates the discrimination usually found in electrokinetic injection. Hjertén (16) employed capillary cooling to cause fluid compression, drawing the sample into the capillary. Yin *et al.* (17) described a device for handling very small sample volumes. Linhares and Kissinger (18) used a fractured capillary to generate an electrosyphon to overcome electrokinetic discrimination. Tsuda *et al.* (19) described a rotary valve injector. Tsuda and Zare (20) reported a method that permits sample injection without interruption of the voltage. Wallingford and Ewing (21) developed a microinjector capable of sampling cytoplasmic material from a single neuron. None of these techniques have been incorporated into commercial instrumentation at this time.

9.6 Stacking

A. Introduction

Compression of or *stacking* of the injection zone is important for two reasons: (i) as Table 9.2 indicates, the injection size can severely limit the number of theoretical plates obtained for the separation; (ii) the problems with detector sensitivity (Chapter 10) can be solved in part by adequate trace enrichment of dilute solutes.

The problem posed by the first point is illustrated in Fig. 9.4 using an MECC separation of urinary porphyrins (22). The injection size is only 10 nL into a 55 cm × 50 μm i.d. capillary. In the electropherogram at left, the injection was performed in the run buffer comprising 100 mM SDS and 20 mM CAPS, pH 11. An injection buffer with only 20 mM CAPS was used to generate the separation at right. Substantial improvements in resolution were found with the nonmicellar injection buffer, even with small injections.[1] Removing the surfactant from the injection buffer reduced the ionic strength and the conductivity of that solution.

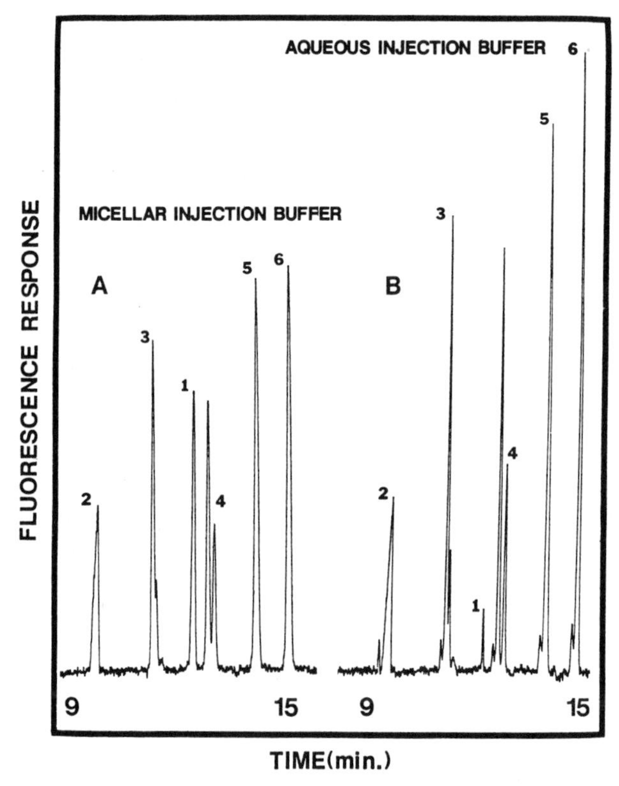

Figure 9.4. Impact of injection buffer on electrophoretic resolution. Injection buffer, (left): 20 mM CAPS, pH 11, 100 mM SDS; (right): 20 mM CAPS, pH 11. See Fig. 7.15 for other conditions. Reprinted with permission from *J. Chromatogr.* **516,** 271, copyright ©1990 Elsevier Science Publishers.

[1] Peak #1, mesoporphyrin, is not very soluble in the absence of the surfactant, accounting for its diminished peak height.

The second point can be addressed as well with a stacking electrolyte. Figure 9.5 shows the separations of some porphyrins with injection sizes ranging from 5 to 100 nL. A 100 nL injection represents about 10% of the volume of the capillary. Compared to a 5 nL injection, the limit of detection is improved by a factor of 13. Even with a stacking buffer, some injection-mediated bandbroadening can occur. While a 20-fold LOD improvement was expected, the improvement was only 13-fold. For a 50 nL injection, the improvement is eight-fold (ten-fold expected).

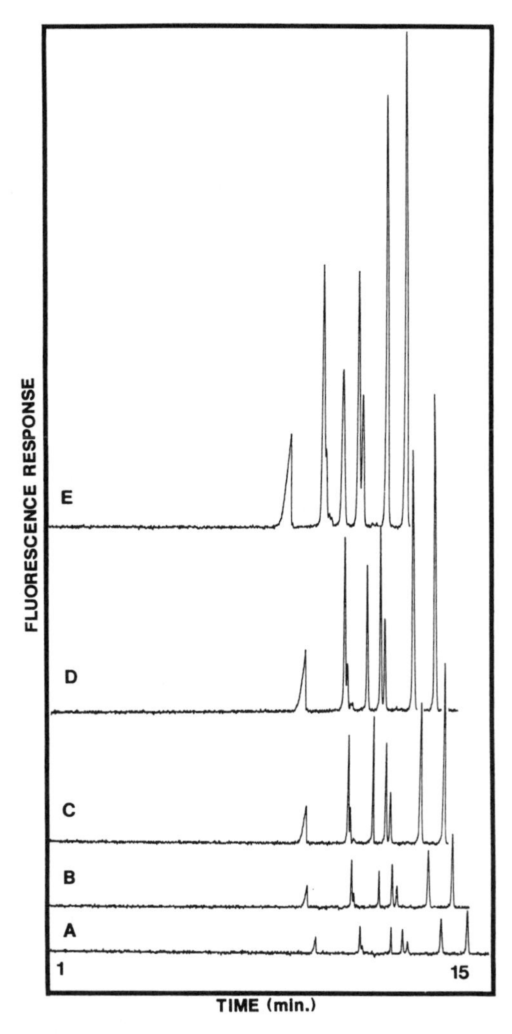

Figure 9.5. Impact of injection size on resolution and sensitivity. Injection buffer, 20 mM CAPS, pH 11. Approximate injection size: (A) 5 nL; (B) 10 nL; (C) 25 nL; (D) 50 nL, and (E) 100 nL. Reprinted with permission from *J. Chromatogr.* **516,** 271, copyright ©1990 Elsevier Science Publishers.

Clearly, the LODs can be improved at the expense of electrophoretic resolution. With proper optimization (see below), it is possible that both resolution and sensitivity can be further improved.

Stacking is not new in electrophoresis or even in HPCE. In 1979, Mikkers *et al.* (23) described stacking in CZE. In that work, they referenced analogous techniques using disc electrophoresis as far back as 1964. Stacking gels are widely used for many biochemical separations. In a gel-free medium such as CZE, stacking electrolytes can be employed to accomplish solute enrichment.

B. Ionic-Strength Mediated Stacking

Differences between the conductivity of the injection zone and that of the carrier electrolyte can have an impact on the field strength that is expressed over each zone. This is simply a consequence of Ohm's Law. Since a solute's electrophoretic velocity is proportional to the field strength, differing velocities can be realized within each compartment. The basis of stacking is to provide a high field strength over the injection zone. This is readily accomplished by injection in low-conductivity buffers.

The phenomenologic situation at low pHs is illustrated in Fig. 9.6. If a low-conductivity injection buffer is employed, the field strength must be higher over that zone compared to the balance of the length of the capillary. When the positively charged solutes begin to migrate out of the injection zone and encounter the separating electrolyte, the field strength abruptly drops, and thus the solute's electrophoretic velocity slows. Meanwhile, the solutes at the middle to rear of the injection zone are still exposed to the high field strength and continue to move forward at "full" speed. As a result, the ions in the injection band continue to narrow until all

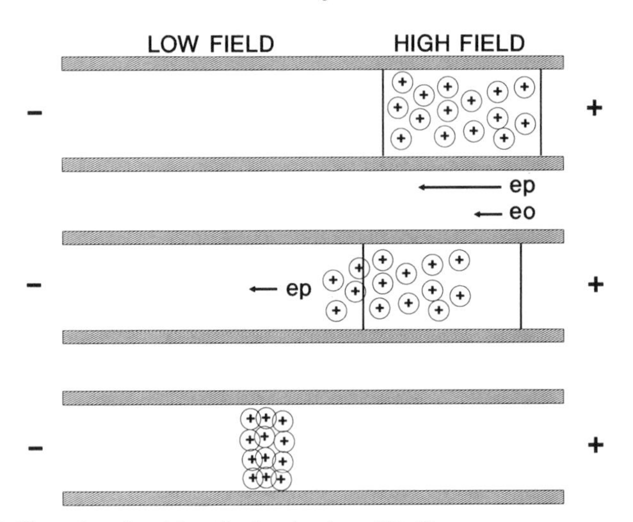

Figure 9.6. Illustration of stacking of cations in a low-pH buffer.

have migrated into the carrier electrolyte. The negatively charged counterions stack up as well, but at the rear of the injection zone (not shown). At high pHs, as shown in Fig. 9.7, a more complex situation occurs because of the presence of the strong EOF. Anionic solutes now compress or stack up at the rear of the injection zone.

Despite these differences in the pH-mediated direction of electrophoresis, the net result is band compression, the degree of which is related by the ratio of conductivities between the injection and separation electrolytes. These differences can be quantitatively expressed as (24)[2].

$$\frac{E_1}{E_2} = \frac{\rho_1}{\rho_2} = \frac{C_2}{C_1} = \gamma, \tag{9.13}$$

where E_1 and E_2 are the field strengths over the injection zone and balance of the capillary, respectively; ρ_1 and ρ_2 are the respective resistivities in these regions; C_1 and C_2 are the respective buffer concentrations; and γ is the field enhancement factor (24). Since the current or ion flux that passes through the capillary must be constant through each zone, the steady state concentration of solutes $[S_1]$ and $[S_2]$ is inversely proportional to the field strength, and thus the electrophoretic velocity therefore

$$[S_1]v_1 = [S_2]v_2 \tag{9.14}$$

then

$$\frac{[S_2]}{[S_1]} = \frac{C_2}{C_1} = \gamma \tag{9.15}$$

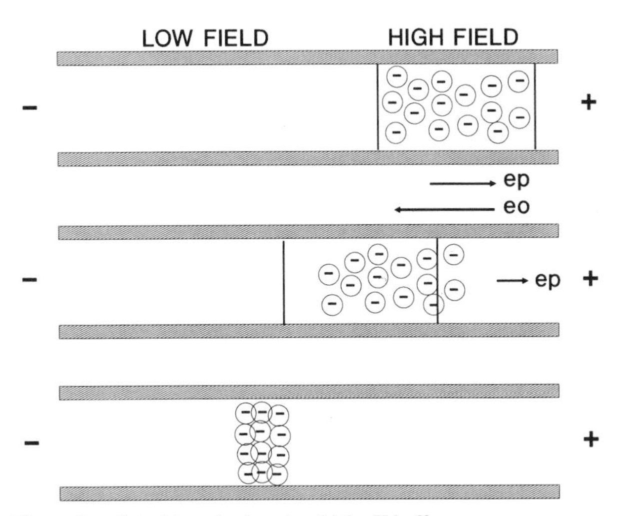

Figure 9.7. Illustration of stacking of anions in a high-pH buffer.

[2] The symbols describing the parameters given in (24) have been simplified for clarity.

The concentration of ions entering the high concentration buffer region must increase. This can only be accomplished by compression of the ions in the injection zone thus

$$x_s = \frac{x_i}{\gamma}, \tag{9.16}$$

where x_s is the effective sample plug length after stacking and x_i is the initial length of the injection plug. The similarities to CITP are clear; however, this is a non-steady state process: The solute ions migrate out of the injection buffer zone into the homogeneous carrier electrolyte. The low ionic strength injection buffer still lingers within the capillary, causing problems when very large injections are made (25).

Based on Eq. (9.16), the sample should always be prepared in water. Unfortunately, this is not optimal because of the generation of electroosmotic pressures. The measured EOF is based on the average electroosmotic contribution of each zone adjusted by the zone length. Since fluids can be considered incompressible, a hydrodynamic component is introduced whenever localized zones contribute differently to the total EOF. This effect generates bandbroadening due to hydrodynamic flow. Calculation and plotting of these effects lead to Fig. 9.8, which shows optimal results for field enhancement factors from 10 to 20 (24). When γ

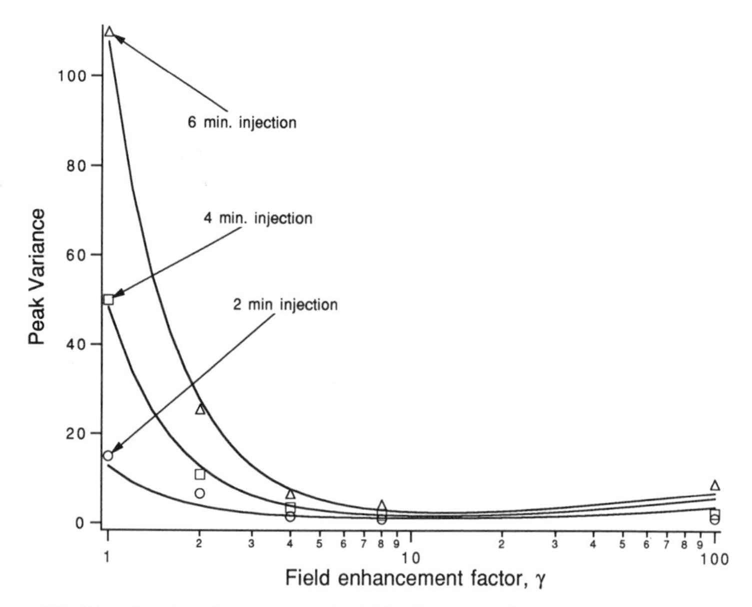

Figure 9.8. Plot of peak variance versus the field enhancement factor γ for several gravity-based (15 cm) injection times using a 100 cm × 75 μm i.d. capillary. Reprinted with permission from *Anal. Chem.* **63,** 2042, copyright ©1991 Am. Chem. Soc.

is small, the peak variance is proportional to the injection time. At high γ values, the laminar-flow induced broadening becomes significant.

These theoretical calculations are supported by experimental data shown in Fig. 9.9. The optimal procedure for stacking is to prepare the sample in an injection buffer that is 10-fold more dilute than the background electrolyte. The length of the injection is also important. For generating a maximum of 10% broadening, the sample plug length should be no more than 1% of the capillary length (24). According to Figs. 9.5 and 9.9, it is possible to trade off some plates for sensitivity. Performed properly, ionic-strength mediated stacking can improve sensitivity by a factor of 10.

C. Whole-Capillary Injection

Very large injections can be performed by removing the injection solution from the capillary after stacking (25). For example, the entire capillary or a large portion thereof can be filled with an anionic sample dissolved in water (Fig. 9.10, top). When a negative voltage is applied, the anions migrate toward the positive

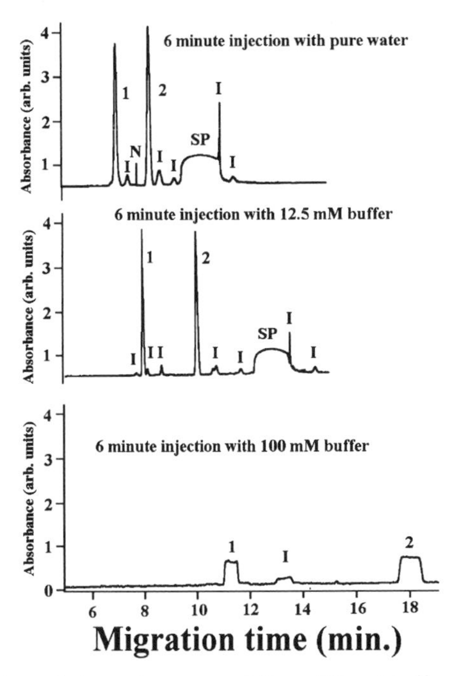

Figure 9.9. Comparison of electropherograms of three different buffer concentrations in the sample plug. Capillary: 100 cm × 75 μm i.d.; buffer: 100 mM MES, pH 6.13; injection: gravity (15 cm) for 6 min (about 400 nL or 5% of the total capillary volume); injection buffer: as specified on figure; voltage: 30 kV. Reprinted with permission from *Anal. Chem.* **63,** 2042, copyright ©1991 Am. Chem. Soc.

electrode and stack at the boundary between the injection zone and the entering background electrolyte (Fig. 9.10, middle). The EOF naturally drives the overall migration velocity toward the negative electrode. Simultaneously, the water electroosmotically exits the capillary at the negative electrode injection end. As water exits, the background electrolyte enters the capillary at the opposite end. As the analyte encounters the support buffer, migration ceases because there is virtually no field strength over the high ionic buffer strength zone; the voltage drop occurs over the water. When the water is nearly out of the capillary, as shown by the current reading, the power supply polarity is reversed and the analytes are quickly separated and eluted (Fig. 9.10, bottom).

This process is illustrated in Fig. 9.11 using some PTH amino acids prepared in water at the CLOD for a normal run. Since the entire capillary is filled with sample, the baseline is elevated slightly as the run begins, indicating the presence of the analyte. Stacking occurs at the rear of the zone, and a peak is seen as the stacked but unseparated anions pass the detector for the first time, driven by the EOF. When the current rises to 99% of the value for the background electrolyte (previously determined), the electrode polarity is reversed. The separation occurs on the segment of capillary that was originally beyond the detector window before switching the power supply polarity. A similar process can be performed for cations, provided a surfactant is added to reverse the EOF (25). While there is no data on the precision for this technique, it may be useful for micropreparative separations. A high ionic-strength sample matrix can badly interfere with this process.

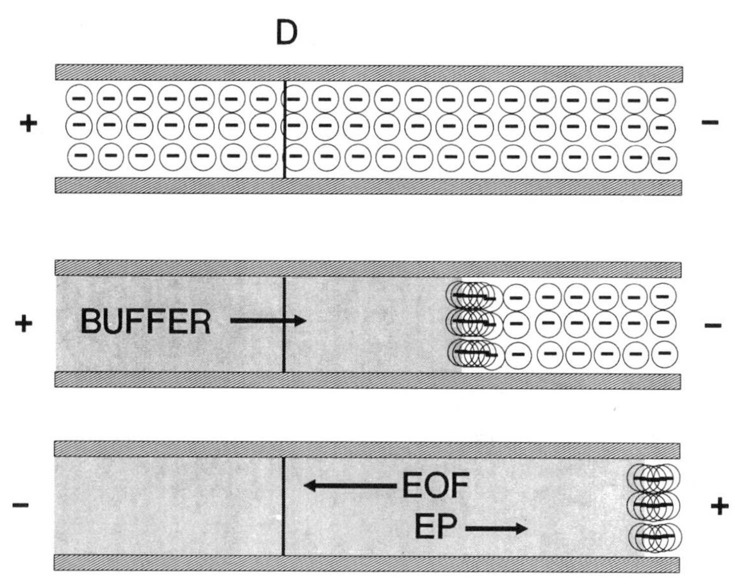

Figure 9.10. Illustration of the whole-capillary injection process for anions.

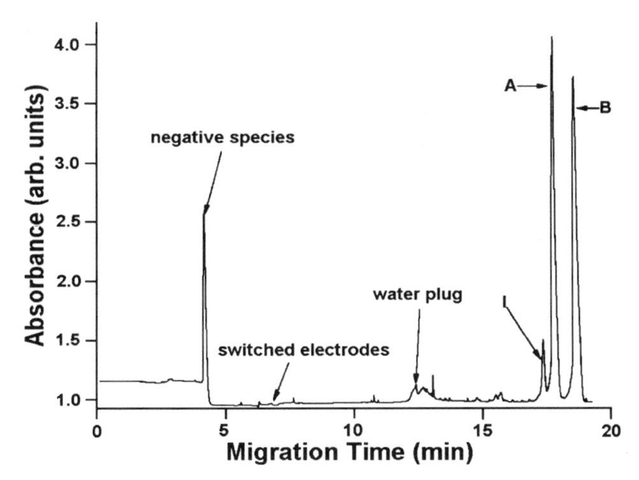

Figure 9.11. Electropherogram from a whole-column injection. The whole capillary (100 cm (65 cm to detector) × 50 μm i.d.) is filled with sample, (A) PTH–Arg, 4.6×10^{-5}M, (B) PTH–Glu, 3.4×10^{-5}M, dissolved in water. The ends of the capillary are dipped in support buffer, 100 mM MES adjusted to pH 6.1 with 100 mM histidine. At 4.5 min, all anions pass the detector as a single band. At 7 min, the polarity is reversed and the anions migrate back in the direction of the injection end and pass the detector as separated analytes. Reprinted with permission from *Anal. Chem.* **64,** 1046, Copyright©1992 Am. Chem. Soc.

D. pH-Mediated Stacking

For separations of zwitterions, the pH of the injection buffer can be advantageously selected to enhance peak compression (26). The stacking mechanism is shown in Fig. 9.12. Peptides are injected using 10 mM ammonium hydroxide into the capillary containing a 10 mM citrate, pH 2.5, supporting electrolyte. Upon application of the voltage, the negatively charged peptides migrate toward the anode. As the peptides at the rear of the band enter the acidic buffer, the charge flips and the direction of migration reverses. Meanwhile, the peptides at the front of the band are still migrating toward the anode. The result is that the band collapses upon itself. Finally, all of the material in the narrowed band becomes positively charged and migrates toward the cathode. The small amount of injected ammonium hydroxide is dissipated and buffered.

The impressive stacking ability of this technique is shown in Fig. 9.13. The injection size was approximately 150 nL into a 72 cm × 50 μm i.d. capillary. Without the pH shift, no separation was obtained. With the pH shift, the resolution is comparable to a 5 nL injection (data not shown). The limit of detection for some dynorphins can be less than 1 ng/mL with a 300 nL injection (data not shown). This technique has been applied toward screening of narrow-bore LC fractions of peptide maps. No data have been published to date on the repeatability of this technique.

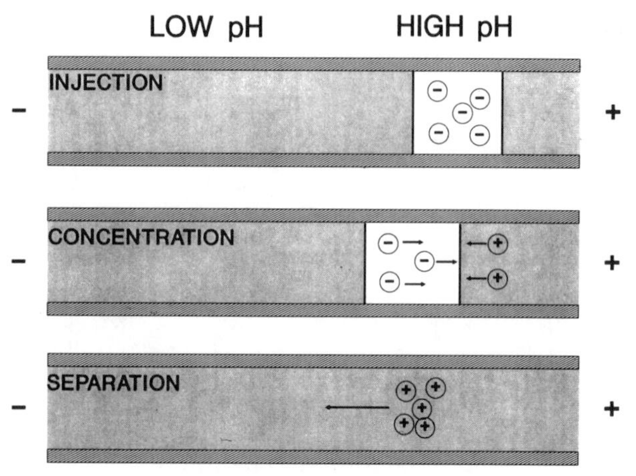

Figure 9.12. Illustration of pH-mediated stacking of zwitterions. Reprinted with permission from *J. Chromatogr.* **516,** 89, copyright ©1990 Elsevier Science Publishers.

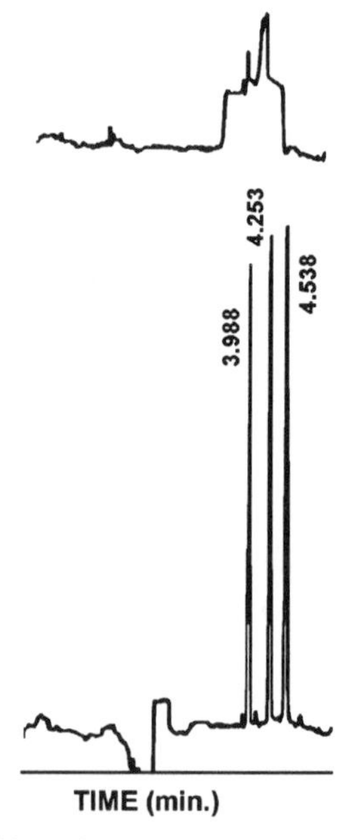

Figure 9.13. Separation of peptides using pH-mediated stacking. Capillary: 72 cm (50 cm to detector) × 50 μm i.d.; separation buffer: 10 mM citrate, pH 2.5; injection buffer: 10 mM ammonium hydroxide; injection: vacuum, 30 s (about 150 nL); detection: UV, 215 nm; temperature: 30°C; solutes: dynorphins, 10 μg/mL. Top: injected in separation buffer; bottom: injected in injection buffer. Redrawn with permission from *J. Chromatogr.* **516,** 79, copyright ©1990 Elsevier Science Publishers.

E. Electrokinetic Stacking

Chien and Burgi (27–29) introduced the term “field amplified injection (FAI)” for describing stacking with an electrokinetic injection. This is quite descriptive, because the electric field is indeed amplified over the injection zone when low-conductivity injection solutions are employed compared with the supporting electrolyte.

To perform FAI, the sample must be dilute and dissolved in water. The capillary is filled with the supporting electrolyte, and a small plug of water is hydrodynamically injected at the head of the capillary. This water plug is to ensure a high electric field at the point of injection. Sampling bias characteristic of electrokinetic injection is reduced, though not eliminated, with the water plug (27). Following introduction of the water plug, the capillary–electrode assembly is immersed in the sample, and a voltage is applied for injection. Injection voltages as high as 30 kV can be employed.

The mechanism of peak compression is similar to ionic-strength mediated stacking and CITP. As the front of the zone enriches, the local ionic strength increases, dropping the field strength in that region. Solute at the rear of the band is still exposed to a high field strength and continues to migrate at full speed until it encounters the higher ionic-strength region. The water plug differentiates this form of peak compression from conventional stacking; the stacked solute provides the higher ionic-strength region. In this fashion, this mode of stacking resembles peak compression by CITP.

The length of the injection plug using FAI can be estimated from (27)

$$X_{\mathrm{i}} = \left(\frac{\mu_{\mathrm{eo}}}{r} + \mu_{\mathrm{ep}} \right) Et \, , \tag{9.17}$$

where r is the ratio of the resistivities of the supporting electrolyte and injecting solution, respectively. For conventional electrokinetic injection, the expression simplifies because r=1. Thus, in FAI, the impact of the EOF on the band width is reduced, so higher voltages or longer injection times can be employed without excessive bandbroadening.

Representative electropherograms are given in Fig. 9.14 a–c for (a) hydrodynamic injection; (b) electrokinetic injection with the sample prepared in water; and (c) electrokinetic injection using a water plug in front of the aqueous sample. An LOD of 10^{-8}M for PTH–Arg has been reported (27).

A consequence of FAI is selective ion injection (28). It is possible to enhance selectivity and speed separations by introduction of either cationic or anionic species exclusively.

Characteristically for FAI, a short plug of water is injected into the capillary (gravity injection with a height of 15 cm for 30 s has been reported) to ensure a high field strength at the point of injection. For positive-ion electroinjection,

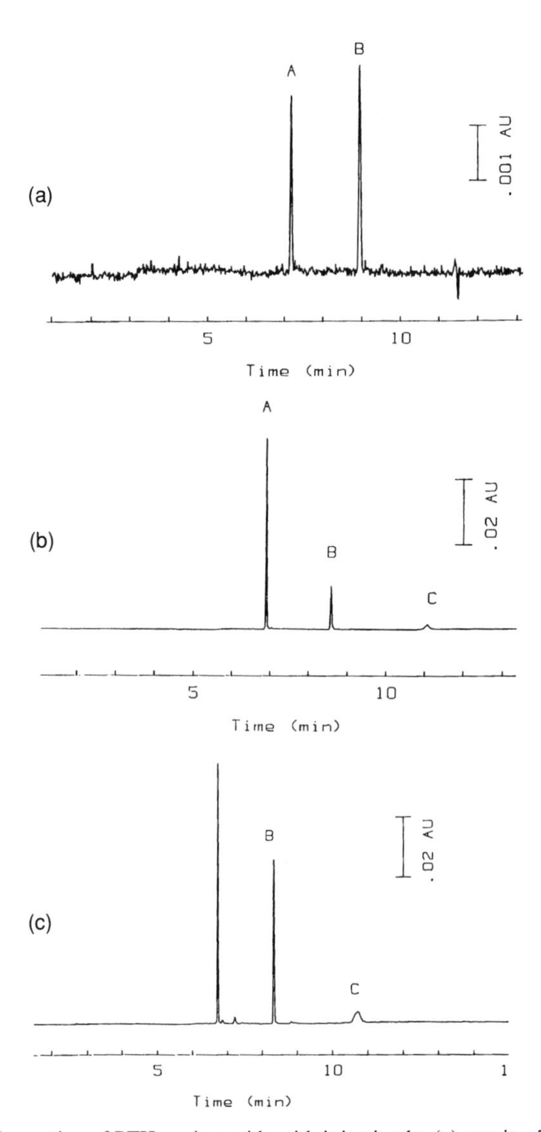

Figure 9.14. Separation of PTH–amino acids with injection by (a) gravity, 7.6 cm for 10 s; (b) electrokinetic injection with sample dissolved in water; (c) field-amplified injection where a small plug of water is introduced into the capillary prior to electrokinetic injection. All electrokinetic injections were at 5 kV for 10 s Capillary: 100 cm (75 cm to detector) × 75 μm i.d.; buffer: 100 mM MES-HIS, pH 6.2; voltage: 30 kV. Solutes: (A) PTH–Arg; (B) PTH–His; (C) neutral marker. Reprinted in part with permission from *J. Chromatogr.* **559,** 141, copyright ©1991 Elsevier Science Publishers.

the sample side of the capillary is held at positive polarity. Nearly all of the electric field is placed over the water and sample zones. Thus, only cations electrophorese into the capillary; the anions migrate toward the positive electrode and never enter the capillary. Since the field strength is carried over the narrow injection zone, the EOF that is averaged over the entire capillary is very small. In other words, the electrophoretic migration of a solute of any charge is greater than the total EOF.

For injections of negative ions, a water plug is injected, followed by the sample, using negative polarity. After injection, the capillary is returned to run buffer and the polarity is reversed to sample-side positive. The anions quickly encounter the run buffer as they electrophorese toward the positive electrode. The field strength over the sample zone decreases, so the anions do not reenter the electrode reservoir. As the water plug mixes with buffer, the field strength redistributes evenly over the entire capillary, and the separation proceeds as usual.

To inject both cations and anions, inject the water plug, then, with positive polarity, electroinject the cations. Switch the polarity to negative to electroinject the anions, return the capillary to the run buffer, and reverse the polarity again to run the separation. For all these injection methods, FAI stacking is in full force because of the differences between the ionic strengths of the running electrolyte and the injection solution.

The merits of these FAI techniques are illustrated in Table 9.5 and compared to gravity injection and conventional electrokinetic injection. Substantial discrimination is obtained with both the positive- and the negative-ion modes. If the polarity switching technique is used for the introduction of both cations and anions, the technique is biased toward cations. Like most of these stacking modes of injection, the precision has not been adequately studied. The sample matrix can have an extreme effect on both the precision and the quantitative accuracy of these techniques. It is extremely important that a consistent injection matrix be employed.

Table 9.5. Study Peak Heights[a] for Anions and Cations Using FAI Compared to Conventional Injection Techniques.

	PTH–Arg	PTH–His	PTH–Asp	PTH–Glu
Gravity injection	1	1	1	1
Electroinjection[b]	0.311	0.225	0.025	0.022
FAI—positive ion mode	28.04	13.44	0	0
FAI—negative ion mode	0	0	13.66	12.58
FAI—dual ion mode	32.03	9.28	2.58	1.96

[a]Peak heights normalized to gravity injection.

[b]Injected using the run buffer, 100 mM MES-HIS. Data from *J. Chromatogr.* **559,** 153 (1991).

9.7 Other Enrichment Techniques

There are several other on-line enrichment techniques that have been studied to address the general detection problem. The two discussed next are multidimensional processes employing either isotachophoresis or liquid chromatography for trace enrichment coupled directly to CZE.

A. LC/CZE

Merion *et. al.* (30) reported on the use of a 1–2 mm plug of a polymeric reverse-phase packing contained at the head of a 75 μm i.d. capillary to effect trace enrichment of peptides. This capillary, depicted in Fig. 9.15, is commercially available.[3] Large volumes of aqueous injection buffer can be loaded into the capillary by electrokinetic injection. Enrichment occurs through binding of hydrophobic solutes to the polymeric packing. After loading, a small volume of organic solvent is injected to elute the solutes into the capillary, after which CZE is performed in the usual manner. CLODs for peptides as low as 1 ng/mL have been reported, with migration time and peak area precision of better than 1.5%.

A separation of angiotensin I at two different concentration levels is shown in Fig. 9.16. With a starting peptide concentration of 1 μg/mL, trace impurities down to 0.02% of the parent are easily detected. A disadvantage of this technique is the long injection time required to introduce sufficient material into the system.

In a related development, Cai and El Rassi (31) employed an open-tubular 20 cm × 50 μm i.d. C_{18}-treated capillary to on-line enrich some triazine herbicides before CZE on a second untreated capillary. Enhancements of 10–35 were reported without affecting separation efficiency. The use of the open-tubular configuration permits the use of either electrokinetic or hydrodynamic injection.

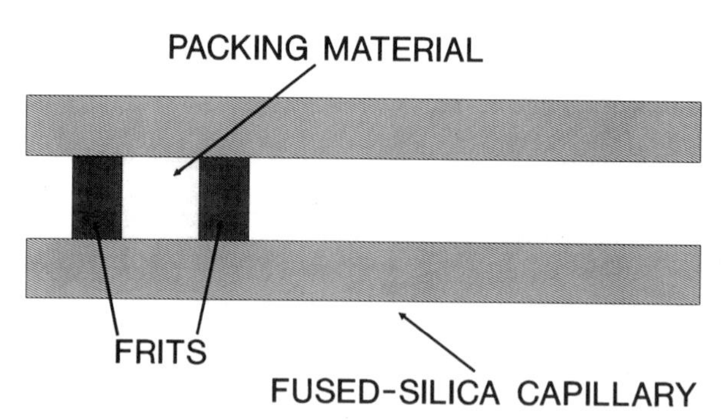

Figure 9.15. Fused-silica capillary with a preconcentration zone of chromatographic material.

[3] Waters Division of Millipore, Milford, MA.

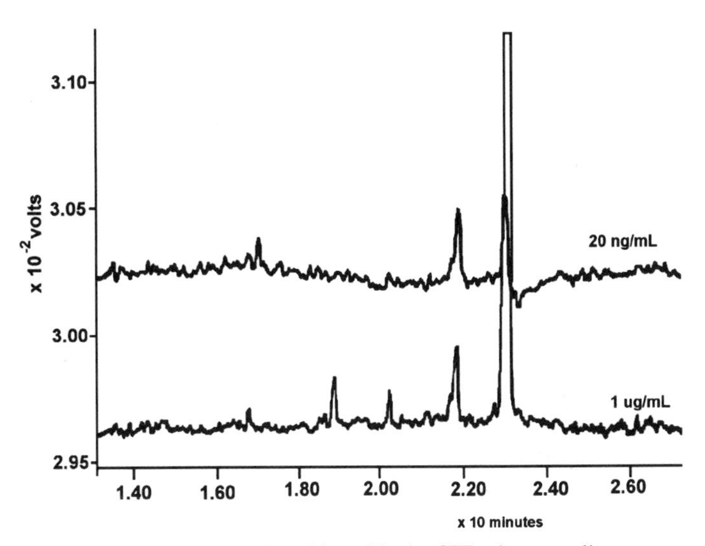

Figure 9.16. Separation of angiotensin and impurities by CZE using an on-line preconcentration capillary. Capillary: 60 cm × 75 μm i.d. with 1 mm of polymeric packing; buffer: 25 mM citrate, pH 4.0; eluting buffer: 25 mM citrate, pH 4.0/acetonitrile, 25/75; injection: 999 s at 15 kV; elution: 10 s at 7.5 kV; run voltage: 15 kV; angiotensin at 20 mg/mL (top) and 1 ng/mL (bottom). Reprinted with permission from Waters Associates.

B. CITP/CZE

Coupling the advantages of CITP and CZE has been reported by several researchers (32–37). The CITP capillary can be relatively wide to permit injection of a sufficient quantity of sample for sensitive detection. Following the CITP preconcentration, the focused analytes are then cut over to the CZE capillary. A potential instrument design is shown in Fig. 9.17. The multiple-electrode arrangement facilitates timed electrokinetic transfer between the two capillaries.

There are several limitations to this technique:

1. CITP is a displacement process, so solutes do not have absolute migration times. The sample matrix can influence the migration times of various zones. Detection of the CITP zones can be difficult because dilute samples are generally employed.
2. A compromise must be made in buffer selection. The CITP leading electrolyte ultimately becomes the CZE supporting electrolyte.
3. It may not be possible to cut all solutes from the CITP separation over to the CZE capillary. This is particularly true when solutes of widely varying mobilities are being separated.
4. Only cations or anions can be separated in a single run. This is a consequence of the CITP process.

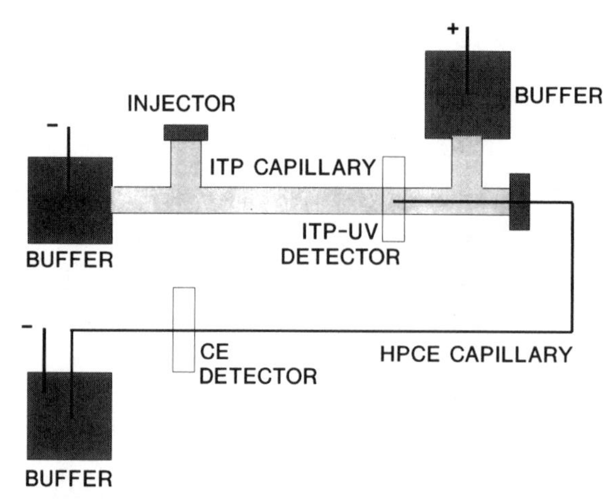

Figure 9.17. An CITP/CZE instrument diagram. Redrawn as described from *J. Chromatogr.* **538,** 393 (1991).

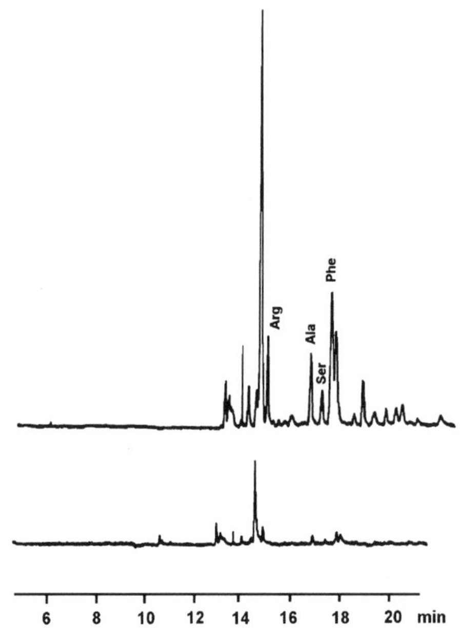

Figure 9.18. Electropherograms of FITC-derivatized amino acids by CZE (lower trace) and CITP/CZE (upper trace). CZE conditions: capillary: 75 cm × 50 μm i.d.; buffer: 5 mM borate, pH 9.5; injection: electrokinetic, 5 kV for 5 s; voltage: 25 kV; solutes: 10 mM amino acid–FITC derivative mixture. CITP/CZE conditions: capillary: 61 cm × 500 μm i.d.; leader: 5 mM borate, pH 9.5; terminator: 5 mM ACES, pH 10.0; CITP injection: 25 μL; CZE cut: 5 kV for 5 s; CITP voltage: 10 kV. Detection: laser fluorescence, 488/515 nm. Reprinted with permission from *J. Chromatogr.* **538,** 393, copyright ©1991 Elsevier Science Publishers.

5. Coupling CITP to MECC may not be possible. This would prove limiting for many small-molecule applications.

Despite these problems, CITP/CZE has the potential of providing 2–3 orders of magnitude of improved sensitivity over CZE alone. This is illustrated in Fig. 9.18 for the separation of some FITC amino acids. Problems with complex sample matrices such as biological fluids are addressed in part by this technique, because only the analytes are cut onto the CZE capillary. This technique shows great promise for the future once problems with both sample handling and sensitivity are addressed. There is no commercial equipment available at this time.

Another mode of CITP/CZE, otherwise entitled "transient CITP," does not require any special instrumentation (37). A 50-fold improvement in pre-concentration was reported utilizing a single capillary. Experimental details are described in Ref. 37.

References

1. F. Foret, M. Deml, and P. Boček, *J. Chromatogr.* **452,** 601 (1988).
2. K. Otsuka and S. Terabe, *J. Chromatogr.* **480,** 91 (1989).
3. H. K. Jones, N. T. Nguyen, and R. D. Smith, *J. Chromatogr.* **504,** 1 (1990).
4. E. Grushka and R. M. McCormick, *J. Chromatogr.* **471,** 421 (1989).
5. X. Huang, W. F. Coleman, and R. N. Zare, *J. Chromatogr.* **480,** 95 (1989).
6. D. J. Rose and J. W Jorgenson, *Anal. Chem.* **60,** 642 (1982).
7. D. E. Burton, M. J. Sepaniak, and M. P. Maskarinec, *Chromatographia* **21,** 583 (1986).
8. J. Tehrani, R. Macomber, and L. Day, *J. High Res. Chromatogr.* **14,** 10 (1991).
9. X. Huang, M. J. Gordon, and R. N. Zare, *Anal. Chem.* **60,** 375 (1988).
10. D. Demorest and R. Dubrow, *J. Chromatogr.* **559,** 43 (1991).
11. H. E. Schwartz, M. Melera, and R. G. Brownlee, *J. Chromatogr.* **480,** 129 (1989).
12. T. Satow, A. Machida, K. Funakushi, and R. Palmieri., *J. High Res. Chromatogr.* **14,** 276 (1991).
13. J. A. Lux, H.-F. Yin, and G. Schomburg, *Chromatographia* **30,** 7 (1990).
14. E. V. Dose and G. Guiochon, *Anal. Chem.* **64,** 123 (1992).
15. R. A. Hartwick, paper presented at HPCE '89, Boston.
16. S. Hjertén, paper presented at HPCE '92, Amsterdam.
17. H. F. Yin, S. R. Motsch, J. A. Lux, and G. Schomberg, *J. High Res. Chromatogr.* **14,** 282 (1991).
18. M. C. Linhares and P. T. Kissinger, *Anal. Chem.* **63,** 2076 (1991).
19. T. Tsuda, T. Mizuno, and J. Akiyama, *Anal. Chem.* **59,** 799 (1987).
20. T. Tsuda and R. N. Zare, *J. Chromatogr.* **559,** 103 (1991).
21. R. A. Wallingford and A. G. Ewing, *Anal. Chem.* **59,** 678 (1987).
22. R. Weinberger, E. Sapp, and S. Moring, *J. Chromatogr.* **516,** 271 (1990).
23. F. M. Everaerts, T. P. E. M. Verheggen, and F. E. P. Mikkers, *J. Chromatogr.* **169,** 11 (1979).
24. D. S. Burgi and R. L. Chien, *Anal. Chem.* **63,** 2042 (1991).
25. R. L. Chien and D. S. Burgi, *Anal. Chem.* **64,** 1046 (1992).

26. R. Aebersold and H. Morrison, *J. Chromatogr.* **516,** 79 (1990).
27. R. L. Chien and D. S. Burgi, *J. Chromatogr.* **559,** 141 (1991).
28. R. L. Chien and D. S. Burgi, *J. Chromatogr.* **559,** 153 (1991).
29. R. L. Chien and D. S. Burgi, *Anal. Chem.* **64,** 489A (1992).
30. M. Merion, R. H. Aebersold, and M. Fuchs, paper presented at HPCE '91, San Diego.
31. J. Cai and Z. El Rassi, *J. Liq. Chromatogr.* **15,** 1179 (1992).
32. V. Dolník, K. A. Cobb, and M. Novotny, *J. Microcol. Sep.* **2,** 127 (1990).
33. F. Foret, V. Sustacek, P. Boček, *J. Microcol. Sep.* **2,** 229 (1990).
34. D. Kaniansky and J. Marák, *J. Chromatogr.* **498,** 191 (1990).
35. D. S. Stegehuis, H. Irth, U. R. Tjaden, and J. Van Der Greef, *J. Chromatogr.* **538,** 393 (1991).
36. D. S. Stegehuis, U. R. Tjaden, and J. Van der Greef, *J. Chromatogr.* **591,** 341 (1992).
37. F. Foret, E. Szoko, and B. L. Karger, *J. Chromatogr.* **608,** 3 (1992).

Chapter 10

Detection

10.1 On-Capillary Detection

On-capillary detection is generally required for HPCE because construction of a postcapillary flow cell is not yet practical considering the minuscule dimensions of the capillaries. The characteristics of on-capillary detection differ dramatically from those of postcolumn detection in HPLC.

In chromatography, solutes move through the chromatographic packing at velocities determined by the mobile-phase flow rate and the overall retention characteristics of each analyte. On the column, the peak velocities depend on each solute's retention characteristics. Once off the column, all solutes are swept past the detector by the mobile phase at uniform flow rates. The detected peak widths are a function of chromatographic processes and are not related to the peak velocity past the flow cell.

In HPCE, a different set of rules applies (1). The migration velocity of each solute through the capillary is a function of its electrophoretic mobility in conjunction with the EOF. Since detection occurs on-capillary, these forces are operative as the solute is transiting the detection window. As a result, slower-moving components spend more time migrating past the detector window than do their more rapidly moving counterparts.

Fig. 10.1a illustrates the separation of a three-component mixture recorded directly from the detector output. Fig. 10.1b gives the same separation, corrected for the zonal velocity. This correction is easily calculated by

$$w_s = (L_d / t_m)\, w_t - w_d\,, \qquad (10.1)$$

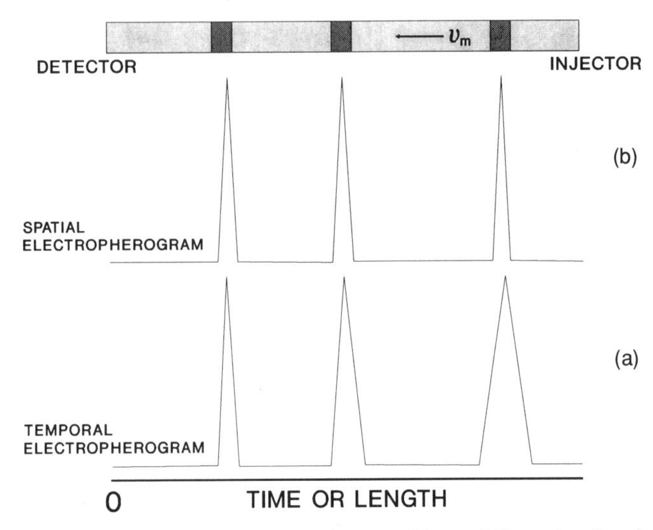

Figure 10.1. Plots of the detector response (a) as function of time and (b) as a function of zone length within the capillary. Redrawn with permission from *J. Chromatogr.* **480,** 95, copyright ©1989 Elsevier Science Publishers.

where w_s is the *spatial* width of the sample in units of length; L_d is the effective capillary length; t_m is the migration time; w_t is the recorded *temporal* width in time units; and w_d is the spatial width of the detector window.

There are practical implications to this phenomenon, because more slowly moving zones produce an increase in the peak area counts. When quantifying solutes by a response factor, a correction factor must be applied to normalize the peak area irrespective of the migration velocity:

$$A_{corr} = \frac{A_{raw}}{t_m}, \tag{10.2}$$

where A_{raw} is the measured peak area, t_m is the migration time, and A_{corr} is the corrected peak area.

Response factors are frequently employed when standards are unavailable, as in oligonucleotide separations (2). For that particular application, it is assumed that all oligonucleotides have the same detector response per unit of mass. In addition to response factors, area % calculations with the assumption of uniform molar absorptivities are used for the determination of impurities in bulk pharmaceuticals, using low-UV detection.

When standards are used, it is unnecessary to provide this correction, since it is assumed that the standard behaves identically to the solute, if the migration times are constant.

10.2 The Detection Problem

Because of the minute amounts of material that can be injected into the capillary, extremely high-sensitivity detection is generally required for all forms of HPCE. The problem is exacerbated by the desire to dilute samples to rid separations of troublesome matrix effects. The instrumental problems for optical detection are two fold: (i) the short optical pathlength as defined by the capillary i.d.; (ii) the poor optical surface of the cylindrical capillary. Perhaps these problems may be solved in part by using square or rectangular capillaries (3), which are now becoming commercially available.[1]

Because of the absence of sensitive detectors, it is often necessary to perform tedious sample-preparation steps to place the solutes in a CE-friendly matrix. Insensitive detection also reduces the linear dynamic range of the technique (see Section 11.4), since high solute concentrations that cause ohmic disturbances must be employed.

Many commercial instruments use absorption detectors that are modifications of standard HPLC detectors. One instrument manufacturer (Dionex) markets a fluorescence detector, and two firms (Beckman and Europhor) market laser fluorescence in conjunction with HPCE. The absorption detectors are modestly sensitive, giving limits of detection of 10^{-6} M for solutes with very high molar absorptivity. While laser fluorescence detection can improve sensitivity down to 10^{-12} M with commercially available equipment and approach single-molecule detection in more sophisticated apparatus, derivatization is usually required to tag a solute with an optimized fluorophore.

New invention is still required to solve the general detection problem in HPCE. Improvements of 2–3 orders of magnitude in absorption detection would solve many problems relating to matrix effects and linear dynamic range. With the high resolving power of CE, less selective detectors will prove useful, providing the sensitivity requirements are fulfilled.

10.3 Classes of Detectors

A tabulation of detectors that have been used in HPCE is given in Table 10.1.[2] Presently, only absorption and fluorescence detectors are commercially available. The interface to the mass spectrometer is slowly being perfected. Most of the literature cited here deals with detection schemes developed in academic laboratories. Several of these new techniques, e.g., conductivity detection, are potential commercial entities in the near future.

[1] Polyimide-coated rectangular capillaries are available from R&S Medical, P.O. Box 344, 91 North Pocono Road, Mountain Lakes, NJ 07046.

[2] The literature describing capillary LC detection may be consulted for further information.

Table 10.1. Detectors for HPCE

Technique	References[a]
Absorbance	4–15
Absorbance, energy transfer	16
Absorbance, indirect	17–33
Absorbance, multiwavelength	34–37
Absorbance, thermo-optical	38–41
Chemiluminescence	42
Circular dichroism	43
Concentration gradient	44–46
Electrochemical, conductivity	7, 20, 47–51
Electrochemical, ampeometric	50, 52–63
Electrochemical, indirect	64
Fluorescence	65–67
Fluorescence, indirect	68–73
Fluorescence, laser	74–99
Fluorescence, microscopy	100–103
Fluorescence, multiwavelength	104
Fluorescence, postcapillary derivatization	67, 105–109
Ion mobility spectrometry	110
Mass spectrometry	111–146
Radioactivity	147–149
Raman	150–151
Refractive index	152–154

[a]The literature citations for absorption detection are limited to hardware developments. Other references are for both applications and hardware.

Detectors fall into one of two broad categories: bulk property or solute property detectors. The bulk property detector measures a general property of matter; refractive index and conductivity detection are the most important of the class. These detectors are not selective and tend to be less sensitive than the solute property class. The sensitivity is enhanced if a solute's measured property is maximally differentiated from that of the background electrolyte. This places an additional constraint in methods development.

Solute property detectors measure a physical property that is specific to the solute compared to the background electrolyte. These are represented by absorption, fluorescence, electrochemical, and radioactivity detectors.

Absorption detectors meet the criteria for solute-specific detection only under certain conditions. In the low UV region, the background electrolyte has significant UV absorption. This poses problems in HPCE, particularly in CGE and CIEF. Both gels and carrier ampholytes absorb strongly in the low UV, despite the short optical pathlength defined by the capillary. The optimal wavelength for detection of proteins is 280 nm for these HPCE modes. Unfortunately, the molar absorptivity of peptides is poor at this wavelength and is dependent on the number of tryptophans contained in the peptide chain. In CZE, buffer absorption is less of a problem, so 200 nm detection is practical. Even lower wavelengths can be used under certain conditions.

Fluorescence detection is far more sensitive and selective than absorption detection. All molecules that fluoresce must first absorb light. The converse is not true, since most molecules do not fluoresce. The resultant selectivity is a double-edged sword. Because the technique is selective, derivatization is frequently required to take advantage of the detector's inherent high sensitivity. Sensitivity of detection is high, since the measurement of low signal levels above a very dark background is more easily measured than a small difference between two high-intensity signals, as in absorption detection.

Information-rich detection represents another important group of detectors. Since fraction collection is difficult, it is important to obtain additional information about a solute on-line. The mass spectrometer and multiwavelength absorption detectors are the most important examples of the group.

Another means of categorizing detectors is the nature of the response toward the solute. Most detectors used for HPCE are concentration-sensitive. They respond in proportion to the concentration of the solute as it transits the detection window. Peak area counts obtained are also dependent on the peak velocity. The mass spectrometer is the most notable exception of the group. Sensitivity is based on the number of formed ions, so this instrument responds to the mass of material that enters the source. Extracapillary dispersion from the interface is a small sacrifice for obtaining such information-rich data.

Much of the general information described about LC detectors is also applicable to HPCE. For readers requiring more information on this subject, Scott's text (155) on LC detection will prove a valuable reference source. Yeung's book (156) on the same subject provides more information on miniaturized detector systems.

10.4 Limits of Detection

There are two means of describing the limits of detection of a system: the concentration limit of detection (CLOD) and the mass limit of detection (MLOD). The CLOD relates to the concentration of the individual sample, whereas the MLOD describes what the instrument can measure. For example, the CLOD of many peptides is around 1 μg/mL. If 10 nL of that material is injected and detected at three times the baseline noise, the MLOD is 10 pg.

Another way of considering MLOD is based on the volume of the detector window. If a 1 μg/mL peptide solution is continuously aspirated through a capillary, then for a 1 nL detector window, the amount of material in the window at any given time is 1 pg. Thus, the MLOD can be manipulated by selecting the size of the detection window. It is frequently possible to improve the MLOD by compressing the detection window at the expense of the CLOD. For most analytical problems, the CLOD is the more important parameter, because it relates to the minimum detectable quantity of a solute in the sample of interest. In extreme cases where the amount of available sample is minuscule, the MLOD becomes the more important parameter describing the LOD.

It is easily concluded the HPCE has excellent MLODs but poor CLODs, especially when compared to optical detection in HPLC.

10.5 Bandbroadening

Two detector-related features can contribute to bandbroadening: the length of the detector window, and the detector time constant. For all practical purposes, the detector-related contribution to peak variance is the same as the injection contribution (157,158), so

$$\sigma_{\mathrm{det}}^2 = \frac{l_{\mathrm{d}}}{12}, \tag{10.3}$$

where σ_{det}^2 is the detector related peak variance and l_{d} is the length of the detector window. Data given in Table 9.2 are applicable to determine if the detector volume is sufficiently small. A 500,000 plate peak with $t_{\mathrm{m}} = 10$ min requires $l_{\mathrm{det}} < 0.56$ mm. Most commercial absorption detectors meet this constraint.

All detectors employ smoothing algorithms to decrease detector noise. To the user, the adjustable parameter is the time constant or rise time. Virtually all modern detectors use rise time functionality, normally contained within the system software. The rise time is approximately twice the effective RC time constant. These newer algorithms permit enhanced smoothing without excessive bandbroadening. Insufficient detector time constant settings needlessly sacrifice signal/noise (*S/N*) ratio, whereas higher values can result in bandbroadening. Scott (155) determined that the detector time constant should be no greater than 10% of the temporal peak variance. Table 10.2 provides a guide for selecting the time constant based on the temporal peak variance. Fig. 10.2 illustrates the impact of detector rise time for a 40,000 plate separation with $v_{\mathrm{m}} = 2.1$ mm/s. Note that 0.5 s is slightly higher than optimal, as evidenced by the reduced peak height (and the tabular value for σ_{t}), but *S/N* is still improved upon. Even at a rise time of 1 s, the peak is still Gaussian with further improvement in *S/N*. As with injection, it is possible to trade some plates for improved sensitivity. For high-resolution separations, the trade-off is far more difficult.

Table 10.2. Selection of Detector Time Constant for a 100 cm capillary and $v_m = 2$ mm/s

N	σ_v (mm)[a]	σ_t (s)[b]	Rise Time[c]
50,000	4.47	2.24	0.5
100,000	3.16	1.58	0.2
500,000	1.41	0.70	0.1
1,000,000	1.00	0.50	0.05

[a] $\sigma_v = L/N_{1/2}$.
[b] $\sigma_t = \sigma_v/v_m$.
[c] Converted into units of rise time and rounded to numbers consistent with commercial detector settings.

10.6 Absorption Detection

A. Types of Detectors

Ultraviolet/visible absorption detection is by far the most popular technique used today. Several types of absorption detectors are available on commercial instrumentation, including

1. fixed-wavelength detector using mercury, zinc, or cadmium lamps with wavelength selection by filters (Waters);

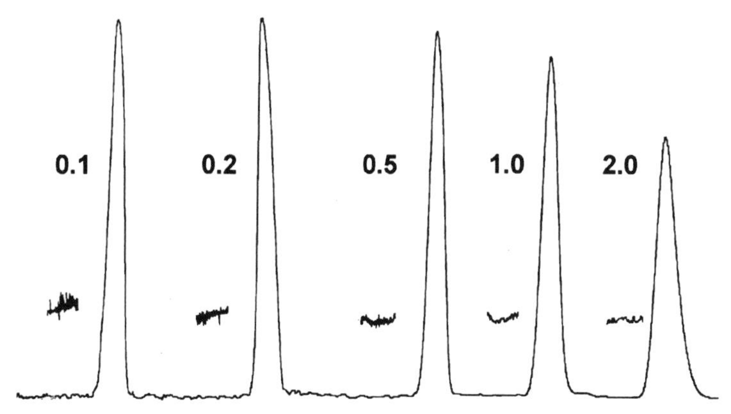

Figure 10.2. Impact of detector rise time on signal, peak shape, and noise. Capillary: 46 cm (35 cm to detector) 50 μm i.d.; buffer: 50 mM SDS, 20 mM borate, pH 9.3; voltage: 22 kV, 33 μA; detection: UV, 220 nm; solute: diflunisal, 13.8 μg/mL; noise traces at left of peak run at chart speed of 1 cm/min, attenuation 4; signal traces run at chart speed of 20 cm/min, attenuation 8. Rise time values in seconds are to the left of each peak.

2. variable-wavelength detector using a deuterium or tungsten lamp with wavelength selection by monochromator (Isco, Applied Biosystems);
3. filter photometer using a deuterium lamp with wavelength selection by filters (Beckman);
4. scanning UV detector (SpectraPhysics, BioRad);
5. photodiode array detector (Hewlett-Packard).

Each of these absorption detectors have certain attributes that are useful in HPCE. Clearly, multiwavelength detectors such as the photodiode array or scanning UV detector are valuable because spectral as well as electrophoretic information can be displayed (Fig. 10.3). These detectors are frequently less sensitive when used in the scanning modes, since signal averaging must be carried out more rapidly than for single-wavelength detection.

Even a filter photometer can be invaluable. Wavelength calibration never varies, so the lines defined by the common atomic vapor lamps such as mercury, zinc, and cadmium can be considered primary standards. Another application is for low-UV detection. The simple optical design reduces the generation of UV-absorbing ozone that causes problems with monochromator-based instruments. The use of the 185 nm mercury line becomes practical in CZE with certain buffers, since the short optical pathlength minimizes the background absorption. Because peptides absorb strongly at 185 nm, sensitivity is frequently enhanced, as shown in Fig. 10.4a,b. Limits of detection are improved by a factor of six for peptides and four for penicillins (9).

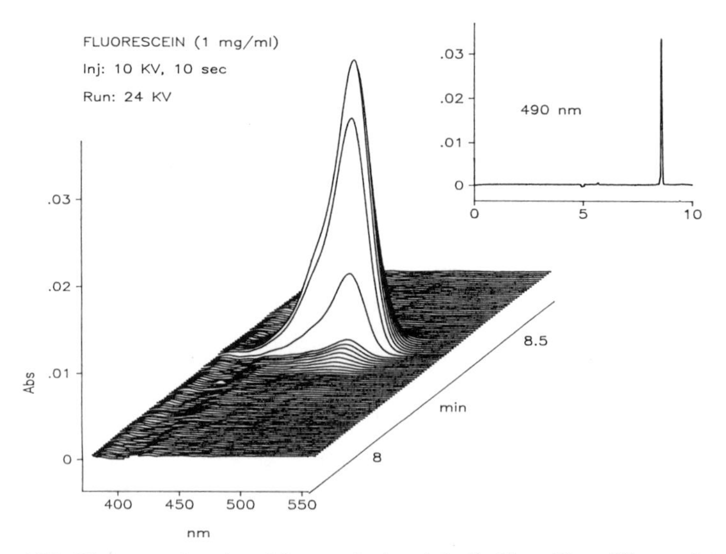

Figure 10.3. Diode-array detection of fluorescein, 1 mg/mL. Capillary: 72 cm (50 cm to detector) 50 μm i.d.; buffer: 20 mM CAPS, pH 11; injection: 10 kV, 10 s.

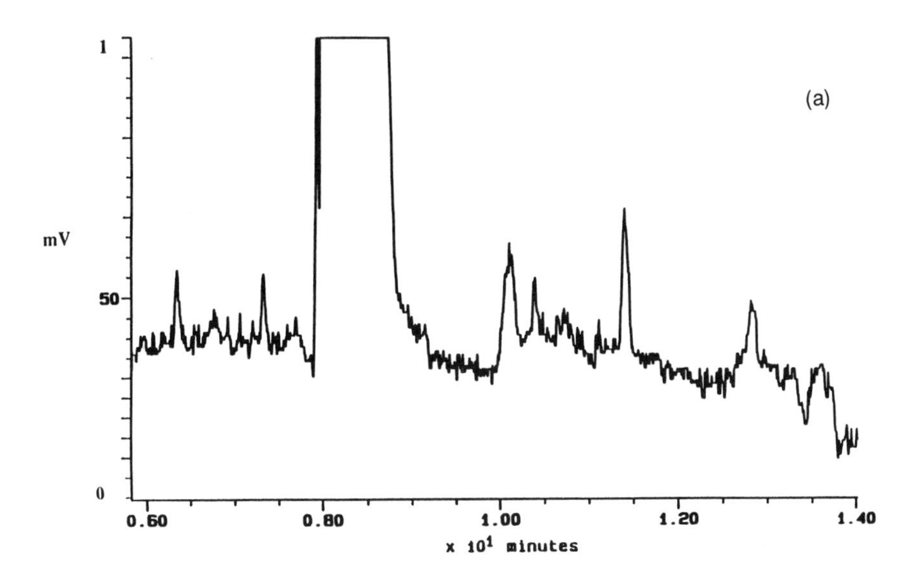

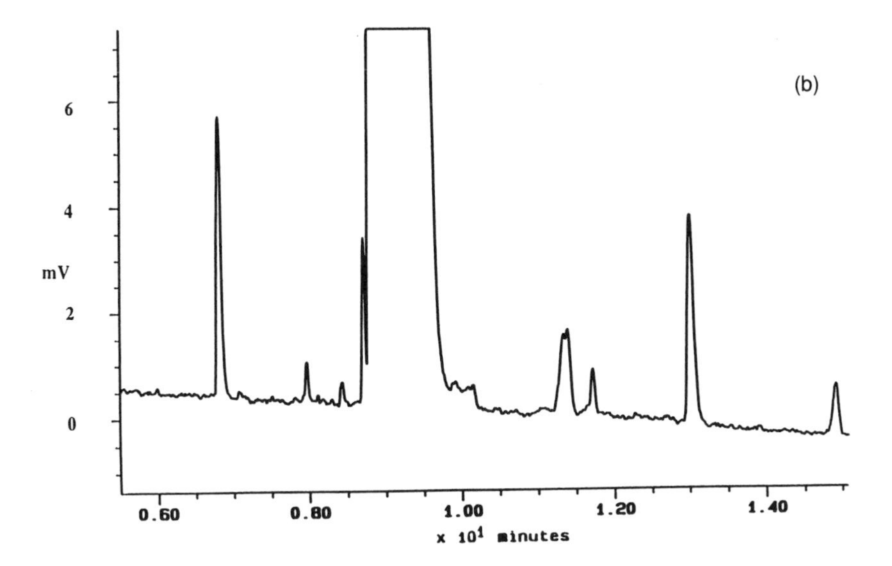

Figure 10.4. Detection of impurities in a synthetic pentapeptide (Asp–Ser–Asp–Pro–Arg) at 214 and 185 nm. Capillary: 60 cm × 75 μm i.d.; buffer: 100 mM phosphoric acid, pH 2; injection: gravity, 10 cm, 10 s; voltage: 12 kV. (a) 214 nm, 1 mg/mL, 12 kV; (b) 185 nm, 335 μg/mL. Courtesy of Waters Division of Millipore, Milford, MA.

B. Optimization

Because of the short optical pathlength, the optimal selection of wavelength is frequently different than in HPLC. With a 1 cm pathlength, HPLC mobile phases absorb too strongly to allow the use of low-UV wavelengths. This is particularly troublesome with gradient elution because substantial baseline drift is generally encountered. In HPCE with a variable-wavelength absorption detector, the optimal signal-to-noise ratio for peptides is found at 200 nm. This is illustrated in Fig. 10.5 with electropherograms of some dynorphins obtained at wavelengths of 200 nm, 214 nm, and 280 nm.

To select the optimal wavelength, it is necessary to plot the signal-to-noise ratio. It is possible to do this in a few minutes without actually performing any real runs. First, determine the effective detector noise at 5 nm intervals using run buffer and applied voltage. Similarly, aspirate some sample through the capillary and repeat each measurement. It is generally necessary to re-autozero the instrument against run buffer after each wavelength change. Finally, calculate the signal-to-noise ratio at each wavelength and select the optimal value.

C. Increasing the Optical Pathlength

Increasing the optical pathlength of the capillary window should increase *S/N* simply as a result of Beer's Law (6). This may be achieved in several ways. One

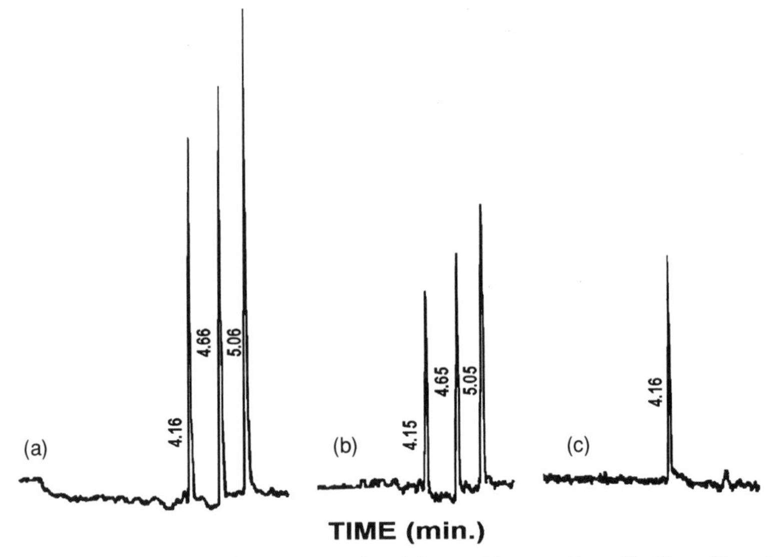

Figure 10.5. Optimization of detector wavelength for peptide separations. Capillary: 65 cm (43 cm to detector) × 50 mm i.d.; buffer: 20 mM citrate, pH 2.5; voltage: 25 kV; temperature: 30°C; sample: dynorphins, 50 μg/mL. (a) 200 nm, attenuation 8×; (b) 214 nm, attenuation 4×; (c) 280 nm, attenuation 4×.

cell is commercially available[3] and is configured as shown in Fig. 10.6. This so-called Z-cell has a pathlength of 3 mm. Electropherograms that are given in Fig. 10.7 showed an *S/N* improvement of only six-fold despite a 60 times longer pathlength. Inadequate focusing of the light is probably responsible for this disproportionate observation. The sensitivity of absorption detectors is related to the amount of light reaching the photodiodes. Since the Z-cell attenuates the light, the *S/N* improvements are not proportional to the increased pathlength. The increased volume of the flow-cell gives rise to 20–30% bandbroadening.

Another means of increasing the detector optical pathlength is to use rectangular capillaries (3). Fig. 10.8 compares the sensitivity viewing across the 50 μm capillary axis versus the 1,000 μm axis. The theoretical improvement should be a factor of 20 by viewing along the long axis, and 15× is actually realized. Bandbroadening that is proportional to the optical pathlength is usually observed.

While increasing the optical pathlength improves sensitivity, efficiency is always degraded. Frequently it is practical to trade off a few plates for better

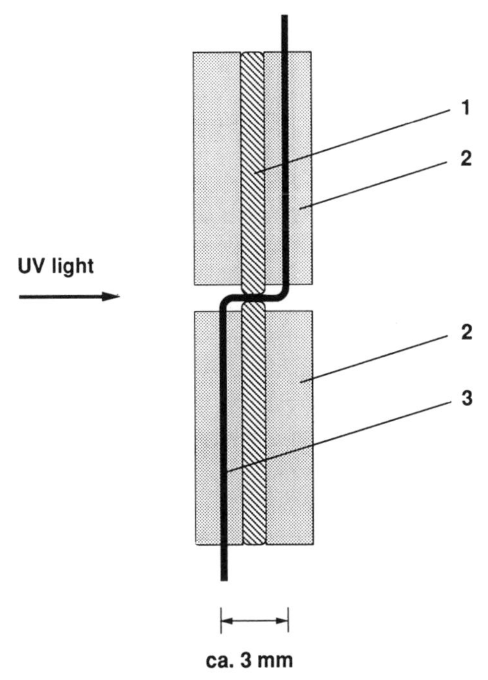

Figure 10.6. Schematic of a 3 mm Z-shaped capillary flow cell. 1 = shim with centered 300 μm i.d. hole; 2 = plastic disks; 3 = fused-silica capillary. Reprinted with permission from *J. Chromatogr.* **543,** 439 (1991).

[3] LC Packings, 80 Caroline Street, San Francisco, CA 94103. A Z-cell optimized for the ABI 270 HT is available from Applied Biosystems, Inc.

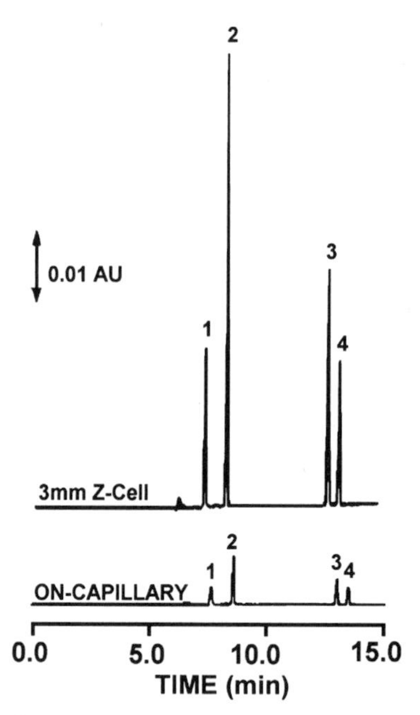

Figure 10.7. Separation of nucleosides by MECC with the capillary Z-cell. Capillary: 60 cm (40 cm to detector) 75 μm i.d.; buffer: 6 mM borate–10 mM phosphate, 75 mM SDS, pH 8.5; voltage: 11 kV; injection: gravity, 10 cm, 5 s; detection: UV, 254 nm. Solutes: 50 μg/mL each: (1) 2′-deoxyxytidine; (2) 2′-deoxyguanosine; (3)2′-deoxyguanosine-5′-monophosphate; (4) 2′-deoxy-cytidine-5′-monophosphate. Reprinted with permission from *J. Chromatogr.* **543,** 439, copyright ©1991 Elsevier Science Publishers.

detectability. Further improvements in the design of absorption detectors should be possible through better management and focusing of the light. Simple focusing optics are already used on several commercial detectors, as illustrated in Fig. 10.9. It is important that the light pass through the i.d. of the capillary without leakage around the outside. Refractive index effects must also be limited by careful optical design. Perhaps more intense and stable deuterium lamps will become available in the future.

10.7 Fluorescence Detection

A. Basic Concepts

Instrumentation for fluorescence detection is slowly reaching the commercial marketplace. Fluorescence detection, even with noncoherent light sources, can improve the limits of detection by several orders of magnitude compared to

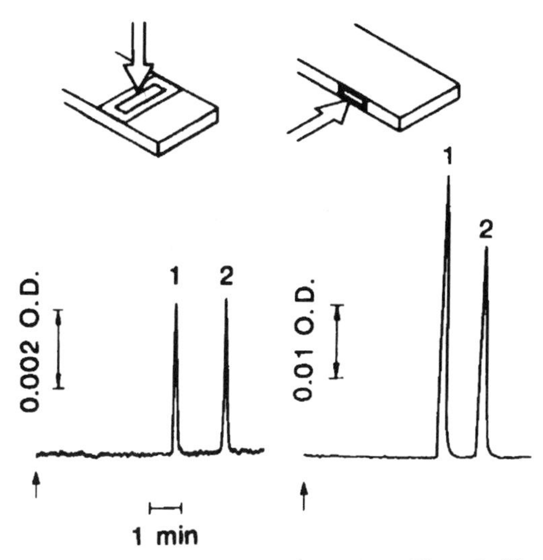

Figure 10.8. Enhancements in detection using a rectangular capillary. Capillary: 57 cm × 50 μm 1000 μm i.d.; buffer: 5 mM phosphate, pH 6.8, with 5% ethylene glycol; solutes: (1) pyridoxine and (2) dansyl-*L*-serine, 4.2×10^{-5} M. Reprinted with permission from *Anal. Chem.* **62,** 2149, copyright ©1990 Am. Chem. Soc.

absorbance detection. Unfortunately, there have been few applications reported for non-laser detection (65–67) and only one commercial instrument (Dionex) is available. The advantage of conventional lamp sources is tunability; lasers operate only at discrete wavelengths.

The fluorescence advantage is the result of detection against a very dark background. An optical transducer, the photomultiplier tube (PMT), is very sensitive to

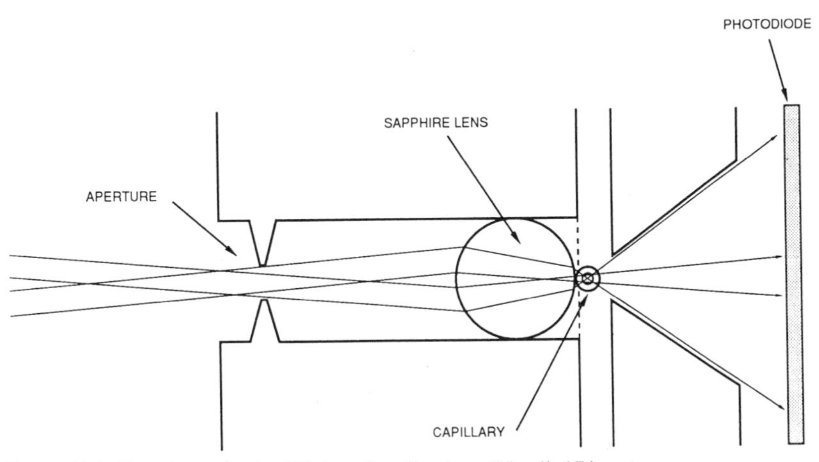

Figure 10.9. Focusing optics for UV detection. Courtesy of Applied Biosystems.

minuscule amounts of emitted light. Unlike absorbance detection, fluorescence detection is extremely dependent on the properties of the instrumental design. The fundamental equation governing fluorescence is

$$I_f = \Phi_f I_o abc E_x E_{fc} E_m E_{pmt}, \tag{10.4}$$

where I_f is the measured fluorescence intensity, Φ is the quantum yield (photons emitted/photons absorbed), I_o is the excitation power of the light source, *abc* are the Beer's Law terms, and the E terms are the efficiencies of the excitation monochromator or filter, the flow cell, the emission monochromator or filter, and the PMT, respectively. The situation is further obscured because I_o and all of the efficiency terms show wavelength dependence. This means that instrumental parameters must be carefully considered when developing an assay. Clearly, the optimization scheme for fluorescence is more complicated than for absorption detection (159), but the extra sensitivity and selectivity can be well worth it.

Various light sources are useful for fluorescence. Deuterium is useful for low-UV excitation, while the xenon arc is superior in the near UV to visible region. Xenon lamps can be very powerful, and limits of detection by CZE of 2 ng/mL for fluorescein is possible with simple fiber-optic collection of fluorescence emission (67). This is a 15-fold improvement compared to absorbance detection with a tungsten lamp at 490 nm and 100-fold improved over deuterium based absorption at 240 nm.

The design of a fluorescence detector for CE is relatively simple. One such design, which uses fiber optics to collect emitted radiation, is shown in Fig. 10.10.

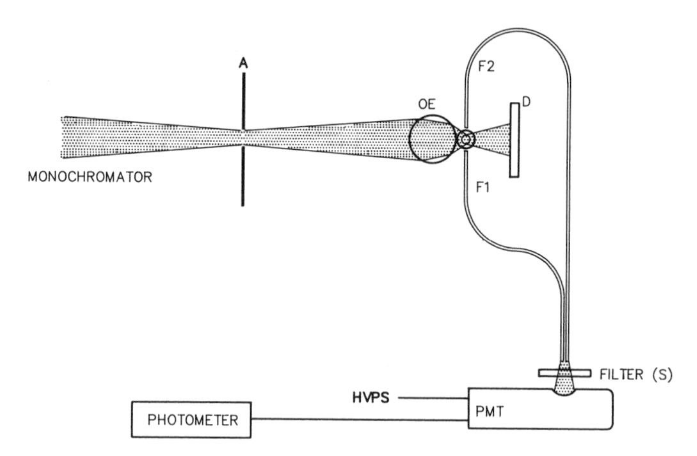

Figure 10.10. Schematic of a tuneable combination absorption–fluorescence detector for HPCE. Key: (A) slit; (OE) sapphire lens; (F1), (F2) optical fibers; (D) photodiode; (S) emission filter. This device is not commercially available. Reprinted with permission from *Anal. Chem.* **63**, 417, copyright ©1991 Am. Chem. Soc.

Many other ingenious designs are also possible, including epi-illumination microscopy (100–102).

Native fluorescence is useful for the determination of many drug substances and tryptophan-containing peptides. The specificity of fluorescence can result in high selectivity as illustrated in Fig. 10.11 for a peptide map. Only tryptophan-containing peptides have significant fluorescence.

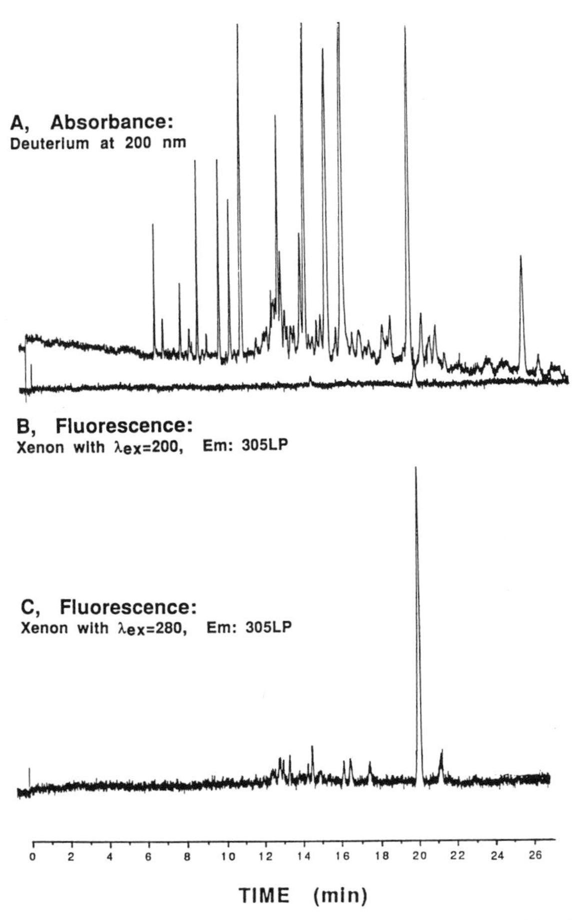

Figure 10.11. Native fluorescence detection of tryptophan-containing peptides from a tryptic digest of β-lactoglobulin. (a) Absorbance detection at 200 nm; (b) fluorescence detection with xenon lamp excitation, emission selected with a 305 nm long wavepass filter; (c) fluorescence detection with xenon arc excitation, emission selected with a 305 nm long wavepass filter. Capillary: 72 cm (50 cm to detector) × 50 μm i.d.; buffer: 20 mM sodium phosphate, 50 mM hexane sulfonic acid, pH 2.5; field strength: 278 V/cm; temperature: 30°C; injection: vacuum, 2 s; initial protein concentration: 20 nmol/mL. Reprinted with permission from *Anal. Chem.* **63,** 417, copyright ©1991 Am. Chem. Soc.

B. Optimization

Since the intensity of fluorescence is directly proportional to the lamp energy at the excitation wavelength, the proper wavelength must be selected. The optimal excitation wavelength is equal to the product of the lamp energy output and the strength of the absorption band. When using a deuterium lamp for excitation, this wavelength frequently corresponds to the absorption maxima of solutes in the low UV. The xenon arc source is better suited to solutes that absorb in the near-UV to the visible spectral region. For the tungsten lamp, only absorption bands in the visible wavelength region are useful.

After the selection of an excitation wavelength, the emission wavelengths must be selected. First obtain a fluorescence emission scan of the solute using a scanning spectrofluorometer. Cutoff or bandpass filters are usually employed to select emission wavelengths in HPCE. Bandpass filters with a 10–25 nm bandwidth are easiest to use because they can be matched to the emission maximum. Interference from specular, Raman, and Rayleigh scattering is less likely with bandpass filters. If a monochromator is used for excitation, never select a bandpass filter that is a whole-number multiple of the excitation wavelength. The monochromator passes higher orders of wavelengths that will raise the background. For example, if 250 nm is used for excitation, do not select a 500 nm bandpass filter.

It is useful to perform an emission scan of the background electrolyte using the proposed excitation wavelength. Search the emission spectrum for the Raman band. The Raman band is red-shifted 10–100 nm (or more) from the obvious Rayleigh band. This shift is excitation-wavelength and solvent-dependent. Using 254 nm excitation, the shift is approximately 20 nm; at 365 nm, the shift reaches 60 nm; while at 546 nm excitation, a Raman shift of 100 nm is normal. If the Raman band overlaps with your proposed emission filter, adjust the excitation wavelength or emission filter to eliminate the interference. Failure to do so can reduce detector sensitivity by an order of magnitude. Any fluorescent impurities in the background electrolyte should be avoided as well. Paying careful attention to these details will optimize the limit of detection. Since excitation and emission bands are quite broad, there is considerable latitude for selecting the operating conditions. Maximizing the distance between excitation and emission, avoidance of Raman bands, and optimizing wavelengths to maximal absorption and emission provide the best results. Optimal *S/N* can usually be obtained empirically after a few experiments.

C. Laser-Induced Fluorescence (LIF)

The high photon flux and spatial coherence (focusing capability) of laser light provide excellent properties for a fluorescence excitation source. Since laser lines are monochromatic, background elevations from Rayleigh and Raman scattering are easily avoided by selection of the appropriate emission wavelengths. There are some problems with lasers, including high cost, instability, lack of durability, high scattering, poor tuneability, and the lack of useful lines in the low UV.

Low-power argon-ion lasers address cost and ruggedness issues effectively. Very clever designs to reduce scattering have been described in the literature. These include complex flow cells that use sheath fluids that are refractive-index matched to the run buffer (78, 79, 90). Changes in the refractive index between abutting surfaces are the root cause of light scattering. In the absence of scattering and with pL volume flowcells, MLODs approach single-molecule detection. Simple designs such as that of Drossman *et al.* (76), shown in Fig. 10.12, can provide LODs of 10^{-11} M for fluorescein. This corresponds to 60,000 molecules, a 10 nL injection of a 10^{-11}M solution.

Commercial LIF instrumentation is available from two sources, Beckman and Europhor. The Beckman instrument is an accessory to the P/ACE 2000 series that includes a power supply and a 3 mW argon-ion laser. Laser light is transmitted to the capillary via a fiber optic. The company reported an LOD for fluorescein of 10^{-12} M. The Europhor IRIS 2000 is an integrated laser-based system containing an argon-ion and/or helium-cadmium laser.

Semiconductor lasers may soon develop as an alternative laser light source. The advantages herein are based on size, cost, and stability. At present these lasers are available in quantity only for visible and near-IR wavelengths. Frequency doubling is required to reach lower wavelengths. Applications have been reported for the determination of derivatized amino acids with sub-femtomole LODs (98, 99).

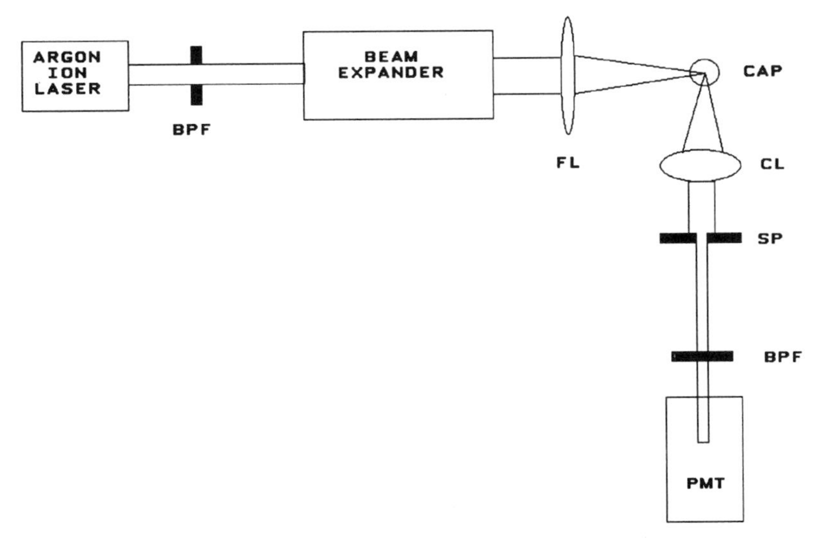

Figure 10.12. Schematic of a simple laser fluorescence detection system for CE. Redrawn with permission from *Anal. Chem.* **62,** 900, copyright ©1990 Am. Chem. Soc.

D. Derivatization

Since few molecules have native fluorescence, derivatization is frequently required. Both pre- and postcapillary derivatization are possible to enhance CE detection. *Precapillary derivatization* can be performed on current instrumentation because no instrumental adaptations are required. Virtually all of the derivatizing agents used for liquid chromatography can be employed in CE. Table 10.3 contains a sampling of some possible reagents. If laser fluorescence were being applied, the tag selection is further dictated by the available laser wavelengths.

The pros and cons of derivatization were covered in Section 7.10G. The problems with validation and the extra sample-handling steps are the major reasons derivatization is not widely used. In commercial environments, corporate culture generally determines whether derivatization schemes can be employed. For those companies forbidding these techniques, pragmatic solutions to difficult problems are being ignored.

Recent work by Liu *et al.* (82, 83) employed the new reagent 3-(4-carboxylbenzoyl)-2-quinolinecarboxaldehyde (CBQCA) to react with primary amines and amino sugars. The absorption maximum of this tag matches well with the 442 nm line of the helium–cadmium laser. This reagent produces MLODs down to 10^{-18} moles, as illustrated in Fig. 10.13 for a tryptic digest of 1.9 pg lysozyme. The design of tags that match laser lines is expected to increase in the future.

When using a non-laser source, the FMOC (2-fluorenylmethyl-*N*-chloroformate) tag is ideal for derivatizing primary and secondary amines (67). Rapid

Table 10.3. Fluorescent Derivatization Reagents

Reagent	Reacts with	Excitation λ (nm)[a]	Emission λ (nm)[b]
Dansyl chloride[c]	1°, 2° amines, phenols	360	520
NBD chloride	1°, 2° amines	420	540
Fluorescamine	1° amines	390	475
o-Phthalaldehyde	1° amines	350	440
Dansyl hydrazine	aldehydes and ketones	340	525
Naphthalenedialdehyde	1° amines	442	490
CBQCA[d]	1° amines	442	550
Fluorescein isothiocyanate	1° amines	488	525
Bromomethylcoumarin	carboxylic acids	325	430
Thiazole Orange	DNA intercalator	488	520

[a]Emission and excitation wavelengths are solute- and solvent-dependent. A number of the excitation wavelengths correspond to laser lines (325, 442, and 488 nm) and do not correspond to the actual excitation maxima.

[b]The emission wavelength is frequently selected to avoid the Raman band and may not correspond to the actual emission maximum.

[c]Dansyl chloride has a poor quantum yield in water.

[d]3-(4-Carboxybenzoyl)-2-quinolinecarboxaldehyde, available from Molecular Probes.

derivatization, stable products, and high sensitivity are features of this reagent. A separation of some amino acids is given in Fig. 10.14.

Postcapillary derivatization is also possible (67, 105–108). The principal requirements are that (i) the derivatizing reagent be invisible to the detector and (ii) rapid reactions occur. A design shown in Fig. 10.15 is not commercially available. The basis for the function of this reactor is differential EOF. The gap junction of this reactor is about 50 μm. With a separating capillary of 50 μm and a reactor capillary of 75 μm, the reagent is drawn into the reactor by differential EOF because the volumetric requirements of the larger-diameter capillary are not being fulfilled. Mixing is accomplished by convection.

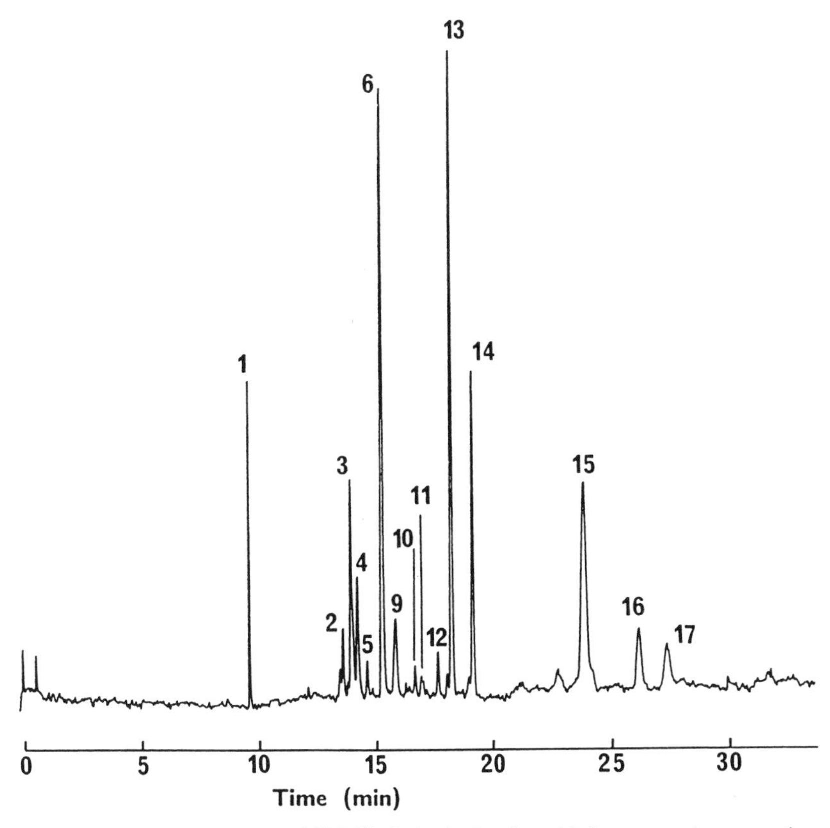

Figure 10.13. Electropherogram of CBQCA-derivatized amino acids from a sample representing 1.9 pg hydrolyzed lysozyme. Capillary: 100 cm (70 cm to detector) 50 μm i.d.; buffer: 50 mM TES, 50 mM SDS, pH 7.02; voltage: 25 kV (14 μA); detection: He–Cd laser, 442 nm. Key: (1) Arg; (2) Trp; (3) Tyr; (4) His; (5) Met; (6) Ile; (7) Gln; (8) Asn; (9) Thr; (10) Phe; (11) Leu; (12) Val; (13) Ser; (14) Ala; (15) Gly; (16) Glu; (17) Asp. Reprinted with permission from *Anal. Chem.* **63,** 408, copyright ©1991 Am. Chem. Soc.

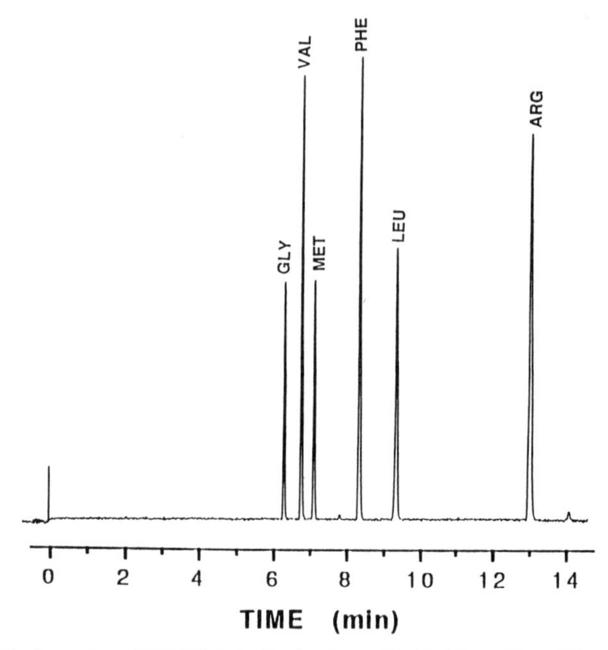

Figure 10.14. Separation of FMOC-derivatized amino acids. Capillary: 62 cm (40 cm to detector) 50 μm i.d.; buffer: 20 mM borate, pH 9.5, 25 mM SDS; field strength: 416 V/cm; temperature: 30°C; detection: xenon arc fluorescence, excitation: 260 nm, emission: 305 nm long wavepass filter. The CLOD is 10 ng/mL. Reprinted with permission from *Anal. Chem.* **63,** 417, copyright ©1991 Am. Chem. Soc..

OPA (*o*-phthalaldehyde) is an ideal postcapillary reagent. The reagent does not fluoresce, is stable, and reacts quickly with primary amines. The optimal conditions are 3.7 mM OPA in run buffer, 0.5% mercaptoethanol, 2% methanol, and 40°C (67). The CLOD for OPA glycine is 60 ng/mL with xenon arc fluorescence, X350 nm, M > 400 nm. The run-to-run reproducibility was about 1% using peak areas. One disadvantage of the postcapillary derivatization system is that only electrokinetic injection can be employed, because vacuum injection would result in drawing the PCRS reagent into the reaction capillary. The sensitivity of peptide mapping can be greatly enhanced relative to absorption detection with the PCRS system, as shown in Fig. 10.16. The concentration of the analytes in the absorbance electropherogram was 40-fold greater than in the fluorescence run.

It is to be hoped that postcapillary derivatization will become commercial in the near future. The disadvantages of precapillary derivatization are overcome because both validation and sample-handling problems are reduced or eliminated.

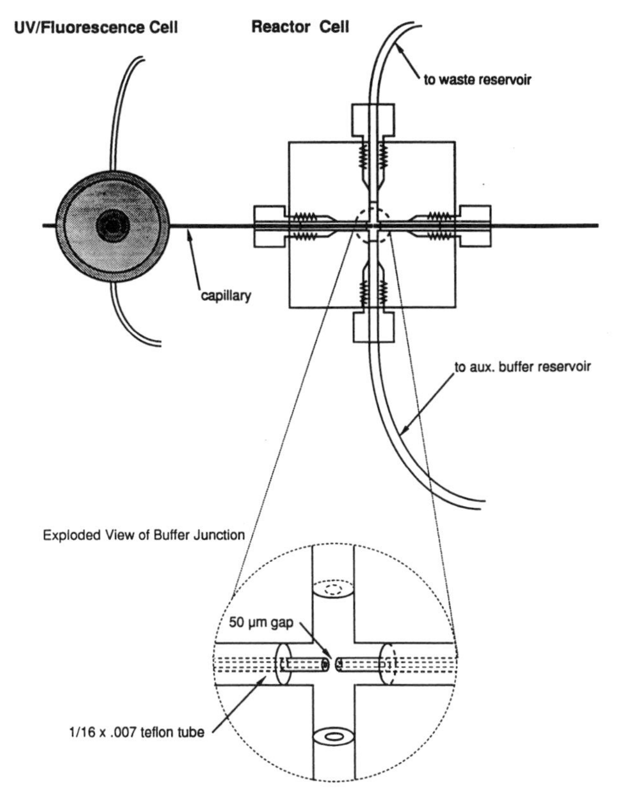

Figure 10.15. Schematic of a liquid junction postcapillary reaction system. Reprinted with permission from *Anal. Chem.* **63**, 417, copyright ©1991 Am. Chem. Soc.

10.8 Indirect Detection

Indirect detection schemes employ an absorbing or fluorescing buffer additive of the identical charge to the solutes being separated. Because of the principle of electroneutrality, the presence of the analyte precludes the additive, so a negative signal is obtained at the detector. These schemes are targeted toward solutes that have low molar absorbtivity at useful wavelengths. Both indirect absorbance and fluorescence can be employed, although the absorbance technique is more common at present owing to the availability of commercial instrumentation. While the primary use is for the determination of small ions, applications for sugars have recently appeared as well. The determination of neutral species via MECC has met only modest success. Waters Associates provides reagent kits, applications support, and operating protocols for separating both anions and cations. Refer to Section 3.6 for factors that contribute to the separations of small ions by CZE.

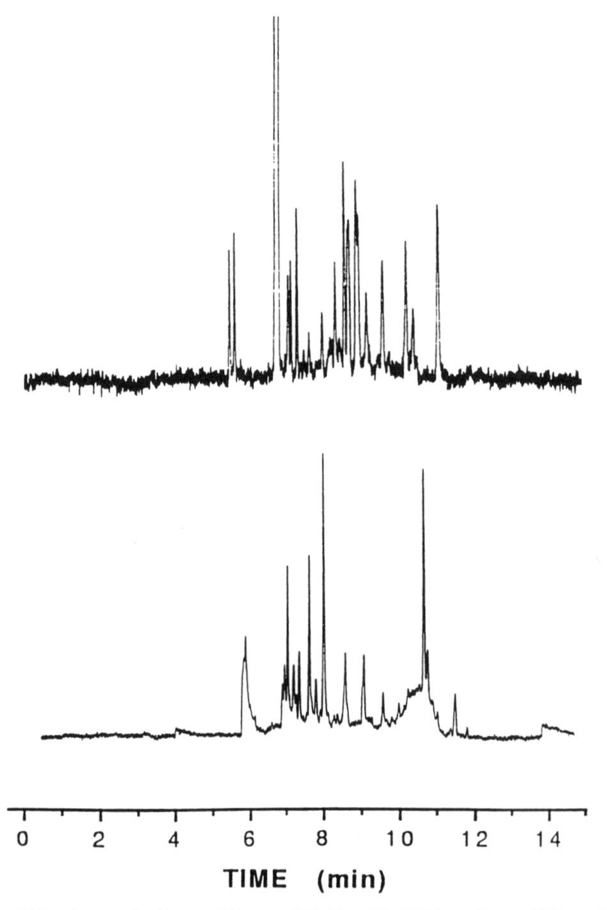

Figure 10.16. CZE of a tryptic digest of β-lactoglobulin with UV detection at 200 nm (bottom) and postcapillary derivatization with OPA (top). Capillary: 62 cm (40 cm to detector) 50 μm i.d.; buffer: 20 mM borate, pH 9.5; field strength: 278 V/cm; postcapillary detection: xenon arc fluorescence, excitation: 390 nm, emission: 450 nm long wavepass filter; sample concentration: absorption, 20 nmol/mL, fluorescence, 0.5 nmol/mL; injection: absorption, vacuum, 1 s, fluorescence, electrokinetic, 7 s at 5 kV. Reprinted with permission from *Anal. Chem.* **63**, 417, copyright ©1991 Am. Chem. Soc..

The principles of indirect detection are illustrated in Fig. 10.17. In indirect absorbance, the additive reduces the amount of light reaching the photodiode. The presence of a solute increases the light throughout, permitting the measurement of a signal. In indirect fluorescence, the reverse occurs. The additive raises the fluorescent background until a nonfluorescing solute transits the detector. The mechanism of separation is identical for both indirect absorption and fluorescence detection; only the spectroscopic properties of the additive differ.

The sensitivity of analysis is determined primarily by the stability of the light source. An unstable source amplifies the noise, because the background

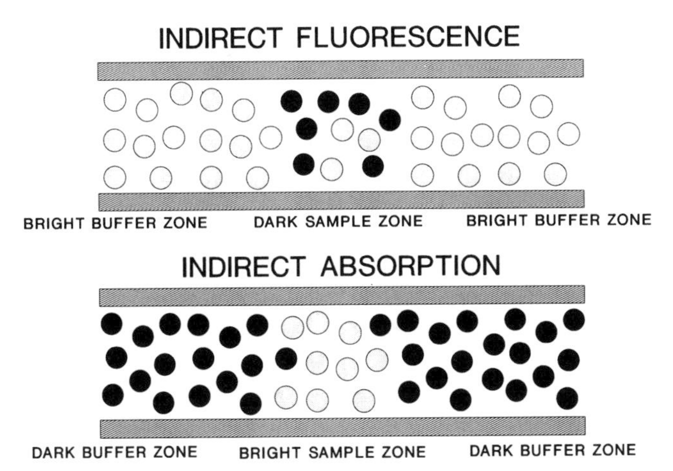

Figure 10.17. Graphic illustration of indirect absorption and fluorescence detection.

is kept artificially high by the presence of the additive. Lasers were first employed by Gross and Yeung in 1988 as a source for indirect fluorescence detection (72, 73). CLODs with laser sources can often be poorer than indirect absorption detection with a highly stable light source (23). LODs of 10^{-6}–10^{-5} M are generally obtained by indirect absorption. Indirect fluorescence with laser excitation usually falls into the same range. Another problem with indirect fluorescence is the low ionic strength of the electrolyte solution. This can result in peak asymmetry at the upper end of the linear dynamic range (see Chapter 11). Because of the intensity of the laser light source, a low additive concentration is required. It may be possible to improve these problems with a conventional light source, but there have been no reports to date on that aspect. Because of these issues, it is difficult to recommend indirect fluorescence with laser sources at this time.

The linear dynamic range is determined by the dynamic reserve of the system, the ratio between the background signal and its noise level. The limit of detection can then be expressed as

$$C_{\text{lim}} = \frac{C_{\text{m}}}{RD}, \tag{10.5}$$

where C_{m} is the concentration of the buffer additive; R is the displacement ratio (divalents can displace two monovalent molecules); and D is the dynamic reserve. While the LOD can be improved by lowering the additive concentration, the linear dynamic range is reduced at low additive concentrations. Table 10.4 describes a series of applications, the mode of detection, and the buffer composition. The additive concentration is usually adjusted to reflect the upper end of the required linear dynamic range.

Table 10.4. Indirect Detection Applications

Application	Mode	Buffer	Reference
Cations	UV	30 mM creatinine-acetate, pH 4.8	22
Lanthanides	UV	30 mM creatinine-acetate, pH 4.8, 4 mM HIBA[a]	22
Anions, carboxylates	UV	25 mM sodium veronal, pH 8.6	33
Anions, carboxylates	UV	20 mM benzoate, pH 6.2 with histidine	21
Carboxylates	UV	250 μM NAA[b]–BTA[c]–Tris, pH 8.09	31
Anions	UV	5 mM chromate, pH 8.1 with EOF modifier[d]	23
Cations	UV	5 mM UVCat-1,[d] 6.5 mM HIBA, pH 4.2	24
Cations	UV	5 mM morpholinoethanesulfonate, pH 6.15	19
Alkyl sulfates	UV	12 mM veronal,[e] pH 8.6	32
Sugars	UV	6 mM sorbate, pH 12.1	18
Amines, metal cations	LIF	380 μM quinine sulfate, pH 3.7	70
Anions	LIF	25 μM salicylate, pH 4.0	68
Nucleotides	LIF	1 mM salicylate, pH 3.5	68
Amino acids	LIF	1 mM salicylate, 200 μM carbonate, pH 9.7	72
Tryptic digest	LIF	500 μM salicylate–CAPS, pH 10.9	71
Sugars, amino acids	LIF	1 mM coumarin 343	69

[a]Hydroxyisobutynic acid, a complexing agent required for separation.
[b]1-Naphthylacetic acid.
[c]1,3,5-Benzenetricarboxylic acid.
[d]Proprietary material from Waters Division of Millipore.
[e]5,5-Diethylbarbituric acid.

10.9 Mass Spectrometry

A. Introduction

Coupling of HPCE to mass spectrometry is developing rapidly since it was first reported by Olivares *et al.* in 1987 (112). The actual interfacing turns out not to be difficult. One problem is sensitivity. Since the mass spectrometer operates as a mass-sensitive detector, providing sufficient amounts of material with HPCE can be problematic. Trace enrichment techniques such as stacking and CITP are helpful here. A second problem is acquisition of the full-scan mass spectrum. Since CE peak widths are generally narrow, it is possible to miss a peak in the full-scan mode (136). Array detectors may become useful in this regard (140).

There are two compatible ionization techniques: electrospray and fast-atom bombardment (FAB). Both techniques generally require makeup flows to elevate the total flowrate to 1–10 μL/min. A syringe pump is typically used for reagent delivery to avoid the pulsations characteristic of reciprocating pumps, particularly at low backpressures and low flow rates. When reciprocating pumps are employed, flow splitters are used to enable a relatively pulse-free stream to be delivered to the interface. Since detection is performed postcapillary and most of the flow is provided from the makeup solution, the problems with peak area normalization described in Section 10.1 are eliminated.

The makeup flow can be added coaxially with a sheath flow or through a liquid junction "T" type of interface. Electrical contact with the "detector-side" electrode is made in this region. Both interfaces are illustrated in Fig. 10.18. The coaxial interface is inserted into another larger-diameter capillary within which the makeup fluid is pumped. This interface appears superior in most respects to the liquid junction, providing less lag (shorter migration times), lower currents, and less bandbroadening (129). The liquid junction can be used without actually pumping a makeup reagent when pneumatically assisted electrospray (124) is employed. In

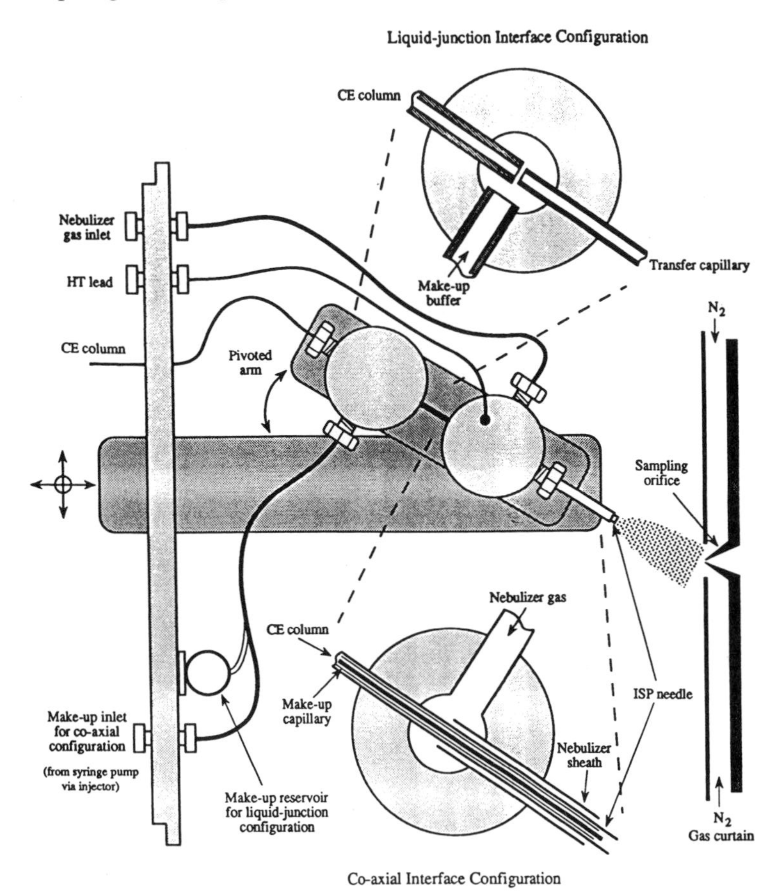

Figure 10.18. Schematic of the liquid junction and coaxial interface between HPCE and MS. Reprinted with permission from *J. Chromatogr.* **591,** 325, copyright ©1991 Elsevier Science Publishers.

this case, the nebulization gas provides a slight vacuum that induces the makeup flow. The junction is simply filled with run buffer.

HPCE instrumentation employed in mass spectrometry has several special considerations. The injection and capillary-filling mechanism should be pressure- rather than vacuum-driven. It is hard to imagine a simple means of interfacing vacuum-driven equipment to any of the interfaces. For pressure-driven systems such as the Beckman P/ACE, this function is easily accomplished. Split-flow syringe loading systems such as the Isco 3850 permit interfacing as well, but with reduced automation capabilities.

The power supply should be capable of providing both polarity switching and a non-grounded negative electrode. The requisite interface potential is generated by the difference between the capillary outlet side voltage and the grounded electrospray. Most instruments are capable of polarity switching but provide only for a fixed ground potential at the cathode. A newly introduced power supply from Beckman accommodates this requirement for the mass spectrometer interface.

For most applications it is useful to employ the on-board UV detector in conjunction with the MS interface. It is important to minimize the length of capillary between the detector and the interface; otherwise, the run time will be unnecessarily prolonged. The capillary length is dictated by the configuration of both the CE instrumentation and the interface. An extra 20–30 cm of capillary is normally used to make this connection. Wider-bore capillaries of 75–100 μm i.d. are frequently used to increase the loading capacity of the system.

To date, CZE, CITP, and most recently CGE have been interfaced to the mass spectrometer. Problems may occur with the interface of MECC and CIEF, since the surfactant and ampholyte additives may not be sufficiently volatile, may form reactive ions, or may interfere with the mass determination of lower molecular weight analytes.

B. Electrospray

The electrospray interface designed by Smith *et al.* (128) illustrated in Fig. 10.19 provides for ion formation at atmospheric pressure (111–133). Since little solvent actually enters the mass spectrometer, the vacuum pumping requirements are minimal compared to those of techniques such as thermospray. The spray is generated in a capillary through a combination of high linear fluid velocity assisted by an electric field of several kilovolts. If the electrospray is maintained at +4 kV, 26 kV is available for the CE separation. The electric field charges the droplets, aiding dispersion via coulombic repulsion of like-charged ions within the droplet. A variant of "pure" electrospray named IonSpray® uses a nebulization gas to further assist spray formation (160). Thermal assists have also been used, but for flow rates that are far more rapid than are ever encountered in HPCE (160).

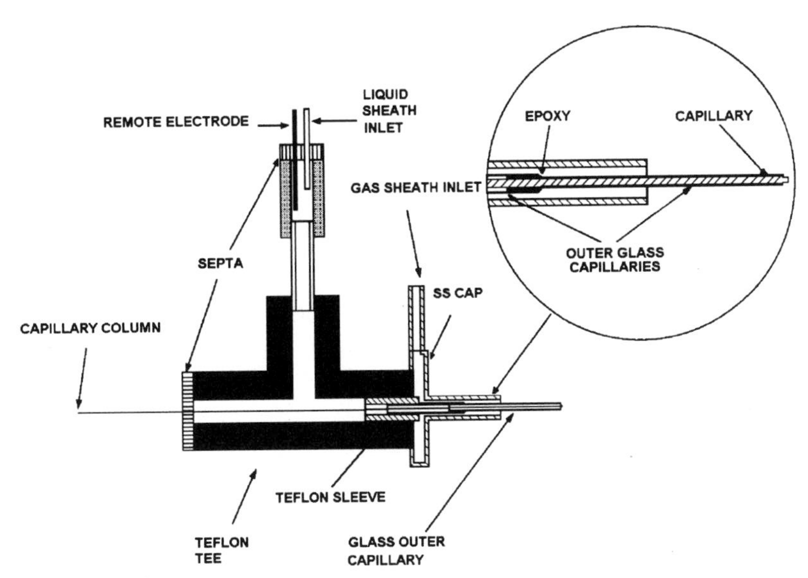

Figure 10.19. Schematic illustration of the electrospray interface for CE/MS. Reprinted with permission from *J. Chromatogr.* **559,** 197, copyright ©1991 Elsevier Science Publishers.

Positively charged ions are produced through complex reactions between the solute and highly reactive ion clusters of water. Negative ions can be produced as well via reactions with O_2^-, O^-, or complex hydrate clusters (160). Nitrogen oxides may also play a role in negative-ion formation. The selection of positive or negative ions is dictated by the polarity of the voltage on the electrospray, the pH of the buffer, and the setup of the mass spectrometer. Nearly all CE/electrospray work has been done in the positive-ion mode, though the negative mode should prove useful for many acidic substances (161). The electrospray is set at a voltage of about –5 kV for negative-ion production.

Elevation of the sample side of a 50 μm i.d. capillary was reported to improve *S/N,* presumably through generation of a modest siphon-produced flow (128). It will probably be necessary to reduce the height differential should larger capillaries be used. A separation comparing CZE/UV with CZE/MS for some myoglobins in shown in Fig. 10.20. The migration times are somewhat longer by CZE/MS because of the extra capillary length that was used. Slightly better resolution and no evidence of significant bandbroadening was found, despite the production of hydrodynamic flow from capillary elevation.

Selection of the buffer and makeup liquid is important. For positive ion formation, 0.2% formic acid, 15 mM ammonium acetate, or 15 mM ammonium formate are useful volatile buffers. Lower buffer concentrations may be insufficient for separation; higher concentrations may lead to a reduction in the total ion current

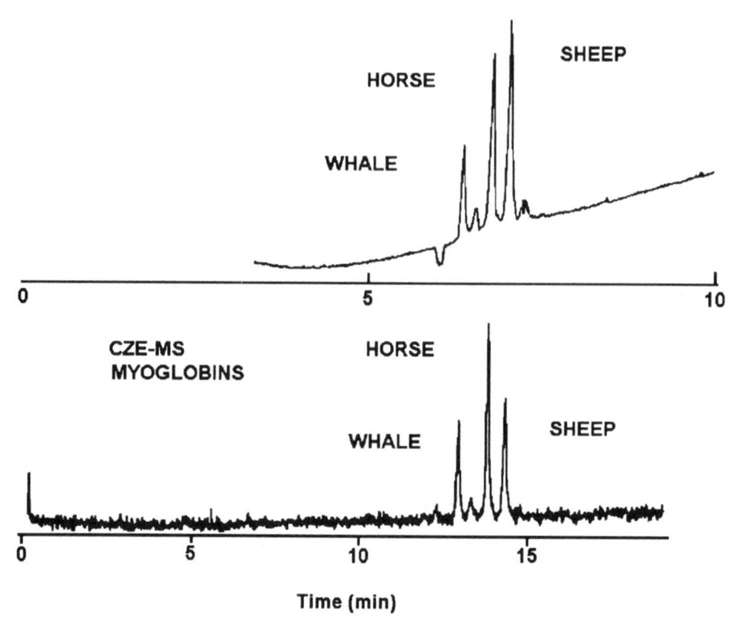

Figure 10.20. Comparison of CE/UV (214 nm) and CE/MS (total ion electropherogram) for a mixture of whale (*M* 17,199.1), horse (*M* 16,950.7), and sheep (*M* 16,923.3) myoglobins. Buffer: 10 mM Tris, pH 8.3; field strength: 120 V/cm; injection size: 100 fmol. Reprinted with permission from *J. Chromatogr.* **559,** 197, copyright ©1991 Elsevier Science Publishers.

(124). Organic modifiers such as 5–25% methanol increase the ion response; methanol is superior to acetonitrile in that regard. Using more than 25% methanol caused bubble formation (124); however, higher concentrations of acetonitrile are possible. In addition, the speed of separation is greater using acetonitrile because the EOF is increased. Tris, ammonium hydrogen carbonate, and acetate buffers have been used as well (128). While negative-ion work with CZE has not been reported, high-pH ammonium buffers should prove workable (161). Relatively nonvolatile buffers such as 20 mM phosphate have also been used with a makeup flow of 0.1% trifluoroacetic acid in methanol pumped at 5 μL/min. (130).

As the solvent evaporates, ions are ejected directly into the gas phase and sampled into the vacuum region of the mass spectrometer through a series of skimmers. A potential difference of 3 kV between the electrospray and the ion-sampling orifice facilitates sampling (124). The spray is best sampled at oblique angles to the MS inlet to reduce the background from the production of cluster ions (129). Nitrogen purge at the atmospheric side of the skimmer further reduces cluster-ion formation (124). Since ions are produced by the electrospray, it is unnecessary to produce chemical or electron-impact ionization inside the instrument. Molecular ions form the predominant species. Fragmentation patterns can be produced and detected with a triple quadrupole instrument via introduction of a collision gas such as argon at the second quadrupole (124).

Nearly all CE/electrospray work has been done using single or triple quadrupole instruments. These instruments have mass ranges of about 2400 D. This is an *m/z* ratio where *m* is the mass and *z* is the charge. Electrospray produces multiply charged ions for proteins and oligonucleotides, so it is possible to detect solutes with molecular weights over 100 kD. The electrospray mass spectrum of myoglobin is shown in Fig. 10.21. The actual mass is determined as follows (160). For adjacent peaks, assume $n_2 = n_1 + 1$, where n = number of charges. The detected mass (m) is given by

$$m_1 = \frac{M + n_1}{n_1}, \tag{10.6}$$

where M is the actual mass and n_n is the number of charges. The measured *m/z* is the sum of the mass plus the mass of the protons forming the positive ion. Then

$$n_2 = \frac{m_1 - 1}{m_2 - m_1}, \tag{10.7}$$

and

$$M = n_2 (m_2 - 1)\,. \tag{10.8}$$

The use of these equations produces a raw n_2 value that is rounded to the nearest integer. The integral n value is used to calculate the actual mass of solute, as given in Table 10.5 for myoglobin. Production of ions in the vapor phase from molecules with very high masses is remarkable, as is the ability to detect and measure them on a quadrupole instrument with a limited mass range. Smith *et al.* (111) reported on the determination of about 100 different peptides and proteins covering a mass range from 409 to 133,000. Most of these materials will be separable by CZE. To overcome wall effects, Thibault *et al.* (126) employed a cationic surfactant coating

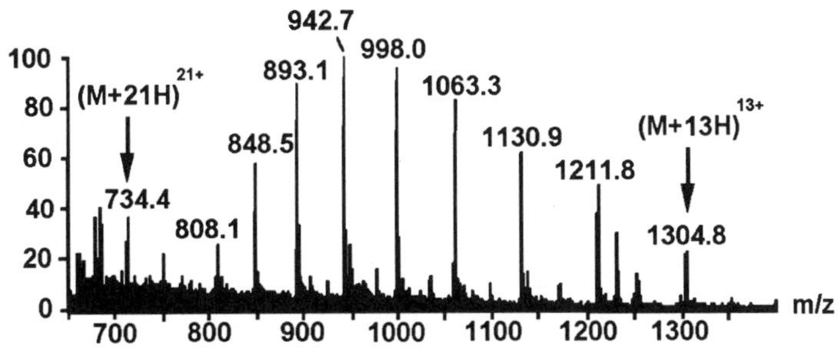

Figure 10.21. Electrospray MS of myoglobin. Reprinted with permission from *The API Book* (Sciex, 1990.)

Table 10.5. Calculation of Molecular Weight of Myoglobin from Multiple Charge Electrospray MS Data[a]

m_2	m_1	Raw n_2	n_2	M
1,304.8	1,211.8	13.03	13	16,949.4
1,211.8	1,130.9	13.97	14	16,951.2
1,130.9	1,060.3	15.00	15	16,948.5
1,060.3	998.0	16.00	16	16,948.5
998.0	942.7	17.03	17	16,949.0
942.7	893.1	17.99	18	16,950.6
893.1	848.5	19.00	19	16,949.5
848.5	808.1	19.98	20	16,950.0
808.1	771.4	20.99	21	16,949.1
771.4	737.7	21.86	22	16,949.8
			Average	16,949.5
			Std dev.	0.9
			% RSD	0.005

[a]Reprinted with permission from *The API Book* (Sciex, 1990).

along with charge reversal for some protein separations. Wall interactions were overcome, and good separations and spectra were obtained. It is expected that this approach, along with the use of other forms of capillary treatments, will permit separation and detection of most macromolecules.

One problem with multiple charges on a solute is loss of sensitivity that depends on molecular weight (122). These data are presented in Table 10.6. There are concentration-dependent issues as well. At lower concentrations, the relative peak intensities shift to higher charge states. The lower charge states include a series of adduct bands (data not shown) on the high *m/z* side of each molecular ion. Despite these problems, high-sensitivity measurements in the full-scan mode are possible, as shown in Fig. 10.22 for a scan of 10^{-8} M (23 fmol injected) cytochrome *c*. In actual practice, the sensitivity will be somewhat lower in the full-scan mode because 90-s scans are impractical.

Separations of small molecules have been reported as well using CZE/electrospray MS. These include quaternary ammonium salts (112), sulfonamides (125), vitamins (114), morphinans (130), paralytic shellfish toxins (131), and benzodiazepines (125). Limits of detection have been as low as 10^{-8} M for small molecules without preconcentration (112, 114).

Garcia and Henion (146) interfaced CGE using the ion spray interface in 1992. They found that the high concentrations of urea typically used here do not elute from the gel. Common buffer systems such as Tris–borate–EDTA are retained as well. The negative-ion mode was employed to separate and monitor benzene and naphthalene sulfonates, dansyl amino acids, and polyacrylic acids. Tandem MS

Table 10.6. Dependence of Signal Intensity on Molecular Mass

Mass	Average Number of Charges[a]	Peak Width (m/z)[b]	Ion Current (Ions/s)[c]
1,000	1	1	1×10^{12}
10,000	10	1	2×10^{10}
40,000	40	3	4×10^{8}
100,000	100	6	3×10^{7}
200,000	200	>6[d]	8×10^{6}[e]

[a]Assumed that the average number of charges increases linearly with M_r and the distribution is centered around *m/z* 1,000.

[b]Peak width due to microheterogeneity typical of large biopolymers and contributions of impurities, solvent adducts, etc.

[c]ESI production before sampling losses assuming an 80% ionization efficiency. Detected ion intensities are 4–5 orders of magnitude lower because of inefficiencies arising from sampling, transmission, and detection.

[d]Peak width of 6 *m/z* units is too large for individual charge states to be resolved; a peak width of ≤4 is required.

[e]For a peak width of 6 *m/z* units.

Data from *J. Chromatogr.* **516,** 157 (1990).

data was produced as well. Since the pore size of the gel is large compared with the first two applications, molecular sieving mechanisms are not operative. Separation by CZE in the presence of the gel matrix is the probable separation mechanism. While it is unlikely that the described applications would be run routinely in the gel matrix, high-viscosity gels should prove useful for oligonucleotide

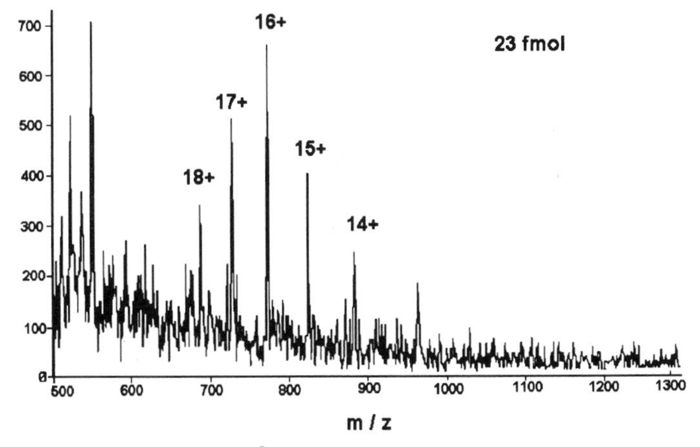

Figure 10.22. ESI-MS of 1.5×10^{-8} M cytochrome *c* obtained by direct infusion at 1 μL/min during a 90 s acquisition period. Reprinted from *J. Chromatogr.* **516,** 157, copyright ©1991 Elsevier Science Publishers.

and DNA separations. Low-viscosity polymeric material such as agarose, hydroxypropylmethylcellulose, or dextrans might elute from the capillary possibly interferring with the MS.

In 1989, Udseth *et al.* (132) reported on the use of CITP interfaced to electrospray MS via the sheath flow. For quaternary phosphonium salt separations, the leader was 1 mM ammonium acetate, and the terminator, 1 mM tetraoctylammonium acetate, both prepared in 50% methanol. Large injections with substantial band compression were reported. For peptides (133), the leader was 10 mM ammonium acetate or potassium acetate/ammonium acetate mixtures, pH 4.5–4.8. The terminator was 10 mM acetic acid. Both uncoated and DB-17-coated capillaries were employed. Better results were obtained on the coated capillary, though it is likely that bare silica is workable under the appropriate experimental conditions. To speed the separations, 3 psi was applied to speed elution (121) with the voltage on. CITP focusing maintained the separation despite the hydrodynamic flow.

The limits of detection can be improved by two orders of magnitude when CITP is used. The use of CITP is indicated when dilute samples must be separated. Otherwise, CZE is the preferred mode of separation. CITP provides the opportunity to employ tandem MS when structural elucidation, including sequence information, is required. Refer to Chapter 6 for additional operational details on performing CITP.

The use of electrospray in combination with ion trap mass spectrometry (ITMS) (162) is expected to develop in the future. Advantages in instrument size and cost are expected, particularly in the area of tandem MS.

C. Fast-Atom Bombardment

Fast-atom bombardment (FAB) is a soft ionization method well suited for the determination of polar and thermally labile large molecules. Samples are introduced into the mass spectrometer, dissolved in glycerol, thioglycerol, or another nonvolatile material. This carrier is known as the FAB matrix. Once inside the high-vacuum region of the instrument, the sample is exposed to a beam of high-energy xenon atoms, causing molecules at the surface of the matrix to be ejected and ionized. High-resolution MS up to a mass of about 6,000 daltons is practical using this technique.

The sheath flow interface (134–139, 141–144), shown in Fig. 10.23, is best employed for introducing the FAB matrix mixed with the effluent from the separation capillary. Flow rates of 1-5 μL/min or less of FAB matrix are generally employed. Since the probe tip is inside the high-vacuum region of the spectrometer, about 50 pL/min of hydrodynamic flow is induced inside a 10 μm i.d. capillary (137). This does not contribute substantially to bandbroadening and can actually be used for sample injection.

Both positive- and negative-ion modes are operative, depending on the nature of the matrix medium. For positive ion formation (Fig. 10.24), heptafluorobutyric

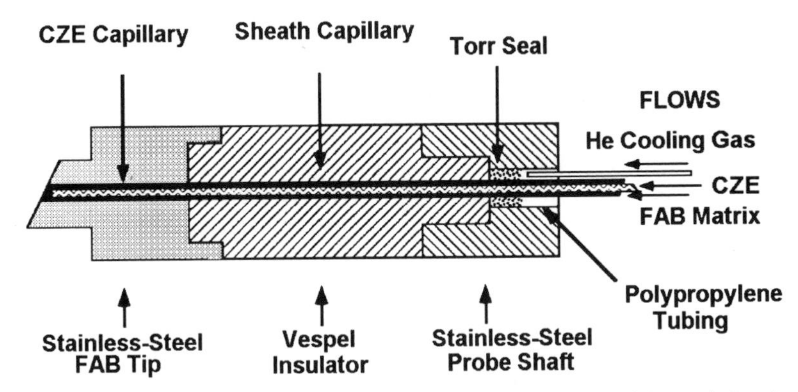

Figure 10.23. Schematic of a CZE coaxial FAB-MS tip. Reprinted with permission from *J. Chromatogr.* **516,** 167, copyright ©1990 Elsevier Science Publishers.

acid, pH 3.5, can be used; for negative ion generation, ammonium hydroxide is useful (144). These modifiers also serve to increase the conductivity of the FAB matrix, which is necessary to maintain electrical contact between the FAB probe tip and the CE capillary (144). As in electrospray, the probe tip also serves as the

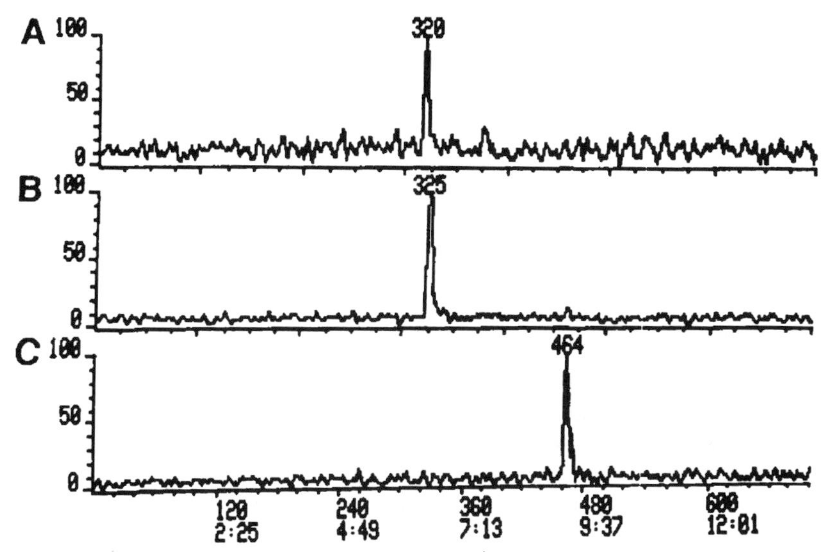

Figure 10.24. Single-ion electropherograms of 2×10^{-4} M solutions of the protonated molecular ions of (a) morphiceptin (30 fmol, $N = 91{,}000$), (b) proctolin (24 fmol, $N = 120{,}000$) and (c) Phe–Leu–Glu–Glu–Ile (24 fmol, $N = 120{,}000$). Capillary: 15 μm i.d., length not specified; buffer: 5 mM ammonium acetate, pH 8.5; FAB matrix: glycerol:water (25:75) adjusted to pH 3.5 with heptafluorobutyric acid; FAB matrix flow rate: 0.5 μL/min; injection: electrokinetic, 6 kV for 2 s. Reprinted with permission from *J. Chromatogr.* **516,** 167, copyright ©1990 Elsevier Science Publishers.

electrical ground of the system (137). The probe tip is maintained at +/–8 kV, leaving +/– 22kV for the electrophoretic separation. Negative ion detection is about an order of magnitude less sensitive compared to the positive-ion mode. Separation as negative ions followed by detection as positive ions can help overcome this problem for zwitterionic peptides and proteins (144). Volatile buffers such as 5 mM ammonium acetate or 10 mM acetic acid are useful for CE/FAB. Phosphate buffers can decrease the total ion current (137). FAB does not tolerate high salt content buffers, so treated capillaries should be used to minimize wall interactions. Alternatively, the salts can be removed from the "detector-side" buffer and replaced with FAB matrix solution (136). This discontinuous buffer system appears to function satisfactorily. The sensitivity of FAB can be as much as 2–3 orders of magnitude poorer than electrospray, especially for smaller molecules. For small peptides, the CLOD is about 10^{-5} M (136, 139). The advantages of FAB over electrospray are (i) easy interface to a high-resolution mass spectrometer, and (ii) freedom from multiple charges, resulting in a simplified mass spectrum. The production of a high-resolution mass spectrum from tandem experiments provides for more clearcut structural assignments compared to those run on a quadrupole instrument. Despite these advantages, it appears that electrospray has greater general utility as a CE/MS interface.

10.10 Other Detection Techniques

As Table 10.1 indicated, detection is a diverse area of research in CE. A number of modes are highly specialized. Some are practical, others not so practical. This section briefly covers a few of these practical or interesting alternative schemes. For example, electrochemical detection is targeted primarily for catecholamine detection in single nerve cells (52, 57–62). It is unlikely that this mode will become commercially available in the near future because of the difficulty of fabricating microelectrodes. That is unfortunate because the LOD can approach that of fluorescence for electroactive solutes.

Non-optical methods for CE are advantageous because the problems with the capillary-defined pathlength are eliminated. Conductivity detection (7, 20, 47–51) is relatively simple, has a good linear dynamic range and a CLOD of about 10^{-7} M for lithium (47). This is an improvement of 1–2 orders of magnitude over indirect absorbance. Considering the interest in small-ion separations by CE, it is realistic to expect a commercial introduction in the near future. Figure 10.25 illustrates monitoring of a 20-fold dilution of blood serum for some cations. The physiological concentration ratio of sodium/potassium is about 53/1. Good peak symmetries are found for all components.

Limits of detection reach 10^{-9} M for chemiluminescence (42), 10^{-8} M for gamma ray detection with a solid scintillator (147), 10^{-7} M for γ or β radiation using a plastic scintillator (148), and 10^{-7} M for resonant Raman spectroscopy (151). With the radioactivity methods, slowing or even stopping the separation increases the counts and improves the sensitivity of the technique (148). It is reasonable to expect a commercialization of radioactivity detection in the near future.

The LOD of absorption detection can be improved more than two orders of magnitude by axial beam illumination with a laser (12) or a conventional light source (13). The beam is light-piped down the capillary to the photodiode. The pathlength becomes dictated by the bandwidth of the sample. A stairstep electropherogram results, since the absorption decreases in a stepwise fashion as the solutes pass the photodiode. While sensitivity is improved, the technique is not appropriate for measuring trace constituents in the presence of a major component. The highly concentrated solute absorbs most of the available light. It is not likely that this device will become commercial in the near future.

Thermooptical methods (38–41) have good potential provided tuneable lasers become available at a reasonable cost. When a molecule absorbs light and does not fluoresce, the excited state energy generates heat when the molecule returns to its ground state. The heat causes refractive index changes in the buffer that can be probed with a laser. Two lasers are required for this process, a pump laser and a probe laser. While it seems complicated, MLOD's can reach the attomole range. Do not expect commercialization for a long time.

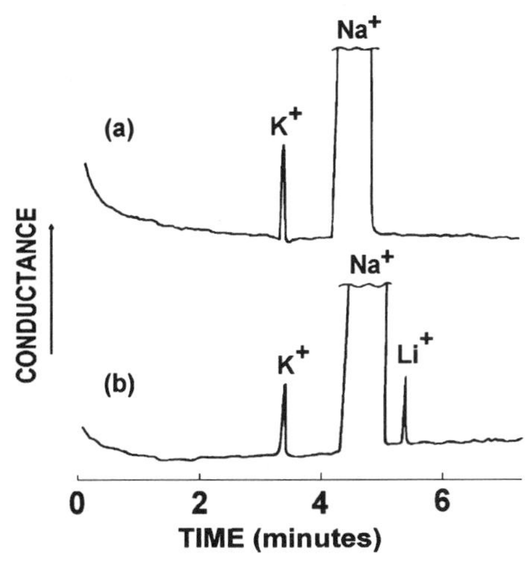

Figure 10.25. Electropherograms of human serum: (a) normal subject, (b) patient on lithium therapy, dilution 1:19 with 20 mM MES/His buffer, pH 6.1; capillary: 75 μm 70 cm; injection: gravity, 10 cm for 10 s; voltage: 25 kV. The sodium peaks are off-scale because sodium's physiologic concentration is about 35 times that of potassium. Reprinted with permission from *Anal. Chem.* **59,** 2747, copyright ©1987 Am. Chem. Soc.

10.11 Fraction Collection

A. Introduction

Fraction collection by CE bears little resemblance to the analogous process employed in HPLC. With the electroosmotic flow rate seldom greater than 100 μL/min, droplets at the end of the capillary do not form, because the rate of evaporation is greater than the rate of droplet formation. The problem of non-uniform migration velocity also exists and is similar to that described in Section 10.1. Both the lag times from the detection point and the appropriate collection time of each fraction are directly related to migration velocity for each component. In addition, one cannot ignore the minute quantities of material that can be separated in a single run. Multiple runs are generally required in order to collect sufficient quantities of material. The potential for electrochemical reactions at the surface of the electrode in the receiving electrolyte must be considered as well.

The first reports employing fraction collection appeared in 1988 in papers by Rose and Jorgenson (163) and Cohen *et al.* (164). These papers described a system where the voltage was interrupted each time the capillary/electrode assembly was moved to a new receiver vial. Most of the commercial instruments use a variation of this procedure.

B. Performing a Run

For instruments containing both an autosampler and "fraction collector" such as the Beckman P/ACE 2000 series or Bio-Rad BioFocus 3000, the following procedure applies:

1. Perform an analytical run to establish the migration times for each solute. Adjust the separation conditions so that the peaks of interest are separated by at least 10–20 s.
2. Calculate the number of runs required to produce a sufficient mass of material. Refer to Eqs. (9.10) and (9.12) and Table 9.3 to make the appropriate calculations. When performing multiple runs, change the buffer solution regularly to minimize buffer depletion. As many as 50 runs may be necessary to collect a sufficient amount of sample. Changing the buffer every 10 runs is generally sufficient.
3. Calculate the time required for the leading and tailing edge of each fraction to transit the length from the detector to the end of the capillary. First determine the migration velocity, υ_m, using

$$\upsilon_m = \frac{L_d}{t_m}. \tag{10.9}$$

Then the lag time for the leading and tailing edge of each component, t_L is given by

$$t_L = \frac{L_f}{\upsilon_m}, \tag{10.10}$$

where L_f = the length of the capillary from the detector to the fraction collector.

4. Fill the receiver containers with 5–50 μL of buffer. Use the smallest-volume vials possible. Both the capillary and electrode must be submerged in buffer solution. Electrochemical reactions become more likely when extremely small collection volumes are employed. When working with small collection volumes, cooling of the fraction collector may be required to lower the evaporation rate.
5. After the run, evaporate or lyophilize to reduce the fraction volume if necessary.

C. Optimizing the Amount of Material Collected

Several combinations and compromises are possible to maximize the mass of material collected with each run.

1. Use large i.d. capillaries such as a 100 μm or even a 200 μm i.d. capillary. Reduce the field strength following an Ohm's Law plot to minimize heating effects.
2. Increase the buffer ionic strength. High-ionic strength buffers permit the loading of more concentrated sample solutions without loss of resolution (see Chapter 11). To compensate for Joule heating, lower the capillary temperature to subambient if possible. Reduce the field strength as well.
3. Lower the EOF to improve resolution. Coated capillaries, linear alcohols, or the use of D_2O as the buffer diluent (165) may be useful here without elevating the current.

D. Instruments Without Fraction Collectors

A clever scheme developed by Albin *et al.* (166) from Applied Biosystems permits the use of the autosampler as a fraction collector. The procedure is illustrated in Fig. 10.26.

1. Perform an analytical run to calculate migration times of all components to be collected.
2. Inject a small plug of water, followed by sample and then another small water plug. Bracketing by water allows the sample to restack (Section 9.6) following hydrodynamic flow to the detector (step 3).
3. Apply vacuum to draw the sample plug just beyond the detector window.

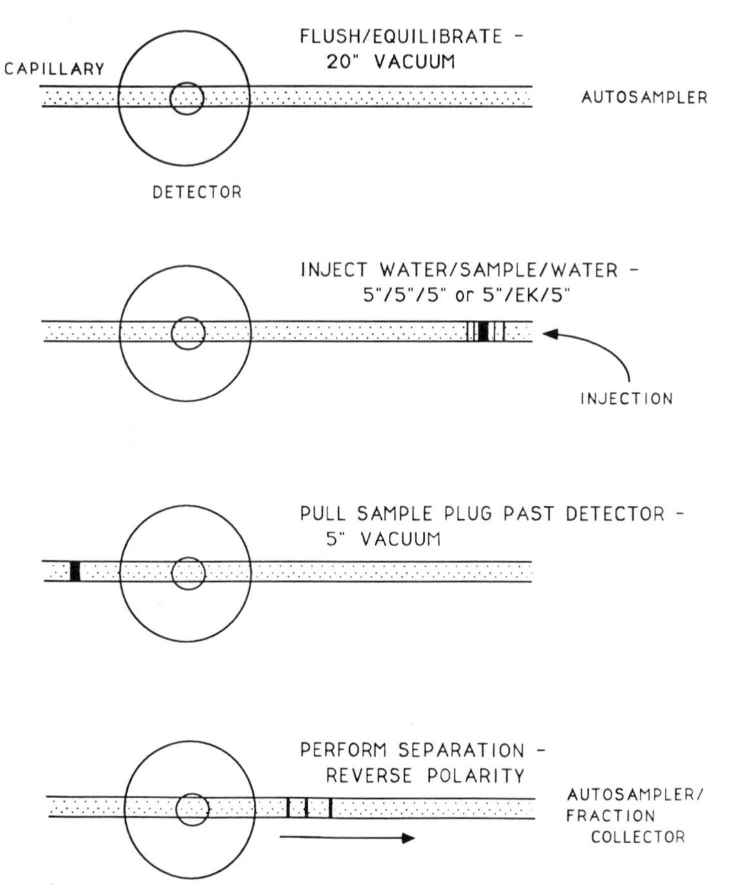

Figure 10.26. Illustration of the procedure for fraction collection in the autosampler of an automated CE instrument. Reprinted with permission from Applied Biosystems.

4. Reverse the power supply polarity and perform the separation in the direction toward the autosampler. Trigger the autosampler for fraction collection as determined by the individual migration times of each component. Collect in 5–10 μL buffer solution. Cool the autosampler to minimize evaporation.

E. Applications

The primary applications reported thus far have been in biotechology-related areas. Camilleri *et al.* (165) prepared calcitonin and, in a single run, generated sufficient material for sequencing. Albin *et al.* (166) recovered 590 pmol of dynorphins at an 85% yield over 30 injections. The fractions were successfully subjected to time-of-flight plasma desorption mass spectrometry and protein sequencing. There is no reason why small molecules could not be collected for spectroscopic analysis by IR, NMR, or other techniques.

F. New Developments

Instrument designs that do not require the voltage to be turned off may represent a future generation of fraction collectors. Collection can take place in a fluid-containing receiver (167–169) or on a moving surface such as a filter paper (170) or a membrane (171, 172). The use of the membrane permits the sample to be transferred directly to a protein sequencer. Multiple capillary assemblies that improve the loading characteristics may become important as well (167, 168).

Many designs employ a grounding junction to complete the electrophoretic circuit before the collection orifice. Such junctions are also used for electrochemical detection, postcapillary reactions, and mass spectrometer interfacing. Pure EOF or the use of a makeup solution can sweep the fractions to the receiving vials. Guzman's (168) approach, which is illustrated in Fig. 10.27, uses a bundle of five 75 μm i.d. capillaries for the separation. Huang and Zare (170) collect on a moving filter paper using a frit to make electrical contact.

Cheng *et al.* (172) from Waters Associates do not use the liquid junction approach. In their rotating membrane fraction collector, illustrated in Fig. 10.28, the ground electrode serves as the base of an assembly kept moist with electrolyte impregnated filter papers. On top is a PVDF membrane, suitable for use in a protein sequencer.

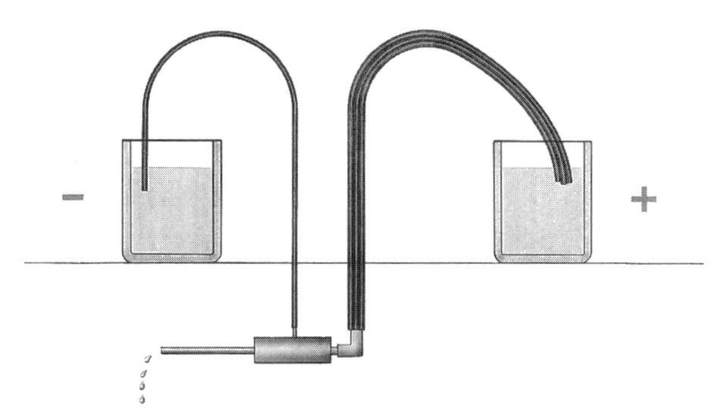

Figure 10.27. Schematic diagram of a continuous fraction collection system for semipreparative capillary electrophoresis. Reprinted with permission from *J. Liq. Chromatogr.* **14,** 997, copyright ©1991 Marcel Dekker.

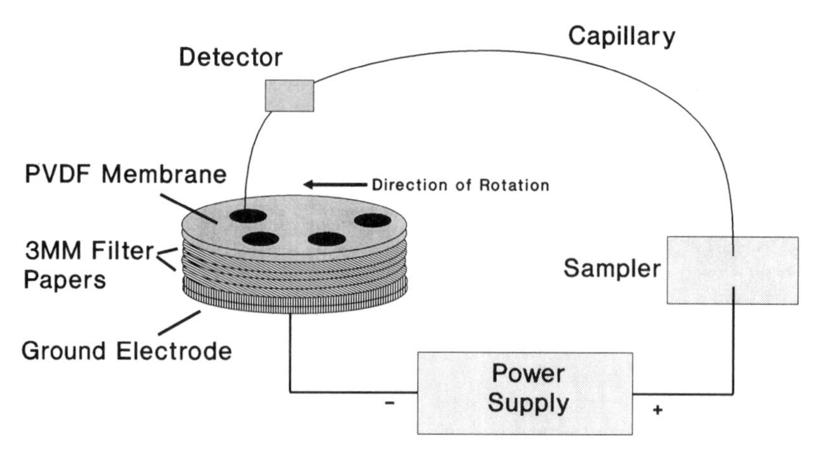

Figure 10.28. Schematic diagram of membrane fraction collection for capillary electrophoresis. Redrawn based on *J. Chromatogr.* **608,** 109, copyright ©1992 Elsevier Science Publishers.

References

1. X. Huang, W. Coleman, and R. N. Zare, *J. Chromatogr.* **480,** 95 (1989).
2. D. Demorest and R. DuBrow, *J. Chromatogr.* **559,** 43 (1991)
3. T. Tsuda, J. V. Sweedler, and R. N. Zare, *Anal. Chem.* **62,** 2149 (1990).
4. G. J. M. Bruin, G. Stegeman, A. C. van Asten, X. Xu, J. C. Kraak, and H. Poppe, *J. Chromatogr.* **559,** 163 (1991).
5. T. Tsuda, J. V. Sweedler, and R. N. Zare, *Anal. Chem.* **62,** 2149 (1990).
6. J. P. Chervet, R. E. J. van Soest, and M. Ursem, *J. Chromatogr.* **543,** 439 (1991).
7. F. Foret, M. Deml, V. Kahle, and P. Boček, *Electrophoresis* **7,** 430 (1986).
8. Y. Walbroehl and J. W. Jorgenson, *J. Chromatogr.* **315,** 135 (1984).
9. M. Fuchs, P. Timmoney, and M. Merion, paper presented at HPCE '91, San Diego.
10. I. H. Grant and W. Steuer, *J. Microcol. Sep.* **2,** 74 (1990).
11. J. S. Green and J. W. Jorgenson, *J. Liq. Chromatogr.* **12,** 2527 (1989).
12. J. A. Taylor and E. S. Yeung, *J. Chromatogr.* **550,** 831 (1991).
13. X. Xi and E. S. Yeung, *Appl. Spectrosc.* **45,** 1199 (1991).
14. T. Wang, J. H. Aiken, C. W. Huie, and R. A. Hartwick, *Anal. Chem.* **63,** 1372 (1991).
15. T. Wang, R. A. Hartwick, and P. B. Champlin, *J. Chromatogr.* **462,** 147 (1989).
16. T. W. Garner and E. S. Yeung, *Anal. Chem.* **62,** 2193 (1990).
17. K. Hargadon and B. McCord, *J. Chromatogr.* **602,** 241 (1992).
18. A. Vorddran, P. Oefner, H. Scherz, and G. Bonn, *Chromatographia* **33,** 163 (1992).
19. A. Weston, P. R. Brown, P. Jandik, W. R. Jones, and A. L. Heckenberg, *J. Chromatogr.* **593,** 289 (1992).
20. M. T. Ackermans, F. M. Everaerts, and J. L. Beckers, *J. Chromatogr.* **549,** 345 (1991).
21. F. Foret, S. Fanali, L. Ossicini, and P. Bocek, *J. Chromatogr.* **470,** 299 (1989).

22. F. Foret, S. Fanali, A. Nardi, and P. Boček, *Electrophoresis* **11,** 780 (1990).
23. P. Jandik and W. R. Jones, *J. Chromatogr.* **546,** 431 (1991).
24. P. Jandik, W. R. Jones, A. Weston, and P. R. Brown, *LC-GC* **9,** 634 (1991).
25. W. R. Jones and P. Jandik, *Amer. Lab.*, June, p. 51 (1990).
26. W. R. Jones, P. Jandik, and R. Pfeifer, *Amer. Lab.*, May, p. 40 (1991).
27. W. R. Jones and P. Jandik, *J. Chromatogr.* **546,** 445 (1991).
28. A. Nardi, S. Fanali, and F. Foret, *Electrophoresis* **11,** 774 (1990).
29. J. Romano, P. Jandik, W. R. Jones, and P. E. Jackson, *J. Chromatogr.* **546,** 411 (1991).
30. B. J. Wildman, P. E. Jackson, W. R. Jones, and P. G. Alden, *J. Chromatogr.* **546,** 459.
31. T. Wang and R. A. Hartwick, *J. Chromatogr.* **589,** 307 (1992).
32. M. W. F. Nielen, *J. Chromatogr.* **588,** 321 (1991).
33. S. Hjertén, K. Elenbring, F. Kilar, J. Liao, A. J. C. Chen, C. J. Siebert and M. Zhu, *J. Chromatogr.* **403,** 47 (1987).
34. S. Kobayashi, T. Ueda, and M. Kikumoto, *J. Chromatogr.* **480,** 179 (1989).
35. W. Thormann, P. Meier, C. Marcolli, and F. Binder, *J. Chromatogr.* **545,** 445 (1991).
36. J. Vindevogel, P. Sandra, and L. C. Verhagen, *J. High Res. Chromatogr.* **13,** 295 (1990).
37. W. Thormann, J. Caslavska, S. Molteni, and J. Chmelik, *J. Chromatogr.* **589,** 321 (1992).
38. A. E. Bruno, A. Paulus, and D. J. Bornhop, *Appl. Spectrosc.* **45,** 462 (1991).
39. C. W. Earle and N. J. Dovichi, *J. Liq. Chromatogr.* **12,** 2575 (1989).
40. M. Yu and N. J. Dovichi, *Mikrochim. Acta* **III,** 27 (1988).
41. M. Yu and N. J. Dovichi, *Anal. Chem.* **61,** 37 (1989).
42. R. Dadoo, L. A. Colon, and R. N. Zare, *J. High Res. Chromatogr.* **15,** 133 (1992).
43. P. L. Christensen and E. S. Yeung, *Anal. Chem.* **61,** 1344 (1989).
44. J. Pawliszyn, *J. Liq. Chromatogr.* **10,** 3377 (1987).
45. J. Wu and J. Pawliszyn, *Anal. Chem.* **64,** 219 (1992).
46. J. Wu and J. Pawliszyn, *Anal. Chem.* **64,** 224 (1992).
47. X. Huang, T. K. J. Pang, M. J. Gordon, and R. N. Zare, *Anal. Chem.* **59,** 2747 (1987).
48. X. Huang, M. J. Gordon, and R. N. Zare, *J. Chromatogr.* **425,** 385 (1988).
49. X. Huang, J. A. Luckey, M. J. Gordon, and R. N. Zare, *Anal. Chem.* **61,** 766 (1989).
50. X. Huang, R. N. Zare, S. Sloss, and A. G. Ewing, *Anal. Chem.* **63,** 189 (1991).
51. X. Huang and R. N. Zare, *Anal. Chem.* **63,** 2193 (1991).
52. J. B. Chien, R. A. Wallingford, and A. G. Ewing, *J. Neurochem.* **54,** 633 (1990).
53. C. E. Engstrom-Silverman and A. G. Ewing, *J. Microcol. Sep.* **3,** 141 (1991).
54. C. D. Gaitonde and P. V. Pathak, *J. Chromatogr.* **514,** 389 (1990).
55. C. Haber, I. Silvestri, S. Roosli, and W. Simon, *Chimia* **45,** 117 (1991).
56. M. D. Oates and J. W. Jorgenson, *Anal. Chem.* **61,** 432 (1989).
57. T. M. Olefirowicz and A. G. Ewing, *J. Neurosci. Meth.* **34,** 11 (1990).
58. T. M. Olefirowicz and A. G. Ewing, *Chimia* **45,** 106 (1991).
59. R. A. Wallingford and A. G. Ewing, *Anal. Chem.* **59,** 1762 (1987).
60. R. A. Wallingford and A. G. Ewing, *Anal. Chem.* **60,** 1972 (1988).
61. R. A. Wallingford and A. G. Ewing, *Anal. Chem.* **60,** 258 (1988).
62. R. A. Wallingford and A. G. Ewing, *Anal. Chem.* **61,** 98 (1989).

63. Y. F. Yik, H. K. Lee, S. F. Y. Lee, and S. B. Khoo, *J. Chromatogr.* **585,** 139 (1991).
64. T. M. Olefirowicz and A. G. Ewing, *J. Chromatogr.* **499,** 713 (1990).
65. J. S. Green and J. W. Jorgenson, *J. Chromatogr.* **352,** 337 (1986).
66. L. N. Amankwa, J. Scholl, and W. G. Kuhr, *Anal. Chem.* **62,** 2189 (1990).
67. M. Albin, R. Weinberger, E. Sapp, and S. Moring, *Anal. Chem.* **63,** 417 (1991).
68. L. Gross and E. Yeung, *J. Chromatogr.* **480,** 169 (1989).
69. T. W. Garner and E. S. Yeung, *J. Chromatogr.* **515,** 639 (1990).
70. L. Gross and E. S. Yeung, *Anal. Chem.* **62,** 427 (1990).
71. B. L. Hogan and E. S. Yeung, *J. Chromatogr. Sci.* **28,** 15 (1990).
72. W. G. Kuhr and E. S. Yeung, *Anal. Chem.* **60,** 1832 (1988).
73. W. G. Kuhr and E. S. Yeung, *Anal. Chem.* **60,** 2642 (1988).
74. D. E. Burton, M. J. Sepanisk, and M. P. Maskarinec, *J. Chromatogr. Sci.* **24,** 347 (1986).
75. D. Y. Chen, H. P. Swerdlow, H. R. Harke, J. Z. Zhang, and N. J. Dovichi, *J. Chromatogr.* **559,** 237 (1991).
76. H. Drossman, J. A. Luckey, A. J. Kostichka, J. D'Cunha, and L. M. Smith, *Anal. Chem.* **62,** 900 (1990).
77. Y. F. Cheng and N. J. Dovichi, *Science* **242,** 562 (1988).
78. Y. F. Cheng and N. J. Dovichi, *SPIE* **910,** 111 (1988).
79. Y. F. Cheng, S. Wu, D. Y. Chen, and N. J. Dovichi, *Anal. Chem.* **62,** 496 (1990).
80. A. E. Karger, J. M. Harris, and F. Gesteland, *Nucl. Acids Res.* **19,** 4955 (1991).
81. T. Lee, E. S. Yeung, and M. Sharma, *J. Chromatogr.* **565,** 197 (1991).
82. J. Liu, Y.-Z. Hsieh, D. Wiesler, and M. Novotny, *Anal. Chem.* **63,** 408 (1991).
83. J. Liu, O. Shirota, and M. Novotny, *Anal. Chem.* **63,** 413 (1991).
84. J. Liu, O. Shirota, D. Wiesler, and M. Novotny, *Proc. Nat. Acad. Sci.* **88,** 2302 (1991).
85. J. Liu, O. Shirota, and M. Novotny, *J. Chromatogr.* **559,** 223 (1991).
86. B. Nickerson and J. W. Jorgenson, *J. High Res. Chromatogr. Chromatogr. Comm.* **11,** 533 (1988).
87. B. Nickerson and J. W. Jorgenson, *J. High Res. Chromatogr. Chromatogr. Comm.* **11,** 878 (1988).
88. M. C. Roach, P. Gozel, and R. N. Zare, *J. Chromatogr.* **426,** 129 (1988).
89. D. F. Swaile and M. J. Sepaniak, *J. Liq. Chromatogr.* **14,** 869 (1991).
90. H. S. Swerdlow, S. Wu, H. Harke, and N. J. Dovichi, *J. Chromatogr.* **516,** 61 (1990).
91. H. P. M. van Villet and H. Poppe, *J. Chromatogr.* **346,** 149 (1985).
92. K. C. Waldron, S. Wu, C. W. Earle, H. R. Harke, and N. J. Dovichi, *Electrophoresis* **11,** 777 (1990).
93. B. W. Wright, G. A. Ross, and R. D. Smith, *J. Microcol. Sep.* **1,** 85 (1989).
94. S. Wu and N. Dovichi, *J. Chromatogr.* **480,** 141 (1989).
95. J. Z. Zhang, D. Y. Chen, S. Wu, H. R. Harke, and N. J. Dovichi, *Clin. Chem.* **37,** 1492 (1991).
96. N. J. Reinhold, U. R. Tjaden, H. Irth, and J. Van der Greef, *J. Chromatogr.* **574,** 327 (1992).
97. K. A. Cobb and M. V. Novotny, *Anal. Biochem.* **200,** 149 (1992).
98. V. Dombek and Z. Stransky, *Anal. Chim. Acta* **256,** 69 (1992).
99. T. Higashijima, T. Fuchigami, T. Imasaka, and N. Ishibashi, *Anal. Chem.* **64,** 711 (1992).
100. N. A. Guzman, M. A. Trebilcock, and J. P. Advis, *Techniques in Protein Chemistry II*, pp. 37–51 (Academic Press, 1991).
101. L. Hernandez, R. Marquina, J. Escalona, and N. A. Guzman, *J. Chromatogr.* **502,** 247 (1990).
102. L. Hernandez, J. Escalona, N. Joshi, and N. A. Guzman, *J. Chromatogr.* **559,** 183 (1991).

103. L. Song and M. F. Maestre, *J. Biomol. Struct. Dyn.* **9,** 525 (1991).
104. J. V. Sweedler, J. B. Shear, H. A. Fishman, R. N. Zarem, and R. H. Scheller, *Anal. Chem.* **63,** 496 (1991).
105. S. L. Pentoney, X. Huang, D. S. Burgi, and R. N. Zare, *Anal. Chem.* **60,** 2625 (1988).
106. D. J. Rose and J. W. Jorgenson, *J. Chromatogr.* **447,** 117 (1988).
107. T. Tsuda, Y. Kobayashi, A. Hori, T. Matsumoto, and O. Suzuki, *J. Chromatogr.* **456,** 375 (1988).
108. B. Nickerson and J. W. Jorgenson, *J. Chromatogr.* **480,** 157 (1989).
109. D. J. Rose, *J. Chromatogr.* **540,** 343 (1991).
110. R. W. Hallen, C. B. Shumate, W. F. Siems, T. Tsuda, and H. H. Hill, *J. Chromatogr.* **480,** 233 (1989).
111. R. D. Smith, J. A. Loo, C. G. Edmonds, C. J. Barinaga, and H. R. Udseth, *Anal. Chem.* **62,** 882 (1990).
112. J. A. Olivares, N. T. Nguyen, C. R. Yonker, and R. D. Smith, *Anal. Chem.* **59,** 1230 (1987).
113. R. D. Smith and H. R. Udseth, *Nature* **331,** 639 (1988).
114. R. D. Smith, J. A. Olivares, N. T. Nguyen, and H. R. Udseth, *Anal. Chem.* **60,** 436 1988.
115. R. D. Smith, C. J. Barinaga, and H. R. Udseth, *Anal. Chem.* **60,** 1948 (1988).
116. E. D. Lee, W. Muck, J. D. Henion, and T. R. Covey, *J. Chromatogr.* **458,** 313 (1988).
117. E. D. Lee, W. Muck, J. D. Henion, and T. R. Covey, *Biomed. & Environ. MS* **18,** 844 (1989).
118. J. A. Loo, H. K. Jones, H. R. Udseth, and R. D. Smith, *J. Microcol. Sep.* **1,** 223 (1989).
119. R. D. Smith, H. R. Udseth, J. A. Loo, B. W. Wright, and G. A. Ross, *Talanta* **36,** 161 (1989).
120. C. G. Edmonds, J. A. Loo, C. J. Barinaga, H. R. Udseth, and R. D. Smith, *J. Chromatogr.* **474,** 21 (1989).
121. R. D. Smith, J. A. Loo, C. J. Barinaga, C. G. Edmonds, and H. R. Udseth, *J. Chromatogr.* **480,** 211 (1989).
122. R. D. Smith, J. A. Loo, C. G. Edmonds, C. J. Barinaga, and H. R. Udseth, *J. Chromatogr.* **516,** 157 (1990).
123. E. C. Huang, T. Wachs, J. J. Conboy, and J. D. Henion, *Anal. Chem.* **62,** 713A (1990).
124. I. M. Johansson, E. C. Huang, J. D. Henion, and J. Zweigenbaum, *J. Chromatogr.* **554,** 311 (1991).
125. I. M. Johansson, R. Pavelka, and J. D. Henion, *J. Chromatogr.* **559,** 515 (1991).
126. P. Thibault, C. Paris, and S. Pleasance, *Rapid Commun. Mass Spectrom.* **5,** 484 (1991).
127. T. Wachs, J. C. Conboy, F. Garcia, and J. D. Henion, *J. Chromatogr. Sci.* **29,** 357 (1991).
128. R. D. Smith, H. R. Udseth, C. J. Barinaga, and C. G. Edmonds, *J. Chromatogr.* **559,** 197 (1991).
129. S. Pleasance, P. Thibault, and J. Kelly, *J. Chromatogr.* **591,** 325 (1992).
130. R. Kostiainen, A. B. L. Lanting, and A. P. Bruins, paper presented at HPCE '92, Amsterdam.
131. S. Pleasance, S. W. Ayer, M. V. Laycock, and P. Thibault, *Rapid Commun. Mass Spectrom.* **6,** 14 (1992).
132. H. R. Udseth, J. A. Loo, and R. D. Smith, *Anal. Chem.* **61,** 228 (1989).
133. R. D. Smith, S. M. Fields, J. A. Loo, C. J. Barinaga, H. R. Udseth, and C. G. Edmonds, *Electrophoresis* **11,** 709 (1990).
134. J. S. M. se Wit, L. J. Deterding, M. A. Mosely, K. B. Tomer, and J. W. Jorgenson, *Rapid Comm. Mass Spectrom.* **2,** 100 (1988).
135. M. A. Mosely, L. J. Deterding, K. B. Tomer, and J. W. Jorgenson, *Rapid Comm. Mass Spectrom.* **3,** 87 (1989).
136. R. M. Caprioli, W. T. Moore, M. Martin, B. B. DaGue, K. Wilson, and S. Moring, *J. Chromatogr.* **480,** 247 (1989).

137. M. A. Moseley, L. J. Deterding, K. B. Tomer, and J. W. Jorgenson, *J. Chromatogr.* **480,** 197 (1989).
138. R. M. Caprioli, *Anal. Chem.* **62,** 477A (1990).
139. M. A. Moseley, L. J. Deterding, K. B. Tomer, and J. W. Jorgenson, *J. Chromatogr.* **516,** 167 (1990).
140. N. J. Reinhold, E. Schroeder, U. R. Tjaden, W. M. A. Niessen, M. C. Ten Noever de Drauw, and J. Van Der Greef, *J. Chromatogr.* **516,** 147 (1990).
141. L. J. Deterding, C. E. Parker, J. R. Perkins, M. A. Moseley, J. W. Jorgenson, and K. B. Tomer, *J. Chromatogr.* **554,** 329 (1991).
142. M. J. F. Suter, B. B. DaGue, W. T. Moore, S. N. Lin, and R. M. Caprioli, *J. Chromatogr.* **553,** 101 (1991).
143. L. J. Deterding, M. A. Moseley, K. B. Tomer, and J. W. Jorgenson, *J. Chromatogr.* **554,** 73 (1991).
144. M. A. Moseley, L. J. Deterding, K. B. Tomer, and J. W. Jorgenson, *Anal. Chem.* **63,** 109 (1991).
145. P. Ferranti, A. Malorni, P. Pucci, S. Fanali, A. Nardi, and L. Ossicini, *Anal. Biochem.* **194,** 1 (1991).
146. F. Garcia and J. D. Henion, *Anal. Chem.* **64,** 985 (1992).
147. K. D. Altria, C. F. Simpson, A. K. Bharih, and A. E. Theobald, *Electrophoresis* **11,** 732 (1990).
148. S. L. Pentoney, R. N. Zare, and J. F. Quint, *Anal. Chem.* **61,** 1642 (1989).
149. S. L. Pentoney, R. N. Zare, and J. F. Quint, *ACS Symp. Ser.* **434,** 60 (1990).
150. C. Y. Chen and M. D. Morris, *Appl. Spectrosc.* **42,** 515 (1988).
151. C. Y. Chen and M. D. Morris, *J. Chromatogr.* **540,** 355 (1991).
152. J. Pawliszyn, *Anal. Chem.* **60,** 2796 (1988).
153. C. Y. Chen, T. Demana, S. D. Huang, and M. D. Morris, *Anal. Chem.* **61,** 1590 (1989).
154. A. E. Bruno, B. Krattiger, F. Maystre, and H. M. Widmer, *Anal. Chem.* **63,** 2689 (1991).
155. R. P. W. Scott, *Liquid Chromatography Detectors* (Elsevier, 1986).
156. E. S. Yeung, ed., *Detectors for Liquid Chromatography* (Wiley-Interscience, 1986).
157. S. Terabe, K. Otsuka, and T. Ando, *Anal. Chem.* **61,** 251 (1989).
158. H. K. Jones, N. T. Nguyen, and R. D. Smith, *J. Chromatogr.* **504,** 1 (1990).
159. R. Weinberger, *Amer. Lab.* May 1984.
160. *The API Book,*(Sciex, 1990).
161. J. A. Loo, R. R. Ogorzalek, K. J. Light, C. G. Edmonds, and R. D. Smith, *Anal. Chem.* **64,** 81 (1992).
162. G. J. Van Berkel, G. L. Glish, and S. A. Mcluckey, *Anal. Chem.* **62,** 1284 (1990).
163. D. J. Rose and J. W. Jorgenson, *J. Chromatogr.* **438,** 23 (1988).
164. A. S. Cohen, D. R. Najarian, A. Paulus, A. Guttman, J. A. Smith, and B. L. Karger, *Proc. Natl. Acad. Sci.* **85,** 9660 (1988).
165. P. Camilleri, G. N. Okafo, C. Southan, and R. Brown, *Anal. Biochem.* **198,** 36 (1991).
166. M. Albin, S.-M. Chen, A. Louie, C. Pairaud, and J. Colburn, *Anal. Biochem.* **206,** 382 (1992).
167. C. Fujimoto, Y. Muramatsu, M. Suzuki, and K. Jinno, *J. High Res. Chromatogr.* **14,** 178 (1991).
168. N. A. Guzman, M. A. Trebilcock, and J. A. Advis, *J. Liq. Chromatogr.* **14,** 997 (1991).
169. N. A. Guzman, M. A. Trebilcock, and J. P. Advis, *Anal. Chim. Acta* **249,** 247 (1991).
170. X. Huang and R. N. Zare, *J. Chromatogr.* **516,** 185 (1990).
171. K. O. Eriksson, A. Palm, and S. Hjertén, *Anal. Biochem.* **201,** 211 (1992).
172. Y.-F. Cheng, M. Fuchs, D. Andrews, and W. Carson, *J. Chromatogr.* **608,** 109 (1992).

Chapter 11

Putting It All Together

11.1 Selecting the Mode of HPCE

Selecting the appropriate separation mode is straightforward for most applications. Table 11.1 contains some guidelines to suggest which technique will yield adequate results in the shortest period. The techniques are listed in descending order from the most favored to least desirable.

The basis for these assignments is primarily ease of methods development and the robustness of the separation process. Rugged and precise methods are generally simple in their design. In this regard, CZE at low pH generally yields the most reproducible separations of all the HPCE modes. CZE at high pH and MECC are also satisfactory, though somewhat less rugged compared to low-pH separations because of the need to control the EOF. Under controlled conditions, expect migration time RSDs as low as 0.5% and externally standardized peak area RSDs as low as 1%. These figures are on the higher side of what is generally acceptable for HPLC. For CIEF and CGE, migration times must frequently be corrected with an internal standard to yield RSDs approaching 1%. Issues concerning precision are covered in more detail in Section 11.6.

Table 11.1. Selecting the Mode of Capillary Electrophoresis

Small Ions	Small Molecules	Peptides	Proteins	Oligonucleotides	DNA
CZE	MECC	CZE	CZE	CGE	CGE
CITP	CZE	MECC	IEF	MECC	
	CITP	IEF	CGE		
		CGE	CITP		
		CITP			

11.2 Requirements for Robust Separations

Poor reproducibility of separations has been a frequent problem in HPCE, particularly among new users. The following guidelines can serve as a checklist that can be applied to all methods.

Temperature control. The capillary must be thermostated, or the instrument placed in a temperature-controlled environment. Since viscosity, and thus mobility, is temperature-dependent, failure to adequately control the capillary temperature will result in increased variability in the migration time.

Solubility. All sample components must be soluble in the supporting electrolyte. This is particularly important whenever the injection solution composition differs from the supporting electrolyte.

Wall effects. Capillary coatings or buffer additives may be required to suppress wall interactions.

Joule heating. An Ohm's Law plot should be performed to optimize the voltage.

Selectivity. The appropriate mobility plots should be performed to select the optimal buffer composition for separation. Variables include pH and additive concentration. Rugged separations are generally obtained when small changes in buffer composition do not dramatically affect mobility. This information is readily discernible from mobility plots.

Buffer refreshment. Buffer depletion can result in drift of both mobility and EOF. The sensitivity of this effect is solute-dependent and may be based on the factors discussed under *Selectivity.* The requisite frequency for buffer replenishment should be determined experimentally. Migration time drift is frequently observed from this effect.

Evaporation control. Use closures provided by the instrument manufacturers for samples and buffers. Reduce the sampler temperature if your instrument has that capability. Evaporation can be particularly troublesome when organic solvents are employed.

Control ionic strength of injection buffer. The conductivity of the injection buffer, including the sample contribution, should be equal to or less than the conductivity of the background electrolyte. Precise intersample control of the ionic strength is required. Beware of sample matrix effects. It may be necessary to perform a sample cleanup to place the solutes in an HPCE-friendly, e.g., low ionic strength, matrix (Section 11.5).

Sample viscosity control. If hydrodynamic injection is employed, the sample viscosity can affect the amount of material introduced into the capillary. Dilution or sample cleanup may be required to ensure that all samples have the same viscosity.

Internal standards. An internal standard can compensate in part for quantitative problems that occur because of sample evaporation or sample-to-sample viscosity differences. The internal standard should be used to adjust for minor experimental effects. Substantive effects should be corrected for experimentally. The use of internal standards will be discussed in Section 11.6.

Capillary washing. To ensure a clean and reproducible inner wall, it is often necessary to wash the capillary between runs with 0.1 N or even 1 N sodium hydroxide. A 1 or 2 min wash is usually adequate, followed by a 1 min water rinse and 1–2 min reequilibration with run buffer. This is only recommended for bare silica capillaries.

11.3 Realistic Compromises

Speed, resolution, and loadability often oppose each other in most separation techniques. The acute sensitivity problem in HPCE must be factored in as well. As a starting point, a reasonable separation must have already been developed following the guidelines given in the individual chapters. Use Table 11.2 as a guide to optimize a particular aspect of a separation.

Do not push these techniques too far. When heroic efforts are called for, the separation scheme is probably not appropriate to begin with.

11.4 Linear Dynamic Range

The contribution that the solute adds to the total ionic strength of the injection solution can dramatically affect both the peak symmetry and the linearity of response. A requirement that the sample not contribute significantly to the overall ionic strength of the injection solution has been recognized for some time (1–3). The problem at high sample concentration is illustrated in Fig. 11.1 for a low-pH separation. In this instance, the ionic strength of the injection solution exceeds that of the carrier electrolyte. At the start of the run, the voltage drop over the injection zone is relatively low compared to that over the separation zone. As cations migrate out of the injection zone, they become exposed to the high field strength expressed over the separation zone (Fig. 11.1, middle). As a result, their electrophoretic velocities increase compared to solutes remaining behind in the injection zone. This anti-stacking effect leads to substantial bandbroadening because of the unstable zone boundary at the front of the zone. The rear of the zone remains stable and sharp.

At high pH, a different mechanism operates, but the result is similar to the low-pH case. The anion's mobility is directed toward the anode in a direction opposed to the EOF. In this instance, the migrating anions encounter the supporting electrolyte at the rear of the injection zone (Fig. 11.2, middle). At this boundary, the rear zone anions accelerate toward the positive electrode, thereby experiencing a net reduction in migration speed toward the cathode. As a result, anions remaining in the injection zone migrate faster than those in the electrolyte zone at the rear of the injection solution.

Table 11.2. Adjusting the Separation Parameters

Goal	Adjustment	Consequence	Compensation	Impact
Speed	Decrease capillary length	Joule heating	Decrease capillary diameter	Loss of sensitivity
	Increase field strength	Loss of resolution	Decrease buffer ionic strength[a]	Increased wall effects
	Increase temperature			Decreased loading capacity
Resolution	Increase capillary length	Longer run times	None[b]	n/a
	Reduce electroosmotic flow	Longer run times		
	Decrease injection size	Loss of sensitivity	None	n/a
Loading	Increase capillary diameter	Joule heating	Decrease field strength	Longer run times
	Increase buffer ionic strength	Loss of resolution	Decrease temperature	
	Increase injection size	Loss of resolution	Increase resolution	Longer run times
Sensitivity	Increase capillary diameter	Joule heating	Decrease field strength	Longer run times
	Increase buffer ionic strength	Loss of resolution	Decrease temperature	Longer run times
	Stacking buffers	Loss of resolution	Increase resolution	Longer run times
	"Z-cell"	Loss of resolution	Increase resolution	Longer run times
	Increase injection size	Loss of resolution	Increase resolution	Longer run times
	Laser fluorescence	Derivatization usually required	None	Validation and sample prep complicated
	On-line concentration (LC/CE)	Limited to CZE	None	n/a

[a]Decreasing the buffer ionic strength also increases both mobility and EOF.

[b]Assuming maximum voltage is already employed.

n/a: not applicable

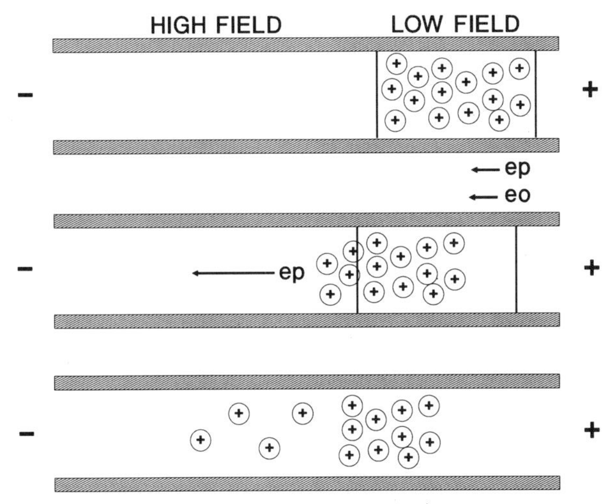

Figure 11.1. Illustration of anti-stacking of cations in a low-pH buffer.

The practical implication of this problem is illustrated in Fig. 11.3 (4). This is an MECC separation of naproxen, an anionic nonsteroidal anti-inflammatory drug. As the serial dilution proceeds, consistent peak shape and migration times are found at concentrations below 156 μg/mL. Above that range, particularly above 1.25 mg/mL, the peaks are badly fronted, as predicted from the aforementioned effects.

The figures of merit are plotted in Fig. 11.4. Above 100 μg/mL, deviations from expected values for migration time, peak width, and peak height are evident. Peak area shows less deviation from linearity. This is quite reasonable, since at high solute concentrations, much of the signal is represented in peak width rather than

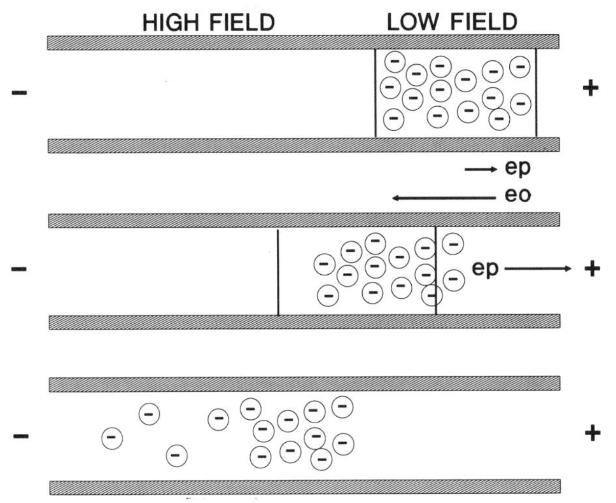

Figure 11.2. Illustration of anti-stacking of anions in a high-pH buffer.

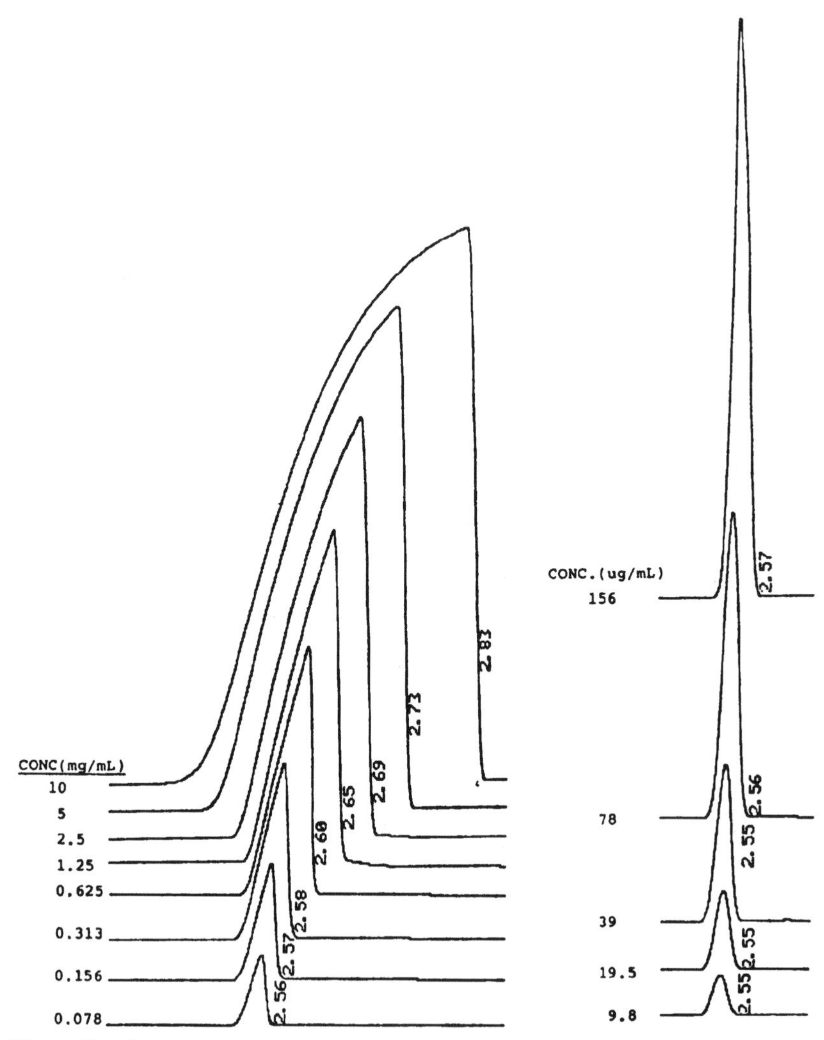

Figure 11.3. Impact of solute concentration on peak shape. Capillary: 60 cm (38 cm to detector) × 50 μm i.d.; buffer: 25 mM SDS, 20 mM borate, pH 9.2; temperature: 50°C; voltage: 25 kV; detection: UV, 230 nm; injection: vacuum, 1 s; solute: naproxen. Reprinted with permission from *J. Liq. Chromatogr.* **14,** 953, copyright ©1991 Marcel Dekker.

height. To maximize the linear dynamic range of the separation, it is important to use peak areas for all calculations. Peak areas are also correctable for migration time variations (Eq. (10.2)).

Increasing the linear range is possible with high ionic-strength buffers (4). Figure 11.5 shows separations of some anti-inflammatory drugs at concentrations of 1 mg/mL and 250 μg/mL using a low ionic-strength buffer. Substantial losses

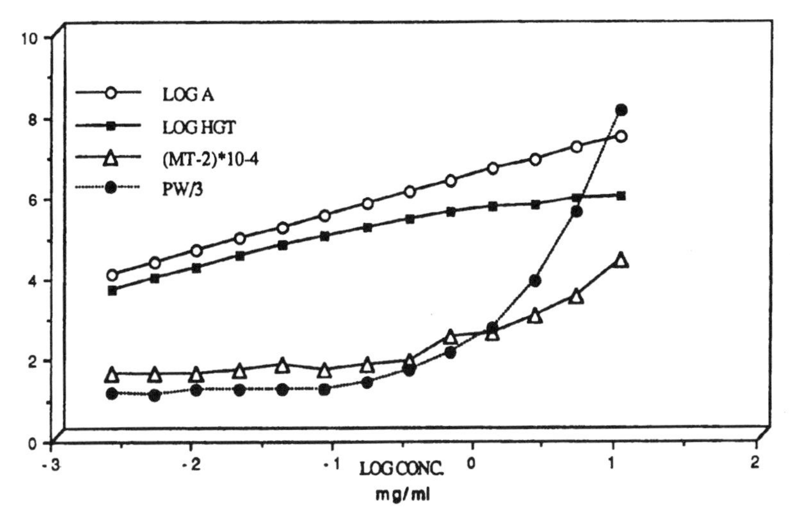

Figure 11.4. Impact of solute concentration on analytical figures of merit. Refer to Fig. 11.3 for experimental conditions. Key: log A = log peak area; log HGT = log peak height; ((MT-2) × 10^{-4}) = scaled migration time (min); PW/3 = scaled peak width (s). Reprinted with permission from *J. Liq. Chromatogr.* **14,** 953, copyright ©1991 Marcel Dekker.

in resolution are found at the higher concentrations. A similar separation in a high ionic-strength buffer is shown in Fig. 11.6. Almost no change in resolution is found between the two electropherograms. Note that a shortened capillary was employed because the EOF was substantially lowered by the high-ionic strength buffer. The broadening of the latter two peaks was probably due to solubility problems in the high ionic-strength buffer. These separations were performed in 25 μm i.d. capillaries to reduce the heating effects. For micropreparative work, it is much better to work with wider-bore capillaries at reduced temperatures.

The overall addressable concentration range of HPCE is illustrated in Fig. 11.7. The use of the laser detector solves most of the compelling problems in HPCE. With this highly sensitive detector, it is possible to perform extreme dilutions of most samples. At high dilution, most ionic-strength mediated effects from the solute and/or the sample matrix become insignificant. Unfortunately, derivatization is required for most laser-fluorescence based applications.

11.5 Sample Preparation

A. Basic Principles

In HPCE, there are several important issues regarding sample preparation. Obviously, interfering components, if they are not separable, must be removed during the sample preparation process. This is true for all analytical techniques. In

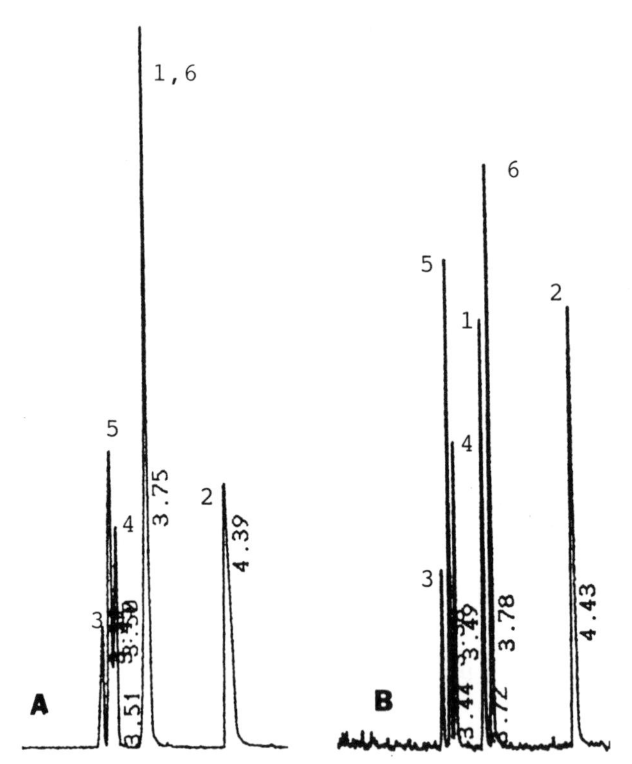

Figure 11.5. Band profile dependence on solute concentration and buffer ionic strength—low ionic-strength buffer. Capillary: 65 cm (43 cm to detector) × 25 μm i.d.; buffer: 20 mM SDS, 20 mM phosphate, pH 9.2; temperature: 30°C; injection: vacuum: 2 s; detection: UV, 230 nm. Key: (a): (1) sulindac, 1 mg/mL; (2) indomethacin, 1 mg/mL; (3) tolmetin, 1 mg/mL; (4) ibuprofen, 1 mg/mL; (5) naproxen, 100 μg/mL; (6) diflunisal, 500 μg/mL. (b): 4× dilution of A. Reprinted with permission from *J. Liq. Chromatogr.* **14,** 953, copyright ©1991 Marcel Dekker.

both chromatography and electrophoresis, the sample matrix can affect the resolution of the separation. Unlike chromatographic techniques, in HPCE the sample matrix may have a profound impact on the amount of material that is injected into the capillary. The sample preparation process must deal with this problem.

For samples containing high concentrations of solutes, e.g., pharmaceutical dosage forms, simple dilution of the dosage form extract in the supporting electrolyte is sufficient since only small injection volumes are generally required. Depending on the strength of the chromophore, final concentrations of 10^{-5}– 10^{-3} M provide adequate sensitivity. Stacking techniques can be useful to improve sensitivity, but matrix effects and artifacts often interfere. For complex samples or when high sensitivity is required, sample preparation to remove interferences and place the solute(s) in a CE-friendly solution is clearly indicated. Centrifugation or filtration to remove particulate matter is usually good practice.

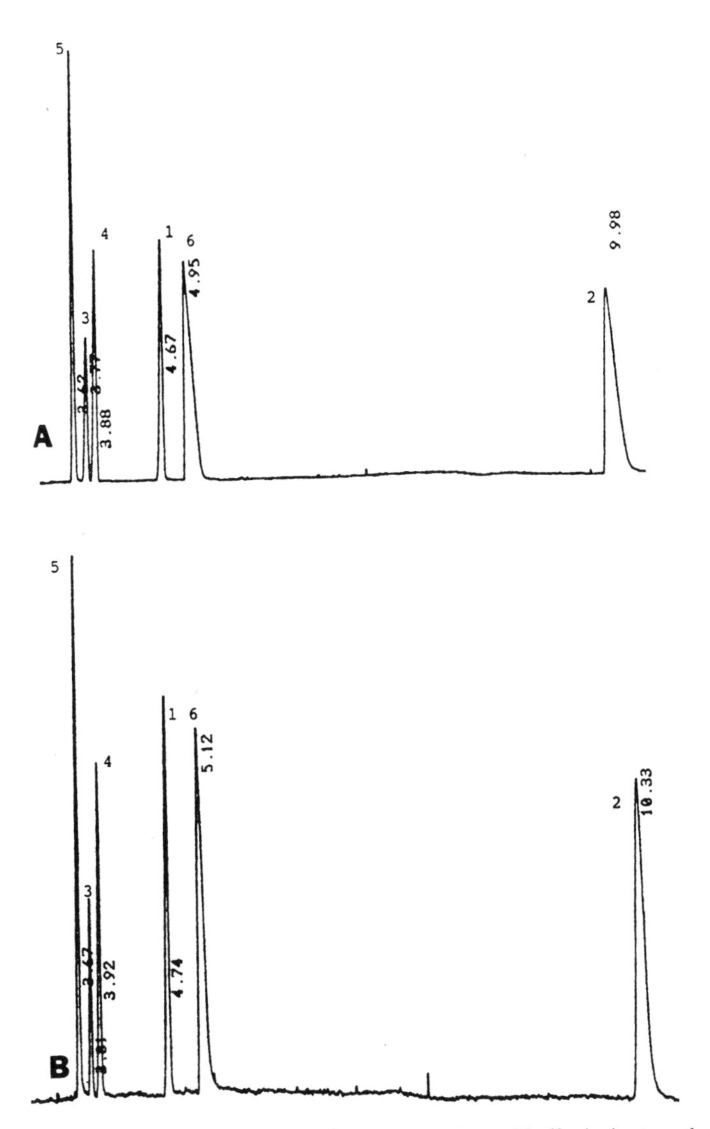

Figure 11.6 Band profile dependence on solute concentration and buffer ionic strength—high ionic-strength buffer. Conditions as in Fig. 11.5, except: capillary: 20 cm to detector; buffer: 100 mM phosphate, 25 mM SDS, pH 7.0. Reprinted with permission from *J. Liq. Chromatogr.* **14,** 953, copyright ©1991 Marcel Dekker.

In reverse-phase LC, it is generally bad practice to prepare the sample in a solvent with greater eluting power than the mobile phase, particularly if large injections are required. The fundamental requirement for HPCE is that the sample should never be prepared to have an ionic strength that is greater than the supporting

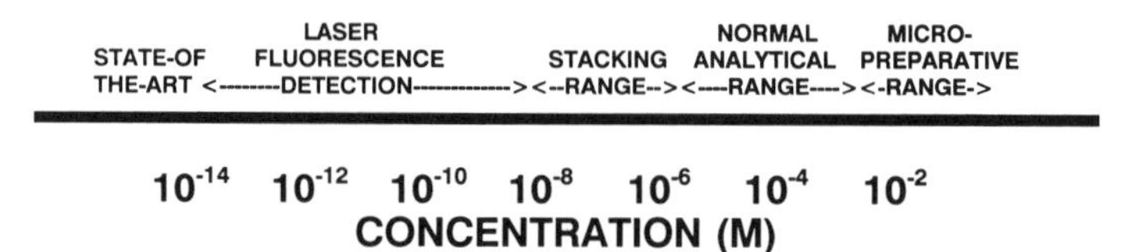

Figure 11.7. The dynamic range of capillary electrophoretic techniques.

electrolyte. This requirement can be loosened only when the injection volume can be kept small.

B. Drugs in Biological Fluids

MECC is generally preferred for separating small synthetic pharmaceuticals. Surfactant solutions are utilized, so direct injection of blood plasma or serum might be feasible because a surfactant such as SDS binds strongly to, and solubilizes, serum proteins. Indeed, micellar liquid chromatography has been shown to be useful for direct injection. In this mode of HPLC, the surfactant solution serves as the mobile-phase modifier. Surfactant-bound serum proteins form an extremely large aggregate that is excluded from the stationary phase. The protein bolus elutes on or about t_0 in a relatively narrow band. The retained drug substance elutes some time later, producing clean chromatograms at the low microgram per milliliter level.

In MECC, the protein–surfactant aggregate has a substantial net negative charge when SDS is used as the additive. The aggregate then elutes relatively late in the separation, leaving only a small window for interference-free monitoring of the drug substance. A typical separation is shown in Fig. 11.8 for aspoxicillin at a concentration of 50 μg/mL (5). Without highly selective detection, it is hard to be optimistic that direct injection will play a major role for the determination of drugs in biological fluids, particularly for trace analysis. An exception may occur when sample sizes are extremely limited, e.g., clinical determination of xanthines from premature infants (6). Other reported applications employing direct injection include creatinine and uric acid (7), porphyrins (8), purines (6), and cefixime (9).

Even more subtle matrix effects can occur. Figure 11.9 illustrates direct urine injection for the determination of urinary porphyrins (8). Clean electropherograms are obtained because selective fluorescence detection with visible excitation is used. With moderately sized injections, e.g., 25 nL, exceptional compression of peak 6 (uroporphyrin) is obtained. For smaller injections, e.g., 10 nL, no such compression is observed. Clearly, some component in the urine is serving as a leading electrolyte invoking CITP over the uroporphyrin zone. The other porphyrins are separated by CZE because their mobilities are lower than the supporting electrolyte's buffer ions (the supporting electrolyte probably acts as a terminator for this high-pH separation). If this effect could be controlled, separations with incredibly high peak capacities could be obtained. For the time being, it is a nuisance to be contended with. Other potential problems with direct injection are

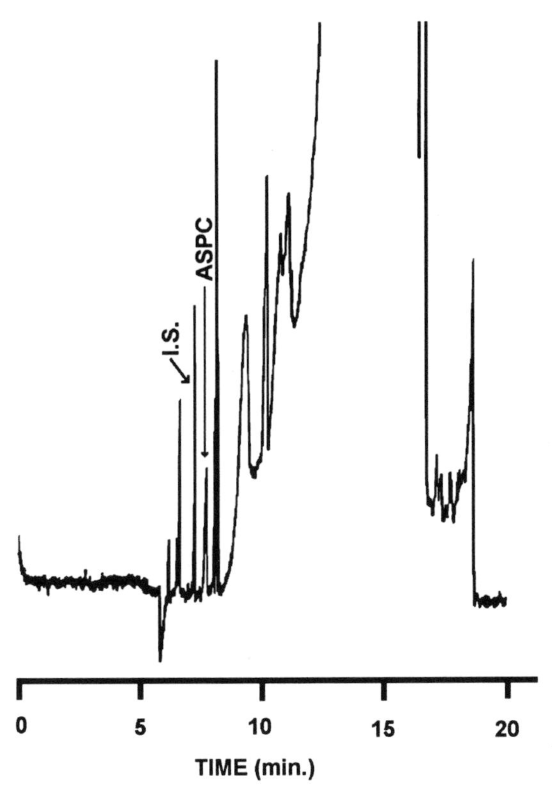

Figure 11.8. Direct plasma injection for the determination of aspoxicillin. Capillary: 65 cm (50 cm to detector) × 50 μm i.d.; buffer: 50 mM SDS, 20 mM phosphate–borate, pH 8.5; volatge: 20 kV; detection: UV, 210 nm; temperature: ambient; solute concentration: 50 μg/mL. Reprinted with permission from *J. Chromatogr.* **515,** 245, copyright ©1990 Elsevier Science Publishers.

ionic-strength effects, protein binding, and impact of endogenous components on electrophoresis and resolution.

Solid-phase extraction is a widely used sample-preparation method for purifying drugs from biological fluids prior to HPLC. Wernly and Thormann (10) employ multistep solid-phase extraction to determine drugs of abuse such as barbiturates, hypnotics, amphetamines, opioids, benzodiazepines, and cocaine metabolites from a single urine specimen. In conjuction with multiwavelength detection, positive confirmation for drugs of abuse in screening urine samples is simple by MECC. The stepwise sample cleanup procedure is illustrated in Fig. 11.10. In this three-step approach, methaqualone is eluted during the first step, morphine, codeine and heroin during the second, and finally, benzoylecognine in the third. There was some carryover between fractions that should be readily eliminated through fine tuning. It is likely that this procedure can be easily adapted for the determination of a wide variety of drug substances in most biological fluid types.

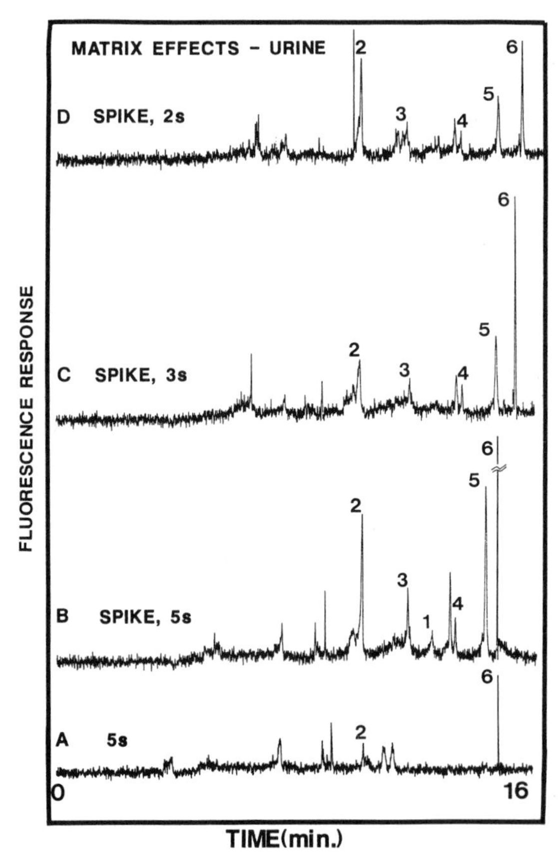

Figure 11.9. Direct urine injection for the determination of urinary porphyrins. Capillary: 77 cm (55 cm to detector) × 50 μm i.d.; buffer: 100 mM SDS, 20 mM CAPS, pH 11; voltage: 20 kV; detection: fluorescence, 400 nm/>550 nm; temperature: 45°C. Samples: (a) normal control urine; (b) control urine spiked with porphyrins to 300 pmol/mL, 5 s vacuum injection; (c) as in (b), 3 s vacuum injection; (d) as in (b), 2 s vacuum injection. Key: (1) mesoporphyrin; (2) coproporphyrin; (3) pentacarboxylporphyrin; (4) hexacarboxylporphyrin isomers; (5) heptacarboxylporphyrin; (6) uroporphyrin. Reprinted with permission from *J. Chromatogr.* **516,** 271, copyright ©1990 Elsevier Science Publishers.

The advantages of this approach are: (i) large sample sizes can be processed and trace-enriched, solving, in part, some of the sensitivity problems; (ii) the organic solvent eluting reagents are easily evaporated, after which the residue can be redissolved in a small quantity of run buffer or a stacking buffer if necessary; (iii) the process produces relatively clean electropherograms.

Liquid–liquid extraction has been used to determine thiopental in serum and plasma (11). Buffered serum was extracted with 5 mL pentane for 10 min and centrifuged. The upper organic layer was removed and evaporated to dryness. The

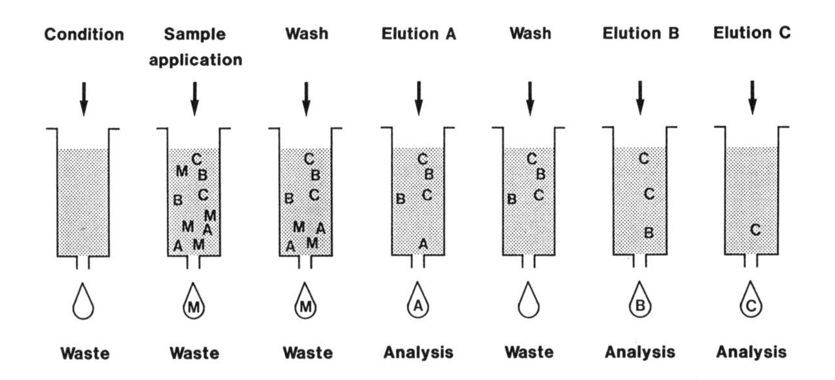

Figure 11.10. Solid-phase extraction process for the determination of drugs in urine samples. (1) Condition with 2 mL each methanol and 100 mM phosphate buffer, pH 6 just before use; (2) load 5 mL urine + 2 mL phosphate buffer via vacuum over a 2 min span; (3) wash sequentially and discard solutions of 1 mL phosphate buffer/methanol (80/20), 1 mL of 1 M acetic acid, and 1 mL hexane; (4) elute with 4 mL methylene chloride; (5) wash with 6 mL methanol; (6) elute with 2 mL of 2% ammonium hydroxide in ethyl acetate; (7) elute with 2 mL methylene chloride/isopropyl alcohol (80/20) containing 2–10% ammonium hydroxide. Reprinted with permission from *Anal. Chem.* **64,** 2155, Copyright©1992 Am. Chem. Soc.

residue was redissolved in run buffer with separation by MECC. The electropherograms were interference-free, and the results from patient samples correlated well to the HPLC assay.

C. DNA and RNA

Salt removal is important prior to CGE of restriction digests or polymerase chain reaction products (PCR). Schwartz *et al.* (12) used ultracentrifugation and desalting with Centricon-30 or 100 filters (W. R. Grace, Amicon Division, Beverly, MA). Many available DNA extraction tubes and filters could be used as well.

Nathakarnkitkool *et al.* (13) purified PCR products using size exclusion chromatography (sample applied to a Sephadex G-150 column and spin-transferred). This procedure was repeated twice. The eluted DNA was precipitated with 0.1 part 3 M sodium acetate and 1 part isopropanol, stored at –30°C for 1 h and centrifuged at 16,000 × *g* for 30 min. The DNA pellet was rinsed twice with 75% ethanol, dried by vacuum, and resuspended in 10 mM Tris, 1 mM EDTA buffer, pH 8.0. Separations were performed using ethidium bromide, hydroxypropylmethylcellulose polymer networks.

The same group (13) isolated mRNA from tissue culture with a commercial mRNA isolation kit (Invitrogen, San Diego, CA). The lysed cells were passed through a 22 gauge needle to shear high molecular weight DNA. mRNA from this 0.5 M lysate solution was isolated using the Invitrogen kit. After elution from the kit's cellulose column, the mRNA was precipitated as described earlier for PCR product isolation.

D. Proteins and Peptides

High salt concentrations routinely found in protein samples cause substantial problems in HPCE (14). Microconcentrators such as those available from Amicon and other vendors are useful for desalting and concentrating samples as required. HPLC is frequently required to isolate fractions from complex samples such as fermentation broths. Only dilution with run buffer is required for simple matrices such as pharmaceutical injectable dosage forms.

Gurley *et al.* (15) separated rat lung proteins from animals exposed to perfluoroisobutylene. The proteins were swept from the lung by lavage with 150 mM sodium chloride. The problem was to concentrate the lavage fluid and remove the salt. The proteins were precipitated overnight at 4°C in 90% acetone. After centrifugation, the precipitate was washed once more with acetone. The protein pellet was dissolved in 1 mL of 0.2% trifluoroacetic acid (TFA). TFA is a good protein solvent and does not interfere with electrophoresis at low pH.

In another approach, Yim (16) used Centriprep concentration followed by dialysis into 10 mM phosphate buffer, pH 2.5, for the preparation of rtPA prior to CIEF. The dialyzed samples were diluted in an ampholyte/urea/buffer blend to a concentration of 0.5 mg/mL.

Gordon *et al.* (17) performed a 40-fold dilution of blood serum with a mixture of 1 mM boric acid, pH 4.5, containing 20% ethylene glycol. The run buffer was 50 mM borate, pH 10. The ethylene glycol appears to be effective in preventing proteins from adhering to the walls of the capillary tubing.

11.6 Quantitative Analysis

Most research on HPCE has focused on instrumentation, capillary, and separation technologies. Only recently has the focus begun to shift into the applied aspects. Table 11.3 summarizes some of the work in quantitation with respect to accuracy and precision. Considering the enormity of the literature in HPCE, it is disturbing but somewhat understandable that so little has been devoted to quantitative analysis. Virtually all of the pre-1990 literature in HPCE contained descriptions of homemade CE instruments. While these manual instruments were simple to operate and produced superb separations, the precision was not up to today's standards. With automated instrumentation, good precision is possible provided the criteria described in Section 11.2 are followed.

Table 11.3. Accuracy and Precision of HPCE

Application	Migration Time (%RSD)	Peak Area (%RDS)	Peak Height (%RSD)	%Recovery	Reference
CZE and MECC					
Analgesics	not reported	0.8–1.7[c]	1.4–2.6[c]	99–101	(18)
Domperidone and others	0.95–1.45	1.05–2.82	0.46–1.01	97–104[a]	(19)
Anti-inflammatory drugs	0.16–0.54	0.86–1.96	0.49–1.9	101–104[b]	(4)
Insulin	0.36–0.54	1.72–2.41	—	91–103[a]	(20)
Serum albumin	0.43	3.52	—	—	
Salicylamide	0.77	1.95	2.06	—	(21)
Dynorphins	0.57–0.63	0.95–1.39			(22)
CGE					
Proteins (polyacrylamide)	0.9–1.6	3.53	—	—	
Restriction fragments (polyacrylamide)	0.9[c]	—	—	—	(23)
Restriction fragments (HPMC)	0.16–0.22	5.32–9.16	4.79–6.00	—	(12)
CIEF					
Proteins	0.5–2.5[d]	—	—	—	(24)

[a]Compared to HPLC.
[b]% of labeled amount.
[c]Used internal standard.
[d]Corrected with a marker protein.

A. Sampling Rate

With fully automated instrumentation, on-line detection and modern computing power for peak integration, quantitative analysis by HPCE is much simpler than by slab-gel electrophoresis. Most data systems or integrators are suitable for HPCE provided their data acquisition rates can be adjusted to account for the sharp peaks characteristic of HPCE. Sampling rates of 20 Hz or less are suitable for all but the most efficient separations. At that sampling rate, a signal with a 5 s peak width is sampled with 100 points, an amount sufficient for all necessary computations. Data collection of 25–50 points per peak is generally adequate. Oversampling does not substantially improve precision, but it does occupy valuable disk storage capacity.

B. Peak Height or Peak Area

No definitive studies comparing precision of peak height or area in quantitative HPCE have been published. The data in Table 11.3 are ambiguous in that regard. Despite the lack of data, peak area is recommended for two reasons: (i) the linear dynamic range is greater and (ii) area can be corrected for migration time, a feature important for area percent calculations.

C. External Standards

External standard calculations are performed by conventional methods, identical to those used in HPLC or GC. Calibration can be single-point, but multipoint regression is usually preferred. For single-point calibration, the standard should have a solute concentration higher than that of any sample. For concentrations approaching and beyond 1 mg/mL, beware of nonlinear response.

Area corrections for migration time scatter are generally not used. It has not been determined if these corrections actually improve precision for well-controlled separations that have solute migration-time RSDs of less than 1%.

For single-point calibration, the concentration of a solute can simply be determined by

$$C_{\mathrm{SAMP}} = \frac{A_{\mathrm{SAMP}}}{A_{\mathrm{STD}}} C_{\mathrm{STD}}, \tag{11.1}$$

where C_{SAMP} and C_{STD} are the respective concentrations of sample and standard solutions, while A_{SAMP} and A_{STD} are the respective peak areas.

For multipoint calibration, linear regression is used to determine the best line through the calibration data. For nonlinear response, quadratic or polynomial curve fitting can be employed (25), but this is less optimal than a linear fit. Standards should bracket the expected solute concentration range.

D. Internal Standards

Internal standards are added to the sample to correct for quantitative losses during cleanup as well as instrumental imprecision, primarily caused by the injection process. Calibration can be single- or multilevel, with the latter the preferred mode. Using a standard solution, the ratio (R) of peak areas of the standard and internal standard is calculated. If the same amount of internal standard is added to a sample,

$$C_{\mathrm{SAMP}} = \frac{R_{\mathrm{SAMP}}}{R_{\mathrm{STD}}} C_{\mathrm{STD}}, \tag{11.2}$$

where C_{SAMP} and C_{STD} are the respective concentrations of sample and standard solutions, while R_{SAMP} and R_{STD} are the respective peak area ratios.

The internal standard must be used to compensate for sample cleanup losses if quantitative recoveries cannot be made. Concerning instrumental factors, definitive studies have not been made comparing external-standard to internal-standard techniques using modern high-precision automated systems.

For the modes of CIEF and CGE, an internal standard must be added to correct for migration-time variation. This same internal standard can be used for quantitation as well.

E. Area Percentage

Area percentage is frequently used for the purity determination of bulk pharmaceuticals. The areas of all peaks are summed, and the percent contribution of each component is calculated. The method is based on the assumption that potential process impurities and degradation products have identical molar absorptivities at the wavelength of detection, usually in the low UV. For this application, it is critical to correct the peak areas of the parent compound and its impurities for their migration velocities. When uncorrected areas are used, positive or negative errors will occur, depending the relative migration order of the sample's components.

11.7 Correction Factors

Correction factors for chromatographic separations have been used for a long time. The best example is the use of retention indices for correcting intercolumn and instrumental variation in gas chromatography (26). For example, a solute with a retention index of 1,250 elutes between n-C_{12} and n-C_{13} normal linear alkanes. In GC, the number of experimental variables is far fewer than in HPLC and HPCE. The primary variable in GC is the stationary phase. In HPLC, both the stationary and mobile phase are important, and in HPCE, the carrier electrolyte dominates the experimental variables. Since the composition of the mobile phase or carrier electrolyte is infinitely variable, the development of retention indices is difficult in HPLC or HPCE. There has been one report on the development of migration indices for HPCE (27).

In an analogous fashion to retention time in chromatography, migration time is used for qualitative analysis. In HPCE, the migration time is dependent on both mobility and EOF. Since EOF is more prone to drift compared to mobility, it has been argued that mobility should be used as the qualitative parameter. Mobility should be independent of field strength and capillary length, but dependent on buffer composition and temperature. The solute's mobility is simply calculated by subtracting the electroosmotic mobility (determined with a neutral marker) from the apparent mobility.

Moring *et al.* (28) reported data from separations in both low- and high-pH buffers. Oddly enough, the precision of mobility improved by 14–20-fold relative to migration time in the low-pH buffer, where the EOF is small to begin with. In the high-pH buffer, the data were equivocal.

For quantitative analysis, it is sufficient to use migration time as the reportable parameter. If mobilities are required to compensate for migration-time drift, it is best to troubleshoot the separation and correct the source of the drift. For structure–migration studies, mobility is the required parameter because the EOF component is factored out.

Protocols to correct for bias in electrokinetic injection due to solute mobility (29) and injection buffer conductivity (30) have been reported. These factors are best compensated for experimentally through the use of standards and constant ionic-strength injection buffers, respectively.

References

1. W. Thormann, *Electrophoresis* **4,** 383 (1983).
2. F. E. P. Mikkers, F. M. Everaerts, and T. P. E. M. Verheggen, *J. Chromatogr.* **169,** 1 (1979).
3. F. E. P. Mikkers, F. M. Everaerts, and T. P. E. M. Verheggen, *J. Chromatogr.* **169,** 11 (1979).
4. R. Weinberger and M. Albin, *J. Liq. Chromatogr.* **14,** 953 (1991).
5. H. Nishi, T. Fukuyama, and M. Matsuo, *J. Chromatogr.* **515,** 245 (1990).
6. W. Thormann, A. Minger, S. Molteni, J. Caslavska, and P. Gebauer, *J. Chromatogr.* **593,** 275 (1992).
7. M. Miyake, A. Shibukawa, and T. Nakagawa, *HRC & CC* **14,** 181 (1991).
8. R. Weinberger, E. Sapp, and S. Moring, *J. Chromatogr.* **516,** 271 (1990).
9. S. Honda, A. Taga, K. Kakehi, S. Koda, and Y. Okamoto, *J. Chromatogr.* **590,** 364 (1992).
10. P. Wernly and W. Thormann, *Anal. Chem.* **64,** 2155 (1992).
11. P. Meier and W. Thormann, *J. Chromatogr.* **559,** 505 (1991).
12. H. E. Schwartz, K. Ulfelder, F. J. Sunzeri, M. P. Busch, and R. G. Brownlee, *J. Chromatogr.* **559,** 267 (1991).

13. S. Nathakarnkitool, P. J. Oefner, G. Bartsch, M. A. Chin, and G. K. Bonn, *Electrophoresis* **13,** 18 (1992).
14. T. Satow, A. Machida, K. Funakushi, and R. Palmieri, *HRC & CC* **14,** 276 (1991).
15. L. R. Gurley, J. S. Buchanon, J. E. London, D. M. Stavert, and B. E. Lehnert, *J. Chromatogr.* **559,** 411 (1991).
16. K. W. Yim, *J. Chromatogr.* **559,** 401 (1991).
17. M. J. Gordon, K. J. Lee, A. A. Arias, and R. N. Zare, *Anal. Chem.* **63,** 69 (1991).
18. S. Fujiwara and S. Honda, *Anal. Chem.* **59,** 2773 (1987).
19. A. Pluym, W. Van Ael, and M. De Smet, *Trends in Anal. Chem.* **11,** 27 (1992).
20. M. Lookabaugh, M. Biswas, and I. S. Krull, *J. Chromatogr.* **549,** 357 (1991).
21. M. E. Swartz, *J. Liq. Chromatogr.* **14,** 923 (1991).
22. P. D. Grossman, H. H. Lauer, S. E. Moring, D. E. Mead, M. F. Oldham, J. H. Nickel, J. R. P. Goudberg, and A. Krever, *Amer. Biotech. Lab.* Feb., p. 35 (1990).
23. D. N. Heiger, A. S. Cohen, and B. L. Karger, *J. Chromatogr.* **516,** 33 (1990).
24. S.-M. Chen and J. E. Wiktotowicz, *Anal. Biochem.* **206,** 84 (1992).
25. G. I. Ouchi, *LC-GC* **10,** 524 (1992).
26. *J & W Scientific Products Catalog, 1992–1993,* p. 250.
27. T. T. Lee and E. S. Yeung, *Anal. Chem.* **63,** 2842 (1991).
28. S. E. Moring, J. C. Colburn, P. D. Grossman, and H. H. Lauer, *LC-GC* **8,** 34 (1990).
29. X. Huang, M. J. Gordon, and R. N. Zare, *Anal. Chem.* **60,** 375 (1988).
30. T. T. Lee and E. S. Yeung, *Anal. Chem.* **64,** 1226 (1992).

Chapter 12

Special Topics

12.1 Introduction

The purpose of this chapter is to offer a glimpse of technologies that may influence the future of HPCE, without introducing distractions into the logical flow of material necessary for a grasp of the basic concepts. Some material described here may soon be or already is incorporated into commercial instrumentation. Other technologies are years away from practical utility. Since most users will not employ these techniques in the near future, only brief descriptions will be given here. Should your curiosity be kindled, the cited references provide a thorough discussion.

Think of this section as a technological tour of some innovations in HPCE. These far-reaching developments include highly miniaturized, multicapillary, and two-dimensional instruments. Since detector sensitivity is often a limiting factor, a glimpse of immunologic trace enrichment is described as well.

12.2 Field Effect Electroendoosmosis

The electroosmotic flow is an important tool that may used to control the speed and resolution of a separation. In a fused-silica capillary, it is not possible to decouple the EOF from the chemistry of the background electrolyte. A change in the buffer composition is frequently accompanied by a modification in the zeta potential at the capillary wall–electrolyte interface. The accompanying change in the EOF complicates methods development, because multiple effects occur as the buffer composition is varied. Ideally, the EOF would be independently controlled, much like the mobile-phase flow rate in HPLC.

Field effect electroosmosis (FEE) is a novel approach to controlling the EOF with a degree of independence from the experimental conditions. A second voltage is imposed perpendicular to the length of the capillary, as illustrated in Fig. 12.1.

Ghowski and Gale (1) treated this system as an electrokinetic transistor called a metal–insulator–electrolyte–electrokinetic-field-effect device (MIEEKFED). They determined that the applied perpendicular field has an impact on the zeta potential at the wall of the capillary, thereby affecting the EOF. Electrical contact to the system is possible with a metal-coated capillary (resistive layer). The fused-silica capillary wall serves as the insulator. By selecting the appropriate values for V_G and V_D, it is possible to adjust both the field strength for the separation and the strength of the perpendicular field. Thin-walled capillaries are required, or the field strength needed to influence the zeta potential would be unrealistically high. Ghowski and Gale (1) predicted a field of 500 V/μm of wall thickness may be needed, though Lee *et al.* (2) used capillaries with a wall thickness of 50 μm and fields of less than 10 kV.

The control of the zeta potential is dependent both on the chemistry and the background electrolyte and on the strength and direction of the applied perpendicular field. Illustrated in Figs. 12.2 and 12.3 is the impact of buffer concentration and pH on the zeta potential versus the applied external field. Under these experimental conditions, it is possible to adjust both the magnitude and charge (+ or −) of the zeta potential. The magnitude of the voltage given in these figures is small because an insulator thickness of only 1 μm is used for this model.

Lee *et al.* (2, 3) developed a capacitance model to quantitate the impact of the applied external field on the zeta potential, ζ. With an instrument design substantially different from that of Ghowski and Gale (1) (Fig. 12.4), they derived a

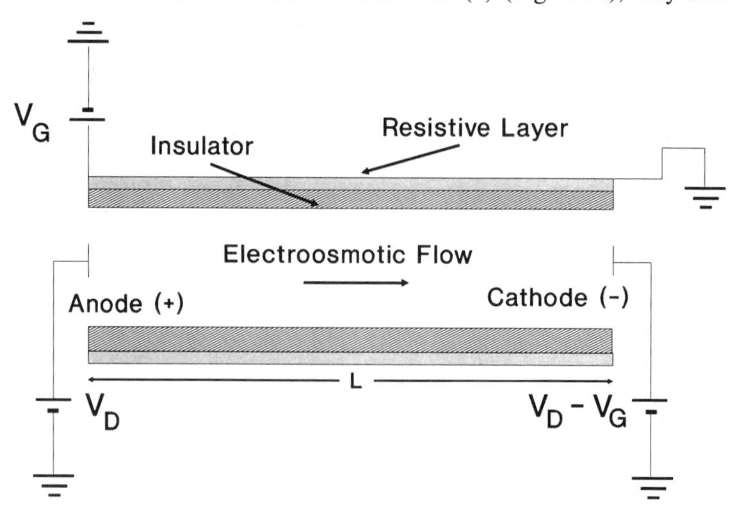

Figure 12.1. Schematic diagram of a MIEEKFED with constant zeta potential across the capillary. Redrawn with permission based on a figure in *J. Chromatogr.* **559,** 95, copyright ©1991 Elsevier Science Publishers.

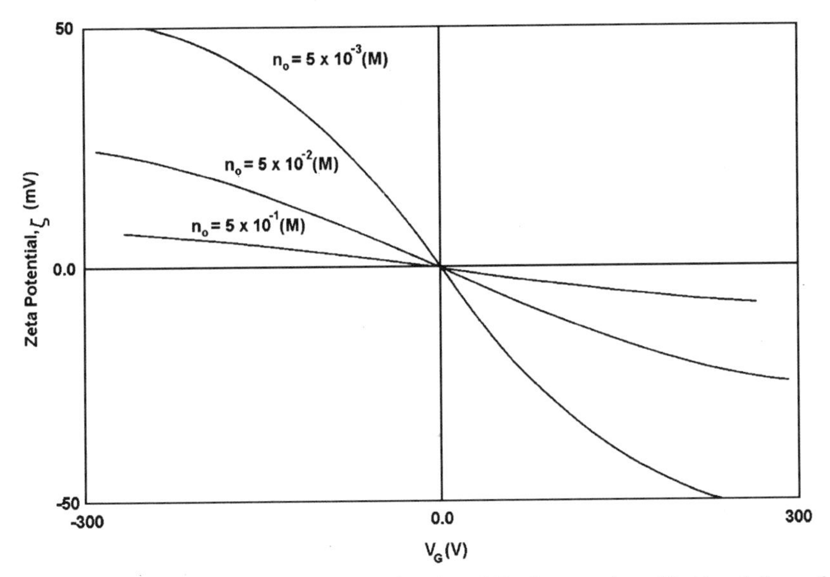

Figure 12.2. Change in zeta potential as a function of V_G for potassium chloride solutions of different concentrations (n_o) and a constant pH of 3.0. The capillary wall thickness is 1 μm. Reprinted with permission from *J. Chromatogr.* **559,** 95, copyright ©1991 Elsevier Science Publishers.

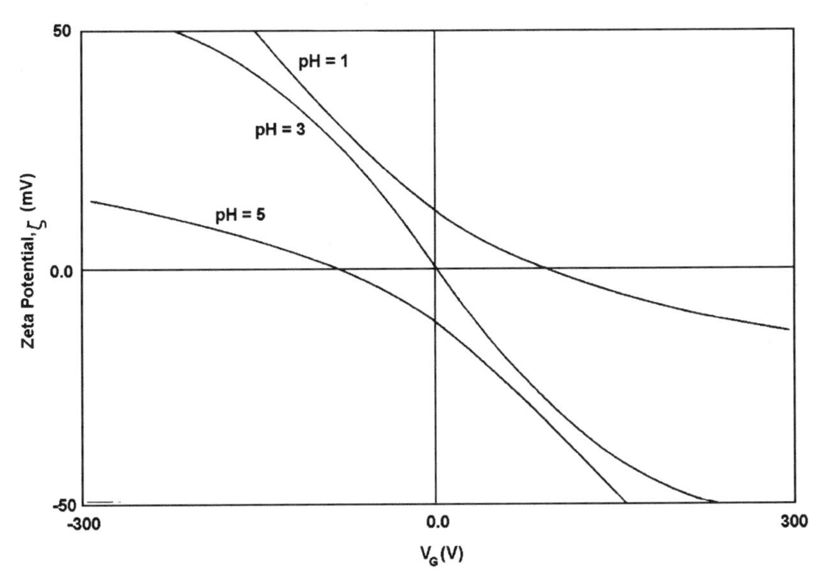

Figure 12.3. Change in zeta potential as a function of V_G for potassium chloride solutions of different pH values at a constant concentration of 5 mM. Reprinted with permission from *J. Chromatogr.* **559,** 95, copyright ©1991 Elsevier Science Publishers.

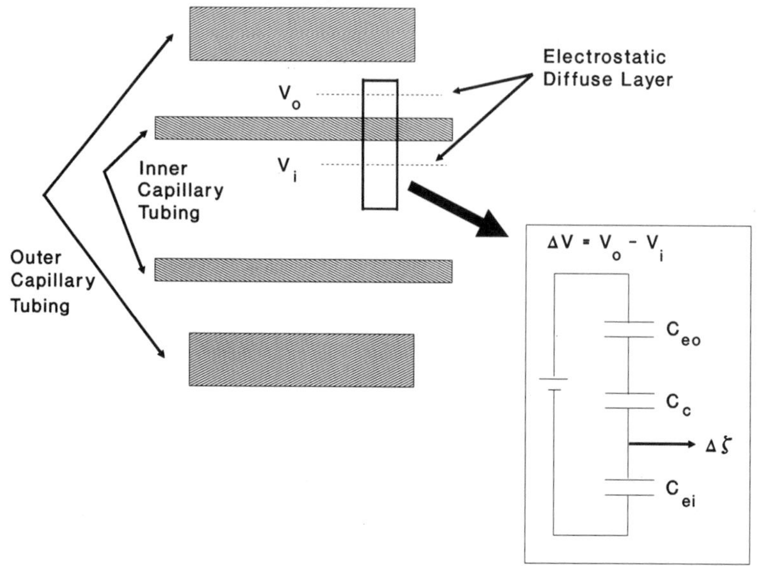

Figure 12.4. Proposed capacitor model for predicting the change in zeta potential due to the application of an applied external field. Redrawn with permission from *Anal. Chem.* **63,** 1550, copyright ©1991 Am. Chem. Soc.

function based on the capacitance of the diffuse layer (Section 2.3) at the capillary wall/buffer interface (C_{ei}) and the capacitance of the fused-silica wall, C_c.

$$\Delta\zeta = \frac{\Delta V}{(C_{ei})/(C_c)}, \tag{12.1}$$

where ΔV is the applied external field. The capacitance of the diffuse layer depends on the square root of the buffer concentration. The zeta potential itself is dependent on both pH and buffer concentration, so the EOF is related to both the solution chemistry and the applied perpendicular field. The capacitance of the fused-silica tubing is inversely proportional to the ln ratio of the capillary o.d./i.d. This is consistent with the work of Ghowski and Gale (1) because the effectiveness of the applied perpendicular field is greatest when thin-walled capillary tubing is used.

Control of the EOF using FEE can be used advantageously for many applications, including

1. increasing the resolution of the separation by reducing or even reversing the direction of the EOF compared with the electrophoretic flow;
2. speeding the separation by setting the EOF in the same direction as the electrophoretic flow;
3. increasing or decreasing the elution range for micellar electrokinetic separations; and

4. lowering the EOF for CIEF, CGE or CITP

One of the first applications illustrating separations tuned via FEE is shown in Figs. 12.5 a and b. At pH 2.7, without an applied external field, both the electrophoretic migration of some peptides and the modest EOF are directed toward the cathode. When the applied perpendicular voltage equals the separation voltage (a 0-kV applied external field), the direction of the EOF is reversed and opposes the electrophoretic migration of the peptides. As predicted by Eq. (2.15), this serves to improve the resolution. Increasing the external field to more positive values further improves the resolution, though the separation times become lengthy. Application of a negative applied external field reduces resolution, because the EOF is increased in the same direction as the electrophoretic mobility of the peptides.

While the advantages of FEE are tantalizing, the technique may increase the cost of operation. Potentially expensive metal-coated thin-walled capillaries may be required, though the technique of Lee *et al.* (3) avoids this problem by inserting the separation capillary inside a buffer-filled outer capillary tube (Fig. 12.4). It is possible that one of these designs will soon be incorporated into a commercial instrument. Alternatively, the technique may be better suited to chip-based microseparation devices, though such instruments are not likely to be fabricated in quantity for many years.

12.3 Ultrafast Separations

There are two fundamental limitations that restrict the speed of separation in HPCE. The first is simply the high voltage capability of the instrument and power supply. While it is possible to incorporate very high-voltage power supplies, the problems of grounding, shielding, and cost are substantial when over 30,000 V is employed. The second problem is Joule heating. When very high fields are incorporated, self-heating quickly causes deterioration in the separation.

Both problems can be solved, but at the expense of injection and detection. For high-speed electrophoresis, injection must be made when the voltage is on. Otherwise, time is lost as the voltage ramps up to its operating value. To generate high fields, short capillaries can be used. To reduce self-heating, very narrow capillaries are useful. Under these circumstances, an extremely small injection is required or the separation efficiency will be harmed. Monnig and Jorgenson (4) developed a system that incorporates short, narrow i.d. capillaries. A new injection process known as on-column sample gating coupled with laser-induced fluorescence (LIF) provides a satisfactory solution for the injection and detection problems of such an extreme system.

In on-column sample gating, the solutes must be derivatized or possess native fluorescence. Unlike conventional injection, the sample solution is continuously

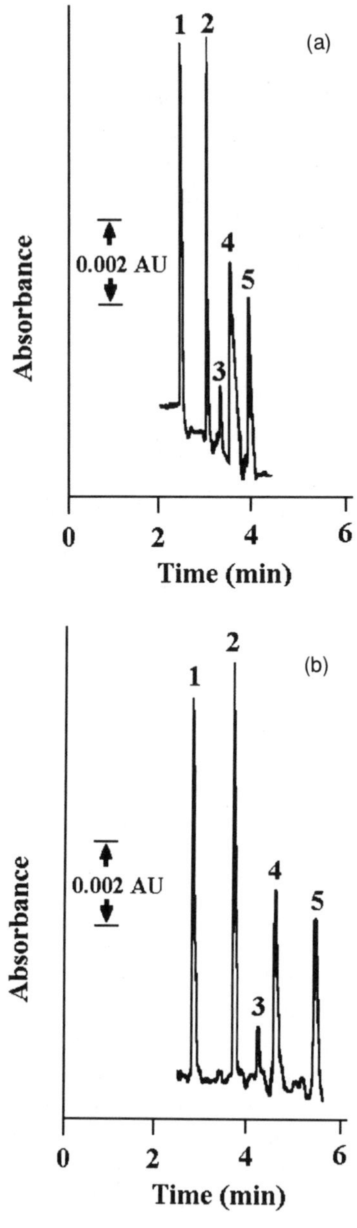

Figure 12.5. CZE of peptides (a) in the absence and (b) in the presence of a 0 kV potential gradient across the capillary. Capillary: 14.5 cm (length to detector) 50 μm i.d., 150 μm o.d.; buffer: 10 mM phosphate buffer, pH 2.7; injection: electrokinetic, 10 s at 1 kV; voltage: 5.5 kV. Solutes: (1) Lys–Trp–Lys; (2) thymopoetin II fragment; (3) adrenocorticotropic hormone fragment; (4) bradykinin; (5) human angiotensin II. Reprinted with permission from *Anal. Chem.* **64,** 886, copyright ©1992 Am. Chem. Soc.

introduced into the separation capillary. As shown in Fig. 12.6, a gating beam (1 W argon-ion laser) is used to continuously photodegrade the fluorophore. The degraded solute migrates through the capillary where it is invisible to the probe beam (an LIF detector). If the gating beam is interrupted for an instant, a small amount of unbleached sample migrates through the capillary and the separated solutes are detected by LIF at the probe beam. Since this "injection" process is optical rather than mechanical, extremely small volumes of undegraded sample can be introduced into the separation capillary.

Another problem was the generation of an extremely high field strength. This was solved by coupling capillaries of different diameters and lengths to "concentrate" the electric field over the separation capillary. As shown in Fig. 12.7, the actual separation capillary consisted of a 4 cm × 10 μm i.d. connected to capillaries of lengths 85 cm and 14 cm, each with an i.d. of 150 μm. Calculated resistances over sections A–C yielded respective voltages of 2.1, 22.7, and 0.2 kV for a 25 kV driving potential. Thus, the field strength over the separation capillary was 5,675 V/cm. This is more than 10 times greater than typical field strengths in HPCE. The narrow i.d. of the separation capillary is effective in minimizing Joule heating. The most efficient separations, due to Joule heating, were found at field strengths of 2,000–3,000 V/cm. The current passed by the narrow separation tube is extremely small. Even at 6,000 V/cm, only 6 μA of current passed through the coupled capillary.

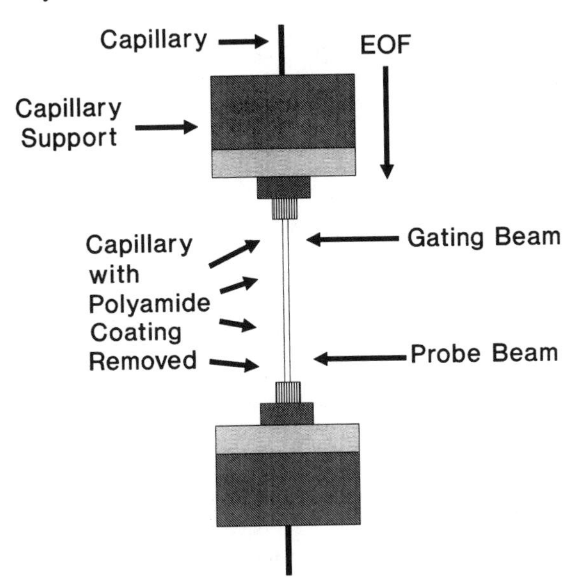

Figure 12.6. Diagram of capillary mount showing the relative position of the capillary and the "gating" and "probe" laser beams. Redrawn with permission from *Anal. Chem.* **63,** 802, copyright ©1991 Am. Chem. Soc.

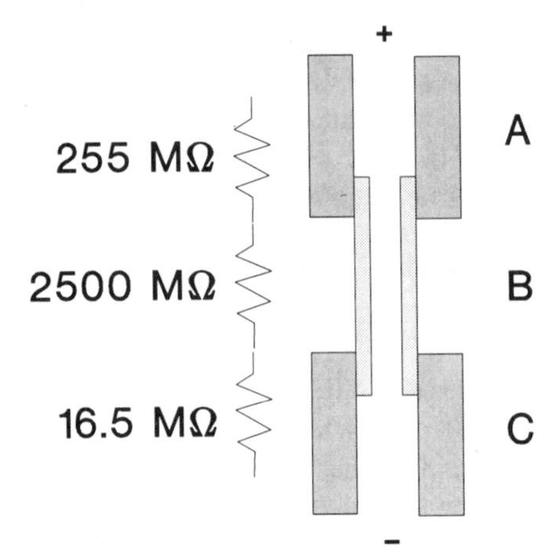

Figure 12.7. Diagram demonstrating the principle of coupling capillaries of different diameters. The resistor equivalent circuit at left lists the resistances calculated for each section of the capillary. Voltage: 25 kV. Key: (a) 85 cm × 150 μm i.d.; (b) 4 cm × 10 μm i.d.; (c) 14 cm × 150 μm i.d. Redrawn with permission from *Anal. Chem.* **63,** 802, copyright ©1991 Am. Chem. Soc.

In running this system, sample is continuously aspirated into the system by electrokinetic migration. With the gating laser set at 400 mW, approximately 80% of fluorescein dye is photobleached. A sample introduction time (pulse-time) of 20 ms produces an injection zone length of 240 μm. Longer zone lengths result in an increase in bandbroadening. A separation of FITC-labeled amino acids using a separation capillary of 1.2 cm with a field strength of 3,300 V/cm is shown in Fig. 12.8. The entire separation is complete in 1.4 s but produces theoretical plate counts of only 5,000–7,000. If the field is dropped to 1,750 V/cm, the efficiencies increase to 70,000–90,000 theoretical plates but the separation time increases to 12 s (separation not shown). It is likely that self-heating causes the reduced plate count for the high field-strength separation, and the usual compromises come into play.

It is not likely that this form of experimental arrangement will be adapted to commercial instruments in their present form. Construction of an entire HPCE instrument on a chip coupled with this technology is another possibility. Such an instrument could be invaluable as a remote sensor in process control, oceanography or space exploration.

Another application for ultrahigh-speed CE is to serve as the second stage in a two-dimensional separation process, e.g., LC–CE (5). Such a system can be required for separation of extremely complex samples. To preserve the speed of separation for the first stage, the second separation process should proceed rapidly. Under this condition, it is possible to perform a sufficient number of CE runs to effectively characterize and separate the chromatographic effluent.

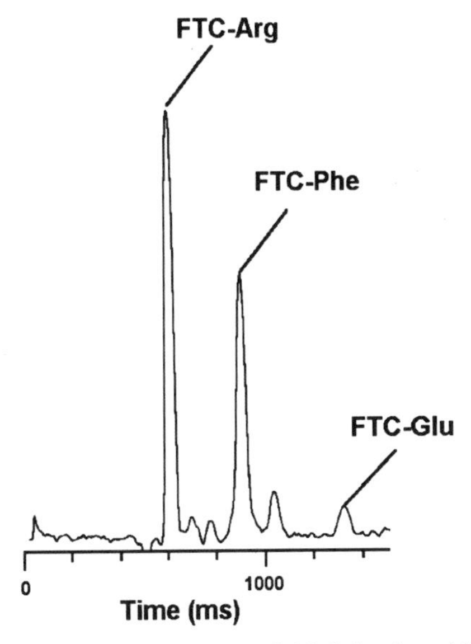

Figure 12.8. Electropherogram of a mixture of FITC-labeled amino acids. Separation capillary: 1.2 cm × 10 μm i.d.; buffer: 30 mM carbonate, pH 9.2; field strength: 3,300 V/cm; sample introduction time: 30 ms; solute concentration: 10 μm each. Redrawn with permission from *Anal. Chem.* **63,** 802, copyright ©1991 Am. Chem. Soc.

12.4 HPCE on a Chip

The techniques described in the preceding two sections may be best implemented in a silicon-wafer or glass microstructure material format. To date, there have been three reports describing the fabrication and performance of these ultraminiaturized CE systems (6–8).

Manz *et al.* (6) described the basis for fabrication based on a photolithography process common to the microelectronics industry. These processes include film deposition (addition of photosensitive material), photolithography (transfer of the layout pattern to the photosensitive film), etching (removal of the oxide and photoresist), and bonding (assembly). Using these technologies, a variety of miniaturized analytical systems, including a gas chromatograph, coulometric titrator, oxygen sensor, and pH sensor, have been fabricated on what is essentially a semiconductor chip (7). Capillary electrophoresis is another technique that is suitable for microfabrication as well.

The layout of the channels on a planar glass substrate is shown in Fig. 12.9. The background electrolyte (designated mobile phase) is introduced in channel 1, sample is contained in channel 2, and the separation occurs in channel 3. To generate high voltage over the separation capillary and sample channel, those

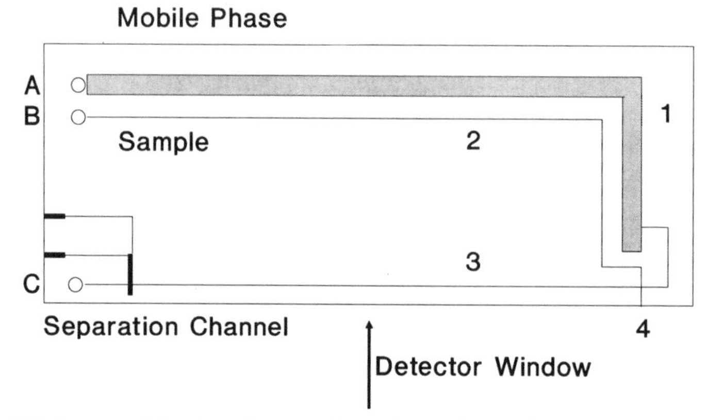

Figure 12.9. Layout of the channels on a planar glass substrate. Channels are referred to by numbers, and inlet points (reservoirs) by letters. Each channel is labeled with its content or its function. Overall dimensions are 14.8 cm × 3.9 cm × 1 cm thick. The location of one pair of platinum electrodes is shown; for clarity, the others are not. Redrawn with permission from *Anal. Chem.* **64,** 1926, copyright ©1992 Am. Chem. Soc.

channels were fabricated to be 30 μm wide by 10 μm deep, whereas channel 1 was 1 mm wide by 10 μm deep. Two bonded glass plates were used, the upper containing platinum electrodes and the lower containing the etched microchannels.

The sample and reagent (buffer) reservoirs are designated A (buffer), B (sample), and C (buffer). Solutions are added to these 3 mm reservoirs with a syringe. Detection was by LIF using a 4 mW argon-ion laser with a fiber optic to direct fluorescence to a photomultiplier tube.

In actual operation of the system (7), Ohm's Law plots were linear for potentials of up to 5,000 V. The slope was dependent on channel length and buffer conductivity. The linearity indicated effective heat dissipation within the system. Fields of up to 350 V/cm were achieved without special electrical isolation techniques. A robust EOF comparable to that in fused-silica was measured as well. Sample injection can be performed electrokinetically by applying a positive potential between channels 2 and 3 (channel 3 is at ground). The potential is then switched between channels 1 and 3 to perform the separation.

An actual separation of two fluorescent dyes, calcein and fluorescein, is shown in Fig. 12.10. Approximately 35,000 theoretical plates were obtained for calcein. It is thought that the unfocused laser beam contributed significantly to the system variance, reducing the theoretical plate count. Both wall interactions and Joule heating were not significant contributors to variance. After correcting for injector and detector variance, the system yielded 15–20 plates/V, compared with 20–25 plates/V for a conventional fused-silica capillary.

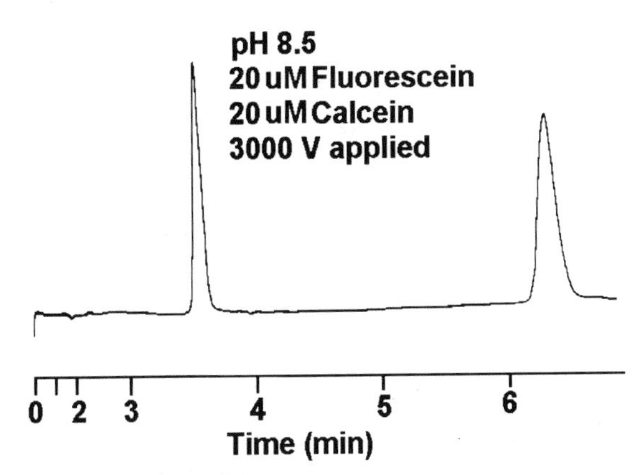

Figure 12.10. Electropherogram of a sample injection from channel 2 into channel 3 with 250 V applied between reservoirs B and C for 30 s. Separation voltage: 3,000 V between reservoirs A and C; buffer: 50 mM tris–borate, pH 8.5; sample concentration: 20 μM. The time scale was expanded 2.7 min after injection. Redrawn with permission from *Anal. Chem.* **64,** 1926, copyright ©1992 Am. Chem. Soc.

It is expected that up to 1,800 V/cm can be achieved using appropriate electrical isolation and channel geometry. Glass is a suitable medium for both microfabrication and capillary electrophoresis. While commercial introduction is in the distant future, it is evident that these technologies will play a major role in the instrumentation of the 21st century.

12.5 Multiple Capillary Systems

Operation of a system with a series of capillaries is desirable to overcome the fundamental limitation of all tubular separation techniques: Only one sample can be processed at a time. In contrast, planar techniques such as slab-gel electrophoresis and thin-layer chromatography process many samples simultaneously. Compared to those techniques described in the preceding sections, multicapillary instruments are on the brink of commercial introduction.[1]

Detection represents the greatest engineering challenge in a multi-capillary system. Injection is conceptually simple and can be considered a multiplexed version of conventional injection techniques. Capillary-to-capillary variation must be carefully controlled as well.

[1] Beckman Instruments is expected to introduce a six-capillary instrument targeted toward the clinical chemist.

Zagursky and McCormick (9) described the first multi-capillary instrument in 1990. They modified a now discontinued DNA sequencing instrument[2] to accept 12 gel-filled capillaries and reported separations of DNA sequencing reaction products up to 512 bases. The laser beam and photomultiplier tube were scanned across the capillary bundle using a digitally controlled stepper motor assembly.

Huang *et al.* (10, 11) employed a laser-excited, confocal-fluorescence scanner to measure DNA sequencing products in a linear array of capillaries. The apparatus is illustrated in Fig. 12.11. A moving stage (translational stage) continuously moves the capillary array through the laser beam. Since epi-illumination is employed (excitation and light collection are performed using 180° geometry), detector alignment problems are minimized. Each capillary is immobilized to an identical height above the translation stage, and all capillaries are precisely in the same plane. As many as 25 capillaries have been run simultaneously (10). Huang *et al.* (11) reported migration time relative standard deviations ranging from 1.9 to 6.2% for a four-capillary (gel-filled) array. Fragments with longer migration times yielded the poorest precision. This precision problem may be solved in part with polymer networks (Chapter 5) instead of rigid gel-filled capillaries.

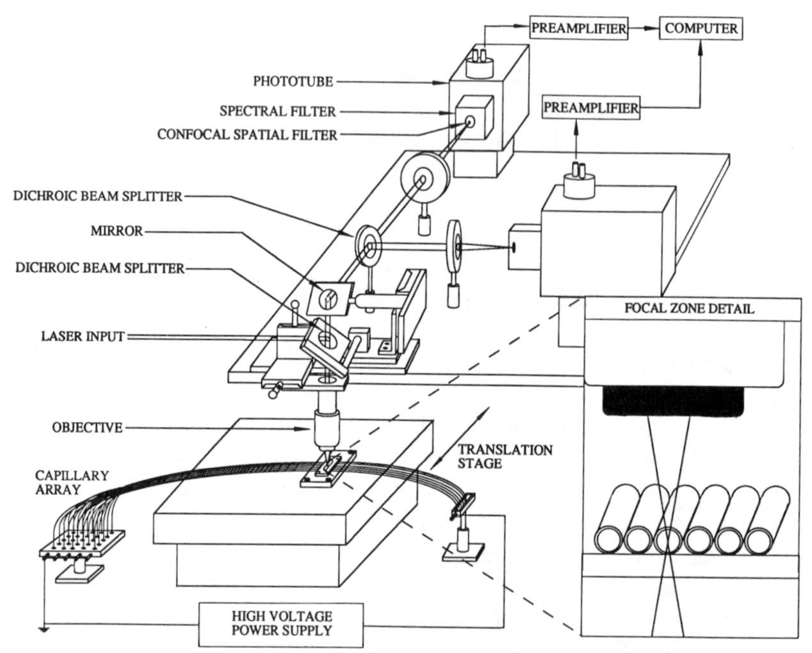

Figure 12.11. Schematic of a multicapillary instrument with laser-excited, confocal-fluorescence detection. Reprinted with permission from *Anal. Chem.* **64**, 2149, copyright ©1992 Am. Chem. Soc.

[2] Genesis 2000, DuPont.

Another problem with these approaches is a reduction in the signal/noise ratio as the number of capillaries is increased.

This can be estimated from

$$S/N = \left(\frac{T}{n}\right)^2, \tag{12.2}$$

where S/N is the signal/noise ratio, T is the bandwidth (s), and n is the number of capillaries. Based on this equation, the signal/noise ratio for a 25-capillary instrument would be 625 times worse than that for a single capillary instrument. This problem may be compensated in part by the high sensitivity of LIF detection. An approach that could eliminate this problem would be to employ a separate detector for each capillary. A form of a diode array detection may prove useful in this regard, because each capillary can be continuously monitored.

12.6 Gradient and Programmed Systems

A. Introduction

Gradient elution is one of the cornerstones of HPLC. The ability to continuously change the solvent composition of the mobile phase is invaluable to solving the general elution problem—that is, the problem of resolving, in the shortest possible time frame, solutes with vastly differing capacity factors. Similarly, in GC, a temperature-programmed system permits the separation of a wide variety of solutes, again in the shortest possible time frame. Programmed systems can be employed in HPCE toward similar goals. These techniques can improve the speed of separations and influence the resolution as well. Among the parameters that can be varied in HPCE are voltage, temperature, pH, and electrolyte composition.

B. Voltage Programming

In 1988, McCormick (12) reported the use of a voltage ramp to improve the separation of proteins by CZE. He found that the separation deteriorates if the voltage is set upon injection to the full operating voltage. It is possible that thermal or nonequilibrum effects are responsible for these observations. However, not all separations exhibit this puzzling effect, and this subject remained dormant for several years.

For most separations, increasing the field in a programmed manner is expected to speed the separation because the electrophoretic velocity is proportional to the voltage. This in itself could be valuable, particularly if there are substantial mobility differences between solutes.

For DNA separations, the shortest run times occur when a decreasing field is employed because the biased reptation mechanism (Section 5.2) may be operative. For example, with the field at a constant potential of 200 V/cm, all ϕX174 restriction fragments are separated, but the run time is 27 min (Fig. 12.12a). Increasing the field speeds the separation at the expense of resolution for the larger

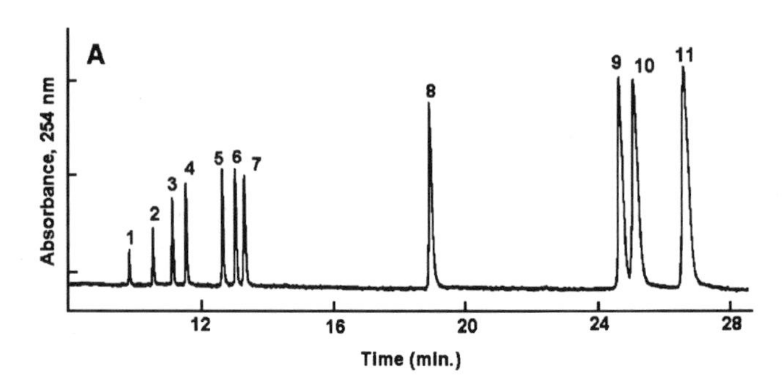

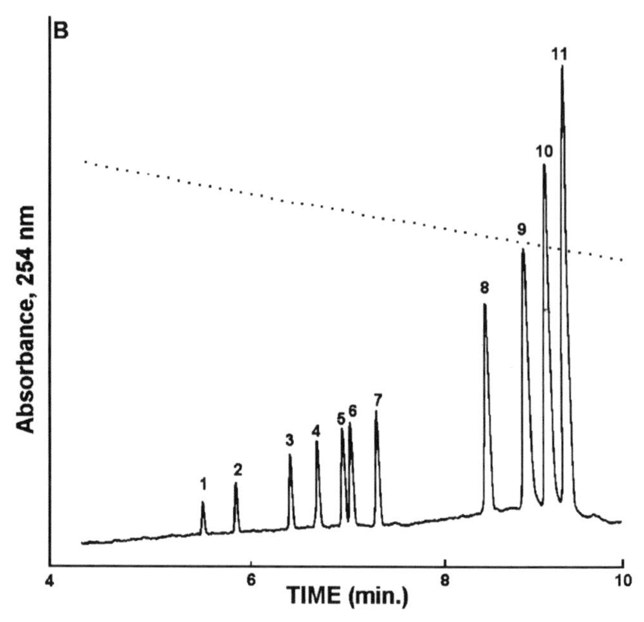

Figure 12.12. Separation of ϕX174 DNA restriction fragments by CGE using (a) a constant applied electric field of 200 V/cm and (b) a linear field strength gradient ramping from 400 V/cm to 100 V/cm in 20 min. Capillary: polyacrylamide gel-filled DB-225 coated 40 cm (length to detector, 27 cm) × 100 μm i.d.; buffer: 100 mM Tris–borate, 2 mM EDTA, pH 8.35. Key: (1) 72; (2) 118; (3) 194; (4) 234; (5) 271; (6) 281; (7) 310; (8) 603; (9) 872; (10) 1,078; (11) 1353 bp. Redrawn with permission from *Anal. Chem.* **64,** 2348, copyright ©1992 Am. Chem. Soc.

fragments. At the high field strength, molecular orientation of DNA with the field prevents adequate resolution of the larger fragments. Guttman *et al.* (13) correctly reasoned that a decreasing field strength (with time) would both speed the separation and improve the resolution, since the various molecular weights could be exposed to a field that is optimal for resolution. The decreasing field reduces biased reptation of larger DNA fragments. Thus, higher fields can be applied over the smaller fragments to speed the separation. Then the field ramps down to ensure resolution of the larger fragments (Fig. 12.12b).

Implementation of voltage programming is simple using a programmable power supply. Unfortunately, not all of the commercial instrumentation is designed with this capability, which is especially valuable for DNA separations as described above.

C. pH Gradients

Generation of pH gradients can be useful for separating a series of solutes having similar ionic mobilities that cover a wide range of p*K* values. For example, a series of purine and pyridine derivatives range in pK_a from 1.9 to 6.0. Separation in a pH 2.2 buffer gives incomplete separation of the fast-migrating components (Fig. 12.13, top). Increasing the pH to 3.5 (data not shown) resolves the early peaks but gives a lengthy and overlapping separation of the last two components. Running a pH gradient from pH 3.5 to pH 2.2 provides separation of all components (Fig. 12.13, bottom).

There are at least two ways of generating a pH gradient: (i) use a system to provide real-time mixing of buffer solutions (14–17); (ii) select a buffer with a large thermal coefficient and temperature-program the system (18).

The temperature-programmed system is a simple approach from the engineering standpoint, but it is somewhat limited since only a narrow pH gradient is achievable. For example, a Tris–borate buffer declines from pH 6.9 to 6.03 when the temperature is increased from 27 to 62°C (18). Another problem is the complicating impact of the temperature effect on the buffer viscosity.

The use of real-time buffer mixing to formulate the gradient on-line is complicated by the necessity of at least one, and possibly two, low-pressure, high-precision pumping systems along with a micromixer. Reproducibility of such a setup has not been extensively studied.

The difficulty in implementing gradient elution overshadows the modest gains illustrated thus far. Perhaps a simpler approach would employ carrier ampholytes to generate an on-line gradient. A step change in ampholyte blend could result in the generation of a new gradient along the capillary. This could be accommodated with virtually any instrumental setup.

D. Solvent Gradients

Organic modifiers are used in MECC to adjust the partition coefficient of a solute between the micelle and the bulk aqueous phase. As in HPLC, it is desirable

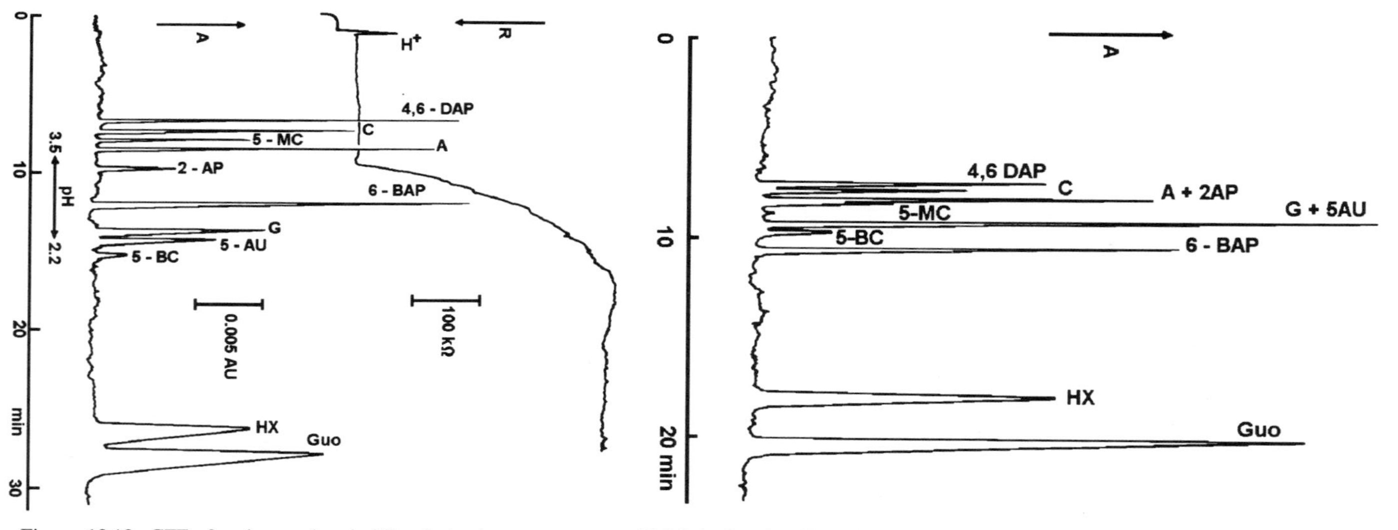

Figure 12.13. CZE of purines and pyrimidine derivatives at a constant pH 2.2 (top) and a pH gradient from pH 3.5 to pH 2.2 (bottom). Capillary: 40 cm 130 μm i.d.; buffer: 10 mM Tris titrated with trichloroacetic acid (TCA) to the appropriate pH with 0.25% PEG-6000 added to suppress the EOF; voltage: 10 kV. The gradient was formed by addition of 0.5 M TCA. Key: 4,6-DAP = 4,6-diaminopyrimidine; C = cytosine; 5-MC = 5-methylcytosine; A = adenine; 2-AP = 2-aminopurine; 6-BAP = 6-benzylaminopurine; G = guanine; 5-AU = 5-aminouracil; 5-BC = 5-bromocytosine; HX = hypoxanthine; Gue = guanosine. Redrawn with permission from *J. Chromatogr.* **480**, 271, copyright ©1989 Elsevier Science Publishers.

to perform gradients to optimize the run time. Using an on-line system consisting of two syringe pumps and a micromixer, Powell and Sepaniak (19) and Sepaniak *et al.* (20) produced functional gradients and provided a model for predicting the solute migration time under gradient conditions. The problems of such a setup were discussed earlier, and it is not expected that this system will be incorporated into commercial instrumentation in the near future.

12.7 Two-Dimensional Systems

Separation of extremely complex systems often requires more than one separation mechanism to remove peak overlap. When two orthogonal[3] separation mechanisms are coupled, the peak capacity is the product of the peak capacity for each dimension, and the resolution is the square root of the sum of the squares of the resolution in each separation mode (5). Bushey and Jorgenson (5, 21) designed an instrument that interfaces a miniaturized HPLC system to CZE. The interface is a six-port valve that intermittently directs the LC effluent past the CZE capillary.

Figure 12.14 illustrates a 3-D plot of a fluorescamine-labeled tryptic digest of horse heart myoglobin C. Figures such as this or its corresponding contour or grey-scale plots are useful to better ensure separations of complex mixtures such as tryptic digests.

Despite the relatively short CZE run time, the LC separation had to be slowed to permit a sufficient number of CZE runs to be made. Otherwise, the effluent from the LC column would be undersampled. Note that the ultrahigh-speed separation system described in Section 12.3 would permit the LC separation to run faster. While this application is somewhat complex, adaptation of commercial instrumentation should be possible. However, since ultrafast CZE is difficult to implement, long run times should be expected.

12.8 Immunologic Trace Enrichment

The final stop of this technological tour deals with trace enrichment, often a necessity in HPCE to obtain sufficient sensitivity. Trace enrichment can be performed either off-line during sample preparation, or on-line. The on-line techniques are often preferable because sample handling is minimized. When the sample size is extremely limited, on-line sample preparation is mandatory.

The problem with trace enrichment is that impurities and endogenous components are frequently concentrated along with the analytes of interest.

[3] Orthogonal separations are generated by unrelated separation mechanisms. A plot comparing migration (retention) times by both methods gives no significant correlation.

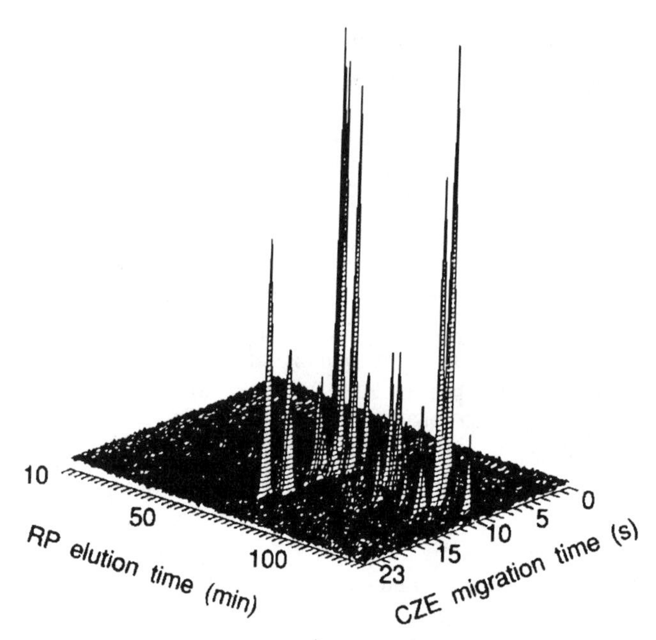

Figure 12.14. Three-dimensional LC/CZE plot of fluorescamine-labeled horse heart myoglobin C tryptic digest. CZE conditions: capillary: 26 cm (6.5 cm to detector) × 15 μm i.d.; buffer: 12 mM potassium phosphate, pH 7.15–7.25; voltage: 28 kV; injection: 1 kV for 5 s. HPLC conditions: column: Brownlee Aquapore RP-300, 25 cm × 1.0 mm i.d.; gradient: complex ramps from 0 to 100% acetonitrile: flow rate: 20 μL/min; detection: fluorescence, excitation: 365 nm; emission: >470 nm. Reprinted with permission from *J. Microcol. Sep.* **2,** 293, copyright ©1990 Microseparations Inc.

This often places severe demands on the conditions of separation and detection. Usually, prolonged run times are required to isolate the minor component from the sample matrix.

The immune response is one of the most selective and tuneable biochemical interactions. Guzman *et al.* (22) immobilized monoclonal antibodies to glass beads and filled a 3–5 mm section of capillary with this material. The beads were secured within the capillary with glass frits (Fig. 12.15).

After electrokinetic or hydrodynamic injection, the analyte concentrator was washed with 50 mM borate buffer, pH 8.3, followed by an elution buffer composed of 50 μM acetate, pH 3.5. After collection of 5 mL aliquots, samples were separated by CZE for purity confirmation (data not shown). This concentrator has been employed for the micropreparative isolation of methamphetamine from urine samples. While there are many problems to be worked out in the fabrication and utilization of immunoconcentrators, the technique has great promise when one is searching for the proverbial needle in the haystack.

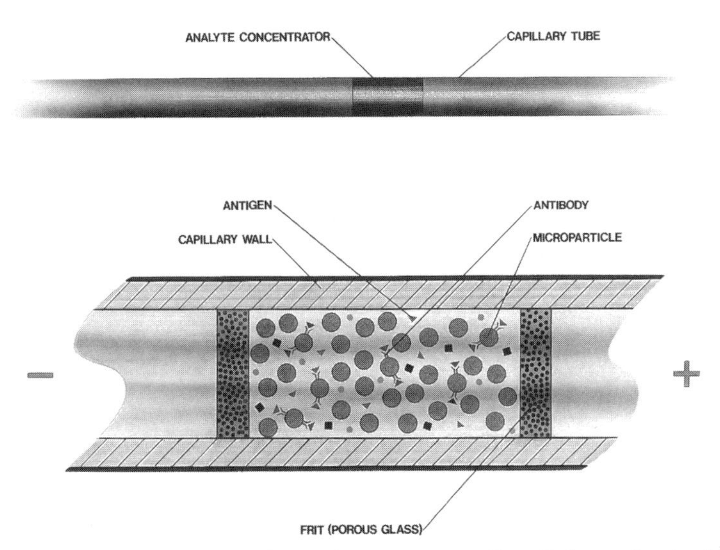

Figure 12.15. Schematic diagram of an analyte concentrator. Controlled-pore glass beads were activated, conjugated with a purified antibody, and installed inside a portion of a fused-silica capillary. Reprinted with permission from *J. Liq. Chromatogr.* **14**, 997, copyright ©1991 Marcel Dekker.

References

1. K. Ghowsi and R. J. Gale, *J. Chromatogr.* **559,** 95(1991).
2. C. S. Lee, C.T. Wu, T. Lopes, and B. Patel, *J. Chromatogr.* **559,** 133 (1991).
3. C. S. Lee, D. McManigill, C.T. Wu, and B. Patel, *Anal. Chem.* **63,** 1519 (1991).
4. C. A. Monnig and J. W. Jorgenson, *Anal. Chem.* **63,** 802 (1991).
5. M. M. Bushey and J. W. Jorgenson, *Anal. Chem.* **62,** 978 (1990).
6. A. Manz, D. J. Harrison, E. M. J. Verpoorte, J. C. Fettinger, A. Paulus, H. Ludi, and H. M. Widmer, *J. Chromatogr.* **593,** 253 (1992).
7. D. J. Harrison, A. Manz, Z. Fan, H. Ludi, and H. M. Widmar, *Anal. Chem.* **64,** 1926 (1992).
8. A. Manz, D. J. Harrison, E. M. J. Verpoorte, J. C. Fettinger, H. Ludi, and H. M. Widmer, *Chimia* **45,** 103 (1991).
9. R. J. Zagursky and R. M. McCormick, *BioTechniques* **9,** 74 (1990).
10. X. C. Huang, M. A. Quesada, and R. Mathies, *Anal. Chem.* **64,** 2149 (1992).
11. X. C. Huang, M. A. Quesada, and R. A. Mathies, *Anal. Chem.* **64,** 967 (1992).
12. R. M. McCormick, *Anal. Chem.* **60,** 2322 (1988).
13. A. Guttman, B. Wanders, and N. Cooke, *Anal. Chem.* **64,** 2348 (1992).
14. P. Boček, M. Deml, J. Pospichal, and J. Sudoe, *J. Chromatogr.* **470,** 309 (1989).
15. T. Tsuda, *Anal. Chem.* **64,** 386 (1992).

16. V. Sustacek, F. Foret, and P. Boček, *J. Chromatogr.* **480,** 271 (1989).
17. H.T. Chang and E. S. Yeung, *J. Chromatogr.* **608,** 65 (1992).
18. C.W. Whang and E. S. Yeung, *Anal. Chem.* **64,** 502 (1992).
19. A. C. Powell and M. J. Sepaniak, *J. Microcol. Sep.* **2,** 278 (1990).
20. M. J. Sepaniak, D. F. Swaile, and A. C. Powell, *J. Chromatogr.* **480,** 185 (1989).
21. M. M. Bushey and J. W. Jorgenson, *J. Microcol. Sep.* **2,** 293 (1990).
22. N. A. Guzman, M. A. Trebilcock, and J. P. Advis, *J. Liq. Chromatogr.* **14,** 997 (1991).

Index

C

D

E

F

M

N

O

P

Q

R